Operator Theory: Advances and Applications
Vol. 172

Modern Trends in Pseudo-Differential Operators

Joachim Toft
M.W. Wong
Hongmei Zhu
Editors

Advances in Partial Differential Equations

Birkhäuser
Basel · Boston · Berlin

Editors:

Joachim Toft
School of Mathematics and
Systems Engineering
Växjö University
SE-351 95 Växjö
Sweden
e-mail: Joachim.Toft@msi.vxu.se

M. W. Wong
Hongmei Zhu
Department of Mathematics and Statistics
York University
4700 Keele Street
Toronto, Ontario M3J 1P3
Canada
e-mail: mwwong@mathstat.yorku.ca
hmzhu@mathstat.yorku.ca

2000 Mathematics Subject Classification Primary 22A10, 32A40, 32A45, 35A17, 35A22, 35B05, 35B40, 35B60, 35J70, 35K05, 35K65, 35L05, 35L40, 35S05, 35S15, 35S30, 43A77, 46F15, 47B10, 47B35, 47B37, 47G10, 47G30, 47L15, 58J35, 58J40, 58J50, 65R10; Secondary 22C05, 30E25, 35G05, 35H10, 35J05, 42B10, 42B35, 47A10, 47A53, 47F05, 58J20, 65N30, 92C55, 94A12

Library of Congress Control Number: 2006939565

Bibliographic information published by Die Deutsche Bibliothek
Die Deutsche Bibliothek lists this publication in the Deutsche Nationalbibliografie; detailed bibliographic data is available in the Internet at <http://dnb.ddb.de>.

ISBN 3-7643-7577574-4 7-7r Verlag, Basel – Boston – Berlin

ISBN 978-3-7643-8097-7 Birkhäuser Verlag AG, Basel – Boston – Berlin

Part of Springer Science+Business Media
Printed on acid-free paper produced from chlorine-free pulp. TCF ∞
Cover design: Heinz Hiltbrunner, Basel
Printed in Germany
ISBN-10: 3-7643-8097-7
ISBN-13: 978-3-7643-8097-7

e-ISBN-10: 3-7643-8
e-ISBN-13: 978-3-76

Contents

Preface

The ISAAC Group in Pseudo-differential Operators (IGPDO) was formed at the Fourth ISAAC Congress held at York University in Toronto in 2003 and the first volume entitled *Advances in Pseudo-differential Operators* and devoted to papers focussing on pseudo-differential operators and its diverse applications was then initiated and published in Professor Israel Gohberg's series *Operator Theory: Advances and Applications* in 2004. As a satellite conference to the Fourth Congress of European Mathematics held at Stockholm University in 2004, the International Conference on Pseudo-differential Operators and Related Topics was held at Växjö University in Sweden. Prompted by the enthusiasm of the participants, the second volume with similar scope and entitled *Pseudo-differential Operators and Related Topics* was published in the same series in 2006.

Members of IGPDO met again at the Fifth ISAAC Congress held at Università di Catania in Italy in July 2005. Core members of the group encouraged the publication of a sequel to the Toronto Volume and the Växjö Volume. The vision is to seek new directions for the broad subject on pseudo-differential operators and the strategy is to devote the Catania Volume not only to papers based on lectures given at the special session on pseudo-differential operators, but also invited papers that bear on the themes of IGPDO. In order to reflect the goal and vision of IGPDO, the Catania Volume is entitled *Modern Trends in Pseudo-differential Operators*.

This Catania Volume consists of nineteen peer-reviewed papers representing *modern trends* in pseudo-differential operators. Topics include partial differential equations, global analysis, geometry, quantization, Wigner transforms, Weyl transforms on Lie groups, mathematical physics and time-frequency analysis. The nineteen articles in this volume should be of great interest to graduate students and researchers in analysis, mathematical physics and mathematical sciences.

The sustained growth of the subject on pseudo-differential operators can be testified by the much wider scope and the significantly greater number of contributors in this Catania Volume.

Operator Theory:
Advances and Applications, Vol. 172, 1–31

The Quantization of Edge Symbols

C.-I. Martin and B.-W. Schulze

Abstract. We investigate operators on manifolds with edges from the point of view of the symbolic calculus induced by the singularities. We discuss new aspects of the quantization of edge-degenerate symbols which lead to continuous operators in weighted edge spaces.

Mathematics Subject Classification (2000). 35S15, 35J70, 35S05, 58J40.

Keywords. Operators on manifolds with edges, quantization of edge-degenerate symbols, composition in the edge algebra.

Introduction

This paper is aimed at studying a number of new properties of edge amplitude functions belonging to the calculus of operators on a manifold with edges. (The notation 'manifold' is used here for convenience; in general, our spaces are manifolds only outside their subset of singularities). The investigation is motivated by the program of establishing operator algebras on configurations with higher singularities, e.g., when the cones of local wedges have cross sections with singularities, based on an iterative definition of such 'corner manifolds', cf. [2], [3]. The hope is that the ideas of the cone and edge pseudo-differential calculus with smooth model cones, cf. [13], [15], are more or less iterative. However, the symbolic structures in the higher floors of the calculus are by no means straightforward. Recall that already for manifolds with smooth edges there are different ways of constructing quantizations of corresponding edge-degenerate symbols; they reflect different aspects of the calculus. Compared with [14] an alternative quantization is given in [5]; remainders belong to the class of flat Green edge symbols. The result of [5] is applied in [11] and [6] to the construction of holomorphic representatives of corner Mellin symbols, given in terms of parameter-dependent families of edge operators of first generation (i.e., for smooth edges).

Here we study edge symbols in another way, namely, with an additional localisation near the diagonal with respect to the cone axis variable at infinity, depending on the edge covariable. In higher corner algebras localisations of a similar kind seem to be necessary, cf. [1], [3].

The (operator-valued) symbols of the edge algebra from [12] or [15] are specific families of pseudo-differential operators on infinite cones $X^\wedge := \mathbb{R}_+ \times X \ni (r, x)$ with $r \to 0$ representing the tip and $r \to \infty$ a conical exit to infinity, where (in the simplest case) the cross section X is a closed compact C^∞ manifold. These symbols may be interpreted as a specific parameter-dependent quantization of edge-degenerate scalar symbols (or, alternatively, of an edge-degenerate family of classical pseudo-differential operators on X). Quantizations are usually not canonical, even in simpler situations. The choice of the edge quantization in the above mentioned expositions (which we briefly recall below) gives rise to an algebra of continuous operators in weighted edge spaces, with all the expected features, such as ellipticity with respect to a principal symbolic hierarchy and existence of a parametrix within the calculus.

However, some aspects of the theory remained complicated, for instance, the rigorous proof of the composition behavior and other necessary elements. For that reason the authors of [5] proposed another edge quantization which entails the composition result in a very simple way. It would be desirable to apply a similar strategy also for manifolds with singularities of higher order. Unfortunately, the attempt to generalise [5], for instance, for singularities of order 2 (i.e., when the base X of the model cone itself has singularities of conical or edge type) leads to an enormous blow-up of technicalities, and the construction of corresponding alternative quantizations along the lines of [5] compared with the 'usual' ones seems to be more complicated than a direct approach. Moreover, the analogue of the usual edge quantization for the edge calculus of second generation as studied in [1] gives rise to difficulties; so there was to be invented a modification. Since this is new in an analogous form also in the edge calculus of first generation, we discuss here such an edge quantization and characterise the remainders compared with the former one. This is one of the points of the present paper, and we prove that the remainders are again flat Green symbols. Moreover, we give a new relatively elementary proof for the composition theorem of edge symbols of first generation.

Our considerations are also motivated by the expectation that analogous conclusions simplify and unify the structure of operator algebras on manifolds with higher corners, according to the program of successive conifications and edgifications of the operator algebras, opened in [16] and continued in [17], [2], [3]. By a manifold with corners we understand a topological space M that contains a subspace M' of singularities such that $M \setminus M'$ is a C^∞ manifold, and near every $m \in M'$ the space M is modelled on a cone $X^\Delta := (\overline{\mathbb{R}}_+ \times X)/(\{0\} \times X)$ or a wedge $X^\Delta \times \Omega$, for an open set $\Omega \subseteq \mathbb{R}^q$, where X is again a manifold with corners (of 'lower order' than M). In addition we require specific properties of the transition maps belonging to different such local representations, cf. [2].

Examples are manifolds with conical singularities or edges; in this case we assume the base spaces X of the model cones to be closed, compact C^∞ manifolds. Manifolds with higher singularities can be obtained by iteratively forming cones or wedges and pasting them together to corresponding global spaces.

In order to illustrate the nature of operator-valued amplitude functions on a singular manifold we first consider the case of differential operators on a manifold B with conical singularities. Assume, for convenience, that B has one conical point v, and let $\mathbb{B}$ denote the associated stretched manifold which is a C^∞ manifold with compact boundary $\partial\mathbb{B} \cong X$ and int $\mathbb{B} \cong B \setminus \{v\}$. Then B is equal to the quotient space $\mathbb{B}/\partial\mathbb{B}$ (with $\partial\mathbb{B}$ collapsed to the point v), and $\mathbb{B}$ is locally near $\partial\mathbb{B}$ identified with $\overline{\mathbb{R}}_+ \times X$.

A differential operator A with smooth coefficients on int$\mathbb{B}$ is said to be of Fuchs type, if locally near $\partial\mathbb{B}$ in the splitting of variables $(r, x) \in \mathbb{R}_+ \times X$ the operator has the form

$$A = r^{-\mu} \sum_{j=0}^{\mu} a_j(r)\Big(-r\frac{\partial}{\partial r}\Big)^j \tag{0.1}$$

with coefficients $a_j \in C^\infty(\overline{\mathbb{R}}_+, \mathrm{Diff}^{\mu-j}(X))$ (here $\mathrm{Diff}^\nu(\cdot)$ denotes the space of all differential operators of order ν with smooth coefficients on the C^∞ manifold in the parentheses). In this case the principal symbolic structure consists of a pair

$$\sigma(A) = (\sigma_\psi(A), \sigma_\mathrm{c}(A)),$$

where $\sigma_\psi(A)$ is the standard homogeneous principal symbol of order μ and

$$\sigma_\mathrm{c}(A)(z) := \sum_{j=0}^{\mu} a_j(0)z^j \tag{0.2}$$

the principal conormal symbol. The function (0.2) is operator-valued and acts between the standard Sobolev spaces on X as a family of continuous operators

$$\sigma_\mathrm{c}(A)(z) : H^s(X) \to H^{s-\mu}(X)$$

depending on the variable $z \in \mathbb{C}$.

Another example is the case of a C^∞ manifold with boundary, locally near the boundary modelled on $\overline{\mathbb{R}}_+ \times \Omega \ni (r, y)$, $\Omega \subseteq \mathbb{R}^q$ open. Then, if A is a differential operator of order μ with smooth coefficients up to the boundary, the principal symbolic hierarchy of A is a pair

$$\sigma(A) = (\sigma_\psi(A), \sigma_\partial(A)).$$

The first component is again the standard homogeneous principal symbol of order μ. Moreover, writing A near the boundary as

$$A = \sum_{|\alpha| \le \mu} a_\alpha(r, y) D^\alpha_{(r,y)}, \tag{0.3}$$

we set

$$\sigma_\partial(A)(y,\eta) := \sum_{\substack{|\alpha|=\mu \\ \alpha=(k,\gamma)}} a_\alpha(0,y) D_r^k \eta^\gamma, \tag{0.4}$$

$(y,\eta) \in T^*\Omega \setminus 0$, called the homogeneous principal boundary symbol of A. The symbol (0.4) is operator-valued and acts between the standard Sobolev spaces on $\mathbb{R}_+$ as a family of continuous operators

$$\sigma_\partial(A)(y,\eta) : H^s(\mathbb{R}_+) \to H^{s-\mu}(\mathbb{R}_+), \tag{0.5}$$

depending on $(y,\eta) \in T^*\Omega \setminus 0$; here $H^s(\mathbb{R}_+) := H^s(\mathbb{R})\big|_{\mathbb{R}_+}$.

A third category of examples are operators on manifolds with edges (cf. also Section 1 below). Manifolds with (smooth) edges can be regarded as a generalisation of manifolds with (smooth) boundary. In this case the local model near the edge is equal to $X^\Delta \times \Omega$ for a closed compact C^∞ manifold X. A manifold with boundary just corresponds to the case $\dim X = 0$.

We often employ the stretched manifolds rather than the manifolds with singularities themselves. In the case of a wedge $X^\Delta \times \Omega$ the associated stretched space is equal to $\overline{\mathbb{R}}_+ \times X \times \Omega$. In the corresponding splitting of variables (r,x,y) the typical differential operators A are assumed to be of the form

$$A = r^{-\mu} \sum_{j+|\alpha|\le\mu} a_{j\alpha}(r,y)\Big(-r\frac{\partial}{\partial r}\Big)^j (rD_y)^\alpha, \tag{0.6}$$

with coefficients $a_{j\alpha} \in C^\infty(\overline{\mathbb{R}}_+ \times \Omega, \mathrm{Diff}^{\mu-(j+|\alpha|)}(X))$. Such operators will also be called edge-degenerate. Their principal symbolic hierarchy is a pair

$$\sigma(A) = (\sigma_\psi(A), \sigma_\wedge(A)) \tag{0.7}$$

with $\sigma_\psi(A)$ being the homogeneous principal symbol of A of order μ in the usual sense, and

$$\sigma_\wedge(A)(y,\eta) := r^{-\mu} \sum_{j+|\alpha|\le\mu} a_{j\alpha}(0,y)\Big(-r\frac{\partial}{\partial r}\Big)^j (r\eta)^\alpha \tag{0.8}$$

the so-called homogeneous principal edge symbol. It represents a family of continuous operators

$$\sigma_\wedge(A)(y,\eta) : \mathcal{K}^{s,\gamma}(X^\wedge) \to \mathcal{K}^{s-\mu,\gamma-\mu}(X^\wedge), \tag{0.9}$$

$(y,\eta) \in T^*\Omega \setminus 0$, acting between weighted Sobolev spaces on the (open infinite stretched) cone $X^\wedge := \mathbb{R}_+ \times X$, cf. Section 1.3 below. Comparing (0.8) with (0.1) we see that there is a family of subordinate conormal symbols

$$\sigma_c\sigma_\wedge(A)(y,z) = \sum_{j=0}^{\mu} a_{j0}(0,y) z^j \tag{0.10}$$

from the interpretation of (0.9) as an operator of Fuchs type for every fixed $(y,\eta) \in T^*\Omega \setminus 0$. In the case of ellipticity the adequate weights $\gamma \in \mathbb{R}$ for (0.9) are to be fixed in connection with the non-bijectivity points of (0.10) in the complex plane.

The difference between the notation 'boundary' and 'edge' symbol is motivated by the fact that the choice of spaces is different.

If we write the operator (0.3) in edge-degenerate form (0.6) (with coefficients $a_{j\alpha}(r,y) \in C^\infty(\overline{\mathbb{R}}_+ \times \Omega)$; in this case we have $\dim X = 0$) we also obtain an edge symbol (0.9).

In the present paper we concentrate on edge-degenerate operators (0.6) and their pseudo-differential versions. The analogy with boundary value problems will not play a major role, but it is worth to recall that the edge calculus for $\dim X = 0$ corresponds to the pseudo-differential calculus of boundary value problems without the transmission property at the boundary, cf. [18].

1. Operators on manifolds with edges

1.1. Manifolds with conical singularities and edges

Let X be a topological space. As in the introduction we set

$$X^\wedge := (\overline{\mathbb{R}}_+ \times X)/(\{0\} \times X) \quad \text{and} \quad X^\wedge := \mathbb{R}_+ \times X.$$

In the following discussion, for convenience, topological spaces are assumed to be countable unions of compact sets; C^∞ manifolds are assumed to be oriented and equipped with a Riemannian metric.

The definition of a manifold W with edge Y is based on a certain kind of (locally trivial) X^Δ-bundles L over Y; here Y and X are C^∞ manifold.

In order to describe the specific structure we first consider a (locally trivial) $\overline{\mathbb{R}}_+ \times X$-bundle $\mathbb{L}$ over Y, the stretched space associated with L.

The transition maps

$$l : \overline{\mathbb{R}}_+ \times X \times \Omega \to \overline{\mathbb{R}}_+ \times X \times \widetilde{\Omega}$$

between trivialisations of $\mathbb{L}$, $\Omega, \widetilde{\Omega} \subseteq \mathbb{R}^q$ open, $q = \dim Y$, are assumed to be restrictions of transition maps $\mathbb{R} \times X \times \Omega \to \mathbb{R} \times X \times \widetilde{\Omega}$ belonging to an $\mathbb{R} \times X$ bundle $2\mathbb{L}$ over Y to $\overline{\mathbb{R}}_+ \times X \times \Omega$. Then $\mathbb{L}$ contains a subspace $\mathbb{L}_{\text{sing}}$ which is an X-bundle over Y, represented by the trivialisations $\{0\} \times X \times \Omega$. Let us set $\mathbb{L}_{\text{reg}} := \mathbb{L} \setminus \mathbb{L}_{\text{sing}}$; this is an $X^\wedge$-bundle over Y with the trivialisations $X^\wedge \times \Omega$.

From the construction of $\mathbb{L}$ and $\mathbb{L}_{\text{sing}}$ it follows that we obtain an X^Δ-bundle L by passing to the quotient space $\overline{\mathbb{R}}_+ \times X \to X^\Delta = (\overline{\mathbb{R}}_+ \times X)/(\{0\} \times X)$ in every fibre of $\mathbb{L}$. The bundle L contains Y as a subspace, interpreted as an edge, and $L \setminus Y$ is a C^∞ manifold which can be identified with $\mathbb{L}_{\text{reg}}$ in a natural way. More precisely, we have a projection

$$\pi : \mathbb{L} \to L, \tag{1.1}$$

fibrewise defined by $\overline{\mathbb{R}}_+ \times X \to X^\Delta$, and π restricts to the bundle projection

$$\pi_{\text{sing}} : \mathbb{L}_{\text{sing}} \to Y \tag{1.2}$$

and to an isomorphism of $X^\wedge$-bundles

$$\pi_{\text{reg}} : \mathbb{L}_{\text{reg}} \to L \setminus Y. \tag{1.3}$$

Definition 1.1. Let W be a topological space and $Y \subset W$ a subspace. Then W is called a manifold with edge Y, if

(i) $W \setminus Y$ and Y are C^∞ manifolds;
(ii) there exists a neighborhood V of Y in W and a homeomorphism

$$\chi : V \to L$$

to an X^Δ-bundle L over Y for a C^∞ manifold X, such that χ restricts to diffeomorphisms

$$\chi_0 : V \cap Y \to Y, \quad \chi_{\text{reg}} : V \setminus Y \to L \setminus Y.$$

Incidentally we will write

$$W' = Y \tag{1.4}$$

From W we can pass to a so-called stretched manifold $\mathbb{W}$ when we first replace V by the stretched set $\mathbb{V}$ that is defined to be a C^∞ manifold with boundary, diffeomorphic to $\mathbb{L}$, such that $\mathbb{V} \setminus \partial\mathbb{V}$ is identified with $V \setminus Y$ and $\partial\mathbb{V}$ isomorphic to $\mathbb{L}_{\text{sing}}$ as an X-bundle over Y.

In other words, $\mathbb{W}$ is a C^∞ manifold with boundary $\partial\mathbb{W} =: \mathbb{W}_{\text{sing}}$ which is an X-bundle over Y. Let us set $\mathbb{W} \setminus \partial\mathbb{W} =: \mathbb{W}_{\text{reg}}$. We then have a canonical continuous map

$$\pi : \mathbb{W} \to W$$

which restricts to the bundle projection

$$\pi_{\text{sing}} : \mathbb{W}_{\text{sing}} \to Y$$

and to a diffeomorphism

$$\pi_{\text{reg}} : \mathbb{W}_{\text{reg}} \to W \setminus Y.$$

With $\mathbb{W}$ we can associate the double $2\mathbb{W}$ which is a C^∞ manifold (without boundary) by gluing together two copies $\mathbb{W}_\pm$ of $\mathbb{W}$ along the common boundary $\partial\mathbb{W}$.

Example. For $W = X^\Delta \times \Omega$ we have $\mathbb{W} = \overline{\mathbb{R}}_+ \times X \times \Omega$ and $2\mathbb{W} = \mathbb{R} \times X \times \Omega$.

Remark 1.2. In the case $\dim Y = 0$ we speak about manifolds with conical singularities.

Remark 1.3. Let W be a manifold with edge Y. Then for every C^∞ manifold M the Cartesian product $W \times M$ is a manifold with edge $Y \times M$.

Definition 1.4. Let W_i be manifolds with edges Y_i, $i = 1, 2$, and let X_i be the base of the model cone for W_i, $i = 1, 2$. A continuous map

$$T : W_1 \to W_2$$

is called an $\mathfrak{M}_1$-morphism, if there is a differentiable map

$$\mathbb{T} : \mathbb{W}_1 \to \mathbb{W}_2$$

between the respective stretched manifolds as manifolds with C^∞ boundary, such that

$$\mathbb{T}|_{\partial\mathbb{W}_1} : \partial\mathbb{W}_1 \to \partial\mathbb{W}_2$$

is a homomorphism between the corresponding X_i-bundles (in particular, $T|_{Y_1}$: $Y_1 \to Y_2$ is then a differentiable map). $T : W_1 \to W_2$ is called an $\mathfrak{M}_1$-isomorphism if there is an $\mathfrak{M}_1$-morphism $T^{-1} : W_2 \to W_1$ which is a two-sided inverse to T.

In this case we also write $W_1 \cong_{\mathfrak{M}_1} W_2$.

In this way, the manifolds with edges form a category $\mathfrak{M}_1$ with the subcategory of manifolds with conical singularities.

Let W be a manifold with edge Y of dimension $q > 0$. Then the above mentioned bijection $\mathbb{V} \underset{\cong}{\to} \mathbb{L}$ allows us to define stretched wedge neighborhoods $\mathbb{U} \subset \mathbb{V}$ that correspond to a trivialisation of $\mathbb{L}$, i.e.,

$$\mathbb{U} \cong_{\mathfrak{M}_1} \overline{\mathbb{R}}_+ \times X \times \Omega \tag{1.5}$$

for open sets $\Omega \subseteq \mathbb{R}^q$. Set $\mathbb{U}_{\text{reg}} := \mathbb{U} \setminus \partial\mathbb{W}$.

Let W be a manifold with edge Y and $\mathbb{W}$ its stretched manifold. By

$$\text{Diff}^{\mu}_{\text{deg}}(W)$$

we denote the space of all differential operators $A \in \text{Diff}^{\mu}(W \setminus Y)$ such that for every (stretched) wedge neighborhood $\mathbb{U}_{\text{reg}}$ in the splitting of variables $(r, x, y) \in \mathbb{R}_+ \times X \times \Omega$ the operator A has the form

$$A = r^{-\mu} \sum_{j+|\alpha|\le\mu} a_{j\alpha}(r, y)\Big(-r\frac{\partial}{\partial r}\Big)^j (rD_y)^{\alpha} \tag{1.6}$$

with coefficients $a_{j\alpha} \in C^{\infty}(\overline{\mathbb{R}}_+ \times \Omega, \text{Diff}^{\mu-(j+|\alpha|)}(X))$.

In local coordinates $(r, x, y) \in \mathbb{R}_+ \times \Sigma \times \Omega$, $\Sigma \subseteq \mathbb{R}^n$, $\Omega \subset \mathbb{R}^q$ open ($n = \dim X$, $q = \dim Y$) the homogeneous principal symbol $\sigma_{\psi}(A)$ of A of order μ has the form

$$\sigma_{\psi}(A)(r, x, y, \varrho, \xi, \eta) = r^{-\mu}\tilde{\sigma}_{\psi}(A)(r, x, y, r\varrho, \xi, r\eta)$$

for a function $\tilde{\sigma}_{\psi}(A)(r, x, y, \tilde{\varrho}, \xi, \tilde{\eta})$ which is smooth up to $r = 0$ (and homogeneous of order μ in $(\tilde{\varrho}, \xi, \tilde{\eta})$). In addition, as noted in the introduction, we have the homogeneous principal edge symbol $\sigma_{\wedge}(A)(y, \eta)$, $(y, \eta) \in T^*Y \setminus 0$, the second component of the principal symbolic hierarchy (0.7). Note that for

$$a(y, \eta) := r^{-\mu} \sum_{j+|\alpha|\le\mu} a_{j\alpha}(r, y)(-r\frac{\partial}{\partial r})^j (r\eta)^{\alpha}, \tag{1.7}$$

and $(\kappa_{\lambda}u)(r, x) := \lambda^{\frac{n+1}{2}} u(\lambda r, x)$, $\lambda \in \mathbb{R}_+$, we have

$$\sigma_{\wedge}(A)(y, \eta) = \lim_{\lambda\to\infty} \lambda^{-\mu}\kappa_{\lambda} a(y, \lambda\eta)\kappa_{\lambda}^{-1},$$

and

$$\sigma_{\wedge}(A)(y, \lambda\eta) = \lambda^{\mu}\kappa_{\lambda}\sigma_{\wedge}(A)(y, \eta)\kappa_{\lambda}^{-1},$$

for all $\lambda \in \mathbb{R}_+$.

1.2. Weighted Sobolev spaces

We now formulate Sobolev spaces on stretched cones $X^\wedge = \mathbb{R}_+ \times X \ni (r, x)$ and wedges $X^\wedge \times \mathbb{R}^q \ni (r, x, y)$, first for the case of a smooth compact manifold X. To this end we first consider the cylindrical Sobolev space

$$H^s(\mathbb{R} \times X)$$

defined as the set of all $u(t, .) \in H^s_{\mathrm{loc}}(\mathbb{R}_t \times X)$ such that

$$(\varphi u) \circ (1 \times \chi)^{-1} \in H^s(\mathbb{R}_t \times \mathbb{R}^n)$$

for every chart $\chi : U \to \mathbb{R}^n$ on X and every $\varphi \in C_0^\infty(U)$; here $(1 \times \chi)(r, \cdot) := (r, \chi(\cdot))$. Let

$$(S_\beta u)(t) := e^{-(\frac{1}{2}-\beta)t} u(e^{-t}),\ t \in \mathbb{R}$$

for any $\beta \in \mathbb{R}$. Then we set

$$\mathcal{H}^{s,\gamma}(X^\wedge) := \left(S_{\gamma-\frac{n}{2}}\right)^{-1} H^s(\mathbb{R} \times X)$$

for $n = \dim X$.

Let us interpret these spaces in connection with the Mellin transform on $\mathbb{R}_+ \ni r$,

$$Mu(z) = \int_0^\infty r^{z-1} u(r) dr.$$

For $u \in C_0^\infty(\mathbb{R}_+)$ the function, $Mu(z)$ is an entire function, and we have $Mu\big|_{\Gamma_\beta} \in \mathcal{S}(\Gamma_\beta)$ for

$$\Gamma_\beta := \{z \in \mathbb{C} : \mathrm{Re} z = \beta\} \tag{1.8}$$

for every $\beta \in \mathbb{R}$, uniformly in compact intervals. Here and in the sequel the 'weight line' Γ_β is treated as a real axis in connection with the spaces, e.g., the Schwartz space, Sobolev spaces, etc., or amplitude functions, given with respect to the corresponding covariable.

The Mellin transform will also be applied to vector- or operator-valued functions depending on $r \in \mathbb{R}$; then the covariable z will often vary on a certain weight line. We have $-r\frac{\partial}{\partial r} = M^{-1} z M$, and $\mathcal{H}^{s,\gamma}(X^\wedge)$ for $s \in \mathbb{N}$ is equal to the subspace of all $u(r, x) \in r^{-\frac{n}{2}+\gamma} L^2(X^\wedge)$ such that $(-r\partial_r)^k D_x^\alpha u(r, x) \in r^{-\frac{n}{2}+\gamma} L^2(X^\wedge)$ for all $k + |\alpha| \le s$. Here D_x^α runs over all elements of $\mathrm{Diff}^{|\alpha|}(X)$, and $L^2(X^\wedge)$ refers to the measure $drdx$. From this definition we can recover $\mathcal{H}^{s,\gamma}(X^\wedge)$ for arbitrary $s \in \mathbb{R}$ by duality and interpolation. It will be adequate to modify the spaces $\mathcal{H}^{s,\gamma}(X^\wedge)$ at infinity by setting

$$\mathcal{K}^{s,\gamma}(X^\wedge) := \{\omega u + (1 - \omega) v : u \in \mathcal{H}^{s,\gamma}(X^\wedge)\ v \in H^s_{\mathrm{cone}}(X^\wedge)\}$$

for a space $H^s_{\mathrm{cone}}(X^\wedge)$ that is defined as follows.

Set $B := \{x \in \mathbb{R}^n : |x| < 1\}$ and $B^\vee := \{(r, rx) \in \mathbb{R}^{1+n} : (r, x) \in \mathbb{R}_+ \times B\}$ which is a conical set in $\mathbb{R}^{1+n}$. Let $\chi : U \to B$ be a chart on X, and consider $1 \times \chi : \mathbb{R}_+ \times U \to \mathbb{R}_+ \times B$. Together with

$$\beta : (r, x) \to (r, rx), \quad \mathbb{R}_+ \times B \to B^\vee,$$

we have the composition

$$\beta \circ (1 \times \chi) : \mathbb{R}_+ \times U \to B^\vee.$$

Then $H^s_{\text{cone}}(X^\wedge)$ is defined to be the subspace of all $u \in H^s_{\text{loc}}(\mathbb{R} \times X)|_{\mathbb{R}_+ \times X}$ such that $(1-\omega)\varphi u \circ (1 \times \chi)^{-1} \circ \beta^{-1} \in H^s(\mathbb{R}^{1+n})$ for every chart $\chi : U \to B$ and arbitrary $\varphi \in C_0^\infty(U)$. Another equivalent characterization of $H^s_{\text{cone}}(X^\wedge)$ is given in Remark 2.1 below.

Let us introduce some convenient terminology in connection with the variety of spaces that will occur in our calculus.

If E_0, E_1 are Fréchet spaces, embedded in a Hausdorff topological vector space H, we set

$$E_0 + E_1 = \{e_0 + e_1 : e_0 \in E_0, e_1 \in E_1\}. \tag{1.9}$$

There is then an algebraic isomorphism $E_0 + E_1 \cong E_0 \oplus E_1 / \Delta$ for $\Delta := \{(e, -e) : e \in E_0 \cap E_1\}$, and we endow (1.9) with the Fréchet topology of the quotient space. We then call (1.9) the non-direct sum of E_0 and E_1.

Moreover, let E be a Fréchet space that is a left module over an algebra A. Then $[a]E$ for $a \in A$ will denote the closure of the space $\{ae : e \in E\}$ in E. In a similar sense we employ the notation $E[b]$ or $[a]E[b]$ for $a, b \in E$ if E is a right or two-sided A-module.

Example. The spaces $\mathcal{H}^s_{\text{cone}}(X^\wedge)$ are two-sided modules over the algebra of all $\varphi(r, x) \in C^\infty(\overline{\mathbb{R}}_+ \times X)$ that do not depend on r for $r > R$ for some $R > 0$. In particular, for every cut-off function $\omega(r)$ we can form the spaces $[\omega]\mathcal{H}^{s,\gamma}(X^\wedge)$ and $[1-\omega]H^s_{\text{cone}}(X^\wedge)$, and we have

$$\mathcal{K}^{s,\gamma}(X^\wedge) = [\omega]\mathcal{H}^{s,\gamma}(X^\wedge) + [1-\omega]H^s_{\text{cone}}(X^\wedge). \tag{1.10}$$

Clearly the space (1.10) is independent of the specific choice of ω.

Remark 1.5. If E_0 and E_1 are Hilbert spaces, also $E_0 + E_1$ becomes a Hilbert space by the identification with the orthogonal complement of Δ in $E_0 \oplus E_1$.

Thus, if we fix the cut-off function ω, we get a Hilbert space structure in $\mathcal{K}^{s,\gamma}(X^\wedge)$ via (1.10). For $s = \gamma = 0$ we take the scalar product from the identification

$$\mathcal{K}^{0,0}(X^\wedge) = r^{-\frac{n}{2}} L^2(\mathbb{R}_+ \times X),$$

where $L^2(\mathbb{R}_+ \times X)$ refers to $drdx$ with dx being connected with a fixed Riemannian metric on X.

For purposes below we tacitly identify a coordinate neighborhood U on X with $B \subset \mathbb{R}^n$, with the coordinates x via a chart $\chi : U \to B$. Then, for the above characterization of the space $H^s_{\text{cone}}(X^\wedge)$ for large r it is enough to look at $(1-\omega)\varphi H^s_{\text{cone}}(B^\wedge)$ for any $\varphi \in C_0^\infty(B)$ and a cut-off function ω.

We then have

$$u \in (1-\omega)\varphi H^s_{\text{cone}}(X^\wedge) \Longleftrightarrow (\beta^* u)(r, \frac{\tilde{x}}{r}) \in H^s(\mathbb{R}^{1+n}_{r,\tilde{x}}) \tag{1.11}$$

for the map $\beta : B^\wedge \to B^\vee \subset \mathbb{R}^{1+n}$.

Remark 1.6. Let $A \in \mathrm{Diff}^{\mu}_{\deg}(\overline{\mathbb{R}}_+ \times X \times \Omega)$ be given in the form (1.6), and set

$$a(y,\eta) := r^{-\mu} \sum_{j+|\alpha|\le\mu} a_{j\alpha}(r,y)\Big(-r\frac{\partial}{\partial r}\Big)^j (r\eta)^\alpha. \tag{1.12}$$

Assume that there is an $R > 0$ such that the coefficients $a_{j\alpha}$ are independent of r for $r > R$. Then

$$D_y^\alpha D_\eta^\beta a(y,\eta) : \mathcal{K}^{s,\gamma}(X^\wedge) \to \mathcal{K}^{s-\mu+|\beta|,\gamma-\mu+|\beta|}(X^\wedge),$$

$(y,\eta) \in \Omega\times\mathbb{R}^q$, is a family of continuous operators for every $s, \gamma \in \mathbb{R}$ and $\alpha, \beta \in \mathbb{N}^q$.

In fact, writing $a(y,\eta) = \tilde{\omega}a(y,\eta) + (1-\tilde{\omega})a(y,\eta)$, for some cut-off function $\tilde{\omega}(r)$ we first have

$$\tilde{\omega}a(y,\eta) : \mathcal{H}^{s,\gamma}(X^\wedge) \to \tilde{\omega}\mathcal{H}^{s-\mu,\gamma-\mu}(X^\wedge)$$

which is fairly obvious. On the other hand, to obtain

$$(1-\tilde{\omega})a(y,\eta) : H^s_{\mathrm{cone}}(X^\wedge) \to (1-\tilde{\omega})H^s_{\mathrm{cone}}(X^\wedge) \tag{1.13}$$

we express $a(y,\eta)$ in local coordinates on X via a chart $\chi : U \to B$. If we write

$$a_{j\alpha}(r,y) = \sum_{|\gamma|\le\mu-(j+|\alpha|)} a_{j\alpha;\gamma}(r,x,y)D_x^\gamma$$

with coefficients $a_{j\alpha;\gamma} \in C^\infty(\overline{\mathbb{R}}_+ \times U \times \Omega)$ for a coordinate neighborhood $U \subset X$, it suffices to consider the summands

$$r^{-\mu}a_{j\alpha;\gamma}(r,x,y)D_x^\gamma\Big(-r\frac{\partial}{\partial r}\Big)^j (r\eta)^\alpha \tag{1.14}$$

separately. For simplicity, let us interpret x as coordinates in the unit ball $B \subset \mathbb{R}^n$. Then (1.14) is a family of operators in $\mathbb{R}_+ \times B \ni (r,x)$. Applying the substitution $\tilde{x} = rx$ and the characterization (1.11), for (1.13) it is enough to observe that

$$(1-\tilde{\omega})\varphi\Big(\frac{\tilde{x}}{r}\Big)r^{-\mu}a_{j\alpha}\Big(r,\frac{\tilde{x}}{r},y\Big)r^{|\gamma|}D_{\tilde{x}}^\gamma\Big(-r\frac{\partial}{\partial r}\Big)^j(r\eta)^\alpha$$

for $\varphi(x) \in C_0^\infty(U)$ respects the standard Sobolev spaces up to $r = \infty$.

1.3. Abstract edge spaces and symbols with twisted homogeneity

A Hilbert space E is said to be equipped with a group action if there is given a strongly continuous group of isomorphisms $\kappa_\lambda : E \to E$, $\lambda \in \mathbb{R}_+$, such that $\kappa_\lambda\kappa_{\lambda'} = \kappa_{\lambda\lambda'}$ for all $\lambda, \lambda' \in \mathbb{R}_+$. An example is the space $E = \mathcal{K}^{s,\gamma}(X^\wedge)$ with

$$(\kappa_\lambda u)(r,x) = \lambda^{\frac{n+1}{2}}u(\lambda r,x),\ \lambda \in \mathbb{R}_+$$

for $n = \dim X$. More generally, if E is a Fréchet space, written as a projective limit $\varprojlim_{j\in\mathbb{N}} E^j$ of Hilbert spaces E^j with continuous embeddings $E^{j+1} \hookrightarrow E^j$ for all j, and if $\{\kappa_\lambda\}_{\lambda\in\mathbb{R}_+}$ is a group action on E^0 which restricts to group actions on E^j for every j we say that E is equipped with a group action.

An example is the space

$$\mathcal{S}^{\gamma}_{\varepsilon}(X^{\wedge}) := \varprojlim_{j\in\mathbb{N}} E^j \tag{1.15}$$

for $E^j := \langle r\rangle^{-j}\mathcal{K}^{j,\gamma+\varepsilon-(1+j)^{-1}}(X^{\wedge})$, $\varepsilon > 0$. Let us define

$$\mathcal{S}_{\mathcal{O}}(X^{\wedge}) := \varprojlim_{\varepsilon>0} \mathcal{S}^{\gamma}_{\varepsilon}(X^{\wedge}) \tag{1.16}$$

(the notation indicates that the left-hand side is independent of γ).

Definition 1.7. Let E be a Hilbert space with group action $\{\kappa_\lambda\}_{\lambda\in\mathbb{R}_+}$. Then $\mathcal{W}^s(\mathbb{R}^q, E)$ (the 'abstract' edge Sobolev space on $\mathbb{R}^q$ of smoothness $s \in \mathbb{R}$) is defined to be the completion of $\mathcal{S}(\mathbb{R}^q, E)$ with respect to the norm

$$\Big\{\int \langle\eta\rangle^{2s}\big\|\kappa^{-1}_{\langle\eta\rangle}\hat{u}(\eta)\big\|^2_E d\eta\Big\}^{\frac{1}{2}};$$

here $\hat{u}(\eta) = F_{y\to\eta}u(\eta)$ is the Fourier transform in $\mathbb{R}^q$, and $\kappa^{-1}_{\langle\eta\rangle}$ acts on the values of $\hat{u}(\eta)$ for every η.

Definition 1.8.

(i) Let E and $\widetilde{E}$ be Hilbert spaces with group actions $\{\kappa_\lambda\}_{\lambda\in\mathbb{R}_+}$ and $\{\tilde{\kappa}_\lambda\}_{\lambda\in\mathbb{R}_+}$, respectively. Then the space of (operator-valued) symbols $S^\mu(U \times \mathbb{R}^q; E, \widetilde{E})$ for an open set $U \subseteq \mathbb{R}^p$, $\mu \in \mathbb{R}$, is defined to be the set all $a(y,\eta) \in C^\infty(U \times \mathbb{R}^q, \mathcal{L}(E, \widetilde{E}))$ such that

$$\sup\Big\{\langle\eta\rangle^{-\mu+|\beta|}\big\|\tilde{\kappa}^{-1}_{\langle\eta\rangle}\{D^\alpha_y D^\beta_\eta a(y,\eta)\}\kappa_{\langle\eta\rangle}\big\|_{\mathcal{L}(E,\widetilde{E})} : (y,\eta) \in K \times \mathbb{R}^q\Big\}$$

is finite for every $K \subset U$ and all multi-indices $\alpha \in \mathbb{N}^p$, $\beta \in \mathbb{N}^q$.

(ii) $S^{(\mu)}(U \times (\mathbb{R}^q \setminus \{0\}); E, \widetilde{E})$ denotes the set of all functions $f_{(\mu)} \in C^\infty(U \times (\mathbb{R}^q \setminus \{0\}), \mathcal{L}(E, \widetilde{E}))$ such that

$$f_{(\mu)}(y, \lambda\eta) = \lambda^\mu \tilde{\kappa}_\lambda f_{(\mu)}(y,\eta)\kappa^{-1}_\lambda$$

for all $(y,\eta) \in U \times (\mathbb{R}^q \setminus \{0\})$, $\lambda \in \mathbb{R}_+$.

(iii) The space $S^\mu_{\mathrm{cl}}(U \times \mathbb{R}^q; E, \widetilde{E})$ of classical symbols is defined as the set of all $a(y,\eta) \in S^\mu(U \times \mathbb{R}^q; E, \widetilde{E})$ such that there are elements $a_{(\mu-j)}(y,\eta) \in S^{(\mu-j)}(U \times (\mathbb{R}^q \setminus \{0\}); E, \widetilde{E})$, $j \in \mathbb{N}$, such that

$$a(y,\eta) - \sum_{j=0}^{N} \chi(\eta)a_{(\mu-j)}(y,\eta) \in S^{\mu-(N+1)}(U \times \mathbb{R}^q; E, \widetilde{E})$$

for every $N \in \mathbb{N}$. Here $\chi(\eta)$ is any excision function, i.e., $\chi \in C^\infty(\mathbb{R}^q)$, $\chi(\eta) = 0$ for $|\eta| < c_0$, $\chi(\eta) = 1$ for $|\eta| > c_1$ for certain $0 < c_0 < c_1$.

In the case $E = \mathbb{C}$ we always set $\kappa_\lambda = \mathrm{id}_{\mathbb{C}}$ for every $\lambda \in \mathbb{R}_+$. Then for $E = \widetilde{E} = \mathbb{C}$ the Definition 1.8 reproduces the standard spaces of scalar symbols. If a notation or a relation is valid both in the classical and the general case we also write '(cl)' as subscript. Let $S^{\mu}_{(\mathrm{cl})}(\mathbb{R}^q; E, \widetilde{E})$ denote the corresponding spaces of symbols $a(\eta)$ that are independent of y, i.e., with constant coefficients.

Definition 1.8 extends to Fréchet spaces $E, \widetilde{E}$, that are projective limits of Hilbert spaces with group action, cf. [13].

Remark 1.9. The symbol spaces of Definition 1.8 depend on the choice of the group actions $\kappa = \{\kappa_\lambda\}_{\lambda\in\mathbb{R}_+}$ and $\widetilde{\kappa} = \{\widetilde{\kappa}_\lambda\}_{\lambda\in\mathbb{R}_+}$ in the spaces E and $\widetilde{E}$, respectively. If necessary we use the notation

$$S^{\mu}_{(\mathrm{cl})}(U \times \mathbb{R}^q; E, \widetilde{E})_{\kappa,\widetilde{\kappa}} \tag{1.17}$$

instead of $S^{\mu}_{(\mathrm{cl})}(U \times \mathbb{R}^q; E, \widetilde{E})$.

Example. For the operator function (1.7) we have

$$S^{\mu}(\Omega \times \mathbb{R}^q; \mathcal{K}^{s,\gamma}(X^\wedge), \mathcal{K}^{s-\mu,\gamma-\mu}(X^\wedge)),$$

provided that the coefficients $a_{j\alpha}$ satisfy a suitable condition for $r \to \infty$, e.g., to be independent of r for large r.

Example. Let us set

$$E := \mathcal{K}^{s,\gamma}(X^\wedge) \oplus \mathbb{C}^{j_-}, \ \widetilde{E}_\varepsilon := \mathcal{S}^{\gamma-\mu}_\varepsilon(X^\wedge) \oplus \mathbb{C}^{j_+}$$

and

$$F := \mathcal{K}^{s,-\gamma+\mu}(X^\wedge) \oplus \mathbb{C}^{j_+}, \ \widetilde{F}_\varepsilon := \mathcal{S}^{-\gamma}_\varepsilon(X^\wedge) \oplus \mathbb{C}^{j_-}$$

for $\gamma, \mu \in \mathbb{R}$ and $\varepsilon > 0$. A family of operators

$$g(y,\eta) : \mathcal{K}^{s,\gamma}(X^\wedge) \oplus \mathbb{C}^{j_-} \to \mathcal{K}^{\infty,\gamma}(X^\wedge) \oplus \mathbb{C}^{j_+}$$

(continuous for every $s \in \mathbb{R}$) is called a Green symbol, if there is an $\varepsilon = \varepsilon(g) > 0$ such that

$$g(y,\eta) \in S^{\mu}_{\mathrm{cl}}(U \times \mathbb{R}^q; E, \widetilde{E}_\varepsilon), g^*(y,\eta) \in S^{\mu}_{\mathrm{cl}}(U \times \mathbb{R}^q; F, \widetilde{F}_\varepsilon) \tag{1.18}$$

for all $s \in \mathbb{R}$. A Green symbol is flat (of infinite order) if the conditions (1.18) hold for all $\varepsilon > 0$. Especially, if $g(y,\eta)$ is an upper left corner, that means $g(y,\eta), g^*(y,\eta) \in S^{\mu}_{\mathrm{cl}}(U \times \mathbb{R}^q; \mathcal{K}^{s,\beta}(X^\wedge), \mathcal{S}_O(X^\wedge))$ for all $s, \beta \in \mathbb{R}$.

Note that Green operators (for fixed y, η) admit beautiful kernel characterizations, cf. [22]. Those imply corresponding descriptions of the symbols themselves, cf. [20].

Remark 1.10. The operator of multiplication by a function $\varphi \in C_0^\infty(\overline{\mathbb{R}}_+)$ belongs to $S^0(\mathbb{R}^q; \mathcal{K}^{s,\gamma}(X^\wedge), \mathcal{K}^{s,\gamma}(X^\wedge))$ for every $s, \gamma \in \mathbb{R}$. If $g(y,\eta)$ is a Green symbol in the sense of Example 1.3, also

$$\mathrm{diag}(\varphi, 1) g(y,\eta) \mathrm{diag}(\tilde{\varphi}, 1)$$

is a Green symbol for every $\varphi, \tilde{\varphi} \in C_0^\infty(\overline{\mathbb{R}}_+)$

Parallel to the spaces of symbols of Definition 1.8 we have vector-valued analogues of Sobolev spaces, based on a Hilbert space E with group action $\{\kappa_\lambda\}_{\lambda\in\mathbb{R}_+}$. Recall the corresponding definition from [12]. By $\mathcal{W}^s(\mathbb{R}^q, E)$ we denote the completion of $\mathcal{S}(\mathbb{R}^q, E)$ with respect to the norm

$$\left\{\int \langle\eta\rangle^{2s} \|\kappa^{-1}_{\langle\eta\rangle}\widehat{u}(\eta)\|^2_E d\eta\right\}^{1/2},$$

$\widehat{u}(\eta) := (F_{y\to\eta}u)(\eta)$. For an open set $\Omega \subseteq \mathbb{R}^q$ we also have the spaces

$$\mathcal{W}^s_{\mathrm{comp}}(\Omega, E) \quad \text{and} \quad \mathcal{W}^s_{\mathrm{loc}}(\Omega, E)$$

of vector-valued analogues of the spaces $H^s_{\mathrm{comp}}(\Omega)$ and $H^s_{\mathrm{loc}}(\Omega)$, respectively.

Concerning useful functional analytic properties of the abstract edge spaces, see [13] or [8]. There is an immediate generalisation of these spaces to the case of Fréchet spaces E that are projective limits of Hilbert spaces with group action.

Let us recall the following general continuity result.

Let $a(y, y', \eta) \in S^\mu(\Omega \times \Omega \times \mathbb{R}^q; E, \widetilde{E})$ be a symbol in the sense of Definition 1.8, here with $U := \Omega \times \Omega$ for an open set $\Omega \subseteq \mathbb{R}^q$. Then, setting $\mathrm{Op}(a)u(y) := \iint e^{i(y-y')\eta}a(y, y', \eta)u(y')dy'đ\eta$ for $đ\eta = (2\pi)^{-n}d\eta$, the operator

$$\mathrm{Op}(a) : \mathcal{W}^s_{\mathrm{comp}}(\Omega, E) \to \mathcal{W}^{s-\mu}_{\mathrm{loc}}(\Omega, \widetilde{E}) \tag{1.19}$$

is continuous for all $s \in \mathbb{R}$. A proof may be found in [13] under some natural conditions on $E, \widetilde{E}$, which are satisfied, for instance, for the spaces $E := \mathcal{K}^{s,\gamma}(X^\wedge)$, $\widetilde{E} := \mathcal{K}^{s-\mu,\gamma-\mu}(X^\wedge)$, occurring in the edge calculus. The continuity (1.19) for $\Omega = \mathbb{R}^q$, $\mu = 0$ and $s = 0$ for $E, \widetilde{E}$ without any extra condition is proved in Seiler [21], as a generalisation of the Calderón-Vaillancourt theorem in the version of [9]. This entails other continuity results, for instance, for arbitrary $\mu, s \in \mathbb{R}$, when we specify the symbol a by imposing a certain exit property with respect to y, cf. the following section in the scalar case; vector-valued variants are studied in [10]. Similarly as (1.17) the spaces $\mathcal{W}^s(\mathbb{R}^q, E)$ depend on the choice of $\kappa = \{\kappa_\lambda\}_{\lambda\in\mathbb{R}_+}$. If necessary we write

$$\mathcal{W}^s(\mathbb{R}^q, E)_\kappa \tag{1.20}$$

instead of $\mathcal{W}^s(\mathbb{R}^q, E)$. It is reasonable, cf. Tarkhanov [23], or Schulze and Tarkhanov [19], also to take for E and $\widetilde{E}$ the spaces

$$K^{s,\gamma}(X^\wedge) := \langle r\rangle^{s-\gamma}\mathcal{K}^{s,\gamma}(X^\wedge) \quad \text{and} \quad K^{s-\mu,\gamma-\mu}(X^\wedge) := \langle r\rangle^{s-\gamma}\mathcal{K}^{s-\mu,\gamma-\mu}(X^\wedge), \tag{1.21}$$

respectively, with the group action $(\kappa^{s-\gamma}_\lambda u)(r, x) := \lambda^{s-\gamma+\frac{n+1}{2}}u(\lambda r, x)$, $\lambda \in \mathbb{R}_+$. In that case we can form

$$W^{s,\gamma}(X^\wedge \times \mathbb{R}^q) := \mathcal{W}^s(\mathbb{R}^q, K^{s,\gamma}(X^\wedge))_{\kappa^{s-\gamma}}, \tag{1.22}$$

and, similarly, spaces of the kind $W^{s,\gamma}_{\mathrm{comp}(y)}(X^\wedge \times \Omega)$ and $W^{s,\gamma}_{\mathrm{loc}(y)}(X^\wedge \times \Omega)$, $\Omega \subseteq \mathbb{R}^q$ open, where 'comp(y)' and 'loc(y)' means 'comp' and 'loc' with respect to the variables $y \in \Omega$.

1.4. Conical exits to infinity

In this section we want to deepen the information of Remark 1.6 in the sense of a more systematic discussion between edge symbols, standard operators in polar coordinates, and symbols within the exit pseudo-differential calculus.

Let us first consider $\mathbb{R}^{n+1}_{\tilde{x}}$ regarded as a space with 'conical exit' to infinity $|\tilde{x}| \to \infty$.

Definition 1.11. The space

$$S^{\mu;\nu}(\mathbb{R}^{n+1}_{\tilde{x}} \times \mathbb{R}^{n+1}_{\tilde{\xi}}), \tag{1.23}$$

$\mu, \nu \in \mathbb{R}$, is defined to be the set of all $a(\tilde{x}, \tilde{\xi}) \in C^\infty(\mathbb{R}^{n+1}_{\tilde{x}} \times \mathbb{R}^{n+1}_{\tilde{\xi}})$ such that

$$\sup\{\langle\tilde{x}\rangle^{-\nu+|\alpha|}\langle\tilde{\xi}\rangle^{-\mu+|\beta|}|D^\alpha_{\tilde{x}} D^\beta_{\tilde{\xi}} a(\tilde{x}, \tilde{\xi})| : (\tilde{x}, \tilde{\xi}) \in \mathbb{R}^{n+1} \times \mathbb{R}^{n+1}\} < \infty$$

for every $\alpha, \beta \in \mathbb{N}^{n+1}$. We also say that a symbol $a(\tilde{x}, \tilde{\xi}) \in S^\mu(\mathbb{R}^{n+1} \times \mathbb{R}^{n+1})$ has the exit property, if it belongs to the space (1.23); then μ is called the pseudo-differential order, ν the exit order of a.

More generally,

$$S^{\mu;\nu,\nu'}(\mathbb{R}^{n+1}_{\tilde{x}} \times \mathbb{R}^{n+1}_{\tilde{x}'} \times \mathbb{R}^{n+1}_{\tilde{\xi}})$$

denotes the set of all $a(\tilde{x}, \tilde{x}', \tilde{\xi}) \in C^\infty(\mathbb{R}^{3(n+1)}_{\tilde{x},\tilde{x}',\xi})$ such that

$$\sup\{\langle\tilde{x}\rangle^{-\nu+|\alpha|}\langle\tilde{x}'\rangle^{-\nu'+|\alpha'|}\langle\tilde{\xi}\rangle^{-\mu+|\beta|}|D^\alpha_{\tilde{x}} D^{\alpha'}_{\tilde{x}'} a(\tilde{x}, \tilde{x}', \tilde{x})| : (\tilde{x}.\tilde{x}', \tilde{\xi}) \in \mathbb{R}^{3(n+1)}\} < \infty$$

for every $\alpha, \alpha', \beta \in \mathbb{N}^{n+1}$.

Moreover, we set

$$S^{\mu;\nu}_{\mathrm{cl}_{\tilde{\xi},\tilde{x}}}(\mathbb{R}^{n+1}_{\tilde{x}} \times \mathbb{R}^{n+1}_{\tilde{\xi}}) := S^\nu_{\mathrm{cl}}(\mathbb{R}^{n+1}_{\tilde{x}}) \hat{\otimes}_\pi S^\mu_{\mathrm{cl}}(\mathbb{R}^{n+1}_{\tilde{\xi}})$$

which is the space of classical symbols in $\tilde{\xi}$ and $\tilde{x}$.

Example. Let $\omega(t) \in C^\infty_0(\overline{\mathbb{R}}_+)$ be a cut-off function such that $\omega(t) = 1$ for $t < \frac{1}{2}$, $\omega(t) = 0$ for $t > \frac{2}{3}$, and set

$$\psi(r, r') := \omega\left(\frac{(r - r')^2}{1 + (r - r')^2}\right).$$

Then ψ represents an element of $S^{0;0,0}(\mathbb{R}_r \times \mathbb{R}_{r'} \times \mathbb{R}_\varrho)$ (which is independent of the covariable ϱ).

Let us set

$$H^{s;g}(\mathbb{R}^{n+1}) := \langle\tilde{x}\rangle^{-g} H^s(\mathbb{R}^{n+1}).$$

Theorem 1.12. *For every $a(\tilde{x}, \tilde{\xi}) \in S^{\mu;\nu}(\mathbb{R}^{n+1} \times \mathbb{R}^{n+1})$ the associated pseudo-differential operator* $\mathrm{Op}(a)$ *induces continuous operators*

$$\mathrm{Op}(a) : H^{s;g}(\mathbb{R}^{n+1}) \to H^{s-\mu;g-\nu}(\mathbb{R}^{n+1})$$

for all $s, g \in \mathbb{R}$.

Remark 1.13. By virtue of

$$\mathcal{S}(\mathbb{R}^{n+1}) = \varprojlim_{N\in\mathbb{N}} H^{N;N}(\mathbb{R}^{n+1})$$

it follows that $\mathrm{Op}(a)$ also induces a continuous operator

$$\mathrm{Op}(a) : \mathcal{S}(\mathbb{R}^{n+1}) \to \mathcal{S}(\mathbb{R}^{n+1}).$$

Let us set

$$L^{\mu;\nu}_{(\mathrm{cl})}(\mathbb{R}^{n+1}) := \{\mathrm{Op}(a) : a(\tilde{x},\tilde{\xi}) \in S^{\mu;\nu}_{(\mathrm{cl}_{\tilde{\xi};\tilde{x}})}(\mathbb{R}^{n+1}\times\mathbb{R}^{n+1})\}.$$

As is known, $\mathrm{Op}(\cdot)$ induces an isomorphism

$$\mathrm{Op} : S^{\mu;\nu}_{(\mathrm{cl}_{\tilde{\xi};\tilde{x}})}(\mathbb{R}^{n+1}\times\mathbb{R}^{n+1}) \to L^{\mu;\nu}_{(\mathrm{cl})}(\mathbb{R}^{n+1}) \tag{1.24}$$

for every $\mu,\nu \in \mathbb{R}$, including $\mu = -\infty$ or $\nu = -\infty$. Note that $L^{-\infty;-\infty}(\mathbb{R}^{n+1}) = \bigcap_{\mu,\nu} L^{\mu;\nu}(\mathbb{R}^{n+1})$ is equal to the space of all integral operators with kernel in $\mathcal{S}(\mathbb{R}^{n+1}\times\mathbb{R}^{n+1})$.

Definition 1.14. An element $a(\tilde{x},\tilde{\xi}) \in S^{\mu;\nu}(\mathbb{R}^{n+1}\times\mathbb{R}^{n+1})$ is called elliptic, if there exists a $p(\tilde{x},\tilde{\xi}) \in S^{-\mu;-\nu}(\mathbb{R}^{n+1}\times\mathbb{R}^{n+1})$ such that

$$a(\tilde{x},\tilde{\xi})p(\tilde{x},\tilde{\xi}) = 1 \bmod S^{\mu-1;\nu-1}(\mathbb{R}^{n+1}\times\mathbb{R}^{n+1}).$$

An operator $A \in L^{\mu;\nu}(\mathbb{R}^{n+1})$ is called elliptic if the associated symbol $a(\tilde{x},\tilde{\xi}) \in S^{\mu;\nu}(\mathbb{R}^{n+1}\times\mathbb{R}^{n+1})$ is elliptic (cf. the bijection (1.24)).

For classical symbols $a(\tilde{x},\tilde{\xi}) \in S^{\mu;\nu}_{\mathrm{cl}_{\tilde{\xi};\tilde{x}}}(\mathbb{R}^{n+1}\times\mathbb{R}^{n+1})$ we have a triple of principal homogeneous components,

$$\sigma(a) := (\sigma_\psi(a),\, \sigma_\mathrm{e}(a),\, \sigma_{\psi\mathrm{e}}(a))$$

given on $\mathbb{R}^{n+1}\times(\mathbb{R}^{n+1}\setminus\{0\})$, $(\mathbb{R}^{n+1}\setminus\{0\})\times\mathbb{R}^{n+1}$, and $(\mathbb{R}^{n+1}\setminus\{0\})\times(\mathbb{R}^{n+1}\setminus\{0\})$, respectively. The homogeneous principal symbol $\sigma_\psi(a)(\tilde{x},\tilde{\xi})$ in $\tilde{\xi}$ order μ is as usual, namely,

$$\sigma_\psi(a)(\tilde{x},\lambda\tilde{\xi}) = \lambda^\mu\sigma_\psi(a)(\tilde{x},\tilde{\xi}) \text{ for all } (\tilde{x},\tilde{\xi}) \in \mathbb{R}^{n+1}\times(\mathbb{R}^{n+1}\setminus\{0\}), \lambda\in\mathbb{R}_+.$$

Analogously, the principal exit symbol $\sigma_\mathrm{e}(a)(\tilde{x},\tilde{\xi})$ in $\tilde{x}$ of order ν has the property

$$\sigma_\mathrm{e}(a)(\delta\tilde{x},\tilde{\xi}) = \delta^\nu\sigma_\mathrm{e}(a)(\tilde{x},\tilde{\xi}) \text{ for all } (\tilde{x},\tilde{\xi}) \in (\mathbb{R}^{n+1}\setminus\{0\})\times\mathbb{R}^{n+1}, \delta\in\mathbb{R}_+.$$

The third component is the homogeneous principal part of $\sigma_\mathrm{e}(a)(\tilde{x},\tilde{\xi})$ in $\tilde{\xi}$ of order μ and has the homogeneity

$$\sigma_{\psi\mathrm{e}}(a)(\delta\tilde{x},\lambda\tilde{\xi}) = \delta^\nu\lambda^\mu\sigma_{\psi\mathrm{e}}(a)(\tilde{x},\tilde{\xi})$$

for all $(\tilde{x},\tilde{\xi}) \in (\mathbb{R}^{n+1}\setminus\{0\})\times(\mathbb{R}^{n+1}\setminus\{0\})$, $\lambda,\delta\in\mathbb{R}_+$.

Remark 1.15. A symbol $a(\tilde{x},\tilde{\xi}) \in S^{\mu;\nu}_{\mathrm{cl}_{\tilde{\xi};\tilde{x}}}(\mathbb{R}^{n+1} \times \mathbb{R}^{n+1})$ is elliptic in the sense of Definition 1.14 if and only if

$$\begin{aligned}\sigma_\psi(a)(\tilde{x},\tilde{\xi}) &\neq 0 \text{ for all } (\tilde{x},\tilde{\xi}) \in \mathbb{R}^{n+1} \times (\mathbb{R}^{n+1} \setminus \{0\}),\\ \sigma_{\mathrm{e}}(a)(\tilde{x},\tilde{\xi}) &\neq 0 \text{ for all } (\tilde{x},\tilde{\xi}) \in (\mathbb{R}^{n+1} \setminus \{0\}) \times \mathbb{R}^{n+1},\\ \sigma_{\psi\mathrm{e}}(a)(\tilde{x},\tilde{\xi}) &\neq 0 \text{ for all } (\tilde{x},\tilde{\xi}) \in (\mathbb{R}^{n+1} \setminus \{0\}) \times (\mathbb{R}^{n+1} \setminus \{0\}).\end{aligned}$$

These three conditions are independent.

The following results are well known.

Theorem 1.16. *Let $A \in L^{\mu;\nu}(\mathbb{R}^{n+1})$, $\mu,\nu \in \mathbb{R}$. Then the following conditions are equivalent:*

(i) *the operator A is elliptic;*
(ii) *A induces a Fredholm operator*

$$A : H^{s;g}(\mathbb{R}^{n+1}) \to H^{s-\mu;g-\nu}(\mathbb{R}^{n+1}) \tag{1.25}$$

for some fixed $s = s_0$, $g = g_0 \in \mathbb{R}$.

Theorem 1.17.

(i) *Let $A \in L^{\mu;\nu}_{(\mathrm{cl})}(\mathbb{R}^{n+1})$ be elliptic. Then A has a parametrix $P \in L^{-\mu;-\nu}_{(\mathrm{cl})}(\mathbb{R}^{n+1})$ in the sense*

$$PA - 1,\ AP - 1 \in L^{-\infty;-\infty}(\mathbb{R}^{n+1}).$$

(ii) *The Fredholm property of the operator (1.25) for some $s = s_0$, $g = g_0 \in \mathbb{R}$ entails the Fredholm property of (1.25) for all $s, g \in \mathbb{R}$.*
(iii) *Let $A \in L^{\mu;\nu}(\mathbb{R}^{n+1})$ be elliptic. Then* $\ker A$ *and* $\operatorname{coker} A$ *of the Fredholm operator (1.25) are independent of s, g, and there are subspaces of finite dimension $V, W \subset \mathcal{S}(\mathbb{R}^{n+1})$ such that* $\ker A = V$ *and*

$$\operatorname{im} A + W = H^{s-\mu;g-\nu}(\mathbb{R}^{n+1}),\ W \cap \operatorname{im} A = \{0\}.$$

We now interpret the stretched cone $X^\wedge \ni (r,x)$ as a space with conical exit $r \to \infty$. In order to avoid clumsy precautions for $r \to 0$ we first pass to the cylinder $\mathbb{R} \times X$ and later on localise the operators on the plus side.

Definition 1.18. We say that the cylinder $\mathbb{R} \times X$ is equipped with the structure of a manifold with conical exits $r \to \pm\infty$ if there is given a locally finite atlas of charts

$$\chi_\iota : \mathbb{R} \times U_\iota \to \Gamma_\iota,\ \iota \in I,$$

for coordinate neighborhoods U_ι on X and open sets $\Gamma_\iota \subset \mathbb{R}^{n+1}$ such that for a constant $R > 0$ independent of ι for every $\iota, \tilde{\iota} \in I$ we have:

(i) $\Gamma_\iota \cap \{\tilde{x} \in \mathbb{R}^{n+1} : |\tilde{x}| \geq R\} = \{\lambda\tilde{x} : \tilde{x} \in \Gamma_{\iota,R}, \lambda \geq 1\}$ for $\Gamma_{\iota,R} := \Gamma_\iota \cap \{|\tilde{x}| = R\}$;
(ii) $\chi_\iota(\lambda r, x) = \lambda\chi_\iota(r,x)$ for every $\lambda \geq 1$, $|r| \geq R$;

(iii) the transition maps

$$\tau_{\tilde{\iota},\iota} := \chi_{\tilde{\iota}}\chi_{\iota}^{-1} : \chi_{\iota}(\mathbb{R} \times (U_{\iota} \cap U_{\tilde{\iota}})) \to \chi_{\tilde{\iota}}(\mathbb{R} \times (U_{\iota} \cap U_{\tilde{\iota}}))$$

have the property

$$\tau_{\tilde{\iota},\iota}(\lambda\tilde{x}) = \lambda\tau_{\tilde{\iota},\iota}(\tilde{x})$$

for every $|\tilde{x}| \geq R$, $\lambda \geq 1$.

If $\mathbb{R} \times X$ is equipped with a structure in that sense we also write $X_{\asymp}$ instead of $\mathbb{R} \times X$.

Let us fix a locally finite partition of unity $\{\varphi'_{\iota}\}_{\iota \in I}$ on X subordinate to $\{U_{\iota}\}_{\iota \in I}$, moreover, let $\{\psi'_{\iota}\}_{\iota \in I}$ be a system of functions $\psi'_{\iota} \in C_0^{\infty}(U_{\iota})$ such that $\psi'_{\iota} \equiv 1$ on $\operatorname{supp}\varphi'_{\iota}$. Moreover, set

$$\varphi_{\iota}(r,x) := \varphi'_{\iota}(x),\ \psi_{\iota}(r,x) := \psi_{\iota}(x)$$

for all $(r,x) \in \mathbb{R} \times X$.

Let us endow $X_{\asymp}$ with a Riemannian metric that is equal to a cone metric $g_{X_{\asymp}} := dr^2 + r^2 g_X$ for $|r| > R$ for a Riemannian metric g_X on X. If dx denotes a corresponding measure on X the associated measure on $X_{\asymp}$ for $|r| > R$ is of the form

$$|r|^n dr dx \tag{1.26}$$

for $n = \dim X$.

Now let $L^{\mu;\nu}_{(\mathrm{cl})}(\Gamma_{\iota})$ denote the restriction of $L^{\mu;\nu}_{(\mathrm{cl})}(\mathbb{R}^{n+1})$ to Γ_{ι} (in the sense of operators $A : C_0^{\infty}(\Gamma_{\iota}) \to C^{\infty}(\Gamma_{\iota})$), and let

$$L^{\mu;\nu}_{(\mathrm{cl})}(X_{\asymp})$$

denote the space of all operators $\sum_{\iota\in I}\varphi_{\iota}\{(\chi_{\iota}^{-1})_* A_{\iota}\}\psi_{\iota} + C$ for arbitrary $A_{\iota} \in L^{\mu;\nu}_{(\mathrm{cl})}(\Gamma_{\iota})$ and C having a kernel in $\mathcal{S}((\mathbb{R}\times X)\times(\mathbb{R}\times X)) := \mathcal{S}(\mathbb{R}\times\mathbb{R}, C^{\infty}(X\times X))$.

The concept of operators on manifolds with conical exits to infinity will be necessary also in the set-up of operator-valued symbols, cf. Definition 1.1. Let us recall some basic technicalities (the formulations will be slightly more general than in the scalar case; so we also give additional information for $E = \widetilde{E} = \mathbb{C}$ and trivial $\kappa_{\lambda} = \tilde{\kappa}_{\lambda}$).

Let E and $\widetilde{E}$ be Hilbert spaces with group actions $\{\kappa_{\lambda}\}_{\lambda\in\mathbb{R}_+}$ and $\{\tilde{\kappa}_{\lambda}\}_{\lambda\in\mathbb{R}_+}$, respectively. Then

$$S^{\mu;\nu,\nu'}(\mathbb{R}^q \times \mathbb{R}^q \times \mathbb{R}^q; E, \widetilde{E}) \tag{1.27}$$

for $\mu,\nu,\nu' \in \mathbb{R}$ denotes the set of all $a(y,y',\eta) \in C^{\infty}(\mathbb{R}^q \times \mathbb{R}^q \times \mathbb{R}^q, \mathcal{L}(E,\widetilde{E}))$ such that

$$\left\|\tilde{\kappa}^{-1}_{\langle\eta\rangle}\{D_y^{\alpha}D_{y'}^{\alpha'}D_{\eta}^{\beta}a(y,y',\eta)\}\kappa_{\langle\eta\rangle}\right\|_{\mathcal{L}(E,\widetilde{E})} \leq c\langle\eta\rangle^{\mu-|\beta|}\langle y\rangle^{\nu-|\alpha|}\langle y'\rangle^{\nu'-|\alpha'|} \tag{1.28}$$

for all $(y,y',\eta) \in \mathbb{R}^{3q}$ and $\alpha,\alpha',\beta \in \mathbb{N}^q$, with constants $c = c(\alpha,\alpha',\beta) > 0$. Similarly we have spaces of the kind

$$S^{\mu;\nu}(\mathbb{R}^q \times \mathbb{R}^q; E, \widetilde{E}), \tag{1.29}$$

where the elements $a(y,\eta)$ satisfy analogues of the estimates (1.28) (obtained by simply omitting y').

The elements of (1.27) are regarded as double symbols of corresponding pseudo-differential operators $\mathrm{Op}(a)$. There are then unique left and right symbols

$$a_L(y,\eta) \quad \text{and} \quad a_R(y',\eta)$$

belonging to $S^{\mu;\nu+\nu'}(\mathbb{R}^q\times\mathbb{R}^q;E,\widetilde{E})$ such that $\mathrm{Op}(a)=\mathrm{Op}(a_L)=\mathrm{Op}(a_R)$. For instance, $a_L(y,\eta)$ can be calculated as an oscillatory integral $a_L(y,\eta)=\iint e^{-iz\zeta}a(y, y+z,\eta+\zeta)dz đ\zeta$; a similar expression holds for $a_R(y',\eta)$. Setting

$$L^{\mu;\nu}(\mathbb{R}^q;E,\widetilde{E}):=\{\mathrm{Op}(a):S^{\mu;\nu}(\mathbb{R}^q\times\mathbb{R}^q;E,\widetilde{E})\}$$

we have an isomorphism

$$\mathrm{Op}:S^{\mu;\nu}(\mathbb{R}^q\times\mathbb{R}^q;E,\widetilde{E})\to L^{\mu;\nu}(\mathbb{R}^q;E,\widetilde{E}).$$

2. Edge quantization

2.1. Pseudo-differential edge symbols

We now pass to the algebra of pseudo-differential edge symbols. These will be parameter-dependent families of operators on the infinite (stretched) cone $X^\wedge$ with a specific dependence on edge variables and covariables $(y,\eta)\in\Omega\times\mathbb{R}^q$. In this section X is assumed to be a closed compact C^∞ manifold.

In order to model families of edge-degenerate pseudo-differential operators we start from the space

$$L^\mu_{\mathrm{cl}}(X;\mathbb{R}^{1+q}_{\varrho,\eta}) \tag{2.1}$$

of classical parameter-dependent pseudo-differential operators on X which is a Fréchet space in a natural way. With $p(r,y,\varrho,\eta)\in C^\infty(\mathbb{R}_+\times\Omega,L^\mu_{\mathrm{cl}}(X;\mathbb{R}^{1+q}_{\varrho,\eta})$ we can associate a family of pseudo-differential operators

$$\mathrm{op}_r(p)(y,\eta)\in C^\infty(\Omega,L^\mu_{\mathrm{cl}}(X^\wedge;\mathbb{R}^q)).$$

Remark 2.1. Let $\tilde{p}(\tilde{\varrho},\tilde{\eta})\in L^\mu_{\mathrm{cl}}(X;\mathbb{R}^{1+q}_{\tilde{\varrho},\tilde{\eta}})$ be parameter-dependent elliptic of order μ and set

$$b(r,\varrho,\eta):=r^{-\mu}\tilde{p}(r\varrho,r\eta).$$

Then for every $\eta\neq 0$ there is an $R(\eta)>0$ such that

$$b(r,\varrho,\eta):H^s(X)\to H^{s-\mu}(X)$$

is a family of isomorphisms for all $r\in\mathbb{R}$, $r\geq R$ and $\varrho\in\mathbb{R}$. In addition for a suitable choice of cut-off functions $\omega(r)$ and $\tilde{\omega}(r)$ the expression

$$\|u\|_{H^s_{\mathrm{cone}}(X^\wedge)}:=\|\omega_{|\eta|}u\|_{H^s(\mathbb{R}_+\times X)}+\|(1-\omega_{|\eta|})\mathrm{op}_r(b)(\eta)(1-\tilde{\omega}_{|\eta|})u\|_{L^2(\mathbb{R}_+\times X)}$$

is an equivalent norm in the space $H^s_{\mathrm{cone}}(X^\wedge)$.

Incidentally the parameter ϱ plays the role of Imz for a complex variable $z = \beta + i\varrho$ with fixed $\beta \in \mathbb{R}$. In this case instead of (2.1) we also write

$$L^\mu_{\mathrm{cl}}(X; \Gamma_\beta \times \mathbb{R}^q), \tag{2.2}$$

cf. also the notation (1.8). Elements $f(r, y, z, \eta) \in C^\infty(\mathbb{R}_+ \times \Omega, L^\mu_{\mathrm{cl}}(X; \Gamma_{\frac{1}{2}-\gamma} \times \mathbb{R}^q))$ will be regarded as amplitude functions of Mellin pseudo-differential operators

$$\mathrm{op}^\gamma_M(f)(y,\eta)u(r) := \int \Big(\frac{r}{r'}\Big)^{-(\frac{1}{2}-\gamma+i\varrho)} f(r, y, \frac{1}{2} - \gamma + i\varrho, \eta)u(r')\frac{dr'}{r'} đ\varrho, \tag{2.3}$$

$đ\varrho = (2\pi)^{-1}d\varrho$. Also in this case we have

$$\mathrm{op}^\gamma_M(f)(y,\eta) \in C^\infty(\Omega, L^\mu_{\mathrm{cl}}(X^\wedge; \mathbb{R}^q)).$$

Definition 2.2. By $L^\mu_{\mathrm{cl}}(X; \mathbb{C}_z \times \mathbb{R}^q_\eta)$ we denote the space of all operator families $f(z,\eta) \in \mathcal{A}(\mathbb{C}, L^\mu_{\mathrm{cl}}(X; \mathbb{R}^q_\eta))$ such that

$$f(\beta + i\varrho, \eta) \in L^\mu_{\mathrm{cl}}(X; \mathbb{R}^{1+q}_{\varrho,\eta})$$

for every $\beta \in \mathbb{R}$, uniformly in compact β-intervals. Moreover, let $\mathcal{M}^{-\infty}(X; \Gamma_\delta)_\varepsilon$ for any $\beta \in \mathbb{R}$, $\varepsilon > 0$, denote the set of all

$$f(z) \in \mathcal{A}(\{\delta - \varepsilon < \mathrm{Re}z < \delta + \varepsilon\}, L^{-\infty}(X))$$

such that

$$f(\beta + i\varrho) \in L^{-\infty}(X; \mathbb{R}_\varrho)$$

for every $\beta \in (\delta - \varepsilon, \delta + \varepsilon)$, uniformly in compact β-intervals. Finally, we set

$$\mathcal{M}^{-\infty}(X; \Gamma_\delta) := \bigcup_{\varepsilon > 0} \mathcal{M}^{-\infty}(X; \Gamma_\delta)_\varepsilon.$$

The spaces $L^\mu_{\mathrm{cl}}(X; \mathbb{C} \times \mathbb{R}^q)$ and $\mathcal{M}^{-\infty}(X; \Gamma_\delta)_\varepsilon$ are Fréchet with natural semi-norm systems that immediately follow from the definition.

In the sequel, if $\varphi(r) \in C^\infty(\overline{\mathbb{R}}_+)$ is any function and $\beta \in \mathbb{R}_+$, we set $\varphi_\beta(r) := \varphi(\beta r)$ (for instance, $\omega_{[\eta]}(r) = \omega(r[\eta])$ when ω is a cut-off function).

Remark 2.3. (i) If $\omega(r), \tilde{\omega}(r)$ are arbitrary cut-off functions and $f(y, z) \in C^\infty(\Omega, \mathcal{M}^{-\infty}(X; \Gamma_{\frac{n+1}{2}-\gamma}))$, $\Omega \subseteq \mathbb{R}^q$ open, we have

$$\begin{aligned} m(y,\eta) &:= r^{-\nu}\omega_{[\eta]}\mathrm{op}^{\gamma-\frac{n}{2}}_M(f)(y)\tilde{\omega}_{[\eta]} \\ &\in S^\nu_{\mathrm{cl}}(\Omega \times \mathbb{R}^q; \mathcal{K}^{s,\gamma}(X^\wedge), \mathcal{K}^{\infty,\gamma-\nu}(X^\wedge)) \end{aligned} \tag{2.4}$$

for every $s \in \mathbb{R}$. For the principal part of order ν we have

$$\sigma_\wedge(m)(y,\eta) = r^{-\nu}\omega_{|\eta|}\mathrm{op}^{\gamma-\frac{n}{2}}_M(f)(y)\tilde{\omega}_{|\eta|}. \tag{2.5}$$

(ii) Let $h(r, y, z, \eta) := \widetilde{h}(r, y, z, r\eta)$ for $\widetilde{h}(r, y, z, \widetilde{\eta}) \in C^\infty(\overline{\mathbb{R}}_+ \times \Omega, L^\mu_{\mathrm{cl}}(X; \mathbb{C}_z \times \mathbb{R}^q_{\widetilde{\eta}}))$, and let $\varphi_0(r), \varphi_1(r) \in C^\infty_0(\overline{\mathbb{R}}_+)$, and $\varphi_1 \equiv 0$ on supp φ_0 (e.g., φ_1 may be a cut-off function, $\varphi_0 \in C^\infty_0(\mathbb{R}_+)$). Then

$$g(y,\eta) := \varphi_1(r[\eta])r^{-\mu}\mathrm{op}^{\gamma-\frac{n}{2}}_M(h)(y,\eta)\varphi_0(r[\eta])$$

is a flat Green symbol of order μ.

The following result may be interpreted as a Mellin quantization.

Theorem 2.4. *For every $p(r,y,\varrho,\eta)$ of the form*

$$p(r,y,\varrho,\eta)=\tilde{p}(r,y,\tilde{\varrho},\tilde{\eta})\big|_{\tilde{\varrho}=r\varrho,\tilde{\eta}=r\eta}$$

for a $\tilde{p}(r,y,\tilde{\varrho},\tilde{\eta})\in C^\infty(\overline{\mathbb{R}}_+\times\Omega,L^\mu_{\mathrm{cl}}(X;\mathbb{R}^{1+q}_{\tilde{\varrho},\tilde{\eta}}))$ there exists an $h(r,y,z,\eta)$ of the form

$$h(r,y,z,\eta)=\tilde{h}(r,y,z,\tilde{\eta})|_{\tilde{\eta}=r\eta}$$

for an $\tilde{h}(r,y,z,\tilde{\eta})\in C^\infty(\overline{\mathbb{R}}_+\times\Omega,L^\mu_{\mathrm{cl}}(X;\mathbb{C}\times\mathbb{R}^q))$ such that

$$\mathrm{op}_r(p)(y,\eta)=\mathrm{op}^\gamma_M(h)(y,\eta)\ \mathrm{mod}\ C^\infty(\Omega,L^{-\infty}(X^\wedge;\mathbb{R}^q)),$$

for every $\gamma\in\mathbb{R}$, and $\tilde{h}(r,y,z\tilde{\eta})$ is unique $\mathrm{mod}\,C^\infty(\overline{\mathbb{R}}_+\times\Omega,L^{-\infty}(X;\mathbb{C}\times\mathbb{R}^q))$. *Moreover, setting*

$$p_0(r,y,\varrho,\eta):=\tilde{p}(0,y,r\varrho,r\eta),\ \ h_0(r,y,z,\eta):=\tilde{h}(0,y,z,r\eta) \tag{2.6}$$

we also have

$$\mathrm{op}_r(p_0)(y,\eta)=\mathrm{op}^\gamma_M(h_0)(y,\eta)\ \mathrm{mod}\ C^\infty(\overline{\mathbb{R}}_+\times\Omega,L^{-\infty}(X^\wedge;\mathbb{R}^q))$$

for every $\gamma\in\mathbb{R}$.

A proof may be found in [13], see also [15], or [5].

Let us fix cut-off functions

$$\omega(r),\ \tilde{\omega}(r)$$

such that $\tilde{\omega}\equiv 1$ on $\mathrm{supp}\,\omega$; in that case we write $\omega\prec\tilde{\omega}$. Set

$$\chi(\eta)=1-\omega(r),\ \tilde{\chi}(\eta)=1-\tilde{\tilde{\omega}}(r) \tag{2.7}$$

for another cut-off function $\tilde{\tilde{\omega}}(r)$ such that $\tilde{\tilde{\omega}}\prec\omega$. Moreover, choose cut-off functions $\sigma(r)$, $\tilde{\sigma}(r)$.

Let

$$p(r,y,\varrho,\eta),\ h(r,y,z,\eta) \tag{2.8}$$

be operator families related to each other as in Theorem 2.4, and set

$$a(y,\eta):=r^{-\mu}\sigma\{\omega_{[\eta]}\mathrm{op}_M^{\gamma-\frac{n}{2}}(h)(y,\eta)\tilde{\omega}_{[\eta]}+\chi_{[\eta]}\mathrm{op}_r(p)(y,\eta)\tilde{\chi}_{[\eta]}\}\tilde{\sigma} \tag{2.9}$$

$n=\dim X$; recall that $\omega_c(r):=\omega(rc)$, $\chi_c(r):=\chi(rc)$, etc., for any $c>0$.

Remark 2.5. Let $a(y,\eta)$ be an operator function of the form (2.9). Then we have

$$a(y,\eta)\in C^\infty(\Omega,L^\mu_{\mathrm{cl}}(X^\wedge;\mathbb{R}^q)).$$

The parameter-dependent homogeneous principal symbol of $a(y,\eta)$

$$\sigma_\psi(a)(r,x,y,\varrho,\xi,\eta)\in C^\infty(T^*(\mathbb{R}_+\times X\times\Omega\times\mathbb{R}^{1+n+q})\setminus 0)$$

has the form

$$\sigma_\psi(a)(r,x,y,\varrho,\xi,\eta)=\sigma(r)\tilde{\sigma}(r)r^{-\mu}\tilde{p}_{(\mu)}(r,x,y,r\varrho,\xi,r\eta),$$

where $\tilde{p}_{(\mu)}(r,x,y,\tilde{\varrho},\xi,\tilde{\eta})$ denotes the parameter-dependent homogeneous principal symbol of $\tilde{p}(r,y,\tilde{\varrho},\tilde{\eta})\in C^\infty(\overline{\mathbb{R}}_+\times\Omega,L^\mu_{\mathrm{cl}}(X;\mathbb{R}^{1+q}_{\tilde{\varrho},\tilde{\eta}}))$ of order μ.

Theorem 2.6. *We have*

$$a(y,\eta) \in S^{\mu}(\Omega \times \mathbb{R}^q; \mathcal{K}^{s,\gamma}(X^\wedge), \mathcal{K}^{s-\mu,\gamma-\mu}(X^\wedge))$$

for every $s \in \mathbb{R}$. *Moreover, for every* $\varepsilon > 0$ *we have*

$$a(y,\eta) \in S^{\mu}(\Omega \times \mathbb{R}^q; \mathcal{S}^{\gamma}_{\varepsilon}(X^\wedge), \mathcal{S}^{\gamma-\mu}_{\varepsilon}(X^\wedge)).$$

A proof of Theorem 2.6 may be found in [4, Section 2.1.3].

Theorem 2.6 can be regarded as a quantization for edge-degenerate families of pseudo-differential operators as in Theorem 2.4. In fact, Theorem 2.6 gives rise to continuous operators

$$\mathrm{Op}_y(a) : \mathcal{W}^s_{\mathrm{comp}}(\Omega, \mathcal{K}^{s,\gamma}(X^\wedge)) \to \mathcal{W}^{s-\mu}_{\mathrm{loc}}(\Omega, \mathcal{K}^{s-\mu,\gamma-\mu}(X^\wedge)) \tag{2.10}$$

for all $s, \gamma \in \mathbb{R}$. In other words, the correspondence $p \to a \to \mathrm{Op}(a)$ represents an operator convention that associates with p continuous operators (2.10).

Let us set

$$\sigma_\wedge(a)(y,\eta) := r^{-\mu}\{\omega_{|\eta|}\mathrm{op}_M^{\gamma-\frac{n}{2}}(h_0)(y,\eta)\tilde{\omega}_{|\eta|} + \chi_{|\eta|}\mathrm{op}_r(p_0)(y,\eta)\tilde{\chi}_{|\eta|}\}$$

for the families h_0 and p_0 as in Theorem 2.4 and $(y,\eta) \in T^*\Omega \setminus 0$. Then we have

$$\sigma_\wedge(a)(y,\lambda\eta) = \lambda^{\mu}\kappa_\lambda \sigma_\wedge(a)(y,\eta)\kappa_\lambda^{-1} \tag{2.11}$$

for all $\lambda \in \mathbb{R}_+$.

Remark 2.7. An inspection of the proof of Theorem 2.6 shows that we also have

$$a(y,\eta) \in S^{\mu}(\Omega \times \mathbb{R}^q; K^{s,\gamma}(X^\wedge), K^{s-\mu,\gamma-\mu}(X^\wedge))_{\kappa^{s-\gamma},\kappa^{s-\gamma}}$$

with the spaces (1.21). Similarly as (2.10) we thus obtain the continuity

$$\mathrm{Op}_y(a) : W^{s,\gamma}_{\mathrm{comp}(y)}(X^\wedge \times \Omega) \to W^{s-\mu,\gamma-\mu}_{\mathrm{loc}(y)}(X^\wedge \times \Omega) \tag{2.12}$$

for all $s \in \mathbb{R}$, cf. the notation (1.22). What we see is that there is a quite different quantization of edge-degenerate families compared with (2.10). It can be proved (cf. [7, Chapter 7]) that (2.10) as well as (2.12) belong to a continuously parametrised family of different (non-equivalent) quantizations. All of them are possible choices for a consistent edge calculus in the (stretched) wedge $X^\wedge \times \Omega$.

2.2. A new edge quantization

The new quantization of edge degenerate symbols consists of taking an operator family of the form

$$a(y,\eta) := r^{-\mu}\sigma\left\{\omega_{[\eta]}\mathrm{op}_M^{\gamma-\frac{1}{2}}(h)(y,\eta)\widetilde{\omega}_{[\eta]} + \chi_{[\eta]}\psi_{[\eta]}\mathrm{op}_r(p)(y,\eta)\widetilde{\chi}_{[\eta]}\right\}\widetilde{\sigma} \tag{2.13}$$

instead of (2.9), where $\omega, \widetilde{\omega}, \widetilde{\omega}, \sigma, \widetilde{\sigma}$ and p, h are as before, while

$$\psi_{[\eta]}(r,r') := \psi(r[\eta], r'[\eta]),$$

with the function $\psi(r,r')$ of Example 1.4.

In order to compare (2.9) and (2.13) we analyse operator families associated with

$$p(r,\varrho,\eta) := \tilde{p}(r, r\varrho, r\eta)$$

for an element $\tilde{p}(r, \tilde{\varrho}, \tilde{\eta}) \in C^\infty(\overline{\mathbb{R}}_+, L^\mu_{\mathrm{cl}}(X; \mathbb{R}^{1+q}_{\tilde{\varrho},\tilde{\eta}}))$ (it is enough to assume the y-independent case). We assume that

$$\tilde{p}(r, \tilde{\varrho}, \tilde{\eta}) \text{ is independent of } r \text{ for } r \geq R$$

some constant $R > 0$.

Theorem 2.8. *Let $\sigma, \tilde{\sigma}, \chi, \tilde{\chi}$ be as in (2.9) and form*

$$g(\eta) := r^{-\mu}\sigma(r)\chi_{[\eta]}(r)(1 - \psi_{[\eta]}(r, r'))\mathrm{op}_r(p)(\eta)\tilde{\chi}_{[\eta]}(r')\tilde{\sigma}(r').$$

Then we have

$$g(\eta), g^*(\eta) \in S^\mu_{\mathrm{cl}}(\mathbb{R}^q; \mathcal{K}^{s,\gamma}(X^\wedge), \mathcal{S}_{\mathcal{O}}(X^\wedge)) \tag{2.14}$$

for every $s, \gamma \in \mathbb{R}$, i.e., $g(\eta)$ is a flat Green symbol of order μ, (cf. Example 1.3 *and the notation* (1.16)).

Proof. We first show the property

$$g(\eta), g^*(\eta) \in C^\infty(\mathbb{R}^q, \mathcal{L}(\mathcal{K}^{s,\gamma}(X^\wedge), \mathcal{S}_{\mathcal{O}}(X^\wedge)).$$

Let us consider $g(\eta)$; the corresponding relation for $g^*(\eta)$ can be obtained in an analogous manner. It is enough to show

$$g(\eta) \in C^\infty(\mathbb{R}^q, \mathcal{L}(\mathcal{K}^{s,\gamma}(X^\wedge)_U, \mathcal{S}_{\mathcal{O}}(X^\wedge)_U)) \tag{2.15}$$

for every coordinate neighborhood U on X where

$$\mathcal{K}^{s,\gamma}(X^\wedge)_U := \{u(r, x) \in \mathcal{K}^{s,\gamma}(X^\wedge) : \mathrm{supp}\, u \subseteq \overline{\mathbb{R}}_+ \times U\}, \tag{2.16}$$

$$\mathcal{S}_{\mathcal{O}}(X^\wedge)_U := \{u(r, x) \in \mathcal{S}_{\mathcal{O}}(X^\wedge) : \mathrm{supp}\, u \subseteq \overline{\mathbb{R}}_+ \times U\}. \tag{2.17}$$

Here we use the fact that there is an atlas $\{U_1, \dots, U_N\}$ on X in such a way that for every two indices $1 \leq j,\ k \leq N$ also $U_j \cup U_k$ is a coordinate neighborhood on X. Without loss of generality we may assume that the coordinate neighborhood U in (2.16), (2.17) are contained in other coordinate neighborhoods $\widetilde{U}$ such that $\overline{U}$ are compact subsets. Now, if we pass to local coordinates on X we can write

$$\begin{aligned} g(\eta)u(r, x) = r^{-\mu}\sigma(r)\chi_{[\eta]}(r) \iint & e^{i(r-r')\varrho + i(x-x')\xi} \\ & \tilde{p}(r, x, x', r\varrho, \xi, r\eta)(1 - \psi_{[\eta]}(r, r'))\tilde{\chi}_{[\eta]}(r')\tilde{\sigma}(r') \\ & u(r', x')dr'dx'đ\varrho đ\xi, \end{aligned} \tag{2.18}$$

where $\mathrm{supp}\, u$ is contained in $\overline{\mathbb{R}}_+ \times K$ for a compact set $K \subset \mathbb{R}^n$. Here $\tilde{p}(r, x, x', \tilde{\varrho}, \xi, \tilde{\eta})$ is a classical symbol in $(\tilde{\varrho}, \xi, \tilde{\eta}) \in \mathbb{R}^{1+n+q}$ of order μ. After multiplying $\tilde{p}(r, x, x', \tilde{\varrho}, \xi, \tilde{\eta})$ from the left and the right by localising functions $\varphi_0(x)$ and $\psi_0(x')$, respectively, belonging $C_0^\infty(K)$, we can assume that $\tilde{p}(r, x, x', \tilde{\varrho}, \xi, \tilde{\eta})$ has compact support with respect to (x, x').

We want to compute the distributional kernel of (2.18). To this end we choose an N and write $e^{i(r-r')\varrho} = |r - r'|^{-2N} D_\varrho^{2N} e^{i(r-r')\varrho}$. Inserting this in (2.18) and integrating by parts under the integral gives us

$$\begin{aligned} g(\eta)u(r,x) =& r^{-\mu}\sigma(x)\chi_{[\eta]}(r)\iint |r[\eta]-r'[\eta]|^{-2N}(1-\psi_{[\eta]}(r,r')) \\ & e^{i(r-r')\varrho+i(x-x')\xi}(r[\eta])^{2N}(D_{\tilde{\varrho}}^{2N}\tilde{p})(r,x,x',r\varrho,\xi,r\eta) \\ & \tilde{\chi}_{[\eta]}(r')\tilde{\sigma}(r')u(r',x')dr'dx'đ\varrho đ\xi. \end{aligned}$$

The kernel of this operator can be written as

$$\begin{aligned} K(g)(r,r',x,x';\eta) = r^{-\mu}\sigma(r)\chi_{[\eta]}(r)\iint & |r[\eta]-r'[\eta]|^{2N}(1-\psi_{[\eta]}(r,r')) \\ & e^{i(r-r')\varrho+i(x-x')\xi}(r[\eta])^{2N}(D_{\tilde{\varrho}}^{2N}\tilde{p}) \\ & (r,x,x',r\varrho,\xi,r\eta)\tilde{\chi}_{[\eta]}(r')\tilde{\sigma}(r')đ\varrho đ\xi. \end{aligned} \tag{2.19}$$

By virtue of the symbolic estimates

$$|D_{\tilde{\varrho}}^{2N}\tilde{p}(r,x,x',r\varrho,\xi,r\eta)| \le c\langle r\varrho,\xi,r\eta\rangle^{\mu-2N} \tag{2.20}$$

for all $x,x' \in K$ and $r \in \operatorname{supp}\sigma\chi_{[\eta]}$ we see that the integral (2.19) converges, together with all derivatives in r,r',x,x' up to some order M when $N = N(M)$ is chosen sufficiently large. This shows that for every fixed $\eta \in \mathbb{R}^q$ the kernel of (2.19) belongs to $C_0^\infty(\mathbb{R}_+ \times \mathbb{R}^n \times \mathbb{R}_+ \times \mathbb{R}^n)$, and this remains true if η varies in a compact set $\subset \mathbb{R}^q$. It is also clear that the kernel smoothly depends on $\eta \in \mathbb{R}^q$. Thus we have verified the relation (2.15). For the formal adjoint we can argue in a similar manner.

For the proof of (2.14) we first generate the homogeneous principal symbol of order μ

$$\sigma_\wedge(y)(\eta) = \lim_{\lambda\to\infty} \lambda^{-\mu}\kappa_\lambda^{-1}g(\lambda\eta)\kappa_\lambda \text{ for } \eta \neq 0.$$

In local coordinates on X it takes the form

$$\begin{aligned} \sigma_\wedge(g)(\eta)u(r,x) = r^{-\mu}\chi_{|\eta|}(r)\iint & e^{i(r-r')\varrho+i(x-x')\xi}(1-\psi_{|\eta|}(r,r')) \\ & \tilde{p}(0,x,x';r\varrho,\xi,r\eta)\tilde{\chi}_{|\eta|}(r')u(r',x')dr'dx'đ\varrho đ\xi. \end{aligned}$$

Integration by parts gives us again

$$\begin{aligned} \sigma_\wedge(g)(\eta)u(r,x) = r^{-\mu}\chi_{|\eta|}(r)\iint & |r|\eta|-r'|\eta||^{-2N}(1-\psi_{|\eta|}(r,r')) \\ & e^{i(r-r')\varrho+i(x-x')\xi}(r|\eta|)^{2N}(D_{\tilde{\varrho}}^{2N}\tilde{p})(0,x,x',r\varrho,\xi,r\eta)\tilde{\chi}_{|\eta|}(r') \\ & u(r',x')dr'dx'đ\varrho đ\xi \end{aligned}$$

for every $N \in \mathbb{N}$. Using the estimate

$$|r-r'|^{-2N}(1-\psi(r,r')) \le \langle r\rangle^{-N}\langle r'\rangle^{-N} \tag{2.21}$$

and the symbolic estimate (2.20) we can easily verify that $\sigma_\wedge(g)(\eta)$ has a kernel in $\mathcal{S}_{\mathcal{O}}(X^\wedge)_U\hat{\otimes}_\pi\mathcal{S}_{\mathcal{O}}(X^\wedge)_U$. Assuming for the moment that $\tilde{p} = \tilde{p}(\tilde{\varrho},\tilde{\eta})$ is independent of r we have

$$g(\eta) - \chi(\eta)\sigma_\wedge(g)(\eta) \in S^{-\infty}(\mathbb{R}^q;\mathcal{K}^{s,\gamma}(X^\wedge),\mathcal{S}_{\mathcal{O}}(X^\wedge))$$

for any excision function $\chi(\eta) \in C^\infty(\mathbb{R}^q)$. Otherwise, we can successively compute the lower order terms by a Taylor expansion

$$\tilde{p}(r, \tilde{\varrho}, \tilde{\eta}) = \sum_{l=0}^{L} r^l \tilde{p}_l(\tilde{\varrho}, \tilde{\eta}) + r^{L+1} \tilde{p}_{(L+1)}(r, \tilde{\varrho}, \tilde{\eta}).$$

The above process applied to $r^l \tilde{p}_l(\tilde{\varrho}, \tilde{\eta})$ then gives us the homogeneous components of $g(\eta)$ of order $\mu - l$. Another elementary calculation with $r^{L+1}\tilde{p}_{(L+1)}(r, \tilde{\varrho}, \tilde{\eta})$ yields a remainder in $S^{\mu-(L+1)}(\mathbb{R}^q; \mathcal{K}^{s,\gamma}(X^\wedge), \mathcal{S}_{\mathcal{O}}(X^\wedge))$ which shows that our symbol is classical. For the formal adjoints we can do the same. □

Proposition 2.9. *Let $\tilde{p}(r, \tilde{\varrho}, \tilde{\eta}) \in C^\infty(\overline{\mathbb{R}}_+, L^{-\infty}(X; \mathbb{R}^{1+q}_{\tilde{\varrho},\tilde{\eta}}))$ be independent of r for $r \geq R$ for some $R > 0$. Then*

$$g(\eta) := r^{-\mu}\sigma(r)\chi_{[\eta]}(r)\mathrm{op}_r(p)(\eta)\tilde{\chi}_{[\eta]}(r')\tilde{\sigma}(r') \tag{2.22}$$

is a flat Green symbol of order μ; in particular, it satisfies the relations (2.14). *The same is true of*

$$g_1(\eta) := r^{-\mu}\sigma(r)\chi_{[\eta]}(r)\psi_{[\eta]}(r, r')\mathrm{op}_r(p)(\eta)\chi_{[\eta]}(r')\tilde{\sigma}(r').$$

Proof. It is evident that $g(\eta)$ has the property (2.14). It remains to show that $g(\eta)$ is a classical symbol. Let us first consider the case that $\tilde{p} = \tilde{p}(\tilde{\varrho}, \tilde{\eta})$ is independent of r. Then for $\eta \neq 0$ we can set

$$g_{(\mu)}(\eta) := r^{-\mu}\chi_{|\eta|}(r)\mathrm{op}_r(p)\tilde{\chi}_{|\eta|}(r').$$

Write

$$\begin{aligned} g_{(\mu)}(\eta) &= r^{-\mu}\chi_{|\eta|}(r)\psi_{|\eta|}(r, r')\mathrm{op}_r(p)(\eta)\tilde{\chi}_{|\eta|}(r') \\ &\quad + r^{-\mu}\chi_{|\eta|}(r)(1 - \psi_{|\eta|}(r, r'))\mathrm{op}_r(p)(\eta)\tilde{\chi}_{|\eta|}(r') \end{aligned} \tag{2.23}$$

with the function $\psi(r, r')$ of Example 1.14, and $\psi_{|\eta|}(r, r') := \psi(r|\eta|, r'|\eta|)$. Observe that we have

$$\psi(r, r')\varphi(r) \in \mathcal{S}(\mathbb{R} \times \mathbb{R}) \tag{2.24}$$

for every $\varphi \in \mathcal{S}(\mathbb{R})$. The first summand on the right of (2.23) is a Schwartz function in r, r' with values in $L^{-\infty}(X)$. In fact, $\chi_{|\eta|}(r)\tilde{p}(\tilde{\varrho}, r\eta)$ belongs to $\mathcal{S}(\mathbb{R}_r, L^{-\infty}(X; \mathbb{R}_{\tilde{\varrho}}))$ for every fixed $\eta \neq 0$; then applying a relation of the kind (2.24) gives us this property. For the second summand on the right of (2.23) we can proceed in a similar manner as in the proof of Theorem 2.8.

We obtain altogether that the kernel of $g_{(\eta)}(\eta)$ belongs to $\mathcal{S}_{\mathcal{O}}(X^\wedge)\widehat{\otimes}_\pi \mathcal{S}_{\mathcal{O}}(X^\wedge)$ for every $\eta \neq 0$. Thus

$$f(\eta) := r^{-\mu}\chi_{[\eta]}\mathrm{op}_r(p)(\eta)\tilde{\chi}_{[\eta]} \tag{2.25}$$

is a Green symbol, since $f(\eta), f^*(\eta)$ have the property (2.14) (with f instead of g) and $f(\eta) = g_{(\mu)}(\eta)$ for large $|\eta|$. Applying Remark 1.10 to $\varphi := |\eta|$, $\tilde{\varphi} := \tilde{\sigma}$, we obtain that (2.22) itself has the desired property.

In the case of an r-dependence of $\tilde{p}(r, \tilde{\varrho}, \tilde{\eta})$ we assume that $\tilde{p}(r, \tilde{\varrho}, \tilde{\eta})$ is independent of r for large r; otherwise we subtract $\tilde{p}(\infty, \tilde{\varrho}, \tilde{\eta})$ which is constant in r

and can be treated as before and then consider the difference $\tilde{p}(r,\tilde{\varrho},\tilde{\eta})-\tilde{p}(\infty,\tilde{\varrho},\tilde{\eta})$. Then we can write

$$\tilde{p}(r,\tilde{\varrho},\tilde{\eta})=\sum_{j=0}^{\infty}\lambda_j\varphi_j(r)\tilde{p}_j(\tilde{\varrho},\tilde{\eta})$$

with $\lambda_j\in\mathbb{C}$, $\sum|\lambda_j|<\infty$, $\varphi_j\in C_0^\infty(\overline{\mathbb{R}}_+)$, $\tilde{p}_j(\tilde{\varrho},\tilde{\eta})\in L^{-\infty}(X;\mathbb{R}^{1+q}_{\tilde{\varrho},\tilde{\eta}})$ tending to 0 in the respective spaces as $j\to\infty$. This easily reduces the general case to r-independent $\tilde{p}$ when we apply other standard arguments, in particular, that $r^{-\mu}\chi_{[\eta]}\mathrm{op}_r(p_j)(\eta)\tilde{\chi}_{[\eta]}$ tends to zero for $j\to\infty$ in the space of Green operators of the kind discussed before.

The assertion on $g_1(\eta)$ is included in the proof, cf. the discussion around the formula (2.23). □

2.3. Edge symbols as parameter-dependent cone operator families

By an edge amplitude function of order μ, referring to the weight data $\boldsymbol{g}=(\gamma,\gamma-\mu)$, we understand an operator family of the form

$$a(y,\eta)=r^{-\mu}\sigma\{\omega_{[\eta]}\mathrm{op}_M^{\gamma-\frac{n}{2}}(h)(y,\eta)\tilde{\omega}_{[\eta]}+\chi_{[\eta]}\mathrm{op}_r(p)(y,\eta)\tilde{\chi}_{[\eta]}\}\tilde{\sigma}+m(y,\eta)+g(y,\eta),$$

where the first expression on the right-hand side is as in (2.9), moreover, $m(y,\eta)$ is a smoothing Mellin edge symbol and $g(y,\eta)$ a Green symbol as in Example 1.3 (for $j_-=j_+=0$).

Let $\mathcal{R}^\mu(\Omega\times\mathbb{R}^q;\boldsymbol{g})$ denote the space of all those $a(y,\eta)$. Let us set

$$\sigma(a):=(\sigma_\psi(a),\sigma_\wedge(a)),\tag{2.26}$$

where $\sigma_\psi(a)$ is defined as in Remark 2.5; we take into account that $\sigma_\psi(m+g)=0$. Moreover, we set

$$\begin{aligned}\sigma_\wedge(a)(y,\eta):=r^{-\mu}\{&\omega_{|\eta|}\mathrm{op}_M^{\gamma-\frac{n}{2}}(h_0)(y,\eta)\tilde{\omega}_{|\eta|}\\&+\chi_{|\eta|}\mathrm{op}_r(p_0)(y,\eta)\tilde{\chi}_{|\eta|}\}+\sigma_\wedge(m+g)(y,\eta)\end{aligned}$$

with the notation (2.6) and $\sigma_\wedge(m+g)(y,\eta)$ as the homogeneous principal symbol of $(m+g)(y,\eta)$ as a classical symbol, cf. also (2.5) for $\mu=\nu$.

Theorem 2.10. *Let $a(y,\eta)\in\mathcal{R}^\mu(\Omega\times\mathbb{R}^q,\boldsymbol{g})$ and assume that $\sigma(a)=0$. Then we have*

$$a(y,\eta)\in S^{\mu-1}(\Omega\times\mathbb{R}^q;\mathcal{K}^{s,\gamma}(X^\wedge),\mathcal{K}^{s-\mu,\gamma-\mu}(X^\wedge)),$$

for every $s\in\mathbb{R}$, and $a(y,\eta)$ takes values in the space of compact operators $\mathcal{K}^{s,\gamma}(X^\wedge)\to\mathcal{K}^{s-\mu,\gamma-\mu}(X^\wedge)$, $s\in\mathbb{R}$.

Proof. The relation $\sigma_\psi(a)=0$ implies that

$$\sigma(r)\tilde{\sigma}(r)p(r,y,\tilde{\varrho},\tilde{\eta})\in C^\infty(\overline{\mathbb{R}}_+\times\Omega,L^{\mu-1}_{\mathrm{cl}}(X;\mathbb{R}^{1+q}_{\tilde{\varrho},\tilde{\eta}}))$$

and

$$\sigma(r)\tilde{\sigma}(r)\tilde{h}(r,y,z,\tilde{\eta})\in C^\infty(\overline{\mathbb{R}}_+\times\Omega,L^{\mu-1}_{\mathrm{cl}}(X;\mathbb{C}\times\mathbb{R}^q_{\tilde{\eta}})).$$

This yields

$$\tilde{p}(0,y,\tilde{\varrho},\tilde{\eta}) \in C^\infty(\Omega, L^{\mu-1}_{\mathrm{cl}}(X;\mathbb{R}^{1+q})),\ \tilde{h}(0,y,z,\tilde{\eta}) \in C^\infty(\Omega, L^{\mu-1}_{\mathrm{cl}}(X;\mathbb{C}\times\mathbb{R}^q)).$$

Now $\sigma_\wedge(a)(y,\eta) = 0$ implies that

$$r^{-\mu}\{\omega_{|\eta|}\mathrm{op}_M^{\gamma-\frac{n}{2}}(h_0)(y,\eta)\tilde{\omega}_{|\eta|} + \chi_{|\eta|}\mathrm{op}_r(p_0)(y,\eta)\tilde{\chi}_{|\eta|}\} = -\sigma_\wedge(m+g)(y,\eta). \quad (2.27)$$

Both sides of the latter relation are operators in the cone algebra on $X^\wedge$. That means their subordinate symbols (interior, conormal, exit) symbols coincide. In particular, we have

$$\sigma_{\mathrm{c}}(r^{-\mu}\omega_{|\eta|}\mathrm{op}_M^{\gamma-\frac{n}{2}}(h_0)(y,\eta)\tilde{\omega}_{|\eta|})(z) = \tilde{h}_0(0,y,z,0) = -\sigma_{\mathrm{c}}\sigma_\wedge(m(y,\eta))(z).$$

In other words, setting $h_{00}(y,z) := \tilde{f}(0,y,z,0)$, we have

$$h_{00}(y,z) = -f(y,z) \quad (2.28)$$

(when $m(y,z)$ is given as $r^{-\mu}\omega_{[\eta]}\mathrm{op}_M^{\gamma-\frac{n}{2}}(f)(y)\tilde{\omega}_{[\eta]}$ for an $f \in C^\infty(\Omega, \mathcal{M}^{-\infty}(X;\Gamma_{\frac{n+1}{2}-\gamma})_\varepsilon)$ which we assume without loss of generality). In addition we may assume that $\sigma \equiv 1$, $\tilde{\sigma} \equiv 1$ on $\mathrm{supp}\,\omega_{[\eta]}$ and $\mathrm{supp}\,\tilde{\omega}_{[\eta]}$. It follows that

$$\delta(\eta)\{r^{-\mu}\{\omega_{[\eta]}\mathrm{op}_M^{\gamma-\frac{n}{2}}(h_0-h_{00})(y,\eta)\tilde{\omega}_{[\eta]} + \chi_{[\eta]}\mathrm{op}_r(p_0)(y,\eta)\tilde{\chi}_{[\eta]}\} + g(y,\eta)\} \quad (2.29)$$

is a Green symbol of order $\mu-1$ for every excision function $\delta(\eta)$ in $\mathbb{R}^q$. In fact, the relation (2.27) shows that

$$r^{-\mu}\{\omega_{|\eta|}\mathrm{op}_M^{\gamma-\frac{n}{2}}(h_0-h_{00})\tilde{\omega}_{|\eta|} + \chi_{|\eta|}\mathrm{op}_r(p_0)(y,\eta)\tilde{\chi}_{|\eta|}\} = -\sigma_\wedge(g)(y,\eta).$$

This entails the identity (2.29) for all η with $|\eta| \geq$ const for a constant > 0. On the other hand, using the technique of proving Theorem 2.10 we see that the left-hand side of (2.29) is an operator-valued symbol, even classical in this situation, with $-\sigma_\wedge(g)(y,\eta)$ as the homogeneous principal symbol. Then the same is true of

$$\begin{aligned} g_0(y,\eta) \quad &:= \quad r^{-\mu}\sigma\delta(\eta)\{\omega_{[\eta]}\mathrm{op}_M^{\gamma-\frac{n}{2}}(h_0-h_{00})(y,\eta)\tilde{\omega}_{[\eta]} + \chi_{[\eta]}\mathrm{op}_r(p_0)(y,\eta)\tilde{\chi}_{[\eta]}\}\tilde{\sigma} \\ &+ \quad \delta(\eta)g(y,\eta), \end{aligned}$$

cf. Remark 1.10.

Now the symbol $a(y,\eta)$ can be written in the form

$$\begin{aligned} &r^{-\mu}\sigma\{\omega_{[\eta]}\mathrm{op}_M^{\gamma-\frac{n}{2}}(h-h_0)(y,\eta)\tilde{\omega}_{[\eta]} + \chi_{[\eta]}\mathrm{op}_r(p-p_0)(y,\eta)\tilde{\chi}_{[\eta]}\}\tilde{\sigma} \\ &\quad + \delta(\eta)\{r^{-\mu}\sigma\{\omega_{[\eta]}\mathrm{op}_M^{\gamma-\frac{n}{2}}(h_0-h_{00})(y,\eta)\tilde{\omega}_{[\eta]} + \chi_{[\eta]}\mathrm{op}_r(p_0)(y,\eta)\tilde{\chi}_{[\eta]}\}\tilde{\sigma}\} \\ &\quad + g(y,\eta) + (1-\delta(\eta))\{r^{-\mu}\sigma\{\omega_{[\eta]}\mathrm{op}_M^{\gamma-\frac{n}{2}}(h_0-h_{00})(y,\eta)\tilde{\omega}_{[\eta]} \\ &\quad + \chi_{[\eta]}\mathrm{op}_r(p_0)(y,\eta)\tilde{\chi}_{[\eta]}\}\tilde{\sigma} + g(y,\eta)\} \qquad (2.30) \\ &= r^{-\mu}\sigma\{\omega_{[\eta]}\mathrm{op}_M^{\gamma-\frac{n}{2}}(h-h_0)(y,\eta)\tilde{\omega}_{[\eta]} + \chi_{[\eta]}\mathrm{op}_r(p-p_0)(y,\eta)\tilde{\chi}_{[\eta]}\}\tilde{\sigma} \\ &\quad + g_0(y,\eta) + (1-\delta(\eta))\{r^{-\mu}\sigma\{\omega_{[\eta]}\mathrm{op}_M^{\gamma-n/2}(h_0-h_{00})(y,\eta)\tilde{\omega}_{[\eta]} \\ &\quad + \chi_{[\eta]}\mathrm{op}_r(p_0)(y,\eta)\tilde{\chi}_{[\eta]}\}\tilde{\sigma}\} \end{aligned}$$

The summand $g_0(y,\eta)$ is a Green symbol of order $\mu-1$ and takes values in compact operators. Also

$$r^{-\mu}\sigma\{\omega_{[\eta]}\mathrm{op}_M^{\gamma-\frac{n}{2}}(h-h_0)(y,\eta)\tilde{\omega}_{[\eta]}\}\tilde{\sigma} \tag{2.31}$$

is compact for every (y,η), since we can write $h-h_0=r\tilde{h}_{(1)}(r,y,z,r\eta)$ for an $\tilde{h}_{(1)}(r,y,z,\tilde{\eta})\in C^\infty(\overline{\mathbb{R}}_+\times\Omega,L_{\mathrm{cl}}^{\mu-1}(X;\mathbb{C}\times\mathbb{R}_\xi^\eta))$ which yields an improvement of the weight at $r=0$, together with the improvement of the order. For a similar reason also

$$r^{-\mu}\sigma\{\chi_{[\eta]}\mathrm{op}_r(p-p_0)(y,\eta)\tilde{\chi}_{[\eta]}\}\tilde{\sigma} \tag{2.32}$$

is compact for every (y,η). Finally, (2.31) and (2.32) are operator-valued symbols of order $\mu-1$, by similar arguments as for the proof of Theorem 2.6, and the last summand on the right of (2.30) takes values in compact operators and is of order $-\infty$ in η. □

3. Composition properties

3.1. Composition of edge symbols

In this section we analyse the composition properties of edge amplitude functions of the form (2.9). It will be more convenient here to employ the cut-off functions $\omega(r),\tilde{\omega}(r),\tilde{\tilde{\omega}}(r)$ rather than the excision functions (2.7).

Moreover, since the presence of the variables $y\in\Omega$ only causes minor modifications of the arguments, we content ourselves with the y-independent case. Let

$$\begin{aligned} a(\eta) &:= r^{-\mu}\sigma\{a_M(\eta)+a_\psi(\eta)\}\tilde{\sigma},\\ b(\eta) &:= r^{-\nu}\sigma\{b_M(\eta)+b_\psi(\eta)\}\tilde{\sigma}, \end{aligned}$$

where $\sigma(r)$ and $\tilde{\sigma}(r)$ are arbitrary cut-off functions, and

$$\begin{aligned} a_M(\eta) &:= \omega_{[\eta]}\mathrm{op}_M^{\gamma-\nu-\frac{n}{2}}(h_1)(\eta)\tilde{\omega}_{[\eta]}, & a_\psi(\eta) &:= (1-\omega_{[\eta]})\mathrm{op}_r(p_1)(\eta)(1-\tilde{\tilde{\omega}}_{[\eta]}),\\ b_M(\eta) &:= \omega_{[\eta]}\mathrm{op}_M^{\gamma-\frac{n}{2}}(h_2)(\eta)\tilde{\omega}_{[\eta]}, & b_\psi(\eta) &:= (1-\omega_{[\eta]})\mathrm{op}_r(p_2)(\eta)(1-\tilde{\tilde{\omega}}_{[\eta]}). \end{aligned}$$

Here $p_j(r,\rho,\eta)=\tilde{p}_i(r,r\rho,r\eta)$ for families $\tilde{p}_i(r,\tilde{\rho},\tilde{\eta})\in C^\infty(\overline{\mathbb{R}}_+,L_{\mathrm{cl}}^{\mu_i}(X;\mathbb{R}_{\tilde{\rho},\tilde{\eta}}^{1+q}))$ with $\mu_1=\mu$, $\mu_2=\nu$ and $h_i(r,z,\eta)=\tilde{h}_i(r,z,r\eta)$ for elements $\tilde{h}_i(r,z,\tilde{\eta})\in C^\infty(\overline{\mathbb{R}}_+,L_{\mathrm{cl}}^{\mu_i}(X;\mathbb{C}\times\mathbb{R}^q))$, $i=1,2$, where we assume that h_i is the Mellin quantization of p_i in the sense of Theorem 2.4.

Theorem 3.1. *We have (with respect to the pointwise composition of operator functions)*

$$(ab)(\eta)=\sigma r^{-(\mu+\nu)}\{c_M(\eta)+c_\psi(\eta)\}\tilde{\sigma}+g(\eta)$$

where

$$c_M(\eta)=\omega_{[\eta]}\mathrm{op}_M^{\gamma-\frac{n}{2}}(h)(\eta)\tilde{\omega}_{[\eta]}+(1-\omega_{[\eta]})\mathrm{op}_r(p)(\eta)(1-\tilde{\tilde{\omega}}_{[\eta]})$$

for operator functions $p(r,\rho,\eta)$ and $h(r,z,\eta)$ which are of the same nature as those in Theorem 2.4, now of order $\mu+\nu$, and $g(\eta)$ is a flat Green symbol of order $\mu+\nu$. We have

$$\sigma_\psi(ab)=\sigma_\psi(a)\sigma_\psi(b),\quad \sigma_\wedge(a)(\eta)\sigma_\wedge(b)(\eta)=\sigma_\wedge(ab)(\eta). \tag{3.1}$$

Proof. For the proof we employ the abbreviations $a=a(\eta)$ for

$$a=\omega a_0\tilde{\omega}+(1-\omega)a_1(1-\tilde{\tilde{\omega}}),$$

where the cut-off functions $\omega,\tilde{\omega},\tilde{\tilde{\omega}}$ have the meaning $\omega=\omega_{[\eta]}$, etc., and

$$a_0:=\sigma r^{-\mu}\mathrm{op}_M^{\gamma-\nu-\frac{n}{2}}(h_1)\tilde{\sigma},a_1:=\sigma r^{-\mu}\mathrm{op}_r(p_1)\tilde{\sigma},$$

and, similarly,

$$b=\omega b_0\tilde{\omega}+(1-\omega)b_1(1-\tilde{\tilde{\omega}})$$

for

$$b_0:=\sigma r^{-\nu}\mathrm{op}_M^{\gamma-\frac{n}{2}}(h_2)\tilde{\sigma},b_1:=\sigma r^{-\mu}\mathrm{op}_r(p_2)\tilde{\sigma}.$$

In the following computations we systematically employ the properties $\tilde{\tilde{\omega}}\prec\omega\prec\tilde{\omega}$ which implies $w\tilde{\omega}=\omega$, $\tilde{\tilde{\omega}}\omega=\tilde{\tilde{\omega}}$. We then obtain $ab=P+\sum_{k=1}^{6}G_k$ after elementary rearrangements of summands in the composition ab for

$$P:=\omega a_0\tilde{\omega}b_0\tilde{\omega}+(1-\omega)a_1(1-\tilde{\tilde{\omega}})b_1(1-\tilde{\tilde{\omega}})$$

and

$$\begin{aligned}
G_1&:=\tilde{\tilde{\omega}}a_0(\tilde{\omega}-\omega)b_1(1-\tilde{\tilde{\omega}})+(\omega-\tilde{\tilde{\omega}})a_0(\tilde{\omega}-\omega)b_1(1-\tilde{\omega}),\\
G_2&:=(1-\tilde{\omega})a_1(\omega-\tilde{\tilde{\omega}})b_0\tilde{\omega}+(\tilde{\omega}-\omega)a_1(\omega-\tilde{\tilde{\omega}})b_0\tilde{\tilde{\omega}},\\
G_3&:=(\omega-\tilde{\tilde{\omega}})a_0(\omega-\tilde{\omega})b_0\tilde{\tilde{\omega}}+\tilde{\tilde{\omega}}a_0(\omega-\tilde{\omega})b_0\tilde{\omega},\\
G_4&:=(1-\tilde{\omega})a_1(\tilde{\tilde{\omega}}-\omega)b_1(1-\tilde{\tilde{\omega}})+(\tilde{\omega}-\omega)a_1(\tilde{\tilde{\omega}}-\omega)b_1(1-\tilde{\omega}),\\
G_5&:=(\tilde{\omega}-\omega)a_0(\tilde{\tilde{\omega}}-\omega)(b_1-b_0)(\tilde{\omega}-\tilde{\tilde{\omega}}),\\
G_6&:=(\omega-\tilde{\tilde{\omega}})(a_1-a_0)(\tilde{\omega}-\omega)b_0(\tilde{\omega}-\tilde{\tilde{\omega}}).
\end{aligned}$$

We now write

$$P=\omega a_0b_0\tilde{\omega}+(1-\omega)c_1(1-\tilde{\tilde{\omega}})+G_7+G_8$$

for

$$G_7:=\omega a_0(\tilde{\omega}-1)b_0\tilde{\omega},\quad G_8:=(1-\omega)\{a_1(1-\tilde{\tilde{\omega}})b_1-c_1\}(1-\tilde{\tilde{\omega}}),$$

and

$$c_1(\eta)=\sigma r^{-(\mu+\nu)}\mathrm{op}_r(p)(\eta)\tilde{\sigma} \tag{3.2}$$

for an operator function $p(r,\rho,\eta)=\tilde{p}(r,r\rho,r\eta)$ which is defined by an element

$$\tilde{p}(r,\tilde{\rho},\tilde{\eta})\in C^\infty(\overline{\mathbb{R}}_+,L_{\mathrm{cl}}^{\mu+\nu}(X;\mathbb{R}^{1+q}_{\tilde{\rho},\tilde{\eta}})), \tag{3.3}$$

obtained by computing the Leibniz product $\#_r$ with respect to r,

$$r^{-(\mu+\nu)}\tilde{p}(r,r\rho,r\eta)=r^{-\mu}\tilde{p}_1(r,r\rho,r\eta)\sharp_r\tilde{\sigma}(r)\sigma(r)r^{-\nu}\tilde{p}_2(r,r\rho,r\eta).$$

In particular, we may assume $\widetilde{p}(r,\widetilde{\rho},\widetilde{\eta})$ to be smooth up to $r=0$. Moreover, we have

$$\begin{aligned} a_0b_0 &= \sigma r^{-\mu}\mathrm{op}_M^{\gamma-\nu-\frac{n}{2}}(h_1)(\eta)\widetilde{\sigma}\sigma r^{-\nu}\mathrm{op}_M^{\gamma-\frac{n}{2}}(h_2)(\eta)\widetilde{\sigma} \\ &= \sigma r^{-(\mu+\nu)}\mathrm{op}_M^{\gamma-\frac{n}{2}}(T^{\nu}h_1)(\eta)\mathrm{op}_M^{\gamma-\frac{n}{2}}(\sigma\widetilde{\sigma}h_2)(\eta)\widetilde{\sigma}. \end{aligned}$$

We therefore obtain a Mellin symbol $h(r,z,\eta)=\widetilde{h}(r,z,r\eta)$ for an $\widetilde{h}(r,w,\widetilde{\eta})\in C^{\infty}(\overline{\mathbb{R}}_+,L_{\mathrm{cl}}^{\mu+\nu}(X;\mathbb{C}\times\mathbb{R}^q))$ such that

$$a_0b_0=\sigma r^{-(\mu+\nu)}\mathrm{op}_M^{\gamma-\frac{n}{2}}(h)(\eta)\widetilde{\sigma}. \tag{3.4}$$

Summing up it follows that

$$ab=\omega c_0\widetilde{\omega}+(1-\omega)c_1(1-\widetilde{\widetilde{\omega}})+g,$$

where $c_0(\eta)$ and $c_1(\eta)$ are given by (3.1) and (3.2), respectively, and

$$g=\sum_{j=1}^{8}G_j. \tag{3.5}$$

It turns out that (3.5) is a flat Green symbol. This will be verified in the following section.

The relation (3.1) follows from the fact that we can apply the above computations for the compositions to corresponding operator functions, where $[\eta]$ is replaced by $|\eta|$, $\eta\neq 0$, and all the first r-variables are frozen at zero. □

3.2. Characterization of remainders

In order to characterise (3.5) as a flat Green symbol we consider the summands separately. Let us write $G_1:=G_1'+G_1''$ for

$$G_1':=\widetilde{\widetilde{\omega}}a_0(\widetilde{\omega}-\omega)b_1(1-\widetilde{\widetilde{\omega}}),G_1'':=(\omega-\widetilde{\widetilde{\omega}})a_0(\widetilde{\omega}-\omega)b_1(1-\widetilde{\omega}).$$

We have

$$G_1'=CD\quad\text{for}\quad C:=\widetilde{\widetilde{\omega}}a_0(\widetilde{\omega}-\omega),D:=\varphi b_1(1-\widetilde{\omega})$$

for any $\varphi\in C_0^{\infty}(\mathbb{R}_+)$ that is equal to 1 on $\operatorname{supp}(\widetilde{\omega}-\omega)$. The function φ is interpreted (similarly as the cut-off functions) as an η-dependent factor, i.e., $\varphi=\varphi_\eta$ for $\varphi_\eta(r):=\varphi(r[\eta])$. The family of operators C is a flat Green symbol, see Example 1.3 and Remark 2.3. It is then easy to verify that the composition with D gives again a Green symbol.

For G_1'' we choose another cut-off function ω_0 that is equal to 1 on $\operatorname{supp}\omega$ and such that $\widetilde{\omega}$ is equal to 1 on $\operatorname{supp}\omega_0$ (this is always possible). Then we can write $G_1''=H+L$ for

$$\begin{aligned} H &:= (\omega-\widetilde{\widetilde{\omega}})a_0(1-\omega_0)(\widetilde{\omega}-\omega)b_1(1-\widetilde{\omega}), \\ L &:= (\omega-\widetilde{\widetilde{\omega}})a_0\omega_0(\widetilde{\omega}-\omega)b_1(1-\widetilde{\omega}). \end{aligned}$$

For any $\varphi\in C_0^{\infty}(\mathbb{R}_+)$ such that $\varphi\equiv 1$ on $\operatorname{supp}(\widetilde{\omega}-\omega)$ we have $H=CD$ for

$$C:=(\omega-\widetilde{\widetilde{\omega}})a_0(1-\omega_0)\varphi,D:=(\widetilde{\omega}-\omega)b_1(1-\widetilde{\omega}),$$

(in this proof, C and D occur in different meaning which will be clear from the context). Since $(1-\omega)$ vanishes on $\operatorname{supp}(\omega-\widetilde{\omega})$, the factor C is smoothing and can be treated in a similar manner as the corresponding factor occurring in G_1'. As above it is again easy to verify that then also H is a flat Green symbol. Moreover, taking some $\varphi \in C_0^\infty(\mathbb{R}_+)$, $\varphi \equiv 1$ on $\operatorname{supp}(\widetilde{\omega}-\omega)$, we can write $L = DC$ for

$$D := (\omega - \widetilde{\widetilde{\omega}})a_0(\widetilde{\omega}-\omega), \quad C := \omega_0\varphi b_1(1-\widetilde{\omega}).$$

Since $1-\widetilde{\omega}$ vanishes on $\operatorname{supp}\omega_0$, the family C is smoothing and a flat Green symbol. This implies the same for L.

The arguments for G_k, $1 < k \leq 8$, are similar and left to the reader.

References

[1] D. Calvo and B.-W. Schulze, Operators on corner manifolds with exits to infinity, *Journal of Differential Equ.* **19**, 2 (2006), 147–192.

[2] D. Calvo, C.-I. Martin and B.-W. Schulze, Symbolic structures on corner manifolds, in *RIMS Conference on Microlocal Analysis and Asymptotic Analysis*, Keio University, Tokyo, 2005, 22–35.

[3] D. Calvo and B.-W. Schulze, *Edge calculus of second generation*, Preprint, Institut für Mathematik, Universität Potsdam, 2005.

[4] N. Dines and B.-W. Schulze, Mellin-edge-representations of elliptic operators, *Math. Meth. Appl. Sci.* **28** (2005), 2133–2172.

[5] J.B. Gil, B.-W. Schulze and J. Seiler, Cone pseudodifferential operators in the edge symbolic calculus, *Osaka J. Math.* **37** (2000), 219–258.

[6] G. Harutjunjan and B.-W. Schulze, The relative index for corner singularities, *Integral Equations Operator Theory* **54**, 3 (2006), 385–426.

[7] G. Harutjunjan and B.-W. Schulze, *Elliptic Mixed, Transmission and Singular Crack Problems*, European Mathematical Soc., Zürich, 2007, to appear.

[8] T. Hirschmann, Functional analysis in cone and edge Sobolev spaces, *Ann. Global Anal. Geom.* **8** (1990), 167–192.

[9] I.L. Hwang, The L^2-boundness of pseudodifferential operators, *Trans. Amer. Math. Soc.* **302** (1987), 55–76.

[10] D. Kapanadze and B.-W. Schulze, *Crack Theory and Edge Singularities*, Kluwer Academic Publ., Dordrecht, 2003.

[11] L. Maniccia and B.-W. Schulze, An algebra of meromorphic corner symbols, *Bull. des Sciences Math.* **127** (2003), 55–99.

[12] B.-W. Schulze, Pseudo-differential operators on manifolds with edges, in *Partial Differential Equations*, Teubner, Teubner-Texte zur Mathematik, Leipzig, 1989, 259–287.

[13] B.-W. Schulze, *Pseudo-differential Operators on Manifolds with Singularities*, North-Holland, Amsterdam, 1991.

[14] B.-W. Schulze, Corner Mellin operators and conormal asymptotics, in *Sem. Equ. aux Dériv. Part.* 1990–1991, Exp. XII, Ecole Polytechnique, Mars 1991.

[15] B.-W. Schulze, *Boundary Value Problems and Singular Pseudo-differential Operators*. J. Wiley, Chichester, 1998.

[16] B.-W. Schulze, Operator algebras with symbol hierarchies on manifolds with singularities, in *Advances in Partial Differential Equations (Approaches to Singular Analysis)*, Editors: J. Gil, D. Grieser and M. Lesch, Birkhäuser, Basel, 2001, 167–207.

[17] B.-W. Schulze, Operators with symbol hierarchies and iterated asymptotics, *Publ. RIMS, Kyoto University* **38** (2002), 735–802.

[18] B.-W. Schulze and J. Seiler, The edge algebra structure of boundary value problems, *Ann. Global Anal. and Geom.* **22** (2002), 197–265.

[19] B.-W. Schulze and N. Tarkhanov, New algebras of boundary value problems for elliptic pseudodifferential operators. Preprint, Institut für Mathematik, Universität Potsdam, 2005.

[20] B.-W. Schulze and A. Volpato, Green operators in the edge Quantisation of elliptic operators. Preprint, Institut für Mathematik, Universität Potsdam, 2004.

[21] J. Seiler, *Continuity of edge and corner pseudodifferential operators*. Math. Nachr., **205**, (1999), 163–182.

[22] J. Seiler, The cone algebra and a kernel characterization of Green operators, in *Advances in Partial Differential Equations (Approaches to Singular Analysis)*, Editors: J. Gil, D. Grieser and M. Lesch, Birkhäuser, Basel, 2001, 1–29.

[23] N. Tarkhanov, Harmonic integrals on domains with edges, Preprint, Institut für Mathematik, Universität Potsdam, 2004.

C.-I. Martin and B.-W. Schulze
Institut für Mathematik
Universität Potsdam
Am Neuen Palais 10
D-14469 Potsdam, Deutschland

e-mail: `martin@math.uni-potsdam.de`
e-mail: `schulze@math.uni-potsdam.de`

Operator Theory:
Advances and Applications, Vol. 172, 33–50

On Rays of Minimal Growth for Elliptic Cone Operators

Juan B. Gil, Thomas Krainer and Gerardo A. Mendoza

Abstract. We present an overview of some of our recent results on the existence of rays of minimal growth for elliptic cone operators and two new results concerning the necessity of certain conditions for the existence of such rays.

2000 Mathematics Subject Classification. Primary 58J50; Secondary 35J70, 47A10.

Keywords. Resolvents, conical singularities, spectral theory.

1. Introduction

The aim of this article is twofold. On the one hand, we present an overview of some of the results contained in [6, 7] on the subject in the title, and of the geometric perspective we developed in the course of the investigations leading to the aforementioned papers. We illustrate the main ideas of our approach by means of examples concerning Laplacians on a compact 2-manifold. Already this simple situation exhibits the structural richness and complexity of the general theory.

On the other hand, we offer some improvements, cf. Theorems 4.3 and 5.5, regarding necessary and sufficient conditions for a closed sector $\Lambda \subset \mathbb{C}$ to be a sector of minimal growth for a certain class of elliptic cone operators A and for the associated model operator $A_\wedge$.

Recall that a closed sector of the form

$$\Lambda = \{\lambda \in \mathbb{C} : \lambda = re^{i\theta} \text{ for } r \geq 0,\ \theta \in \mathbb{R},\ |\theta - \theta_0| \leq a\} \tag{1.1}$$

is called a *sector of minimal growth* (or of maximal decay) for a closed operator

$$A : \mathcal{D} \subset H \to H,$$

where H is a Hilbert space and $\mathcal{D}$ is dense in H, if there is a constant $R > 0$ such that $A - \lambda$ is invertible for every $\lambda \in \Lambda_R = \{\lambda \in \Lambda : |\lambda| \geq R\}$, and the resolvent

$(A-\lambda)^{-1}$ satisfies either of the equivalent estimates

$$\|(A-\lambda)^{-1}\|_{\mathscr{L}(H)} \le C/|\lambda|, \quad \|(A-\lambda)^{-1}\|_{\mathscr{L}(H,\mathcal{D})} \le C \tag{1.2}$$

for some $C > 0$ and all $\lambda \in \Lambda_R$.

We are interested in cone operators on smooth manifolds with boundary. Specifically, let M be a smooth n-manifold with boundary $Y = \partial M$ and let $E \to M$ be a Hermitian vector bundle over M. Fix a defining function x for Y. A differential *cone operator* of order m acting on sections of $E \to M$ is an operator of the form $A = x^{-m}P$ with P in the class $\operatorname{Diff}_b^m(M;E)$ of totally characteristic differential operators of order m, cf. Melrose [14]. We write $A \in x^{-m}\operatorname{Diff}_b^m(M;E)$. More explicitly, in the interior $\mathring{M}$ of M, A is a differential operator with smooth coefficients, and near the boundary, in local coordinates $(x,y) \in (0,\varepsilon) \times Y$, it is of the form

$$A = x^{-m} \sum_{k+|\alpha|\le m} a_{k\alpha}(x,y)(xD_x)^k D_y^\alpha \tag{1.3}$$

with coefficients $a_{k\alpha}$ smooth up to $x = 0$; here $D_x = -i\partial/\partial x$ and likewise D_{y_j}. We will say that A (or P) has *coefficients independent of x near Y*, if the coefficients $a_{k\alpha}$ in (1.3) do not depend on x (this notion depends on the choice of tubular neighborhood map, defining function x, and connection on E. For a precise definition see [6]).

This paper consists of 5 sections. In Section 2 we review some basic properties of cone operators while in Section 3 we discuss the associated model operators. The new results on rays of minimal growth can be found in Sections 4 and 5.

Apart from the works explicitly cited in the text, our list of references contains additional items referring to related works on resolvents and rays of minimal growth for elliptic operators. The investigation of spectral properties for elliptic cone operators was initiated by Cheeger [4] in the seventies and has since undergone major developments. Concerning the large-parameter behavior of resolvents, the interested reader is referred to [3, 5, 12, 13, 15], among others. A more detailed account on the subject is provided in the introduction of the papers [6, 7].

2. Preliminaries on cone operators

Let A be a differential cone operator. As introduced in [6], the principal symbol of A, ${}^c\boldsymbol{\sigma}(A)$, is defined on the c-cotangent ${}^cT^*M$ of M rather than on the cotangent itself. Over $\mathring{M}$ it is essentially the usual principal symbol, and equal to

$$\sum_{k+|\alpha|=m} a_{k\alpha}(x,y)\xi^k\eta^\alpha$$

near the boundary Y, see (1.3).

Example. Let M be a compact 2-manifold with boundary $Y = S^1$. Let $g_Y(x)$ be a smooth family of Riemannian metrics on S^1 such that $g_Y(0)$ is the standard metric, dy^2. We equip M with a "cone metric" g that near Y takes the form

$g = dx^2 + x^2 g_Y(x)$ (g is a regular Riemannian metric in the interior of M). Then, near Y, the Laplace-Beltrami operator Δ has the form

$$x^{-2}\big((xD_x)^2 + a(x,y)(xD_x) + \Delta_Y(x)\big), \tag{2.1}$$

where $a(x,y)$ is a smooth function with $a(0,y) = 0$ and $\Delta_Y(x)$ is the nonnegative Laplacian on S^1 associated with $g_Y(x)$. In this case, near the boundary, we have

$$^c\boldsymbol{\sigma}(\Delta) = \xi^2 + \boldsymbol{\sigma}(\Delta_Y(x)).$$

Ellipticity and boundary spectrum

An operator $A \in x^{-m}\operatorname{Diff}_b^m(M;E)$ is *c-elliptic* if $^c\boldsymbol{\sigma}(A)$ is invertible on $^cT^*M\backslash 0$. Moreover, the family $A - \lambda$ is said to be *c-elliptic with parameter* $\lambda \in \Lambda \subset \mathbb{C}$ if $^c\boldsymbol{\sigma}(A) - \lambda$ is invertible on $(^cT^*M \times \Lambda)\backslash 0$.

Associated with $A = x^{-m}P$ there is an operator-valued polynomial

$$\mathbb{C} \ni \sigma \mapsto \hat{P}(\sigma) \in \operatorname{Diff}^m(Y; E|_Y)$$

called the *conormal symbol* of P (and of A). If we write A as in (1.3), then

$$\hat{P}(\sigma) = \sum_{k+|\alpha|\le m} a_{k\alpha}(0,y)\sigma^k D_y^\alpha.$$

If A is c-elliptic, then $\hat{P}(\sigma)$ is invertible for all $\sigma \in \mathbb{C}$ except a discrete set $\operatorname{spec}_b(A)$, the *boundary spectrum of* A, cf. [14]; $\hat{P}(\sigma)$ is a holomorphic family of elliptic operators on Y and $\sigma \to \hat{P}(\sigma)^{-1}$ is a meromorphic operator-valued function on $\mathbb{C}$.

Example. The Laplacian (2.1) is clearly c-elliptic. If y is the angular variable on S^1, then

$$\hat{P}(\sigma) = \sigma^2 + \Delta_Y(0) = \sigma^2 + D_y^2,$$

and the boundary spectrum of Δ is given by

$$\operatorname{spec}_b(\Delta) = \{\pm ik : k \in \mathbb{N}_0\}.$$

Closed extensions

Let $\mathfrak{m}$ be a positive b-density on M, that is, $x\mathfrak{m}$ is a smooth everywhere positive density on M. Let $L_b^2(M;E)$ be the L^2 space of sections of E with respect to the Hermitian form on E and the density $\mathfrak{m}$. Consider A initially defined on $C_0^\infty(\mathring{M};E)$ and look at it as an unbounded operator on the Hilbert space

$$x^{-m/2}L_b^2(M;E) = L^2(M;E;x^m\mathfrak{m}).$$

The particular weight $x^{-m/2}$ is just a convenient normalization and represents no loss. If we are interested in A on $x^\mu L_b^2(M;E)$ for $\mu \in \mathbb{R}$, we can base all our analysis on the space $x^{-m/2}L_b^2(M;E)$ by considering the operator $x^{-\mu-m/2}A\,x^{\mu+m/2}$.

Typically, A has a large class of closed extensions

$$A_{\mathcal{D}} : \mathcal{D} \subset x^{-m/2}L_b^2(M;E) \to x^{-m/2}L_b^2(M;E). \tag{2.2}$$

There are two canonical closed extensions, namely the ones with domains

$$\mathcal{D}_{\min}(A) = \text{closure of } C_0^\infty(\overset{\circ}{M};E) \text{ with respect to } \|\cdot\|_A,$$
$$\mathcal{D}_{\max}(A) = \{u \in x^{-m/2}L_b^2(M;E) : Au \in x^{-m/2}L_b^2(M;E)\},$$

where $\|u\|_A = \|u\| + \|Au\|$ is the graph norm in $\mathcal{D}_{\max}(A)$. Both domains are dense in $x^{-m/2}L_b^2(M;E)$, and for any closed extension (2.2),

$$\mathcal{D}_{\min}(A) \subseteq \mathcal{D} \subseteq \mathcal{D}_{\max}(A).$$

Let

$$\mathfrak{D}(A) = \{\mathcal{D} \subset \mathcal{D}_{\max}(A) : \mathcal{D} \text{ is a vector space and } \mathcal{D}_{\min}(A) \subset \mathcal{D}\}.$$

The elements of $\mathfrak{D}(A)$ are in one-to-one correspondence with the subspaces of $\mathcal{D}_{\max}(A)/\mathcal{D}_{\min}(A)$. If the operator A is fixed and there is no possible ambiguity, we will omit A from the notation and will write simply $\mathcal{D}_{\min}$, $\mathcal{D}_{\max}$, and $\mathfrak{D}$.

Theorem 2.1 (Lesch [12]). *If $A \in x^{-m}\operatorname{Diff}_b^m(M;E)$ is c-elliptic, then*

$$\dim \mathcal{D}_{\max}/\mathcal{D}_{\min} < \infty$$

and all closed extensions of A are Fredholm. Moreover,

$$\operatorname{ind} A_{\mathcal{D}} = \operatorname{ind} A_{\mathcal{D}_{\min}} + \dim \mathcal{D}/\mathcal{D}_{\min}. \tag{2.3}$$

Modulo $\mathcal{D}_{\min}$, the elements of $\mathcal{D}_{\max}$ are determined by their asymptotic behavior near the boundary of M. The structure of these asymptotics depends on the conormal symbols of A and on the part of $\operatorname{spec}_b(A)$ in the strip $\{|\Im\sigma| < m/2\}$. More details will be discussed in the next section.

Corollary 2.2. *If A is c-elliptic and symmetric (formally selfadjoint), then*

$$\operatorname{ind} A_{\mathcal{D}_{\max}} = -\operatorname{ind} A_{\mathcal{D}_{\min}} \quad \textit{and} \quad \operatorname{ind} A_{\mathcal{D}_{\min}} = -\frac{1}{2}\dim \mathcal{D}_{\max}/\mathcal{D}_{\min}.$$

Example. Consider the cone Laplacian Δ, cf. (2.1). Then (3.2) and (3.6) imply

$$\dim \mathcal{D}_{\max}(\Delta)/\mathcal{D}_{\min}(\Delta) = 2$$

and thus, by the previous corollary,

$$\operatorname{ind} \Delta_{\min} = -1 \quad \text{and} \quad \operatorname{ind} \Delta_{\max} = 1. \tag{2.4}$$

If $A \in x^{-m}\operatorname{Diff}_b^m(M;E)$ is c-elliptic, the embedding $\mathcal{D}_{\max} \hookrightarrow x^{-m/2}L_b^2(M;E)$ is compact. Therefore, for every $\mathcal{D} \in \mathfrak{D}$ and $\lambda \in \mathbb{C}$, the operator $A_{\mathcal{D}} - \lambda$ is also Fredholm with $\operatorname{ind}(A_{\mathcal{D}} - \lambda) = \operatorname{ind} A_{\mathcal{D}}$. Consequently, if $\operatorname{spec}(A_{\mathcal{D}}) \neq \mathbb{C}$, then we necessarily have $\operatorname{ind} A_{\mathcal{D}} = 0$. For this reason, we will primarily be interested in the set of domains

$$\mathfrak{G} = \{\mathcal{D} \in \mathfrak{D} : \operatorname{ind} A_{\mathcal{D}} = 0\} \tag{2.5}$$

which is empty unless $\operatorname{ind} A_{\mathcal{D}_{\min}} \leq 0$ and $\operatorname{ind} A_{\mathcal{D}_{\max}} \geq 0$. Let $d'' = -\operatorname{ind} A_{\mathcal{D}_{\min}}$.

Using that the map $\mathfrak{D} \ni \mathcal{D} \mapsto \mathcal{D}/\mathcal{D}_{\min}$ is a bijection, we identify $\mathfrak{G}$ with the complex Grassmannian of d''-dimensional subspaces of $\mathcal{D}_{\max}/\mathcal{D}_{\min}$.

Example. For Δ we have

$$\mathfrak{G}(\Delta) \cong \mathbb{CP}^1 = S^2.$$

Note that by (2.3) and (2.4), $\operatorname{ind} \Delta_{\mathcal{D}} = 0$ if and only if $\dim \mathcal{D}/\mathcal{D}_{\min} = 1$.

We finish this section with the following proposition that gives a first glimpse of the complexity of the spectrum of elliptic cone operators.

Proposition 2.3. *If A is c-elliptic and $\dim \mathfrak{G} > 0$, then for any $\lambda \in \mathbb{C}$ there is a domain $\mathcal{D} \in \mathfrak{G}$ such that $\lambda \in \operatorname{spec}(A_{\mathcal{D}})$. If, in addition, A is symmetric on $\mathcal{D}_{\min}$, then for any $\lambda \in \mathbb{R}$ there is a $\mathcal{D} \in \mathfrak{G}$ such that $A_{\mathcal{D}}$ is selfadjoint and $\lambda \in \operatorname{spec}(A_{\mathcal{D}})$.*

A proof is given in [6, Propositions 5.7 and 6.7]. A surprising consequence of the second statement is that for any arbitrary negative number λ there is always a selfadjoint extension of A having λ as eigenvalue, even if A is positive on $\mathcal{D}_{\min}$.

3. The model operator

Let $A \in x^{-m} \operatorname{Diff}_b^m(M;E)$ be c-elliptic. The *model operator* $A_\wedge$ associated with A is an operator on N_+Y, the closed inward normal bundle of Y, that in local coordinates takes the form

$$A_\wedge = x^{-m} \sum_{k+|\alpha|\le m} a_{k\alpha}(0,y)(xD_x)^k D_y^\alpha,$$

if A is written as in (1.3). A Taylor expansion in x (at $x=0$) of the coefficients of the operator A induces a decomposition

$$x^m A = \sum_{k=0}^{N-1} P_k x^k + x^N \tilde{P}_N \quad \text{for every } N \in \mathbb{N}, \tag{3.1}$$

where each P_k has coefficients independent of x near Y. Thus the model operator can be written, near Y, as $A_\wedge = x^{-m}P_0$. In other words, $A_\wedge$ can be thought of as the "most singular" part of A.

We trivialize N_+Y as $Y^\wedge = [0,\infty)\times Y$. The operator $A_\wedge \in x^{-m} \operatorname{Diff}_b^m(Y^\wedge;E)$ acts on $C_0^\infty(\mathring{Y}^\wedge;E)$ and can be extended as a densely defined closed operator in $x^{-m/2}L_b^2(Y^\wedge;E)$. The space $L_b^2(Y^\wedge;E)$ is the L^2 space with respect to a density of the form $\frac{dx}{x} \otimes \pi^*\mathfrak{m}_Y$ and the canonically induced Hermitian form on $\pi^*(E|_Y)$, where $\pi : Y^\wedge \to Y$ is the projection on the factor Y. The density $\mathfrak{m}_Y$ is related to $\mathfrak{m}$ and, by abuse of notation, we denote $\pi^*(E|_Y)$ by E, cf. [6]. Again, there are two canonical domains $\mathcal{D}_{\wedge,\min}$ and $\mathcal{D}_{\wedge,\max}$ and we denote by $\mathfrak{D}_\wedge$ the set of subspaces of $\mathcal{D}_{\wedge,\max}$ that contain $\mathcal{D}_{\wedge,\min}$. There is a natural (and useful) linear isomorphism

$$\theta : \mathcal{D}_{\max}/\mathcal{D}_{\min} \to \mathcal{D}_{\wedge,\max}/\mathcal{D}_{\wedge,\min},$$

cf. Section 5. As a consequence we have

$$\dim \mathcal{D}_{\wedge,\max}/\mathcal{D}_{\wedge,\min} = \dim \mathcal{D}_{\max}/\mathcal{D}_{\min} \tag{3.2}$$

which by Theorem 2.1 is finite. It is known (cf. Lesch [12]) that $\mathcal{D}_{\wedge,\max}/\mathcal{D}_{\wedge,\min}$ is isomorphic to a finite-dimensional space $\mathcal{E}_{\wedge,\max}$ consisting of functions of the form

$$\varphi = \sum_{\substack{\sigma\in\operatorname{spec}_b(A)\\ |\Im\sigma|<m/2}} \left(\sum_{k=0}^{m_\sigma} c_{\sigma,k}(y)\log^k x\right) x^{i\sigma} \tag{3.3}$$

where $c_{\sigma,k} \in C^\infty(Y;E)$. More precisely, for every $u \in \mathcal{D}_{\wedge,\max}$ there is a function $\varphi \in \mathcal{E}_{\wedge,\max}$ such that $u(x,y) - \omega(x)\varphi(x,y) \in \mathcal{D}_{\wedge,\min}$ for some (hence any) cut-off function $\omega \in C_0^\infty([0,1))$, $\omega = 1$ near 0. The function φ is uniquely determined by the equivalence class $u + \mathcal{D}_{\wedge,\min}$.

We identify

$$\mathcal{E}_{\wedge,\max} = \mathcal{D}_{\wedge,\max}/\mathcal{D}_{\wedge,\min}$$

and let

$$\pi_{\wedge,\max} : \mathcal{D}_{\wedge,\max} \to \mathcal{E}_{\wedge,\max}$$

be the canonical projection.

Contrary to the situation in Theorem 2.1, the closed extensions of $A_\wedge$ do not need to be Fredholm. However, if $A - \lambda$ is c-elliptic with parameter, then the canonical extensions $A_{\wedge,\min} - \lambda$ and $A_{\wedge,\max} - \lambda$ are both Fredholm for $\lambda \neq 0$, cf. [7, Remark 5.26]. Moreover, we have

$$\operatorname{ind}(A_{\wedge,\min} - \lambda) = \operatorname{ind} A_{\mathcal{D}_{\min}}, \tag{3.4}$$

cf. Corollary 5.35 in [7].

Example. On $Y^\wedge = [0,\infty) \times S^1$ with the cone metric $dx^2 + x^2dy^2$, the Laplace-Beltrami operator is given by

$$\Delta_\wedge = x^{-2}\big((xD_x)^2 + \Delta_Y\big), \tag{3.5}$$

where Δ_Y is the nonnegative Laplacian on S^1. $\Delta_\wedge$ is precisely the model operator associated with the cone Laplacian Δ discussed in the previous section, cf. (2.1). It is easy to check that for any cut-off function $\omega \in C_0^\infty([0,1))$, the functions

$$\omega(x)\cdot 1,\ \Delta_\wedge(\omega(x)\cdot 1),\ \omega(x)\log x,\ \text{ and } \Delta_\wedge(\omega(x)\log x)$$

are all in the space $x^{-1}L_b^2(Y^\wedge)$. Thus $\omega(x)\cdot 1$ and $\omega(x)\log x$ are elements of $\mathcal{D}_{\wedge,\max}$. In fact,

$$\mathcal{E}_{\wedge,\max} = \operatorname{span}\{1, \log x\}. \tag{3.6}$$

Observe that $\Delta - \lambda$ is c-elliptic with parameter $\lambda \in \mathbb{C}\backslash\mathbb{R}_+$ and therefore the closed extensions of $\Delta_\wedge - \lambda$ are Fredholm for every $\lambda \in \mathbb{C}\backslash\overline{\mathbb{R}}_+$.

The model operator has a dilation/scaling property that can be exploited to analyze its closed extensions and their resolvents from a geometric point of view. In order to describe this property we first introduce the one-parameter group of isometries

$$\mathbb{R}_+ \ni \varrho \mapsto \kappa_\varrho : x^{-m/2}L_b^2(Y^\wedge;E) \to x^{-m/2}L_b^2(Y^\wedge;E)$$

which on functions is defined by

$$(\kappa_\varrho f)(x,y) = \varrho^{m/2} f(\varrho x, y). \tag{3.7}$$

It is easily verified that the operator $A_\wedge$ satisfies the relation

$$\kappa_\varrho A_\wedge = \varrho^{-m} A_\wedge \kappa_\varrho.$$

This implies

$$A_\wedge - \lambda = \varrho^m \kappa_\varrho (A_\wedge - \lambda/\varrho^m)\kappa_\varrho^{-1} \tag{3.8}$$

for every $\varrho > 0$ and $\lambda \in \mathbb{C}$. This homogeneity property, called *κ-homogeneity*, will be used systematically to describe the closed extensions of $A_\wedge$ with nonempty resolvent sets.

It is convenient to introduce the set

$$\text{bg-res}\, A_\wedge = \{\lambda \in \mathbb{C} : A_{\wedge,\min} - \lambda \text{ is injective and } A_{\wedge,\max} - \lambda \text{ is surjective}\},$$

the *background resolvent set* of $A_\wedge$, cf. [6].

Lemma 3.1 (Lemma 7.3 in [6]). *If $\lambda \in \text{bg-res}\, A_\wedge$ and $\mathcal{D} \in \mathfrak{D}_\wedge$, then $A_{\wedge,\mathcal{D}} - \lambda$ is Fredholm. The set bg-res $A_\wedge$ is a disjoint union of open sectors,*

$$\text{bg-res}\, A_\wedge = \bigcup_{\alpha \in \mathfrak{I} \subset \mathbb{N}} \mathring{\Lambda}_\alpha.$$

This lemma follows immediately from (3.8).

For $\lambda \in \text{bg-res}\, A_\wedge$ and $\mathcal{D} \in \mathfrak{D}_\wedge$ we have

$$\operatorname{ind}(A_{\wedge,\mathcal{D}} - \lambda) = \operatorname{ind}(A_{\wedge,\min} - \lambda) + \dim \mathcal{D}/\mathcal{D}_{\wedge,\min}. \tag{3.9}$$

Moreover, the map

$$\mathring{\Lambda}_\alpha \ni \lambda \mapsto \operatorname{ind}(A_{\wedge,\mathcal{D}} - \lambda)$$

is constant since the embedding $\mathcal{D} \hookrightarrow x^{-m/2} L_b^2(Y^\wedge; E)$ is continuous. Now, in analogy with (2.5) we define

$$\mathfrak{G}_{\wedge,\alpha} = \{\mathcal{D} \in \mathfrak{D}_\wedge : \operatorname{ind}(A_{\wedge,\mathcal{D}} - \lambda) = 0 \text{ for } \lambda \in \mathring{\Lambda}_\alpha\}$$

and let $d''_\alpha = -\operatorname{ind}(A_{\wedge,\min} - \lambda)$ for $\lambda \in \mathring{\Lambda}_\alpha$. We identify $\mathfrak{G}_{\wedge,\alpha}$ with the complex Grassmannian of d''_α-dimensional subspaces of $\mathcal{E}_{\wedge,\max}$.

The canonical domains $\mathcal{D}_{\wedge,\min}$ and $\mathcal{D}_{\wedge,\max}$ are both κ-invariant. Thus the group action κ_ϱ induces an action on $\mathcal{E}_{\wedge,\max}$. In general, κ_ϱ does not preserve the elements of $\mathfrak{D}_\wedge$. In fact, the set of κ-invariant domains in $\mathfrak{D}_\wedge$ is an analytic variety because it consists of the stationary points of a holomorphic flow, cf. Section 7 in [6]. To better analyze the resolvents of the closed extensions of $A_\wedge$ over the open sector $\mathring{\Lambda}_\alpha$, we will consider the manifold $\mathfrak{G}_{\wedge,\alpha}$ together with the flow generated by the induced action of κ_ϱ given by $\kappa_\varrho(\mathcal{D}/\mathcal{D}_{\wedge,\min}) = \kappa_\varrho(\mathcal{D})/\mathcal{D}_{\wedge,\min}$.

Example. The background resolvent set of $\Delta_\wedge$ is the open sector $\mathbb{C}\backslash\overline{\mathbb{R}}_+$; this is easily seen after noting that $\Delta_\wedge$ is the standard Laplacian in $\mathbb{R}^2$ written in polar coordinates. Moreover, since $\operatorname{ind}(\Delta_{\wedge,\min} - \lambda) = -1$ for every $\lambda \in \mathbb{C}\backslash\overline{\mathbb{R}}_+$, we have that $\mathcal{D} \in \mathfrak{D}_\wedge$ belongs to $\mathfrak{G}_\wedge$ if and only if $\dim \mathcal{D}/\mathcal{D}_{\wedge,\min} = 1$. Thus

$$\mathfrak{G}_\wedge \cong \mathbb{CP}^1 = S^2.$$

We identify $\mathcal{E}_{\wedge,\max}$ with $\mathcal{D}_{\wedge,\max}/\mathcal{D}_{\wedge,\min}$ and use (3.6) to write

$$\mathcal{E}_{\wedge,\max} = \operatorname{span}\{1, \log x\}.$$

For $\mathcal{D} \in \mathfrak{G}_\wedge$ we then have

$$\pi_{\wedge,\max}\mathcal{D} = \operatorname{span}\{\zeta_0 \cdot 1 + \zeta_1 \log x\} \quad \text{for some } \zeta_0, \zeta_1 \in \mathbb{C},\ (\zeta_0,\zeta_1) \neq 0. \tag{3.10}$$

Hence, with κ as defined in (3.7), we get

$$\pi_{\wedge,\max}\kappa_\varrho^{-1}\mathcal{D} = \operatorname{span}\{(\zeta_0 - \zeta_1 \log \varrho)\cdot 1 + \zeta_1 \log x\}. \tag{3.11}$$

Clearly, the only κ-invariant domain in $\mathfrak{G}_\wedge$ is the domain $\mathcal{D}_F$ such that

$$\pi_{\wedge,\max}\mathcal{D}_F = \operatorname{span}\{1\};$$

$\mathcal{D}_F$ is precisely the domain of the Friedrichs extension of $\Delta_\wedge$, cf. [8]. Every domain $\mathcal{D} \in \mathfrak{G}_\wedge$ with $\zeta_1 \neq 0$ in (3.10) generates a nontrivial orbit as given by (3.11). In order to describe the flow of κ on these nonstationary points, rewrite (3.10) as

$$\pi_{\wedge,\max}\mathcal{D} = \operatorname{span}\left\{\tfrac{\zeta_0}{\zeta_1}\cdot 1 + \log x\right\}.$$

Then the projection to $\mathcal{E}_{\wedge,\max}$ of the dilation $\kappa_\varrho^{-1}\mathcal{D}$ is given by

$$\pi_{\wedge,\max}\kappa_\varrho^{-1}\mathcal{D} = \operatorname{span}\left\{\left(\tfrac{\zeta_0}{\zeta_1} - \log\varrho\right)\cdot 1 + \log x\right\}. \tag{3.12}$$

If $[\zeta_0 : \zeta_1] \in \mathbb{CP}^1$ is the point corresponding to $\mathcal{D}$, then $\kappa_\varrho^{-1}\mathcal{D}$ is represented by $[\zeta_0 - \zeta_1 \log\varrho : \zeta_1]$. In other words, in the situation at hand, the flow generated by κ on $\mathfrak{G}_\wedge \cong \mathbb{CP}^1$ consists of curves that in projective coordinates are lines parallel to the real axis, see Figure 1.

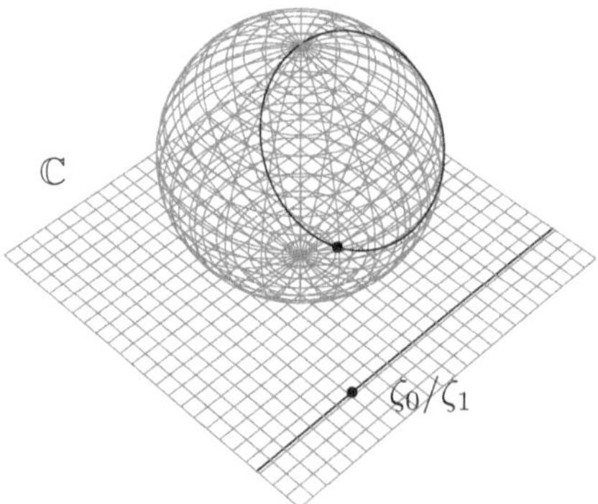

FIGURE 1. Orbit in $\mathfrak{G}_\wedge(\Delta_\wedge)$ generated by $\mathcal{D} \leftrightarrow \zeta_0/\zeta_1 \in \mathbb{C}$.

Observe that the Friedrichs extension corresponds to the point $[1:0] \in \mathbb{CP}^1$. Using

$$\pi_{\wedge,\max}\kappa_\varrho^{-1}\mathcal{D} = \operatorname{span}\left\{1 + \tfrac{\zeta_1}{\zeta_0 - \zeta_1\log\varrho}\log x\right\} \quad \text{if } \varrho \neq e^{\zeta_0/\zeta_1}$$

we see that

$$\pi_{\wedge,\max}\kappa_\varrho^{-1}\mathcal{D} \to \operatorname{span}\{1\} = \pi_{\wedge,\max}\mathcal{D}_F \quad \text{as } \varrho \to \infty \text{ or } \varrho \to 0. \tag{3.13}$$

Let $\mathcal{D}_0 \in \mathfrak{G}_\wedge$ be such that $\pi_{\wedge,\max}\mathcal{D}_0 = \operatorname{span}\{\log x\}$. This domain gives a selfadjoint extension of $\Delta_\wedge$ which on the sphere corresponds to the point $[0:1]$. The circle consisting of the orbit of $\mathcal{D}_0$ together with $\mathcal{D}_F$ is the set of domains of selfadjoint extensions of $\Delta_\wedge$.

4. Ray conditions on the model cone

In this section we will discuss the existence of sectors of minimal growth for the model operator $A_\wedge \in x^{-m}\operatorname{Diff}_b^m(Y^\wedge; E)$ associated with a c-elliptic cone operator. We fix a component $\mathring{\Lambda}_\alpha$ of bg-res $A_\wedge$ and let $\mathfrak{G}_\wedge = \mathfrak{G}_{\wedge,\alpha}$.

Let Λ be a closed sector such that $\Lambda\backslash 0 \subset \mathring{\Lambda}_\alpha$, cf. (1.1), let $\operatorname{res} A_{\wedge,\mathcal{D}}$ be the resolvent set of $A_{\wedge,\mathcal{D}}$. The κ-invariant domains are the simplest domains to analyze.

Proposition 4.1 (Proposition 8.4 in [6]). *Suppose $\mathcal{D} \in \mathfrak{G}_\wedge$ is κ-invariant. If there exists $\lambda_0 \in \mathring{\Lambda}_\alpha$ such that $A_{\wedge,\mathcal{D}} - \lambda_0$ is invertible, then $\mathring{\Lambda}_\alpha \subset \operatorname{res} A_{\wedge,\mathcal{D}}$ and Λ is a sector of minimal growth for $A_{\wedge,\mathcal{D}}$.*

If $\mathcal{D} \in \mathfrak{G}_\wedge$ is not κ-invariant, the situation is more complicated. Nonetheless, in [6] we found a condition necessary and sufficient for a sector Λ to be a sector of minimal growth for $A_{\wedge,\mathcal{D}}$. This condition is expressed in terms of finite-dimensional spaces and projections that we proceed to discuss briefly.

For $\lambda \in$ bg-res $A_\wedge$ we let

$$\mathcal{K}_{\wedge,\lambda} = \ker(A_{\wedge,\max} - \lambda).$$

Then

$$\operatorname{res} A_{\wedge,\mathcal{D}} = \text{bg-res}\, A_\wedge \cap \{\lambda : \mathcal{K}_{\wedge,\lambda} \cap \mathcal{D} = 0\},$$

and for $\lambda \in \operatorname{res} A_{\wedge,\mathcal{D}}$ we have

$$\mathcal{D}_{\wedge,\max} = \mathcal{K}_{\wedge,\lambda} \oplus \mathcal{D}. \tag{4.1}$$

Projecting on $\mathcal{E}_{\wedge,\max}$, this direct sum induces the decomposition

$$\mathcal{E}_{\wedge,\max} = \pi_{\wedge,\max}\mathcal{K}_{\wedge,\lambda} \oplus \pi_{\wedge,\max}\mathcal{D}, \tag{4.2}$$

and the projection on $\pi_{\wedge,\max}\mathcal{K}_{\wedge,\lambda}$ according to (4.2) is given by the map

$$\begin{gathered} \hat{\pi}_{\mathcal{K}_{\wedge,\lambda},\mathcal{D}} : \mathcal{E}_{\wedge,\max} \to \mathcal{E}_{\wedge,\max} \\ (u + \mathcal{D}_{\wedge,\min}) \mapsto \pi_{\mathcal{K}_{\wedge,\lambda},\mathcal{D}}u + \mathcal{D}_{\wedge,\min}, \end{gathered} \tag{4.3}$$

where $\pi_{\mathcal{K}_{\wedge,\lambda},\mathcal{D}}$ is the projection on $\mathcal{K}_{\wedge,\lambda}$ according to (4.1).

The following theorem gives a condition on the operator norm of (4.3) for a sector Λ to be a sector of minimal growth for $A_{\wedge,\mathcal{D}}$. Define

$$\|u\|_\lambda^2 = \|u\|^2 + |\lambda|^{-2}\|A_\wedge u\|^2$$

for $\lambda \neq 0$ and $u \in \mathcal{D}_{\wedge,\max}$.

Theorem 4.2. *Let $\mathcal{D} \in \mathfrak{G}_\wedge$, let Λ be a closed sector with $\Lambda\backslash 0 \subset \mathring{\Lambda}_\alpha$. Then Λ is a sector of minimal growth for $A_{\wedge,\mathcal{D}}$ if and only if there are C, $R > 0$ such that $\Lambda_R \subset \operatorname{res} A_{\wedge,\mathcal{D}}$, and*

$$\left\|\hat{\pi}_{\mathcal{K}_{\wedge,\lambda},\mathcal{D}}\right\|_{\mathscr{L}(\mathcal{E}_{\wedge,\max},\|\cdot\|_\lambda)} \leq C \quad \textit{for } \lambda \in \Lambda_R, \tag{4.4}$$

where $\hat{\pi}_{\mathcal{K}_{\wedge,\lambda},\mathcal{D}}$ is the projection (4.3).

This theorem is a rephrasing of [6, Theorem 8.7]. There the condition (4.4) appears in the equivalent form

$$\left\|\hat{\pi}_{\mathcal{K}_{\wedge,\hat{\lambda}},\kappa^{-1}_{|\lambda|^{1/m}}\mathcal{D}}\right\|_{\mathscr{L}(\mathcal{E}_{\wedge,\max})} \leq C \quad \text{for } \lambda \in \Lambda_R, \tag{4.4$'$}$$

where $\hat{\lambda} = \lambda/|\lambda|$, and $\hat{\pi}_{\mathcal{K}_{\wedge,\hat{\lambda}},\kappa^{-1}_{|\lambda|^{1/m}}\mathcal{D}}$ is the projection on $\mathcal{K}_{\wedge,\hat{\lambda}}$ induced (following the steps (4.1)–(4.3)) by the direct sum

$$\mathcal{D}_{\wedge,\max} = \mathcal{K}_{\wedge,\varrho^{-m}\lambda} \oplus \kappa_\varrho^{-1}\mathcal{D} \tag{4.5}$$

for $\lambda \in \operatorname{res} A_{\wedge,\mathcal{D}}$ and $\varrho > 0$. This decomposition is a consequence of (4.1) and the κ-invariance of $\mathcal{D}_{\wedge,\max}$, as follows. First, the κ-homogeneity of $A_\wedge - \lambda$, cf. (3.8), implies

$$\kappa_\varrho^{-1}(\mathcal{K}_{\wedge,\lambda}) = \mathcal{K}_{\wedge,\varrho^{-m}\lambda} \quad \text{for } \varrho > 0.$$

Furthermore, if $\mathcal{D} \in \mathfrak{G}_\wedge$ and $\lambda \in \text{bg-res}\, A_\wedge$, then

$$\varrho^{-m}\lambda \in \operatorname{res} A_{\wedge,\kappa_\varrho^{-1}\mathcal{D}} \iff \lambda \in \operatorname{res} A_{\wedge,\mathcal{D}}.$$

In particular,

$$\mathcal{K}_{\wedge,\varrho^{-m}\lambda} \cap \kappa_\varrho^{-1}\mathcal{D} = \{0\} \iff \mathcal{K}_{\wedge,\lambda} \cap \mathcal{D} = \{0\},$$

as claimed.

The equivalence of (4.4) and (4.4$'$) follows immediately from the identity

$$\kappa^{-1}_{|\lambda|^{1/m}} \pi_{\mathcal{K}_{\wedge,\lambda},\mathcal{D}}\, \kappa_{|\lambda|^{1/m}} = \pi_{\mathcal{K}_{\wedge,\hat{\lambda}},\kappa^{-1}_{|\lambda|^{1/m}}\mathcal{D}}$$

using the relation (3.8) and the fact that κ is an isometry on $x^{-m/2}L^2_b$. The virtue of (4.4$'$) is that the norm is fixed, while the advantage of (4.4) lies in that it gives a more explicit dependence on λ and deals with a projection on a subspace of $\mathcal{E}_{\wedge,\max}$ with *fixed* complement $\pi_{\wedge,\max}\mathcal{D}$.

In [6, Corollary 8.22] it is proved that Λ is a sector of minimal growth for $A_{\wedge,\mathcal{D}}$ if and only if there are constants C, $R > 0$ such that $\Lambda_R \subset \operatorname{res} A_{\wedge,\mathcal{D}}$ and

$$\left\|\kappa^{-1}_{|\lambda|^{1/m}}(A_{\wedge,\mathcal{D}} - \lambda)^{-1}\right\|_{\mathscr{L}(x^{-m/2}L^2_b,\mathcal{D}_{\wedge,\max})} \leq C/|\lambda|, \quad \lambda \in \Lambda_R.$$

It can be shown that this estimate is equivalent to (4.4) and (4.4$'$).

Example. We consider again the model Laplacian $\Delta_\wedge$ from the previous section. Recall that $\text{bg-res}\,\Delta_\wedge = \mathbb{C}\backslash\overline{\mathbb{R}}_+$. For $\lambda \in \mathbb{C}\backslash\overline{\mathbb{R}}_+$, we have

$$\pi_{\wedge,\max}\mathcal{K}_{\wedge,\lambda} = \operatorname{span}\{-k_0\log(-\lambda) + k_1 \log x\} \quad \text{for some } k_0, k_1 > 0,$$

where log means the principal branch of the logarithm. Moreover, by (3.12),

$$\pi_{\wedge,\max}\kappa_{\varrho}^{-1}\mathcal{D} = \operatorname{span}\Big\{\big(\tfrac{\zeta_0}{\zeta_1} - \log\varrho\big)\cdot 1 + \log x\Big\}.$$

The projection in (4.4′) can be computed explicitly. Namely, if $u = \alpha_0 + \alpha_1 \log x \in \mathcal{E}_{\wedge,\max} = \operatorname{span}\{1, \log x\}$ and $\lambda = \varrho^m\lambda_0$, then

$$\hat{\pi}_{\mathcal{K}_{\wedge,\lambda_0},\kappa_{\varrho}^{-1}\mathcal{D}}u = \frac{-\alpha_0 + \alpha_1(\frac{\zeta_0}{\zeta_1} - \log\varrho)}{k_0\log(-\lambda_0) + k_1(\frac{\zeta_0}{\zeta_1} - \log\varrho)}\,(-k_0\log(-\lambda_0) + k_1\log x). \tag{4.6}$$

Let Λ be a closed sector in $\mathbb{C}\backslash\mathbb{R}_+$ containing the half-plane $\{\Re\lambda < 0\}$. Since the family of projections (4.6) is bounded as $\varrho \to \infty$, uniformly for $|\lambda_0| = 1$ in Λ, regardless of the specific choice of α_0, α_1, Theorem 4.2 implies that every closed extension $\Delta_{\wedge,\mathcal{D}}$, $\mathcal{D} \in \mathfrak{G}_\wedge$, of the model Laplacian admits Λ as a sector of minimal growth.

Equivalent geometric condition

We identify $\mathfrak{G}_\wedge$ with the Grassmannian $\operatorname{Gr}_{d''}(\mathcal{E}_{\wedge,\max})$ where $d'' = -\operatorname{ind}(A_{\wedge,\min}-\lambda)$ for $\lambda \in \mathring{\Lambda}_\alpha \subset \text{bg-res}\, A_\wedge$. Let $d' = \dim \mathcal{K}_{\wedge,\lambda}$. The condition that in the Grassmannian $\operatorname{Gr}_{d''}(\mathcal{E}_{\wedge,\max})$, the curve

$$[R,\infty) \ni \varrho \mapsto \pi_{\wedge,\max}\kappa_{\varrho}^{-1}\mathcal{D}$$

does not approach the set

$$\mathscr{V}_{\mathcal{K}_{\wedge,\lambda}} = \{D \in \operatorname{Gr}_{d''}(\mathcal{E}_{\wedge,\max}) : D \cap \pi_{\wedge,\max}\mathcal{K}_{\wedge,\lambda} \neq 0\}$$

as $\varrho \to \infty$, is sufficient for the validity of (4.4′). This is [6, Theorem 8.28]. The following theorem states that the condition is also necessary.

For $D \in \operatorname{Gr}_{d''}(\mathcal{E}_{\wedge,\max})$ let

$$\Omega^-(D) = \big\{D' \in \operatorname{Gr}_{d''}(\mathcal{E}_{\wedge,\max}) : \exists\,\{\varrho_k\}_{k=1}^\infty \subset \mathbb{R}_+ \text{ such that } \varrho_k \to \infty \text{ and } \kappa_{\varrho_k}^{-1}D \to D' \text{ as } k \to \infty\big\}.$$

Theorem 4.3. *Let $\lambda_0 \in \mathring{\Lambda}_\alpha$. The ray through λ_0 is a ray of minimal growth for $A_{\wedge,\mathcal{D}}$ if and only if $\Omega^-(\pi_{\wedge,\max}\mathcal{D}) \cap \mathscr{V}_{\mathcal{K}_{\wedge,\lambda_0}} = \varnothing$.*

Proof. Let $\lambda_0 \in \mathring{\Lambda}_\alpha$ and $\mathcal{D} \in \mathfrak{G}_\wedge$. For simplicity, we use the notation

$$D = \pi_{\wedge,\max}\mathcal{D}, \quad \mathscr{V} = \mathscr{V}_{\mathcal{K}_{\wedge,\lambda_0}}, \quad \mathcal{K} = \pi_{\wedge,\max}\mathcal{K}_{\wedge,\lambda_0} \text{ and } \pi_{\mathcal{K},D} = \hat{\pi}_{\mathcal{K}_{\wedge,\lambda_0},\mathcal{D}}.$$

Suppose $\Omega^-(D) \cap \mathscr{V} = \varnothing$. Since $\Omega^-(D)$ and $\mathscr{V}$ are closed sets, there are a neighborhood $\mathcal{U}$ of $\mathscr{V}$ and a constant $R > 0$ such that if $\varrho > R$ then $\kappa_{\varrho}^{-1}D \notin \mathcal{U}$. Then Lemma 5.24 in [6] gives that $\|\pi_{\mathcal{K},\kappa_{\varrho}^{-1}D}\|$ is uniformly bounded as $\varrho \to \infty$, and therefore, by Theorem 4.2 the ray through λ_0 is a ray of minimal growth for $A_{\wedge,\mathcal{D}}$.

Assume now that there are C, $R > 0$ such that $\Lambda_R \subset \operatorname{res} A_{\wedge,\mathcal{D}}$ and the condition (4.4′) is satisfied. Suppose $\Omega^-(D) \cap \mathscr{V} \neq \varnothing$ and let $D_0 \in \Omega^-(D) \cap \mathscr{V}$. Since $D_0 \in \mathscr{V}$, we have $D_0 \cap \mathcal{K} \neq \{0\}$. On the other hand, $D_0 \in \Omega^-(D)$ implies that there is a sequence $\{\varrho_k\}_{k=1}^\infty \subset \mathbb{R}_+$ such that $\varrho_k \to \infty$ and $D_k = \kappa_{\varrho_k}^{-1}D \to D_0$

as $k \to \infty$. Note that for ϱ_k large we have $\varrho_k^m \lambda_0 \in \operatorname{res} A_{\wedge,\mathcal{D}}$, so $\lambda_0 \in \operatorname{res} A_{\wedge,\kappa_{\varrho_k}^{-1}\mathcal{D}}$ and therefore, $D_k \notin \mathscr{V}$.

Pick $v \in D_0 \cap \mathcal{K}$ with $\|v\| = 1$. Let π_{D_k} denote the orthogonal projection on D_k. Since $D_k \to D_0$ as $k \to \infty$, we have $\pi_{D_k} \to \pi_{D_0}$, so $v_k = \pi_{D_k} v \to \pi_{D_0} v = v$ as $k \to \infty$. Since $D_k \notin \mathscr{V}$, $v_k - v \neq 0$ and $\pi_{\mathcal{K},D_k} v_k = 0$. Hence

$$\pi_{\mathcal{K},D_k}\left(\frac{v - v_k}{\|v - v_k\|}\right) = \frac{v}{\|v - v_k\|} \to \infty \text{ as } k \to \infty,$$

since $\|v\| = 1$ and $v_k \to v$ as $k \to \infty$. But this implies that $\|\pi_{\mathcal{K},D_k}\| \to \infty$ contradicting the boundedness of the norm in (4.4′). Thus $\Omega^-(D) \cap \mathscr{V} = \varnothing$. □

Example. Let $\Delta_\wedge$ be the model Laplacian and let $\mathcal{D} \in \mathfrak{G}_\wedge$. In this case, the limiting set $\Omega^-(\pi_{\wedge,\max}\mathcal{D})$ consists of the one element of $\mathbb{CP}^1$ corresponding to the Friedrichs extension of $\Delta_\wedge$, cf. (3.13). From this new perspective, it is evident that every closed extension $\Delta_{\wedge,\mathcal{D}}$ of $\Delta_\wedge$ with $\mathcal{D} \in \mathfrak{G}_\wedge$ must admit a sector of minimal growth.

5. Rays of minimal growth

We continue to assume that $A \in x^{-m} \operatorname{Diff}_b^m(M; E)$ is c-elliptic.

Unlike the case of a differential operator with smooth coefficients on a closed manifold, that a ray Γ is a ray of minimal growth for the principal symbol ${}^c\boldsymbol{\sigma}(A)$ of A is not expected to imply that Γ is a ray of minimal growth for A. In this context, it is useful to think of $A_\wedge$ as a symbol (the wedge symbol) associated with A, cf. Schulze [16], so that it is natural to impose ray conditions on $A_\wedge$. For this to work, however, we need a procedure to transfer the information on the given domain $\mathcal{D}$ of A on M to equivalent information for $A_\wedge$ on $Y^\wedge$, and vice versa.

Theorem 5.1 (Theorem 4.12 in [6]). *There is a natural isomorphism*

$$\theta^{-1} : \mathcal{D}_{\wedge,\max}/\mathcal{D}_{\wedge,\min} \to \mathcal{D}_{\max}/\mathcal{D}_{\min}$$

given by a finite iterative procedure that involves the boundary spectrum of A and the decomposition (3.1). *In particular, if A has coefficients independent of x near Y, then θ is the identity map.*

A constructive proof of this theorem can be found in [6]. For illustration purposes, we will give here only a simplified description of θ^{-1} and will make some convenient identifications that can be justified via the Mellin transform, cf. [6, 7].

According to the decompositions (1.3) and (3.1) we define

$$\hat{P}_k(\sigma) = \sum_{j+|\alpha|\le m} \frac{1}{k!}(\partial_x^k a_{j\alpha})(0,y)(\sigma + ik)^j D_y^\alpha.$$

Note that $\hat{P}_0(\sigma)$ is the conormal symbol of A. Again, we identify $\mathcal{D}_{\wedge,\max}/\mathcal{D}_{\wedge,\min}$ with the space $\mathcal{E}_{\wedge,\max}$ of singular functions of the form (3.3). As in loc. cit., we let

$\mathcal{E}_{\wedge,\sigma_0}$ be the space of all functions in $\mathcal{E}_{\wedge,\max}$ that are of the form

$$\psi = \Big(\sum_{k=0}^{m_{\sigma_0}} c_{\sigma_0,k}(y)\log^k x\Big)x^{i\sigma_0}. \tag{5.1}$$

These spaces give a splitting

$$\mathcal{E}_{\wedge,\max} = \bigoplus_{\sigma_0\in\Sigma} \mathcal{E}_{\wedge,\sigma_0},$$

where

$$\Sigma = \operatorname{spec}_b(A) \cap \{\sigma \in \mathbb{C} : -m/2 < \Im(\sigma) < m/2\}.$$

Every function $\psi \in \mathcal{E}_{\wedge,\sigma_0}$ of the form (5.1) corresponds (via the Mellin transform) to a meromorphic function of the form

$$\widehat{\psi}(\sigma) = \sum_{k=0}^{m_{\sigma_0}} \frac{(-i)^{k-1}k!\,c_{\sigma_0,k}(y)}{(\sigma-\sigma_0)^{k+1}}.$$

We use this correspondence $\psi \mapsto \widehat{\psi}$ to define linear operators

$$\mathrm{e}_{\sigma_0,k} : \mathcal{E}_{\wedge,\sigma_0} \to C^\infty(\mathring{Y}^\wedge; E)$$

through the following inductive algorithm. For $\psi \in \mathcal{E}_{\wedge,\sigma_0}$ we let $\mathrm{e}_{\sigma_0,0}(\psi) = \psi$, and given $\mathrm{e}_{\sigma_0,0},\dots,\mathrm{e}_{\sigma_0,\vartheta-1}$ for some $\vartheta \in \mathbb{N}$, we define $\mathrm{e}_{\sigma_0,\vartheta}(\psi)$ to be the unique singular function of the form

$$\Big(\sum_{k=0}^{m_{\sigma_0-i\vartheta}} c_{\sigma_0-i\vartheta,k}(y)\log^k x\Big)x^{i(\sigma_0-i\vartheta)}$$

such that

$$\mathrm{e}_{\sigma_0,\vartheta}(\psi)^\wedge(\sigma) + \hat{P}_0(\sigma)^{-1}\Big(\sum_{k=1}^{\vartheta} \hat{P}_k(\sigma)\mathrm{e}_{\sigma_0,\vartheta-k}(\psi)^\wedge(\sigma+ik)\Big)$$

is holomorphic at $\sigma = \sigma_0 - i\vartheta$. Finally, we define θ^{-1} on $\mathcal{E}_{\wedge,\sigma_0}$ as the sum

$$\theta^{-1}|_{\mathcal{E}_{\wedge,\sigma_0}} = \sum_{k=0}^{N(\sigma_0)} \mathrm{e}_{\sigma_0,k}, \tag{5.2}$$

where $N(\sigma_0) \in \mathbb{N}_0$ is the largest integer such that $\Im\sigma_0 - N(\sigma_0) \geq -m/2$.

Analogous to Section 3, the range $\mathcal{E}_{\max}$ of θ^{-1} in $C^\infty(\mathring{Y}^\wedge; E)$ can be identified with the quotient $\mathcal{D}_{\max}/\mathcal{D}_{\min}$ via the mapping

$$\mathcal{E}_{\max} \ni \psi \mapsto \omega\psi + \mathcal{D}_{\min},$$

where $\omega \in C_0^\infty([0,1))$ is an arbitrary cut-off function near zero.

Example. Let M be a compact 2-manifold with boundary $Y = S^1$. Let A be a differential operator in $x^{-2}\operatorname{Diff}_b^2(M)$ that over the interior of M coincides with some Laplacian, and near Y, is of the form

$$A = x^{-2}\big((xD_x)^2 + q(x)\Delta_Y\big),$$

where Δ_Y is the standard nonnegative Laplacian on S^1 and q is a smooth function. We assume q to have the form

$$q(x) = \alpha^2 + \beta x + x^2\gamma(x),$$

where α, β are constants with $\frac{1}{2} < \alpha < 1$, $\beta \neq 0$, and γ is smooth up to $x = 0$. The associated model operator is then given by

$$A_\wedge = x^{-2}\big((xD_x)^2 + \alpha^2\Delta_Y\big),$$

and $\operatorname{spec}_b(A) = \{\pm i\alpha k : k \in \mathbb{N}_0\}$. Since $\frac{1}{2} < \alpha < 1$, only the set $\{-i\alpha, 0, i\alpha\}$ is relevant for the spaces $\mathcal{E}_{\max}$ and $\mathcal{E}_{\wedge,\max}$, cf. (3.3).

If y denotes the angular variable on S^1, then

$$\begin{aligned}\mathcal{E}_{\wedge,-i\alpha} &= \operatorname{span}\{e^{-iy}x^{\alpha}, e^{iy}x^{\alpha}\},\\ \mathcal{E}_{\wedge,0} &= \operatorname{span}\{1, \log x\},\\ \mathcal{E}_{\wedge,i\alpha} &= \operatorname{span}\{e^{-iy}x^{-\alpha}, e^{iy}x^{-\alpha}\},\end{aligned}$$

and $\mathcal{E}_{\wedge,\max} = \mathcal{E}_{\wedge,-i\alpha} \oplus \mathcal{E}_{\wedge,0} \oplus \mathcal{E}_{\wedge,i\alpha}$. Moreover, we have

$$\hat{P}_0(\sigma) = \sigma^2 + \alpha^2\Delta_Y \quad\text{and}\quad \hat{P}_1(\sigma) = \beta\Delta_Y.$$

Since every $\psi \in \mathcal{E}_{\wedge,0}$ is independent of y, we have $\hat{P}_1(\sigma)\hat{\psi}(\sigma) = 0$. Thus $\mathrm{e}_{0,1}(\psi) = 0$ and so $\theta^{-1}|_{\mathcal{E}_{\wedge,0}} = \mathrm{e}_{0,0}$ is the identity. On the other hand, $N(-i\alpha) = 0$ and therefore $\theta^{-1}|_{\mathcal{E}_{\wedge,-i\alpha}} = \mathrm{e}_{-i\alpha,0}$ is also the identity. Finally, for $\psi = e^{\pm iy}x^{-\alpha}$ we have

$$\mathrm{e}_{-i\alpha,1}(\psi) = -\frac{\beta}{2\alpha - 1}x\psi, \qquad \text{so} \qquad \theta^{-1}\psi = \Big(1 - \frac{\beta}{2\alpha-1}x\Big)\psi.$$

In other words,

$$\mathcal{E}_{\max} = \operatorname{span}\Big\{1, \log x, e^{\pm iy}x^{\alpha}, e^{\pm iy}x^{-\alpha}\big(1 - \tfrac{\beta}{2\alpha-1}x\big)\Big\}.$$

The map θ induces an isomorphism

$$\Theta : \mathfrak{D} \to \mathfrak{D}_\wedge$$

that we use to define $\mathcal{D}_\wedge = \Theta\mathcal{D}$ for any given $\mathcal{D} \in \mathfrak{D}$. The operator $A_{\wedge,\mathcal{D}_\wedge}$ is the closed extension of $A_\wedge$ in $x^{-m/2}L^2_b(Y^\wedge; E)$ uniquely associated with $A_{\mathcal{D}}$.

As in [7, Section 6], and motivated by the importance of κ_ϱ in studying the model operator $A_\wedge$, we introduce on $\mathcal{D}_{\max}(A)/\mathcal{D}_{\min}(A)$ the one-parameter group

$$\tilde{\kappa}_\varrho = \theta^{-1}\kappa_\varrho\theta \ \text{ for } \varrho > 0.$$

Similar to the situation on the model cone, the spectrum and resolvent of the closed extensions of A can be geometrically analyzed by considering the manifold $\mathfrak{G}$, cf. (2.5), together with the flow generated by $\tilde{\kappa}_\varrho$.

An interesting consequence of Theorem 5.1 is the following.

Proposition 5.2. *If $A - \lambda$ is c-elliptic with parameter $\lambda \neq 0$, then*

$$\operatorname{ind}(A_{\wedge,\mathcal{D}_\wedge} - \lambda) = \operatorname{ind} A_{\mathcal{D}}.$$

Proof. The existence of θ implies $\dim \mathcal{D}_\wedge/\mathcal{D}_{\wedge,\min} = \dim \mathcal{D}/\mathcal{D}_{\min}$. Now, the proposition follows by combining this identity with the relative index formulas (2.3) and (3.9), together with the equation (3.4). □

The following theorem describes the pseudo-differential structure of the resolvent of a cone operator A and gives tangible conditions over a given sector Λ on the symbols ${}^c\boldsymbol{\sigma}(A)$ and $A_\wedge$ for A to have Λ as a sector of minimal growth.

Theorem 5.3 (Theorem 6.9 in [7]). *Let $A \in x^{-m}\operatorname{Diff}_b^m(M;E)$ be such that $A-\lambda$ is c-elliptic with parameter $\lambda \in \Lambda$. If Λ is a sector of minimal growth for $A_{\wedge,\mathcal{D}_\wedge}$, then it is a sector of minimal growth for $A_\mathcal{D}$. Moreover,*

$$(A_\mathcal{D} - \lambda)^{-1} = B(\lambda) + G_\mathcal{D}(\lambda),$$

where $B(\lambda)$ is a parametrix of $A_{\mathcal{D}_{\min}} - \lambda$ with $B(\lambda)(A_{\mathcal{D}_{\min}} - \lambda) = 1$ for λ sufficiently large, and $G_\mathcal{D}(\lambda)$ is a pseudo-differential regularizing operator of finite rank.

The following lemma gives further information about the behavior at large of the resolvent along a sector of minimal growth.

Given two cut-off functions ω_0 and ω_1, the notation $\omega_1 \prec \omega_0$ will indicate that $\omega_0 = 1$ in a neighborhood of the support of ω_1.

Lemma 5.4. *Let $A \in x^{-m}\operatorname{Diff}_b^m(M;E)$ be c-elliptic and let Λ be a sector of minimal growth for $A_\mathcal{D}$. For every pair of cut-off functions $\omega_1 \prec \omega_0$, supported near the boundary, we have*

$$(1-\omega_0)(A_\mathcal{D} - \lambda)^{-1}\omega_1 \in \mathscr{S}\big(\Lambda, \mathscr{L}(x^{-m/2}L_b^2, \mathcal{D}_{\max})\big),$$

where $\mathscr{S}$ stands for Schwartz (rapidly decreasing as $|\lambda| \to \infty$).

Proof. Since Λ is a sector of minimal growth for $A_\mathcal{D}$, the family $A-\lambda$ must be c-elliptic with parameter $\lambda \in \Lambda$, and $A_{\wedge,\min} - \lambda$ must be injective for every $\lambda \in \Lambda$, $\lambda \neq 0$. A proof of this can be found in [7, Theorem 4.1].

As a consequence (cf. [7, Section 5]), there is a parametrix $B(\lambda)$ such that $B(\lambda)(A_{\mathcal{D}_{\min}} - \lambda) = 1$ for large $\lambda \in \Lambda$, and

$$(1-\omega_0)B(\lambda)\omega_1 \in \mathscr{S}\big(\Lambda, \mathscr{L}(x^{-m/2}L_b^2, \mathcal{D}_{\max})\big) \tag{5.3}$$

for all cut-off functions $\omega_1 \prec \omega_0$ supported near the boundary. We now make use of the identity

$$(A_\mathcal{D} - \lambda)^{-1} = B(\lambda) + (1 - B(\lambda)(A-\lambda))(A_\mathcal{D} - \lambda)^{-1}.$$

Multiplying by $(1-\omega_0)$ from the left and by ω_1 from the right, (5.3) proves the assertion for the first term involving $B(\lambda)$. On the other hand, since $1-B(\lambda)(A-\lambda)$ vanishes on $\mathcal{D}_{\min}$ for large λ, we have for such λ,

$$\begin{aligned}(1-\omega_0)(1-B(\lambda)(A-\lambda)) &= (1-\omega_0)(1-B(\lambda)(A-\lambda))\omega_2 \\ &= -(1-\omega_0)B(\lambda)(A-\lambda)\omega_2 \\ &= -(1-\omega_0)B(\lambda)\omega_1(A-\lambda)\omega_2\end{aligned}$$

whenever $\omega_2 \prec \omega_1$. Thus, by (5.3),

$$(1-\omega_0)(1-B(\lambda)(A-\lambda)) : \mathcal{D}_{\max} \to \mathcal{D}_{\max}$$

is rapidly decreasing as $|\lambda| \to \infty$. Finally, the assertion of the lemma can be completed using the fact that $(A_{\mathcal{D}}-\lambda)^{-1}\omega_1 : x^{-m/2}L^2_b \to \mathcal{D}_{\max}$ is uniformly bounded. □

Necessity of the conditions

The converse of Theorem 5.3 involves proving that the minimal growth of the resolvent $(A_{\mathcal{D}}-\lambda)^{-1}$ over a sector Λ implies a corresponding behavior for the inverse of ${}^c\boldsymbol{\sigma}(A)-\lambda$ and for the resolvent $(A_{\wedge,\mathcal{D}_\wedge}-\lambda)^{-1}$.

While in [7, Theorem 4.1] we established the necessity of the condition on ${}^c\boldsymbol{\sigma}(A)$, we did not address the question whether Λ must necessarily be a sector of minimal growth for $A_{\wedge,\mathcal{D}_\wedge}$. In the next theorem we prove that this is indeed the case when A has coefficients independent of x near $Y=\partial M$.

Theorem 5.5. *Let $A \in x^{-m}\operatorname{Diff}^m_b(M;E)$ be c-elliptic with coefficients independent of x near Y. If Λ is a a sector of minimal growth for $A_{\mathcal{D}}$, then $A-\lambda$ is c-elliptic with parameter $\lambda\in\Lambda$, and Λ is a sector of minimal growth for $A_{\wedge,\mathcal{D}_\wedge}$.*

Proof. As stated in the proof of Lemma 5.4, the assumption on the resolvent of $A_{\mathcal{D}}$ implies that $A-\lambda$ is c-elliptic with parameter $\lambda\in\Lambda$ and that $A_{\wedge,\min}-\lambda$ is injective for every $\lambda\neq 0$. Thus we only need to prove the statement about $A_{\wedge,\mathcal{D}_\wedge}$.

By Proposition 5.2, and since $\operatorname{ind} A_{\mathcal{D}}=0$, we have $\operatorname{ind}(A_{\wedge,\mathcal{D}_\wedge}-\lambda)=0$ for $\lambda\neq 0$. For this reason, in order to show that Λ is a sector of minimal growth for $A_{\wedge,\mathcal{D}_\wedge}$, it suffices to find (for large $\lambda\in\Lambda$) a right-inverse of $A_{\wedge,\mathcal{D}_\wedge}-\lambda$ that is uniformly bounded in $\mathscr{L}(x^{-m/2}L^2_b,\mathcal{D}_\wedge)$ as $|\lambda|\to\infty$.

Since A is assumed to have coefficients independent of x near the boundary, there is a cut-off function ω_0 such that

$$A\omega_0 = A_\wedge\omega_0 \quad\text{and}\quad \omega_0\mathcal{D}=\omega_0\mathcal{D}_\wedge.$$

Let ω_1, ω_2 be cut-off functions with $\omega_2\prec\omega_1\prec\omega_0$. Then the operator

$$B(\lambda)=\omega_1(A_{\mathcal{D}}-\lambda)^{-1}\omega_2$$

can be regarded as an operator on M with values in $\mathcal{D}$ or as an operator on $Y^\wedge$ with values in $\mathcal{D}_\wedge$. Depending on the context we will write $B(\lambda)$ as

$$B_{\mathcal{D}}(\lambda) : x^{-m/2}L^2_b(M;E)\to\mathcal{D} \quad\text{or}\quad B_{\mathcal{D}_\wedge}(\lambda) : x^{-m/2}L^2_b(Y^\wedge;E)\to\mathcal{D}_\wedge.$$

On M we consider

$$\begin{aligned}(A_{\mathcal{D}}-\lambda)B_{\mathcal{D}}(\lambda) &= \omega_0(A_{\mathcal{D}}-\lambda)\omega_1(A_{\mathcal{D}}-\lambda)^{-1}\omega_2\\ &= \omega_2-\omega_0(A_{\mathcal{D}}-\lambda)(1-\omega_1)(A_{\mathcal{D}}-\lambda)^{-1}\omega_2\\ &= \omega_2+R(\lambda)\end{aligned}$$

with $R(\lambda)=-\omega_0(A_{\mathcal{D}}-\lambda)(1-\omega_1)(A_{\mathcal{D}}-\lambda)^{-1}\omega_2$. By Lemma 5.4, $R(\lambda)$ is rapidly decreasing in the norm as $|\lambda|\to\infty$.

Because of the presence and nature of the cut-off functions ω_0 and ω_2, $R(\lambda)$ can also be regarded as an operator on $Y^\wedge$, say $R_\wedge(\lambda) \in \mathscr{S}\big(\Lambda, \mathscr{L}(x^{-m/2}L_b^2)\big)$. Now, using that $(A_\mathcal{D} - \lambda)\omega_1 = (A_{\wedge,\mathcal{D}_\wedge} - \lambda)\omega_1$, we get on $Y^\wedge$ the identity

$$(A_{\wedge,\mathcal{D}_\wedge} - \lambda)B_{\mathcal{D}_\wedge}(\lambda) = \omega_2 + R_\wedge(\lambda). \tag{5.4}$$

Furthermore, we have

$$\|B_{\mathcal{D}_\wedge}(\lambda)\|_{\mathscr{L}(x^{-m/2}L_b^2, \mathcal{D}_{\wedge,\max})} = O(1) \quad \text{as } |\lambda| \to \infty,$$

since, by assumption, $\|B_\mathcal{D}(\lambda)\|_{\mathscr{L}(x^{-m/2}L_b^2, \mathcal{D}_{\max})}$ has the same asymptotic behavior.

On the other hand, as $A - \lambda$ is c-elliptic with parameter, by [7, Theorem 5.24] there is a family of pseudo-differential operators $B_{2,\wedge}(\lambda) : x^{-m/2}L_b^2 \to \mathcal{D}_{\wedge,\min}$ (uniformly bounded in λ) such that $(A_\wedge - \lambda)B_{2,\wedge}(\lambda) - 1$ is regularizing, and for $\omega_3 \prec \omega_2$, the families $\omega_3 B_{2,\wedge}(\lambda)(1-\omega_2)$ and $\big[(A_\wedge - \lambda)B_{2,\wedge}(\lambda) - 1\big](1-\omega_2)$ are rapidly decreasing in the norm as $|\lambda| \to \infty$. Thus, as $A_{\wedge,\mathcal{D}_\wedge}(1-\omega_3) = A_\wedge(1-\omega_3)$,

$$(A_{\wedge,\mathcal{D}_\wedge} - \lambda)(1-\omega_3)B_{2,\wedge}(\lambda)(1-\omega_2) = (1-\omega_2) + S_\wedge(\lambda) \tag{5.5}$$

with $S_\wedge(\lambda) \in \mathscr{S}\big(\Lambda, \mathscr{L}(x^{-m/2}L_b^2)\big)$. Finally, the operator family

$$Q_\wedge(\lambda) = B_{\mathcal{D}_\wedge}(\lambda) + (1-\omega_3)B_{2,\wedge}(\lambda)(1-\omega_2) : x^{-m/2}L_b^2 \to \mathcal{D}_{\wedge,\max}$$

is bounded in the norm as $|\lambda| \to \infty$ and by (5.4) and (5.5) we have

$$(A_{\wedge,\mathcal{D}_\wedge} - \lambda)Q_\wedge(\lambda) - 1 \in \mathscr{S}\big(\Lambda, \mathscr{L}(x^{-m/2}L_b^2)\big).$$

By a Neumann series argument, it follows that $A_{\wedge,\mathcal{D}_\wedge} - \lambda : \mathcal{D}_\wedge \to x^{-m/2}L_b^2$ has a uniformly bounded right-inverse for large $\lambda \in \Lambda$. $\square$

Acknowledgment

The new results contained herein reflect part of work carried out by the three authors at the Mathematisches Forschungsinstitut Oberwolfach under their "Research in Pairs" program. They gratefully acknowledge the Institute's support and hospitality.

References

[1] M. Agranovich and M. Vishik, Elliptic problems with a parameter and parabolic problems of general type, *Russ. Math. Surveys* **19** (1963), 53–159.

[2] S. Agmon, On the eigenfunctions and on the eigenvalues of general elliptic boundary value problems, *Comm. Pure Appl. Math.* **15** (1962), 119–147.

[3] J. Brüning and R. Seeley, The expansion of the resolvent near a singular stratum of conical type, *J. Funct. Anal.* **95** (1991), 255–290.

[4] J. Cheeger, On the spectral geometry of spaces with cone-like singularities, *Proc. Nat. Acad. Sci. U.S.A.* **76** (1979), 2103–2106.

[5] J. Gil, Full asymptotic expansion of the heat trace for non-self-adjoint elliptic cone operators, Math. Nachr. **250** (2003), 25–57.

[6] J. Gil, T. Krainer, and G. Mendoza, Geometry and spectra of closed extensions of elliptic cone operators, *Canad. J. Math.*, to appear.

[7] J. Gil, T. Krainer, and G. Mendoza, Resolvents of elliptic cone operators, *J. Funct. Anal.* **241** (2006), no. 1, 1–55.

[8] J. Gil and G. Mendoza, Adjoints of elliptic cone operators, *Amer. J. Math.* **125** (2003), no. 2, 357–408.

[9] G. Grubb, *Functional calculus of pseudodifferential boundary problems*, Second Edition, Birkhäuser, Basel, 1996.

[10] V. Kondrat'ev, Boundary problems for elliptic equations in domains with conical or angular points, *Trans. Mosc. Math. Soc.* **16** (1967), 227–313.

[11] T. Krainer, Resolvents of elliptic boundary problems on conic manifolds, to appear in *Communications in PDE*.

[12] M. Lesch, *Operators of Fuchs type, conical singularities, and asymptotic methods*, B.G. Teubner, Stuttgart, Leipzig, 1997.

[13] P. Loya, On the resolvent of differential operators on conic manifolds, *Comm. Anal. Geom.* **10** (2002), no. 5, 877–934.

[14] R. Melrose, Transformation of boundary value problems, *Acta Math.* **147** (1981), 149–236.

[15] E. Schrohe and J. Seiler, The resolvent of closed extensions of cone differential operators, *Canad. J. Math.* **57** (2005), 771–811.

[16] B.-W. Schulze, Pseudo-differential operators on manifolds with edges, in *Proc. Symp. Partial Differential Equations, Holzhau 1988* (*Leipzig*), Teubner-Texte zur Math. Vol. 112, Teubner, 1989, 259–288.

[17] R. Seeley, Complex powers of an elliptic operator, in *Singular Integrals*, Amer. Math. Soc., Providence, 1967, 288–307.

Juan B. Gil
Department of Mathematics and Statistics
Penn State Altoona
3000 Ivyside Park
Altoona, PA 16601, USA
e-mail: jgil@psu.edu

Thomas Krainer
Institut für Mathematik
Universität Potsdam
D-14415 Potsdam, Germany
e-mail: krainer@math.uni-potsdam.de

Gerardo A. Mendoza
Department of Mathematics
Temple University
Philadelphia, PA 19122, USA
e-mail: gmendoza@math.temple.edu

Operator Theory:
Advances and Applications, Vol. 172, 51–66

Symbolic Calculus of Pseudo-differential Operators and Curvature of Manifolds

Chisato Iwasaki

Abstract. The method of construction of the fundamental solution for heat equations using pseudo-differential operators with parameter time variable is discussed, which is applicable to calculate traces of operators. This gives extensions of both the Gauss-Bonnet-Chern Theorem and the Riemann-Roch Theorem.

Mathematics Subject Classification (2000). Primary 58J35; Secondary 58J40, 58J20.

Keywords. Symbolic calculus, fundamental solution, curvature, local index.

1. Introduction

In this paper we give, by means of symbolic calculus of pseudo-differential operators, both an extension theorem of a local version of the Gauss-Bonnet-Chern theorem given in [8] and that of a local version of the Riemann-Roch theorem given in [9].

Let M be a Riemannian manifold of dimension n without boundary. The Gauss-Bonnet-Chern theorem is stated as follows:

$$\sum_{p=0}^{n}(-1)^p \dim H_p(M) = \int_M C_n(x,M)dv,$$

where H_p is the set of harmonic p-forms, $C_n(x,M)dv$ is the Euler form if n is even and $C_n(x,M)dv = 0$ if n is odd. Its analytical proof is based on the formula

$$\sum_{p=0}^{n}(-1)^p \dim H_p(M) = \int_M \sum_{p=0}^{n}(-1)^p \operatorname{tr} e_p(t,x,x)dv,$$

where $e_p(t,x,y)$ denotes the kernel of the fundamental solution $E_p(t)$ of the Cauchy problem for the heat equation of Δ_p on differential p-forms $\Gamma(\wedge^p T^*(M))$;

$$E_p(t)\varphi(x) = \int_M e_p(t,x,y)\varphi(y)dv_y, \quad \varphi \in \Gamma(\wedge^p T^*(M))$$

satisfies

$$\begin{aligned}\left(\frac{d}{dt}+\Delta_p\right)E_p(t) &= 0 \qquad \text{in } (0,T)\times M,\\ E_p(0) &= I \qquad \text{in } M.\end{aligned}$$

So, we may say that a local version of the Gauss-Bonnet-Chern theorem holds, if

$$\sum_{p=0}^{n}(-1)^p \operatorname{tr} e_p(t,x,x) = C_n(x,M) + 0(\sqrt{t}) \tag{1.1}$$

as t tends to 0.

The author proved (1.1) in [8], using an algebraic theorem on linear spaces stated in [2], and the method of construction of the fundamental solution by the technique of pseudo-differential operators with new weights on symbols. In this paper, a generalization of a local version of the Gauss-Bonnet-Chern theorem is obtained. Before stating our theorems, we introduce the notation.

We denote by $\mathcal{I}$ the index set

$$\mathcal{I} = \{I = (i_1, i_2, \ldots, i_r) : 0 \le r \le n, 1 \le i_1 < \cdots < i_\ell \le n\},$$

and

$$\binom{a}{b} = 0 \quad \text{if } a < b, \text{ or } b < 0, \quad \binom{0}{0} = 1.$$

In the rest of this paper, fix an integer ℓ such that $0 \le \ell \le n$.

Define the constants $\{f_p\}_{p=0,1,\ldots,n}$

$$f_p = \binom{n-p}{n-\ell} + \sum_{j=\max\{p,\ell+1\}}^{n} k_j \binom{n-p}{n-j} \quad (0 \le p \le n) \tag{1.2}$$

with arbitrary constants $\{k_j\}_{j=\ell+1,\ldots,n}$.

Theorem 1.1 (Main Theorem I). *Let M be a Riemannian manifold of dimension n without boundary and let $E_p(t)$ be the fundamental solution on $\Gamma(\wedge^p T^*(M))$. Suppose the f_p are of the form (1.2). Then we have*

$$\sum_{p=0}^{n}(-1)^p f_p \operatorname{tr} e_p(t,x,x) = C_\ell(x) t^{-\frac{n}{2}+\frac{\ell}{2}} + 0(t^{-\frac{n}{2}+\frac{\ell}{2}+\frac{1}{2}}) \qquad \text{as } t \to 0,$$

where $C_\ell(x)$ is given as follows:

(1) *If ℓ is odd, $C_\ell(x) = 0$.*
(2) *If ℓ is even ($\ell = 2m$), $C_\ell(x) = \sum_{I\in\mathcal{I},\sharp(I)=\ell} C_I(x)$, for $I = (i_1, i_2, \ldots, i_\ell) \in \mathcal{I}$*

$$C_I(x) = \left(\frac{1}{2\sqrt{\pi}}\right)^n \frac{1}{m!}\left(\frac{1}{2}\right)^m \sum_{\pi,\sigma\in S_\ell} \operatorname{sign}(\pi)\operatorname{sign}(\sigma) \times R_{i_{\pi(1)}i_{\pi(2)}i_{\sigma(1)}i_{\sigma(2)}} \cdots R_{i_{\pi(\ell-1)}i_{\pi(\ell)}i_{\sigma(\ell-1)}i_{\sigma(\ell)}}$$

for $I = (i_1, i_2, \ldots, i_\ell) \in \mathcal{I}$.

Remark 1.2. Assume $\ell = n$. Then $f_p = 1$ of (1.2) for all p. Theorem 1.1 is a local version of the Gauss-Bonnet-Chern theorem.

Remark 1.3. Assume $k_j = 0$ for all j. Then $f_p = \binom{n-p}{n-\ell}$ $(0 \le p \le \ell)$, $f_p = 0$ $(\ell + 1 \le p \le n)$. So Theorem 1.1 coincides with the result in [3].

Now consider the similar problem for Dolbeault complex on a Kaehler manifold M, that is, a local version of the Riemann-Roch theorem. Let $e_p(t, x, y)$ denote the kernel of the fundamental solution $E_p(t)$ of the Cauchy problem for the heat equation of $\Box_p$ on differential $(0,p)$-forms $\Gamma(\wedge^p T^{*(0,1)}(M))$;

$$E_p(t)\varphi(x) = \int_M e_p(t,x,y)\varphi(y)dv_y, \quad \varphi \in \Gamma(\wedge^p T^{*(0,1)}(M))$$

satisfies

$$\begin{aligned} \left(\frac{d}{dt} + \Box_p\right) E_p(t) &= 0 \qquad \text{in } (0,T) \times M, \\ E_p(0) &= I \qquad \text{in } M. \end{aligned}$$

In [9] the author has given a proof of a local version of the Riemann-Roch theorem, constructing the fundamental solution according to the method of symbolic calculus for a degenerate parabolic operator in [7]. There are several papers about a local version of the Riemann-Roch theorem. In [11] this formula is proved for manifolds of dimension 1. Then in [13] it is proved for Kaehler manifolds of any dimension. In [6] it is also shown, using invariant theory. In [5] this problem is treated by a different approach. We obtain an extension of this problem as follows.

Theorem 1.4 (Main Theorem II). *Let M be a compact Kaehler manifold with complex dimension n, and let $E_p(t)$ be the fundamental solution on $A^{0,p}(M) = \Gamma(\wedge^p T^{*(0,1)}(M))$. Suppose f_p are of the form (1.2). Then we have*

$$\sum_{p=0}^{n}(-1)^p f_p \operatorname{tr} e_p(t,x,x)dv = \left(\frac{1}{2\pi i}\right)^n C_\ell^D(x) t^{-n+\ell} + 0(t^{-n+\ell+1}) \qquad \text{as } t \to 0,$$

where $D_\ell^D(x)$ are defined as follows:
$C_\ell^D(x) = \sum_{I \in \mathcal{I}, \sharp(I)=\ell} C_I^D(x)$, where for $I = (i_1, i_2, \dots, i_\ell) \in \mathcal{I}$

$$C_I^D(x) = \left[\det\left(\frac{\Omega}{e^\Omega - Id}\right)\right]_{2\ell} \wedge dv^{I^c}.$$

Here Ω is a matrix whose (j,k) element is a 2-form defined as

$$(\Omega)_{jk} = \sum_{a,b=1}^{n} R_{\bar{k}ja\bar{b}}\,\omega^a \wedge \bar{\omega}^b$$

and

$$dv^{I^c} = \bar{\omega}^{j_1} \wedge \omega^{j_1} \wedge \bar{\omega}^{j_2} \wedge \omega^{j_2} \cdots \wedge \bar{\omega}^{j_{n-\ell}} \wedge \omega^{j_{n-\ell}},$$

where $I^c = (j_1, j_2, \dots, j_{n-\ell}) \in \mathcal{I}$ such that $I \cup I^c = \{1, 2, \dots, n\}$.

Remark 1.5. Assume $\ell = n$. Then $f_p = 1$ of (1.2) for all p. In this case Theorem 1.4 is a local version of the Riemann-Roch theorem.

Our point is that one can prove the above theorems by only calculating the main term of the symbol of the fundamental solution, introducing a new weight of symbols of pseudo-differential operators.

The plan of this paper is following. In Section 2 an algebraic theorem, which is the key of the proof, is proved. The sketch of proof is given in Section 3 and Section 4.

2. Algebraic properties for the calculation of the trace

Let V be a vector space of dimension n with an inner product and let $\wedge^p(V)$ be its anti-symmetric p tensors. Set $\wedge^*(V) = \sum_{p=0}^n \wedge^p(V)$. Let $\{v_1, \ldots, v_n\}$ be an orthonormal basis for V. Let a_i^* be a linear transformation on $\wedge^*(V)$ defined by $a_i^* v = v_i \wedge v$ and let a_i be the adjoint operator of a_i^* on $\wedge^*(V)$.

We note that $\{a_i^*, a_j\}_{1 \le i,j \le n}$ satisfy the following relations.

$$\begin{aligned} a_i a_j + a_j a_i &= 0, \\ a_i^* a_j^* + a_j^* a_i^* &= 0, \\ a_i a_j^* + a_j^* a_i &= \delta_{ij}. \end{aligned} \tag{2.1}$$

Definition 2.1. Set $\mathcal{A} = \{(\mu_1, \ldots, \mu_k) : 1 \le k \le 2n, 1 \le \mu_1 < \cdots < \mu_k \le 2n\}$, $\gamma_{2k-1} = a_k + a_k^*$, $\gamma_{2k} = i^{-1}(a_k - a_k^*)$ for $k \in \{1, 2, \ldots, n\}$, $\gamma_A = i^{\frac{k(k-1)}{2}} \gamma_{\mu_1} \cdots \gamma_{\mu_k}$ for $A = (\mu_1, \ldots, \mu_k) \in \mathcal{A}$ and $\gamma_\phi = 1$.

By (2.1) we have

$$\gamma_\mu \gamma_\nu + \gamma_\nu \gamma_\mu = 2\delta_{\mu\nu}, \quad 1 \le \mu, \nu \le 2n$$

and

$$\gamma_A^2 = 1 \quad \text{for any } A \in \mathcal{A}.$$

The following propositions are shown in [2] under the above assumptions.

Proposition 2.2. *We have the following equality for transformation on* $\wedge^*(V)$.

$$\operatorname{tr}(\gamma_A) = \begin{cases} 0, & \text{if } A \ne \phi; \\ 2^n, & \text{if } A = \phi. \end{cases}$$

Corollary 2.3. *For any* $A, B \in \mathcal{A}$

$$\operatorname{tr}(\gamma_A \gamma_B) = \begin{cases} 0, & \text{if } A \ne B; \\ 2^n, & \text{if } A = B. \end{cases}$$

Definition 2.4. Set $\beta_\phi = 1, \beta_j = i\gamma_{2j-1}\gamma_{2j}$ for $1 \le j \le n$ and $\beta_I = \beta_{i_1} \cdots \beta_{i_k}$ for $I = (i_1, \ldots, i_k) \in \mathcal{I}$.

It holds that for $I = (i_1, i_2, \ldots, i_k) \in \mathcal{I}$

$$\beta_I = \gamma_{\tilde{I}},$$

where $\tilde{I} = (2i_1 - 1, 2i_1, 2i_2 - 1, 2i_2, \ldots, 2i_k - 1, 2i_k) \in \mathcal{A}$. It is clear that

$$a_k^* a_k = \frac{1}{2}(1 + \beta_k), \quad \beta_j \beta_k = \beta_k \beta_j, \quad \beta_j^2 = 1$$

by the properties of γ_j.

Proposition 2.5. *We have for any $I = (i_1, \ldots, i_k) \in \mathcal{I}$ the following assertions:*

(1) *If $p < k$*

$$\operatorname{tr}[\beta_I a_{j_1} a_{j_2} \cdots a_{j_p} a_{h_1}^* a_{h_2}^* \cdots a_{h_p}^*] = 0.$$

(2) *Suppose $p = k$ and $\{j_1, j_2, \ldots, j_k\} \neq \{i_1, i_2, \ldots, i_k\}$ or $\{h_1, h_2, \ldots, h_k\} \neq \{i_1, i_2, \ldots, i_k\}$. Then*

$$\operatorname{tr}[\beta_I a_{j_1} a_{j_2} \cdots a_{j_p} a_{h_1}^* a_{h_2}^* \cdots a_{h_p}^*] = 0.$$

(3) *Let π, σ be elements of the permutation group of degree k. Then we have*

$$\operatorname{tr}[\beta_I a_{i_{\pi(1)}}^* a_{i_{\sigma(1)}} a_{i_{\pi(2)}}^* a_{i_{\sigma(2)}} \cdots a_{i_{\pi(k)}}^* a_{i_{\sigma(k)}}] = 2^{n-k} \operatorname{sign}(\pi) \operatorname{sign}(\sigma).$$

Proof. (1) It holds that

$$a_k = \frac{1}{2}(\gamma_{2k-1} + i\gamma_{2k}), a_k^* = \frac{1}{2}(\gamma_{2k-1} - i\gamma_{2k}). \tag{2.2}$$

So $a_{j_1} a_{j_2} \cdots a_{j_p} a_{h_1}^* a_{h_2}^* \cdots a_{h_p}^*$ is represented by a linear combination of γ_A with $A \in \mathcal{I}, \sharp(A) = 2p$. On the other hand we have $\beta_I = \gamma_{\tilde{I}}$ with $\sharp(\tilde{I}) = 2k$. We get the result by Corollary 2.3.

(2) In this case $k = p$ but $A \in \mathcal{A}$ does not coincide with $\tilde{I}$. More precisely $a_{j_1} a_{j_2} \cdots a_{j_p} a_{h_1}^* a_{h_2}^* \cdots a_{h_p}^*$ are represented by a linear combination of the form $\gamma_{2\ell_1 - 1}\gamma_{2\ell_2 - 1} \cdots \gamma_{2\ell_k - 1}\gamma_{2\tilde{\ell}_1}\gamma_{2\tilde{\ell}_2} \cdots \gamma_{2\tilde{\ell}_k}$ with

$$\{j_1, j_2, \ldots, j_k, h_1, h_2, \ldots, h_k\} = \{\ell_1, \ell_2, \ldots, \ell_k, \tilde{\ell}_1, \tilde{\ell}_2, \ldots, \tilde{\ell}_k\}.$$

By the assumption this set does not coincide with $\{i_1, i_2, \ldots, i_k\}$. So, by the above corollary we get (2).

(3) We have for $\pi, \sigma \in S_k$

$$\begin{aligned} &a_{i_{\pi(1)}}^* a_{i_{\sigma(1)}} a_{i_{\pi(2)}}^* a_{i_{\sigma(2)}} \cdots a_{i_{\pi(k)}}^* a_{i_{\sigma(k)}} \\ &= \operatorname{sign}(\pi) \operatorname{sign}(\sigma) a_{i_k}^* a_{i_{k-1}}^* \cdots a_{i_1}^* a_{i_1} \cdots a_{i_k} \\ &\quad + \text{lower order terms}, \end{aligned} \tag{2.3}$$

because

$$\begin{aligned} &a_{i_{\pi(1)}}^* a_{i_{\sigma(1)}} a_{i_{\pi(2)}}^* a_{i_{\sigma(2)}} \cdots a_{i_{\pi(k)}}^* a_{i_{\sigma(k)}} \\ &= (-1)^{\frac{k(k-1)}{2}} a_{i_{\pi(1)}}^* a_{i_{\pi(2)}}^* \cdots a_{i_{\pi(k)}}^* a_{i_{\sigma(1)}} a_{i_{\sigma(2)}} \cdots a_{i_{\sigma(k)}} \\ &\quad + \text{lower order terms}, \end{aligned}$$

$$a^*_{i_{\pi(1)}}a^*_{i_{\pi(2)}}\cdots a^*_{i_{\pi(k)}}=\mathrm{sign}(\pi)a^*_{i_1}a^*_{i_2}\cdots a^*_{i_k}=(-1)^{\frac{k(k-1)}{2}}\mathrm{sign}(\pi)a^*_{i_k}a^*_{i_{k-1}}\cdots a^*_{i_1}.$$

By (2.3) and the statement of (1) we have

$$\begin{aligned} tr\Big(\beta_I a^*_{i_{\pi(1)}}a_{i_{\sigma(1)}}a^*_{i_{\pi(2)}}a_{i_{\sigma(2)}} &\cdots a^*_{i_{\pi(k)}}a_{i_{\sigma(k)}}\Big) \\ &= 2^{-k}\,\mathrm{sign}(\pi)\,\mathrm{sign}(\sigma)tr\Big(\beta_I(1+\beta_{i_1})\cdots(1+\beta_{i_k})\Big), \end{aligned}$$

where we use

$$a^*_{i_k}a^*_{i_{k-1}}\cdots a^*_{i_1}a_{i_1}\cdots a_{i_k}=a^*_{i_1}a_{i_1}a^*_{i_2}a_{i_2}\cdots a^*_{i_k}a_{i_k}=2^{-k}(1+\beta_{i_1})\cdots(1+\beta_{i_k}). \quad (2.4)$$

By the above corollary we get (3), noting $\beta_J=\gamma_{\tilde{J}}$ for any $J\in\mathcal{I}$. □

Let Ψ_p be the projection of $\wedge^*(V)$ on $\wedge^p(V)$. The following proposition is the key algebraic argument of the proof in this paper.

Proposition 2.6. *For any p $(0\le p\le n)$ we have the equation*

$$\Psi_p=\sum_{q=0}^{n}\mathcal{M}_{pq}\Gamma_q,$$

where

$$\mathcal{M}_{pq}=\sum_{p,q\le j\le n}(-1)^{p+j}2^{-j}\binom{j}{p}\binom{n-q}{n-j}$$

and $\Gamma_0=1$, $\Gamma_k=\sum_{I\in\mathcal{I},\sharp(I)=k}\beta_I$.

Proof. Set $M_0=1$ and for $1\le p\le n$

$$M_p=2^{-p}\sum_{\sharp(I)=p,I=(i_1,\dots,i_p)\in\mathcal{I}}(1+\beta_{i_1})(1+\beta_{i_2})\cdots(1+\beta_{i_p}). \quad (2.5)$$

Then it holds that for $0\le p\le n$

$$M_p=\sum_{j=p}^{n}\binom{j}{p}\Psi_j. \quad (2.6)$$

In fact, we have

$$M_p=\sum_{\sharp(I)=p,I=(i_1,\dots,i_p)\in\mathcal{I}}a^*_{i_1}a_{i_1}a^*_{i_2}a_{i_2}\cdots a^*_{i_p}a_{i_p}, \quad (2.7)$$

noting $(1+\beta_j)=2a^*_ja_j$. We denote

$$v_I=v_{i_1}\wedge\cdots\wedge v_{i_p}\quad \text{for } I=(i_1,\dots,i_p).$$

Then

$$a^*_ka_k=\begin{cases}1, & \text{on } v_I \text{ if } k\in I;\\ 0, & \text{on } v_I \text{ if } k\notin I.\end{cases}$$

So we have on $\wedge^j(V)$

$$M_p=\begin{cases}0, & \text{if } j\le p-1;\\ \binom{j}{p}, & \text{if } j\ge p.\end{cases} \quad (2.8)$$

By (2.8) we have (2.6). Owing to (2.6) it holds

$$\Psi_p = \sum_{j=p}^{n} (-1)^{j+p} \binom{j}{p} M_j, \quad (0 \le p \le n). \tag{2.9}$$

On the other hand by the definition of M_p we have

$$M_j = 2^{-j} \sum_{q=0}^{j} \binom{n-q}{n-j} \Gamma_q. \tag{2.10}$$

By (2.9) and (2.10) we get the assertion. □

Note that a $(n+1) \times (n+1)$ matrix $\mathcal{M} = (\mathcal{M}_{pq})_{0 \le p,q \le n}$ is nonsingular because, with the argument of Proposition 2.6, $\mathcal{M} = CD$ where C and D are $(n+1) \times (n+1)$ matrices whose elements c_{pq}, d_{pq} $(0 \le p, q, \le n)$ are given by $c_{pq} = (-1)^{p+q}\binom{q}{p}$, $d_{pq} = 2^{-p}\binom{n-q}{n-p}$. C^{-1} and D^{-1} are matrices whose elements $\tilde{c}_{pq}$, $\tilde{d}_{pq}$ $(0 \le p, q, \le n)$ are given by $\tilde{c}_{pq} = \binom{q}{p}$, $\tilde{d}_{pq} = 2^q(-1)^{p+q}\binom{n-q}{n-p}$. So, the following equation holds;

$$(\mathcal{M}^{-1})_{pq} = \sum_{0 \le j \le p,q} (-1)^{p+j} 2^j \binom{q}{j} \binom{n-j}{n-p}.$$

Then we have

Corollary 2.7. *Let f_j, h_j be constants. Then*

$$\sum_{q=0}^{n} f_q \Psi_q = \sum_{p=0}^{n} h_p \Gamma_p$$

holds if and only if

$$\mathbf{f} = {}^t \mathcal{M}^{-1} \mathbf{h}$$

with $\mathbf{h} = (h_0, h_1, \dots, h_n)$ *and* $\mathbf{f} = (f_0, f_1, \dots, f_n)$.

We are interested in obtaining the exact formula of $\mathbf{f}$ for $\mathbf{h}$ of the following special form.

Proposition 2.8. *Assume $h_j = 0$ $(0 \le j \le \ell - 1)$, $h_\ell = 2^{\ell-n}(-1)^\ell$. Then we obtain the following typical solutions $\mathbf{f} = {}^t(f_0, f_1, \dots, f_n)$ of*

$${}^t\mathcal{M}\mathbf{f} = \mathbf{h},\ \mathbf{h} = {}^t(h_0, h_1, \dots, h_n). \tag{2.11}$$

(1) *If $\mathbf{f}$ satisfies $f_p = 0$ $(\ell + 1 \le p \le n)$, the unique solution of (2.11) is*

$$f_p = (-1)^p \binom{n-p}{n-\ell} \quad (0 \le p \le \ell).$$

In this case $h_p = (-1)^p 2^{\ell-n}\binom{p}{\ell}$ $(\ell + 1 \le p \le n)$.

(2) *If $\mathbf{f}$ satisfies $f_p = 0$ $(0 \le p \le n - \ell - 1)$, the unique solution of (2.11) is*

$$f_p = (-1)^{n-\ell+p} \binom{p}{n-\ell} \quad (n - \ell \le p \le n).$$

In this case $h_p = (-1)^\ell 2^{\ell-n}\binom{p}{\ell}$ $(\ell + 1 \le p \le n)$.

Proof. Fix an integer $q\,(0\le q\le n)$. The following equation holds for arbitrary x.

$$\sum_{j=q}^{n}(-1)^j 2^{-j}\binom{n-q}{n-j}x^j=(-1)^q 2^{-q}x^q(1-\frac{1}{2})^{n-q}. \tag{2.12}$$

Put $x=1-u$ in (2.12). Then we have

$$\sum_{p=0}^{n}\sum_{j\ge p,q}^{n}(-1)^{p+j}2^{-j}\binom{j}{p}\binom{n-q}{n-j}u^p=2^{-n}(u-1)^q(1+u)^{n-q}.$$

The above equation means for any q

$$\sum_{p=0}^{n}\mathcal{M}_{pq}u^p=2^{-n}(u-1)^q(1+u)^{n-q}. \tag{2.13}$$

Put $u=\frac{1}{v-1}$ in (2.13). Then we have

$$\sum_{p=0}^{n}\mathcal{M}_{pq}(v-1)^{n-p}=2^{-n}v^{n-q}(2-v)^q.$$

Comparing the coefficients of $(-1)^r v^{n-r}$ for any $r\,(0\le r\le n)$ of both sides of the above equation, we obtain for any q

$$\sum_{p=0}^{n}\mathcal{M}_{pq}\binom{n-p}{n-r}(-1)^p=\begin{cases}0, & \text{if } r>q;\\ (-1)^q 2^{r-n}\binom{q}{r}, & \text{if } r\le q.\end{cases}$$

This means that for any integer $\ell\,(0\le\ell\le n)$

$$\sum_{p=0}^{n}\mathcal{M}_{pq}\binom{n-p}{n-\ell}(-1)^p=\begin{cases}0, & \text{if } q\le\ell-1;\\ (-1)^q 2^{\ell-n}\binom{q}{\ell}, & \text{if } q\ge\ell.\end{cases}$$

The above equations mean that

$$^t\mathcal{M}\mathbf{F}_\ell=\mathbf{H}_\ell \tag{2.14}$$

if we set a pair of $(n+1)$ vectors

$$\mathbf{F}_\ell={}^t(f_{\ell,0},f_{\ell,1},\dots,f_{\ell,n}) \quad\text{and}\quad \mathbf{H}_\ell={}^t(h_{\ell,0},h_{\ell,1},\dots,h_{\ell,n})$$

as follows:

$$f_{\ell,p}=\begin{cases}(-1)^p\binom{n-p}{n-\ell}, & (0\le p\le\ell);\\ 0, & (\ell+1\le p\le n)\,,\end{cases} \tag{2.15}$$

$$h_{\ell,p}=\begin{cases}0, & (0\le p\le\ell-1);\\ (-1)^p 2^{\ell-n}\binom{p}{\ell}, & (\ell\le p\le n).\end{cases}$$

The assertion (1) is proved.

For the proof of (2), put $u=w-1$ in (2.12). Then we have for any q

$$\sum_{p=0}^{n}\mathcal{M}_{pq}(w-1)^p=2^{-n}w^{n-q}(w-2)^q.$$

Comparing the coefficients of $(-w)^{n-r}$ for any $r\,(0 \le r \le n)$ of both sides of the above equation, we obtain for any q

$$\sum_{p=0}^{n} \mathcal{M}_{pq} \binom{p}{n-r} (-1)^p = \begin{cases} 0, & \text{if } r > q; \\ (-1)^n 2^{r-n} \binom{q}{r}, & \text{if } r \le q. \end{cases}$$

This means that for any integer $\ell\,(0 \le \ell \le n)$

$$\sum_{p=0}^{n} \mathcal{M}_{pq} \binom{p}{n-\ell} (-1)^p = \begin{cases} 0, & \text{if } q \le \ell - 1; \\ (-1)^n 2^{\ell-n} \binom{q}{\ell}, & \text{if } q \ge \ell. . \end{cases}$$

The above equations mean that

$${}^t\mathcal{M}\mathbf{f} = \mathbf{h}$$

if we set

$$f_p = \begin{cases} 0, & (0 \le p \le n - \ell - 1); \\ (-1)^{n-\ell+p} \binom{p}{n-\ell}, & (n - \ell \le p \le n), \end{cases} \tag{2.16}$$

$$h_p = \begin{cases} 0, & (0 \le p \le \ell - 1); \\ (-1)^{\ell} 2^{\ell-n} \binom{p}{\ell}, & (\ell \le p \le n). \end{cases}$$

The assertion (2) is proved. □

Theorem 2.9. *Let α_p $(\ell + 1 \le p \le n)$ be constants. The equation*

$$\sum_{q=0}^{n} f_q \Psi_q = (-1)^{\ell} 2^{\ell-n} \Gamma_\ell + \sum_{p=\ell+1}^{n} \alpha_p \Gamma_p \tag{2.17}$$

has a solution as follows:

$$f_p = (-1)^p \left\{ \binom{n-p}{n-\ell} + \sum_{j=\max(\ell+1,p)}^{n} k_j \binom{n-p}{n-j} \right\} \text{ for any } p \tag{2.18}$$

with constants k_j $(\ell + 1 \le j \le n)$ defined by

$$k_j = (-1)^j 2^{n-j} \left\{ 2^{\ell-n} (-1)^{\ell} \binom{j}{\ell} + \sum_{p=\ell+1}^{j} \binom{j}{p} \alpha_p \right\}. \tag{2.19}$$

Especially:

(1) *If $\ell = n$, then*

$$\sum_{q=0}^{n} f_q \Psi_q = (-1)^n \Gamma_n$$

holds if and only if

$$f_p = (-1)^p \quad \text{for any } p.$$

(2) *If $\alpha_p = (-1)^p 2^{\ell-n} \binom{p}{\ell}$ $(\ell + 1 \le p \le n)$, we have f_p of the following form:*

$$f_p = \begin{cases} (-1)^p \binom{n-p}{n-\ell}, & (0 \le p \le \ell); \\ 0, & (\ell + 1 \le p \le n). \end{cases}$$

(3) *If* $\alpha_p = (-1)^\ell 2^{\ell-n}\binom{p}{\ell}$ $(\ell+1 \le p \le n)$, *we have* f_p *of the following form:*

$$f_p = \begin{cases} 0, & (0 \le p \le n-\ell-1); \\ (-1)^{n-\ell+p}\binom{p}{n-\ell}; & (n-\ell \le p \le n). \end{cases}$$

Proof. It is shown that by (2.14) for any j $(0 \le j \le n)$, ${}^t\mathcal{M}\mathbf{F_j} = \mathbf{H_j}$ hold. $\{\mathbf{F_j}\}_{j=\ell}^n$ are linearly independent because $f_{j,j} = (-1)^j$ and $f_{j,k} = 0$ for $(j < k)$. Put

$$\mathbf{f} =^t (f_0, f_1, \dots, f_n) = \mathbf{F}_\ell + \sum_{j=\ell+1}^n k_j \mathbf{F_j}.$$

Then we have by (2.15) ${}^t\mathcal{M}\mathbf{f} = \mathbf{h}$ with $\mathbf{h} =^t (h_0, h_1, \dots, h_n)$ as follows:

$$f_p = \begin{cases} (-1)^p\Big\{\binom{n-p}{n-\ell} + \sum_{j=\ell+1}^n k_j\binom{n-p}{n-j}\Big\}, & (0 \le p \le \ell); \\ (-1)^p \sum_{j=p}^n k_j\binom{n-p}{n-j}, & (\ell+1 \le p \le n), \end{cases}$$

$$h_p = \begin{cases} 0, & (0 \le p \le \ell-1); \\ (-1)^p\Big\{2^{\ell-n}\binom{p}{\ell} + \sum_{j=\ell+1}^p 2^{j-n}k_j\binom{p}{j}\Big\}, & (\ell \le p \le n). \end{cases}$$

So, $\mathbf{f}$ defined by (2.18) satisfies (2.17) if and only if k_j satisfies

$$\alpha_p = (-1)^p\Big\{2^{\ell-n}\binom{p}{\ell} + \sum_{j=\ell+1}^p 2^{j-n}\binom{p}{j}k_j\Big\}. \tag{2.20}$$

The equation (2.20) means that

$$2^p k_p = (-1)^p 2^n \alpha_p - 2^\ell\binom{p}{\ell} - \sum_{j=\ell+1}^{p-1} 2^j\binom{p}{j}k_j.$$

We get (2.19) solving the above equation inductively.

The other statements are clear by (2.14) and (2.15). □

3. The proof of Main Theorem I (Riemannian manifold)

We give a rough review of the method of construction of the fundamental solution in [8].

Let M be a smooth Riemannian manifold of dimension n with a Riemannian metric g. Let $X_1, X_2, \dots, X_n$ be a local orthonormal frame of $T(M)$ in a local path U. And let $\omega^1, \omega^2, \dots, \omega^n$ be its dual. The differential d and its dual ϑ acting on $\Gamma(\wedge^p T^*(M))$ are written as follows, using the Lévi-Cività connection ∇ (see Appendix A of [12]):

$$d = \sum_{j=1}^n e(\omega^j)\nabla_{X_j}, \qquad \vartheta = -\sum_{j=1}^n \imath(X_j)\nabla_{X_j},$$

where we use the following notation.

Notation.

$$e(\omega^j)\omega = \omega^j \wedge \omega,\ \imath(X_j)\omega(Y_1,\ldots,Y_{p-1}) = \omega(X_j,Y_1,\ldots,Y_{p-1}).$$

Let $R(X,Y)$ be the curvature transformation, that is

$$R(X,Y) = \big[\nabla_X,\nabla_Y\big] - \nabla_{[X,Y]}.$$

Set

$$R(X_i,X_j)X_k = \sum_{\ell=1}^{n} R_{\ell kij}X_\ell \quad 1 \le i,j,k, \le n.$$

The Laplacian $\Delta = d\vartheta + \vartheta d$ on $\sum_{p=0}^{n}\Gamma(\wedge^p T^*(M))$ has the following Weitzenböck formula;

$$\Delta = -\{\sum_{j=1}^{n}\nabla_{X_j}\nabla_{X_j} - \sum_{j=1}^{n}\nabla_{(\nabla_{X_j}X_j)} + \sum_{i,j=1}^{n} e(\omega))\imath(X_j)R(X_i,X_j)\}.$$

We use the following notation in the rest of this section.

$$a_j^* = e(\omega *), \qquad a_k = \iota(X_k).$$

The fundamental solution $E(t)$ has an expansion, due to [8],

$$E(t) \sim \sum_{j=0} u_j(t,x,D),$$

where $u_j(t,x,D)$ are pseudo-differential operators with parameter t.

The following statement is obtained in p. 255 of [8]. The kernel of a pseudo-differential operator with symbol $u_0(t,x,\xi)$ is obtained as

$$\begin{aligned}\tilde{u}_0(t,x,x) &= (2\pi)^{-n}\int_{\mathbf{R}^n} u_0(t,x,\xi)d\xi \\ &= (\frac{1}{2\sqrt{\pi t}})^n\sqrt{\det g}e^{-tR}\Big(1+0(\sqrt{t})\Big),\end{aligned} \tag{3.1}$$

where

$$R = \sum_{i,j,k,q=1}^{n} R_{qkij}a_i^* a_j a_k^* a_q.$$

We shall calculate

$$\mathrm{tr}\ (\beta_I\tilde{u}_0(t,x,x))dx = (\frac{1}{2\sqrt{\pi t}})^n \mathrm{tr}\ (\beta_I e^{-tR})dv(1+0(\sqrt{t})), \tag{3.2}$$

for $I \in \mathcal{I}$, $\sharp(I) = r$.

Using

$$e^{-tR} = \sum_{k=0}^{\infty}\{\frac{(-1)^k}{k!}R^k t^k\},$$

and by Proposition 2.5 we have

$$\mathrm{tr}\ (\beta_I e^{-tR}) = \begin{cases}\mathrm{tr}(\beta_I\frac{(-1)^m}{m!}R^m)t^m + 0(t^{m+1}), & \text{if } r = 2m;\\ 0(t^{\frac{r+1}{2}}), & \text{if } r \text{ is odd.}\end{cases} \tag{3.3}$$

We have the following proposition.

Proposition 3.1. *For $I = (i_1, i_2, \dots, i_r) \in \mathcal{I}$ $(r = 2m)$*

$$\operatorname{tr}(\beta_I(-1)^m R^m) = 2^{n-r-m} \sum_{\pi,\sigma\in S_r} \operatorname{sign}(\pi)\operatorname{sign}(\sigma) \times R_{i_{\pi(1)}i_{\pi(2)}i_{\sigma(1)}i_{\sigma(2)}} \cdots R_{i_{\pi(r-1)}i_{\pi(r)}i_{\sigma(r-1)}i_{\sigma(r)}}. \tag{3.4}$$

Proof. We can write

$$-R = \sum_{i,j,k,q=1}^{n} R_{ijkq} a_i^* a_j a_k^* a_q = \frac{1}{2}\{\sum_{i,j,k,q=1}^{n} R_{ikjq} a_i^* a_j a_k^* a_q - \sum_{j,k,q=1}^{n} R_{jkjq} a_k^* a_q\}.$$

By Proposition 2.5 we have

$$\begin{aligned}\operatorname{tr}(\beta_I(-1)^m R^m) &= \operatorname{tr}\Big(\beta_I\Big(\frac{1}{2}\sum_{i,j,k,q=1}^{n} R_{ikjq} a_i^* a_j a_k^* a_q\Big)^m\Big)\\ &= 2^{-m}\sum_{\pi,\sigma\in S_r}\operatorname{tr}\Big(\beta_I R_{i_{\pi(1)}i_{\pi(2)}i_{\sigma(1)}i_{\sigma(2)}}\cdots R_{i_{\pi(r-1)}i_{\pi(r)}i_{\sigma(r-1)}i_{\sigma(r)}}\\ &\quad a^*_{i_{\pi(1)}}a_{i_{\sigma(1)}}a^*_{i_{\pi(2)}}a_{i_{\sigma(2)}}\cdots a^*_{i_{\pi(r)}}a_{i_{\sigma(r)}}\Big)\\ &= 2^{n-r-m}\sum_{\pi,\sigma\in S_r}\operatorname{sign}(\pi)\operatorname{sign}(\sigma)\\ &\quad R_{i_{\pi(1)}i_{\pi(2)}i_{\sigma(1)}i_{\sigma(2)}}\cdots R_{i_{\pi(r-1)}i_{\pi(r)}i_{\sigma(r-1)}i_{\sigma(r)}}.\end{aligned}$$

□

By (3.2),(3.3) and Proposition 3.1 we have

$$\operatorname{tr}\,(\beta_I\tilde{u}_0(t,x,x))dx = \begin{cases} 2^{n-r}t^{-\frac{n}{2}+\frac{r}{2}}C_I(x)dv + 0(t^{-\frac{n}{2}+\frac{r}{2}+1}), & \text{if} r = 2m;\\ 0(t^{-\frac{n}{2}+\frac{r}{2}+\frac{1}{2}}), & \text{if } r \text{ is odd}\end{cases} \tag{3.5}$$

with $C_I(x)$ defined in Definition 1.1. Similarly we have

$$\operatorname{tr}\,(\beta_I\tilde{u}_j(t,x,x))dx = 0(t^{-\frac{n}{2}+\frac{r}{2}+\frac{j}{2}}). \tag{3.6}$$

By Theorem 2.9 we obtain

$$\begin{aligned}\operatorname{tr}\left(\sum_{p=0}^{n} f_p e_p(t,x,x)\right) &= (-1)^{\ell}2^{\ell-n}\sum_{I\in\mathcal{I},\sharp(I)=\ell}\operatorname{tr}\,(\beta_I e(t,x,x))\\ &\quad + \sum_{p=\ell+1}^{n}\alpha_p\operatorname{tr}\,(\Gamma_p e(t,x,x)).\end{aligned}$$

By (3.5) and (3.6) we have

$$\operatorname{tr}\,(\Gamma_p e(t,x,x)) = 0(t^{-\frac{n}{2}+\frac{\ell}{2}+\frac{1}{2}}).$$

Applying (3.5) (3.6), we have

$$\operatorname{tr}\left(\sum_{p=0}^{n} f_p e_p(t,x,x)\right) = \begin{cases} C_I(x)t^{-\frac{n}{2}+\frac{\ell}{2}} + 0(t^{-\frac{n}{2}+\frac{\ell}{2}+\frac{1}{2}}), & \text{if } \ell \text{ is even;} \\ 0(t^{-\frac{n}{2}+\frac{\ell}{2}+\frac{1}{2}}), & \text{if } \ell \text{ is odd.} \end{cases}$$

4. The proof of Main Theorem II (Kaehler manifolds)

We give a rough review of the method of construction of the fundamental solution in [9].

Let M be a compact Kaehler manifold whose complex dimension is n with a Hermitian metric g. Let $Z_1, Z_2, \ldots, Z_n$ be a local orthonormal frame of $T^{1,0}(M)$ in a local patch of a chart U. And let $\omega^1, \omega^2, \ldots, \omega^n$ be its dual. The differential d'' and its dual ϑ'' acting on $A^{0,p}(M)$ are given as follows, using the Lévi-Cività connection ∇.

$$d'' = \sum_{j=1}^{n} e(\bar{\omega}^j)\nabla_{\bar{Z}_j}, \vartheta'' = -\sum_{j=1}^{n} \imath(\bar{Z}_j)\nabla_{Z_j},$$

where we use the following notation.

Notation.

$$Z_{\bar{j}} = \bar{Z}_j, \quad \omega^{\bar{j}} = \bar{\omega}^j \quad (j = 1, \ldots, n),$$

$$e(\omega^\alpha)\omega = \omega^\alpha \wedge \omega, \quad \imath(Z_\alpha)\omega(Y_1, \ldots, Y_{p-1}) = \omega(Z_\alpha, Y_1, \ldots, Y_{p-1}),$$

$$(\alpha \in \Lambda = \{1, \ldots, n, \bar{1}, \ldots, \bar{n}\}).$$

Let $R(Z_\alpha, \bar{Z}_\beta)$ be the curvature transformation,

$$R(Z_\alpha, Z_\beta) = \left[\nabla_{Z_\alpha}, \nabla_{Z_\beta}\right] - \nabla_{[Z_\alpha, Z_\beta]}.$$

The curvature transformations satisfy

$$R(Z_i, Z_j) = 0, \quad R(\bar{Z}_i, \bar{Z}_j) = 0,$$

because M is a Kaehler manifold. Set

$$R(Z_i, \bar{Z}_j)Z_\beta = \sum_{\gamma \in \Lambda} R_{\bar{\gamma}\beta i\bar{j}} Z_\gamma.$$

The Laplacian $\square = d''\vartheta'' + \vartheta''d''$ on $A^{0,*}(M) = \sum_{p=0}^{n} A^{0,p}(M)$ has the following Bochner-Kodaira formula;

$$\square = -\frac{1}{2}\{\sum_{j=1}^{n}(\nabla_{Z_j}\nabla_{\bar{Z}_j} + \nabla_{\bar{Z}_j}\nabla_{Z_j}) - \sum_{j=1}^{n}\nabla_{(\nabla_{Z_j}\bar{Z}_j + \nabla_{\bar{Z}_j}Z_j)} - \sum_{j=1}^{n} R(Z_j, \bar{Z}_j)\}.$$

We use the following notation in the rest of this section.

$$e(\bar{\omega}^j) = a_j^*, \iota(\bar{Z}_k) = a_k.$$

The fundamental solution has an expansion, due to [9]

$$E(t) \sim \sum_{j=0} u_j(t, x, D),$$

where $u_j(t,x,D)$ are pseudo-differential operators with parameter t and the main part of their symbols $u_0(t,x,\xi)$ is represented by the precise form. The kernel $\tilde{u}_0(t,x,x)$ of a pseudo-differential operator with symbol $u_0(t,x,\xi)$ is obtained as in p. 90 [9]

$$\tilde{u}_0(t,x,x) = (2\pi t)^{-n} \det\left(\frac{t\mathcal{M}_0}{\exp(t\mathcal{M}_0) - Id}\right)\sqrt{\det g}.$$

Note that

$$e(t,x,x)dv = \tilde{u}_0(t,x,x)dx(1+0(t)). \tag{4.1}$$

We shall calculate

$$\operatorname{tr}\,(\beta_I \tilde{u}_0(t,x,x))dx = (2\pi t)^{-n}\operatorname{tr}\left(\beta_I \det(\frac{t\mathcal{M}_0}{\exp(t\mathcal{M}_0) - Id})\right)dv, \tag{4.2}$$

for $I \in \mathcal{I}$, $\sharp(I) = r$, where

$$(\mathcal{M}_0)_{jk} = R(Z_j,\bar{Z}_k) = -\textstyle\sum_{p,q=1}^n R_{p\bar{q}j\bar{k}}a_q^* a_p = \textstyle\sum_{p,q=1}^n R_{j\bar{k}\bar{q}p}a_q^* a_p.$$

Set

$$\det\left(\frac{t\mathcal{M}_0}{\exp(t\mathcal{M}_0) - Id}\right) = \sum_{j=0}^n A_j t^j.$$

Then we have by Proposition 2.5

$$\operatorname{tr}\,\left(\beta_I \det\left(\frac{t\mathcal{M}_0}{\exp(t\mathcal{M}_0) - Id}\right)\right) = \operatorname{tr}\,(\beta_I A_r)t^r + 0(t^{r+1}). \tag{4.3}$$

Set Ω a matrix whose (j,k) element is a 2-form defined by

$$(\Omega)_{jk} = -\sum_{p,q=1}^n R_{j\bar{k}\bar{q}p}\bar{\omega}^q \wedge \omega^p = \sum_{p,q=1}^n R_{\bar{k}jp\bar{q}}\omega^p \wedge \bar{\omega}^q.$$

Then we have

Proposition 4.1.

$$\operatorname{tr}\,(\beta_I A_r)dv = (-1)^r\left(\frac{1}{i}\right)^n 2^{n-r}\left[\det(\frac{\Omega}{\exp\Omega - Id})\right]_{2r} \wedge dv^{I^c} \tag{4.4}$$

Proof. If A_r is of the form

$$A_r = \sum_{\substack{1\le j_1\le n, 1\le j_2\le n,\dots,1\le j_r\le n\\ 1\le j_1\le n, 1\le j_2\le n,\dots,1\le j_r\le n}} B(j_1,k_1,j_2,k_2,\dots,j_r,k_r)a_{j_1}^* a_{k_1} a_{j_2}^* a_{k_2}\cdots a_{j_r}^* a_{k_r},$$

then we have for $I = (i_1, i_2, \dots, i_r)$ by Proposition 2.5

$$\begin{aligned}\operatorname{tr}\,(\beta_I A_r) &= \sum_{\pi,\sigma\in S_r}\operatorname{tr}\,\Big(\beta_I B(i_{\pi(1)}, i_{\sigma(1)}, i_{\pi(2)}, i_{\sigma(2)},\dots,i_{\pi(r)}, i_{\sigma(r)})\\ &\qquad \times a_{i_{\pi(1)}}^* a_{i_{\sigma(1)}} a_{i_{\pi(2)}}^* a_{i_{\sigma(2)}}\cdots a_{i_{\pi(r)}}^* a_{i_{\sigma(r)}}\Big) \qquad (4.5)\\ &= 2^{n-r}\sum_{\pi,\sigma\in S_r}\operatorname{sign}(\pi)\operatorname{sign}(\sigma)B(i_{\pi(1)}, i_{\sigma(1)}, i_{\pi(2)}, i_{\sigma(2)},\dots,i_{\pi(r)}, i_{\sigma(r)}).\end{aligned}$$

On the other hand we have the equations

$$\begin{aligned} dv &= i^n \omega^1 \wedge \bar{\omega}^1 \wedge \omega^2 \wedge \bar{\omega}^2 \cdots \wedge \omega^n \wedge \bar{\omega}^n \\ &= (-i)^n \bar{\omega}^1 \wedge \omega^1 \wedge \bar{\omega}^2 \wedge \omega^2 \cdots \wedge \bar{\omega}^n \wedge \omega^n \\ &= (-i)^n \bar{\omega}^{i_1} \wedge \omega^{i_1} \wedge \bar{\omega}^{i_2} \wedge \omega^{i_2} \cdots \wedge \bar{\omega}^{i_r} \wedge \omega^{i_r} \wedge dv^{I^c}, \end{aligned} \tag{4.6}$$

$$\begin{aligned} &\operatorname{sign}(\pi)\operatorname{sign}(\sigma)\bar{\omega}^{i_1} \wedge \omega^{i_1} \wedge \bar{\omega}^{i_2} \wedge \omega^{i_2} \cdots \wedge \bar{\omega}^{i_r} \wedge \omega^{i_r} \\ &\quad = \bar{\omega}^{i_{\pi(1)}} \wedge \omega^{i_{\sigma(1)}} \wedge \bar{\omega}^{i_{\pi(2)}} \wedge \omega^{i_{\sigma(2)}} \cdots \wedge \bar{\omega}^{i_{\pi(r)}} \wedge \omega^{i_{\sigma(r)}}. \end{aligned} \tag{4.7}$$

We obtain by (4.5), (4.6) and (4.7)

$$\operatorname{tr}\,(\beta_I A_r) dv = \left(\frac{1}{i}\right)^n 2^{n-r} \sum_{\pi,\sigma \in S_r} B(i_{\pi(1)}, i_{\sigma(1)}, i_{\pi(2)}, i_{\sigma(2)}, \dots, \tag{4.8}$$

$$\cdots i_{\pi(r)}, i_{\sigma(r)}) \bar{\omega}^{i_{\pi(1)}} \wedge \omega^{i_{\sigma(1)}} \wedge \bar{\omega}^{i_{\pi(2)}} \wedge \omega^{i_{\sigma(2)}} \cdots \wedge \bar{\omega}^{i_{\pi(r)}} \wedge \omega^{i_{\sigma(r)}} \wedge dv^{I^c}.$$

Then by (4.8) we have

$$\operatorname{tr}\,(\beta_I A_r) dv = (-1)^r \left(\frac{1}{i}\right)^n 2^{n-r} \left[\det\left(\frac{\Omega}{\exp \Omega - Id}\right)\right]_{2r} \wedge dv^{I^c}. \qquad \square$$

From (4.2), (4.3) and Proposition 4.1 the following equation holds.

$$\begin{aligned} &\operatorname{tr}\,(\beta_I \tilde{u}_0(t,x,x)) dx \\ &= (-1)^r \left(\frac{1}{2\pi i}\right)^n 2^{n-r} t^{-n+r} \left[\det\left(\frac{\Omega}{\exp \Omega - Id}\right)\right]_{2r} \wedge dv^{I^c} + 0(t^{-n+r+1}). \end{aligned} \tag{4.9}$$

Now by Theorem 2.9 we obtain

$$\begin{aligned} \operatorname{tr}\left(\sum_{p=0}^n f_p e_p(t,x,x)\right) &= (-1)^\ell 2^{\ell-n} \sum_{I \in \mathcal{I}, \sharp(I)=\ell} \operatorname{tr}\,(\beta_I e(t,x,x)) \\ &\quad + \sum_{p=\ell+1}^n \alpha_p \operatorname{tr}\,(\Gamma_p e(t,x,x)) \end{aligned}$$

with some constants $a_j (\ell+1 \leq j \leq n)$. Applying (4.9) we have

$$\sum_{p=\ell+1}^n \alpha_p \operatorname{tr}\,(\Gamma_p e(t,x,x)) dv = 0(t^{-n+\ell+1}).$$

By (4.1) and (4.9) we obtain

$$\begin{aligned} \operatorname{tr}\left(\sum_{p=0}^n f_p e_p(t,x,x)\right) dv = \left(\frac{1}{2\pi i}\right)^n t^{-n+\ell} \sum_{I \in \mathcal{I}, \sharp(I)=\ell} \left[\det\left(\frac{\Omega}{\exp \Omega - Id}\right)\right]_{2\ell} \wedge dv^{I^c} \\ + 0(t^{-n+\ell+1}). \end{aligned}$$

References

[1] N. Berline, E. Getzler and M. Vergne, *Heat Kernels and Dirac Operators,* Springer-Verlag, 1992.

[2] H.L. Cycon, R.G. Froese, W. Kirsch and B. Simon, *Schrödinger Operators,* Springer, 1987.

[3] P. Günther and R. Schimming, Curvature and spectrum of compact Riemannian manifolds, *J. Diff. Geom.* **12** (1977), 599–618.

[4] E. Getzler, The local Atiyah-Singer index theorem, in *Critical Phenomena, Random Systems, Gauge Theories*, Editors: K. Sterwalder and R. Stora, Les Houches, North-Holland, 1984, 967–974.

[5] E. Getzler, A short proof of the local Atiyah-Singer index theorem, *Topology* **25** (1986), 111–117.

[6] P.B. Gilkey, *Invariance Theory, the Heat Equation, and the Atiyah-Singer Index Theorem,* Publish or Perish, Inc., 1984.

[7] C. Iwasaki and N. Iwasaki, Parametrix for a degenerate parabolic equation and its application to the asymptotic behavior of spectral functions for stationary problems, Publ. Res. Inst. Math. Sci. **17** (1981), 557–655.

[8] C. Iwasaki, A proof of the Gauss-Bonnet-Chern theorem by the symbol calculus of pseudo-differential operators, Japanese J. Math.**21** (1995), 235–285.

[9] C. Iwasaki, Symbolic calculus for construction of the fundamental solution for a degenerate equation and a local version of Riemann-Roch theorem, in *Geometry, Analysis and Applications*, Editor: R.S. Pathak, World Scientific, 83–92.

[10] S. Kobayashi and K. Nomizu, *Foundations of Differential Geometry, I, II*, John Wiley & Sons, 1963.

[11] T. Kotake, An analytic proof of the classical Riemann-Roch theorem, in *Global Analysis, Proc. Symp. Pure Math. XVI* Providence, 1970.

[12] S. Murakami, *Manifolds,* Kyoritsusshuppan, 1969 (in Japanese).

[13] V.K. Patodi, An analytic proof of Riemann-Roch-Hirzebruch theorem for Kaehler manifold, J. Diff. Geom. **5** (1971), 251–283.

Chisato Iwasaki
Department of Mathematics
University of Hyogo
2167 Shosha
Himeji Hyogo 671-2201, Japan
e-mail: `iwasaki@sci.u-hyogo.ac.jp`

Operator Theory:
Advances and Applications, Vol. 172, 67–85

Weyl Transforms, Heat Kernels, Green Functions and Riemann Zeta Functions on Compact Lie Groups

M.W. Wong

Abstract. The Plancherel formula and the inversion formula for Weyl transforms on compact and Hausdorff groups are given. A formula expressing the relationships of the wavelet constant, the degree of the irreducible and unitary representation and the volume of an arbitrary compact and Hausdorff group is derived. The role of the Weyl transforms in the derivation of the formulas for the heat kernels of Laplacians on compact Lie groups is explicated. The Green functions and the Riemann zeta functions of Laplacians on compact Lie groups are constructed using the corresponding heat kernels.

Mathematics Subject Classification (2000). Primary 22A10, 35K05, 43A77, 47G30; Secondary 22C05, 35J05.

Keywords. Weyl transforms, pseudo-differential operators, compact groups, Lie groups, Laplacians, irreducible and unitary representation, wavelet constants, resolution of the identity formula, Plancherel formula, Weyl inversion formula, eigenvalues and eigenvectors, heat kernels, Green functions, Riemann Zeta functions.

1. Introduction

Pseudo-differential operators and their invariants, first systematically developed by Kohn and Nirenberg [8] in 1965 and modified almost immediately by Hörmander [7] and others for problems in partial differential equations, have roots in quantization due to Hermann Weyl [13] more than thirty years earlier. As such, they are known as Weyl transforms. The product formula for these classical Weyl transforms entails a twisted convolution, which is essentially a convolution on the Heisenberg group and can be found in the paper [17] by Wong. This suggests a notion of Weyl

This research has been partially supported by the Natural Sciences and Engineering Research Council of Canada.

transforms on locally compact and Hausdorff topological groups, which we present in the context of compact groups in this paper. These Weyl transforms play a very fundamental role in the formulas for the heat kernels of the Laplacians on compact Lie groups.

We first recall in Section 2 the construction of the Laplacian on a compact Lie group. In Sections 3–6, we have the freedom of working in a more general setting and we study harmonic analysis on compact and Hausdorff topological groups. To be specific, we give in Section 3 a streamlined presentation of irreducible and unitary representations, admissible wavelets and the resolution of the identity formula for compact and Hausdorff groups. An interesting formula relating the wavelet constant, the degree of the representation and the volume of the compact group is derived. The notion of a Weyl transform on a compact and Hausdorff group is introduced. We give a version of the Plancherel formula for Weyl transforms on compact groups in Section 4, the Weyl inversion formula in Section 5 and a formulation of the Weyl inversion formula in terms of characters in Section 6. The technique that underpins the results in Sections 4–6 is the Peter-Weyl theorem and the role played by the Weyl transforms is emphasized. Sections 7–9 are devoted to compact Lie groups in the context of partial differential equations. The eigenvalues and eigenfunctions of Laplacians are computed in Section 7. The heat kernels are constructed Section 8. The ubiquity of the heat kernels is demonstrated in Sections 9 and 10 by constructing, respectively, the Green functions and the Riemann zeta functions for the Laplacians on compact Lie groups. The Riemann zeta function of the Laplacian on a compact Lie group is the Mellin transform of the regularized trace of the heat kernel, and we express the Riemann zeta function in terms of the eigenvalues of the Laplacian. Issues on the regions of convergence of the series defining the Riemann zeta functions are beyond the scope of this paper and hence omitted. The book [3] by Fegan contains discussions on the impact of the trace of the heat kernel and hence the Riemann zeta function on the geometry of compact Lie groups. The unit circle S^1 with center at the origin of $\mathbb{R}^2$ is given as an example of an abelian compact Lie group in Section 11. SU(2) is given as an example of a non-commutative compact Lie group in Section 12. Since SU(2) can be identified as the unit sphere S^3 with center at the origin in $\mathbb{R}^4$ and the Laplacian on S^3 can best be studied using spherical harmonics, it is more convenient and natural to construct the heat kernel and Green function for the Laplacian $\Delta_{S^{n-1}}$ on the unit sphere S^{n-1} with center at the origin in $\mathbb{R}^n$ using spherical harmonics in $\mathbb{R}^n$. This is done in Section 13. It is worth pointing out that with the exception of S^1 and S^3, S^{n-1} is a compact symmetric space instead of a compact Lie group.

2. Laplacians

Let G be a compact Lie group. This means that G is a compact and Hausdorff topological group, which is also an n-dimensional C^∞ manifold. A vector field X

on G is a linear mapping $X : C^\infty(G) \to C^\infty(G)$ such that

$$X(\varphi\psi) = X(\varphi)\psi + \varphi X(\psi), \quad \varphi, \psi \in C^\infty(G).$$

For all vector fields X and Y on G, we can define the bracket $[X, Y]$ of X and Y by

$$[X, Y] = XY - YX.$$

The bracket $[\,,\,]$ is bilinear, anti-symmetric and satisfies the Jacobi identity

$$[X, [Y, Z]] + [Y, [Z, X]] + [Z, [X, Y]] = 0$$

for all vector fields X, Y and Z on G. The vector space of all vector fields on G is infinite-dimensional and hence too big to be useful. Therefore we look at instead the vector space of all left-invariant vector fields on G. A vector field X on G is said to be left-invariant if

$$L(h)X = XL(h), \quad h \in G,$$

where $L(h)$ is the left translation by h defined by

$$(L(h)\varphi)(g) = \varphi(h^{-1}g), \quad g \in G,$$

for all φ in $C^\infty(G)$. It is obvious that the set of all left-invariant vector fields on G is closed with respect to the bracket operation.

Let X be a vector field on G. If we denote the identity element in G by e, then we define the element X_e in the tangent space $T_e(G)$ of G at e by

$$X_e(\varphi) = X(\varphi)(e), \quad \varphi \in C^\infty(G).$$

If we let $\mathrm{LIV}(G)$ be the vector space of all left-invariant vector fields on G, then we have the following proposition.

Proposition 2.1. *The mapping*

$$\mathrm{LIV}(G) \ni X \mapsto X_e \in T_e(G)$$

is a bijective isomorphism.

The isomorphism alluded to in the preceding proposition induces a natural bracket operation on $T_e(G)$. Equipped with this bracket operation, $T_e(G)$ is the Lie algebra $\mathfrak{g}$ of G. The nice thing about $\mathfrak{g}$ is that it is n-dimensional. As such, it has a basis consisting of n elements. We choose a basis for $\mathfrak{g}$, say, $\{\mathbf{v}_1, \mathbf{v}_2, \ldots, \mathbf{v}_n\}$. Then we can get a basis $\{X_1, X_2, \ldots, X_n\}$ for $\mathrm{LIV}(G)$ as follows.

For $j = 1, 2, \ldots, n$, let $\gamma_j : \mathbb{R} \to G$ be a homomorphic curve in G such that $\gamma_j(0) = e$ and $\gamma_j'(0) = \mathbf{v}_j$. Then for $j = 1, 2, \ldots, n$, we define the left-invariant vector field X_j on G by

$$(X_j f)(g) = \frac{d}{ds}\bigg|_{s=0} f(g \cdot \gamma_j(s)), \quad g \in G,$$

for all $f \in C^\infty(G)$.

Here comes the Laplacian Δ_G on G, which we define by

$$\Delta_G = -\sum_{j=1}^{n} X_j^2.$$

It can be proved that Δ_G is positive and bi-invariant.

3. Weyl transforms

Let G be a compact and Hausdorff group on which the left (and right) Haar measure is denoted by μ. Let π be an irreducible and unitary representation of G on a complex and separable Hilbert space X_π. We denote the inner product and the norm in X_π by $(\,,\,)_{X_\pi}$ and $\|\;\|_{X_\pi}$ respectively. Since G is compact, it is well known that X is finite-dimensional. A good reference is Theorem 5.2 in the book [4] by Folland. We let d_π be the dimension of X_π. The number d_π is also known as the degree of the representation π of G on X_π. For every element φ in X, the compactness of G implies that

$$\int_G |(\varphi, \pi(g)\varphi)|^2 d\mu(g) < \infty.$$

Using the terminology in the general theory of wavelet transforms in the book [15] by Wong, we see that every element φ in the unit sphere of X is an admissible wavelet and the wavelet constant $c_{\varphi,\pi}$ for the admissible wavelet φ is defined by

$$c_{\varphi,\pi} = \int_G |(\varphi, \pi(g)\varphi)_{X_\pi}|^2 d\mu(g).$$

The starting point is the following theorem.

Theorem 3.1. *Let $\varphi \in X_\pi$ be such that $\|\varphi\|_{X_\pi} = 1$. Then*

$$(x, y)_{X_\pi} = \frac{1}{c_{\varphi,\pi}} \int_G (x, \pi(g)\varphi)_{X_\pi} (\pi(g)\varphi, y)_{X_\pi} d\mu(g), \quad x, y \in X_\pi. \tag{3.1}$$

See, for instance, Theorem 3.1 in the paper [5] by Grossmann, Morlet and Paul and Theorem 6.1 in the book [15] by Wong for more general results. The essence of the formula is that the identity operator (matrix) on X can be resolved into an infinite sum of rank-one operators on X_π. Thus, it is reasonable to call the formula (3.1) a resolution of the identity formula.

As an immediate consequence of Theorem 3.1 and the unimodularity of G, we have the following interesting result on wavelet constants.

Theorem 3.2. *Let φ and ψ be in X_π such that $\|\varphi\|_{X_\pi} = \|\psi\|_{X_\pi} = 1$. Then*

$$c_{\varphi,\pi} = c_{\psi,\pi}.$$

Proof. By (3.1),

$$c_{\varphi,\pi} = \int_G (\psi, \pi(g)\varphi)_{X_\pi} (\pi(g)\varphi, \psi)_{X_\pi} d\mu(g) \tag{3.2}$$

and

$$c_{\psi,\pi} = \int_G (\varphi, \pi(g)\psi)_{X_\pi} (\pi(g)\psi, \varphi)_{X_\pi} d\mu(g). \tag{3.3}$$

By (3.3) and the unimodularity of G, we get

$$\begin{aligned} c_{\psi,\pi} &= \int_G (\psi, \pi(g^{-1})\varphi)_{X_\pi} (\pi(g^{-1})\varphi, \psi)_{X_\pi} d\mu(g) \\ &= \int_G (\psi, \pi(g)\varphi)_{X_\pi} (\pi(g)\varphi, \psi)_{X_\pi} d\mu(g). \end{aligned} \tag{3.4}$$

So, by (3.2) and (3.4), the proof is complete. □

Another interesting consequence of Theorem 3.1 is the following formula relating the wavelet constant, the degree of the representation and the volume of the compact group.

Theorem 3.3. *Let* $\varphi \in X$ *be such that* $\|\varphi\|_{X_\pi} = 1$. *Then*

$$c_{\varphi,\pi} = \frac{\mu(G)}{d_\pi}.$$

Proof. Let $\{\psi_1, \psi_2, \ldots, \psi_{d_\pi}\}$ be an orthonormal basis for X_π. Then, by Theorem 3.1 and Parseval's identity,

$$\begin{aligned} d_\pi &= \sum_{j=1}^{d_\pi} (\psi_j, \psi_j)_{X_\pi} \\ &= \sum_{j=1}^{d_\pi} \frac{1}{c_{\varphi,\pi}} \int_G |(\psi_j, \pi(g)\varphi)_{X_\pi}|^2 d\mu(g) \\ &= \frac{1}{c_{\varphi,\pi}} \int_G \sum_{j=1}^{d_\pi} |(\psi_j, \pi(g)\varphi)_{X_\pi}|^2 d\mu(g) \\ &= \frac{1}{c_{\varphi,\pi}} \int_G \|\pi(g)\varphi\|_{X_\pi}^2 d\mu(g). \end{aligned}$$

Since π is a unitary representation of G on X_π and $\|\varphi\|_{X_\pi} = 1$, we get

$$d_\pi = \frac{\mu(G)}{c_{\varphi,\pi}},$$

which is the same as asserted. □

From now on, we assume that $\mu(G) = 1$.

Let $F \in L^1(G)$. Then we define the linear operator (matrix) $W_{F,\pi} : X_\pi \to X_\pi$ by

$$(W_{F,\pi}x, y)_{X_\pi} = \int_G F(g)(x, \pi(g)y)_{X_\pi} d\mu(g), \quad x, y \in X_\pi. \tag{3.5}$$

We call $W_{F,\pi} : X_\pi \to X_\pi$ the Weyl transform associated to the symbol F and the representation π. A more transparent formula for $W_{F,\pi}$ than (3.5) is given by

$$W_{F,\pi} = \int_G F(g)\pi(g^{-1})d\mu(g).$$

The terminology is due to the fact that the classical Weyl transform in, say, the book [14] by Wong is in fact given by (3.5) when the compact group and the representation π are replaced, respectively, by the non-compact Weyl-Heisenberg group and the Schrödinger representation. A seasoned harmonic analyst can recognize immediately that this is nothing but the Fourier transform of F on the group G evaluated at π. Such a viewpoint has been exploited in the paper [16] by Wong. Moreover, motivated by the formula of the Weyl transform provided by Theorem 4.3 in [14], we call

$$G \ni g \mapsto (\pi(g)x, y)_{X_\pi} \in \mathbb{C}$$

the Wigner transform of x and y.

Using the realization of Weyl transforms as Fourier transforms on the group G, and the fact that the Fourier transform of the convolution of two functions F and H in $L^1(G)$ at π is the product $\hat{F}(\pi)\hat{H}(\pi)$ of the bounded linear operators $\hat{F}(\pi)$ and $\hat{H}(\pi)$, we have the following result on the product of two Weyl transforms.

Theorem 3.4. *Let F and H be functions in $L^1(G)$. Then the product of the Weyl transforms $W_{F,\pi} : X_\pi \to X_\pi$ and $W_{H,\pi} : X_\pi \to X_\pi$ is the Weyl transform $W_{F*H,\pi} : X_\pi \to X_\pi$, where $F * H$ is the convolution of F and H defined by*

$$(F * H)(g) = \int_G F(gh^{-1})H(h)\, d\mu(h), \quad g \in G.$$

4. The Plancherel formula

Let $\hat{G}$ be the set of all irreducible and unitary representations of G. Let $\nu \in \hat{G}$. Then ν is finite-dimensional. We denote its degree by d_ν and its representation space by X_ν. Let $\{\psi_1, \psi_2, \dots, \psi_{d_\nu}\}$ be an orthonormal basis for X_ν. Then for all $j, k = 1, 2, \dots,$ we define ν_{jk} by

$$\nu_{jk}(g) = (\nu(g)\psi_k, \psi_j)_{X_\nu}, \quad g \in G.$$

The fact that $\{\sqrt{d_\nu}\nu_{jk} : \nu \in \hat{G},\ j, k = 1, 2, \dots\}$ forms an orthonormal basis for $L^2(G)$ is the core of the Peter-Weyl theorem.

Theorem 4.1. (The Plancherel formula) *Let $F \in L^2(G)$. Then*

$$\sum_{\pi \in \hat{G}} d_\pi \|W_{F,\pi}\|^2_{S_2(X_\pi)} = \|F\|^2_{L^2(G)},$$

where $\|W_{F,\pi}\|_{S_2(X_\pi)}$ is the Hilbert-Schmidt norm of $W_{F,\pi} : X_\pi \to X_\pi$.

Proof. By the core of the Peter-Weyl theorem, we can write F as

$$F = \sum_{\nu,j,k} c_{\nu,j,k}\sqrt{d_\nu}\nu_{jk}. \qquad \text{Then} \quad W_{F,\pi} = \sum_{\nu,j,k} c_{\nu,j,k} W_{\sqrt{d_\nu}\nu_{jk},\pi}.$$

By Parseval's identity,

$$\|F\|^2_{L^2(G)} = \sum_{\nu,j,k} |c_{\nu,j,k}|^2.$$

Let $\{\psi_1, \psi_2, \ldots, \psi_{d_\pi}\}$ be an orthonormal basis for X_π. Then for all

$$x = \sum_{l=1}^{d_\pi} x_l \psi_l \qquad \text{and} \qquad y = \sum_{m=1}^{d_\pi} y_m \psi_m$$

in X_π, we get

$$\begin{aligned}(W_{F,\pi}x, y)_{X_\pi} &= \sum_{\nu,j,k} c_{\nu,j,k}(W_{\sqrt{d_\nu}\nu_{jk},\pi}x, y)_{X_\pi} \\ &= \sum_{\nu,j,k} c_{\nu,j,k} \sum_{l,m} x_l \overline{y_m} \int_G \sqrt{d_\nu}\nu_{jk}(g)\overline{\pi_{lm}(g)}\, d\mu(g) \\ &= \sum_{l,m} x_l \overline{y_m} \sum_{\nu,j,k} \sqrt{d_\nu} c_{\nu,j,k}(\nu_{jk}, \pi_{lm})_{L^2(G)} = \sum_{l,m} x_l \overline{y_m} c_{\pi,l,m} \frac{1}{d_\pi}.\end{aligned}$$

Now, for $j = 1, 2, \ldots, d_\pi$,

$$(W_{F,\pi}\psi_j, y)_{X_\pi} = \sum_m \overline{y_m} c_{\pi,j,m} \frac{1}{d_\pi} = (z, y)_{X_\pi},$$

where $z_m = \frac{1}{\sqrt{d_\pi}} c_{\pi,j,m}$, $m = 1, 2, \ldots, d_\pi$. Thus,

$$W_{F,\pi}\psi_j = \frac{1}{d_\pi}(c_{\pi,j,1}, c_{\pi,j,2}, \ldots, c_{\pi,j,n}).$$

Therefore

$$\|W_{F,\pi}\psi_j\|^2_{X_\pi} = \frac{1}{d_\pi}\sum_{k=1}^{d_\pi} |c_{\pi,j,k}|^2$$

and consequently,

$$\sum_{j=1}^{d_\pi} \|W_{F,\pi}\psi_j\|^2_{X_\pi} = \frac{1}{d_\pi}\sum_{j,k=1}^{d_\pi} |c_{\pi,j,k}|^2.$$

Therefore

$$d_\pi \|W_{F,\pi}\|^2_{S_2(X_\pi)} = \sum_{j,k=1}^{d_\pi} |c_{\pi,j,k}|^2$$

and we get

$$\sum_{\pi\in\hat{G}} d_\pi \|W_{F,\pi}\|^2_{S_2(X_\pi)} = \|F\|^2_{L^2(G)},$$

as asserted. □

5. The Weyl inversion formula

The Peter-Weyl theorem states that for every f in $L^2(G)$,

$$f = \sum_{\pi,j,k} (f, \sqrt{d_\pi}\pi_{jk})_{L^2(G)} \sqrt{d_\pi}\pi_{jk} = \sum_{\pi,j,k} \alpha_{\pi,j,k}\pi_{jk},$$

where

$$\alpha_{\pi,j,k} = d_\pi \int_G f(g)\overline{\pi_{jk}(g)}\, d\mu(g).$$

For every π in $\hat{G}$, the entry $(W_{f,\pi})_{kj}$ of the matrix $W_{f,\pi}$ is given by

$$(W_{f,\pi})_{kj} = \int_G f(g)\overline{\pi_{jk}(g)}\, d\mu(g) = \frac{1}{d_\pi}\alpha_{\pi,j,k}.$$

Therefore for $j, k = 1, 2, \ldots, d_\pi$, and $g \in G$,

$$\sum_{j,k} \alpha_{\pi,j,k}\pi_{jk}(g) = d_\pi \sum_{j,k} (W_{f,\pi})_{kj}\pi_{jk}(g) = d_\pi \mathrm{tr}[W_{f,\pi}\pi(g)].$$

So, we have the Weyl inversion formula to the effect that for every $f \in L^2(G)$,

$$f = \sum_{\pi\in\hat{G}} d_\pi \mathrm{tr}[W_{f,\pi}\pi(\cdot)].$$

6. Characters

For every finite-dimensional unitary representation π of G, the character χ_π of π is defined by

$$\chi_\pi(g) = \mathrm{tr}[\pi(g)], \quad g \in G.$$

Then for all g in G,

$$\begin{aligned}
\mathrm{tr}[W_{f,\pi}\pi(g)] &= \mathrm{tr}\left[\int_G f(h)\pi(h^{-1})\, d\mu(h)\, \pi(g)\right] \\
&= \mathrm{tr}\left[\int_G f(h)\pi(h^{-1})\pi(g)\, d\mu(h)\right] \\
&= \mathrm{tr}\left[\int_G f(h)\pi(h^{-1}g)\, d\mu(h)\right] \\
&= \int_G f(h)\mathrm{tr}[\pi(h^{-1}g)]\, d\mu(h) \\
&= (\chi_\pi * f)(g).
\end{aligned}$$

Thus, for every f in $L^2(G)$,

$$f = \sum_{\pi\in\hat{G}} d_\pi(\chi_\pi * f).$$

7. Eigenvalues and eigenfunctions of Laplacians

We prove in this section that for every Laplacian Δ_G on a compact Lie group G, there exists an orthonormal basis for $L^2(G)$ consisting of eigenfunctions of Δ_G. The precise result is the following theorem.

Theorem 7.1. *For all π in $\hat{G}$, there exists a nonnegative number λ_π such that*

$$\Delta_G \pi_{jk} = \lambda_\pi \pi_{jk}, \quad j,k = 1,2,\ldots,d_\pi.$$

Proof. Let $\pi \in \hat{G}$. Then for $j,k = 1,2,\ldots,d_\pi$, we let

$$\nu_{jk} = \Delta \pi_{jk}.$$

Since for all $j,k = 1,2,\ldots,d_\pi$, and all g and h in G,

$$\begin{aligned}\sum_{l=1}^{d_\pi} \pi_{kl}(g)\pi_{lj}(h) &= \sum_{l=1}^{d_\pi} (\pi(g)\psi_l, \psi_k)_{X_\pi} (\psi_j, \pi(h^{-1})\psi_l)_{X_\pi} \\ &= \sum_{l=1}^{d_\pi} (\pi(h)\psi_j, \psi_l)_{X_\pi} (\psi_l, \pi(g^{-1})\psi_k)_{X_\pi} \\ &= (\pi(h)\psi_j, \pi(g^{-1})\psi_k)_{X_\pi} \\ &= (\pi(gh)\psi_j, \psi_k)_{X_\pi} = \pi_{kj}(hg),\end{aligned}$$

it follows from the left-invariance of Δ_G that

$$\nu_{kj}(gh) = \sum_{l=1}^{d_\pi} \pi_{kl}(g)\nu_{lj}(h).$$

Letting $h = e$, we get for $j,k = 1,2,\ldots,d_\pi$ and all g in G,

$$\nu_{kj}(g) = \sum_{l=1}^{d_\pi} \pi_{kl}(g)\nu_{lj}(e).$$

Similarly, if we use the right-invariance of Δ_G, we get

$$\nu_{kj}(g) = \sum_{l=1}^{d_\pi} \nu_{kl}(e)\pi_{lj}(g).$$

Thus, the matrix $[\nu_{jk}(e)]_{j,k=1}^{d_\pi}$ commutes with the irreducible and unitary representation π. So, by Schur's lemma, there exists a constant λ_π such that

$$[\nu_{jk}(e)]_{j,k=1}^{d_\pi} = \lambda_\pi = \lambda_\pi I,$$

where I is the identity matrix. Therefore for $j,k = 1,2,\ldots,d_\pi$ and all g in G,

$$(\Delta_G \pi_{jk})(g) = \nu_{jk}(g) = \sum_{l=1}^{d_\pi} \pi_{jl}(g)\nu_{lk}(e) = \lambda_\pi \pi_{jk}(g).$$

So, we are done if we can show that $\lambda_\pi \geq 0$. But this follows from the fact that

$$(\Delta_G f, f)_{L^2(G)} \geq 0, \quad f \in L^2(G). \qquad \square$$

8. Heat kernels

We give a formula for the heat kernels of Laplacians on compact Lie groups. The formula involves Weyl transforms in a very fundamental way. To this end, let u and v be functions in $L^2(G)$. Then for $t > 0$, Theorem 7.1 gives

$$(e^{-t\Delta_G}u, v)_{L^2(G)} = \sum_{\pi,j,k} e^{-t\lambda_\pi} d_\pi (u, \pi_{jk})_{L^2(G)}(\pi_{jk}, v)_{L^2(G)}.$$

We know from Section 5 that

$$(u, \pi_{jk})_{L^2(G)} = (W_{u,\pi})_{jk}$$

and a simple calculation gives

$$(\pi_{jk}, v)_{L^2(G)} = (W_{v^*,\pi})_{jk},$$

where

$$v^*(g) = \overline{v(g^{-1})}, \quad g \in G.$$

So, if we let $\{\psi_1, \psi_2, \ldots, \psi_{d_\pi}\}$ be an orthonormal basis for X_π, then

$$\begin{aligned}
(e^{-t\Delta_G}u, v)_{L^2(G)} &= \sum_{\pi\in\hat{G}} e^{-t\lambda_\pi} d_\pi \sum_{j,k=1}^{d_\pi} (W_{u,\pi})_{jk}(W_{v^*,\pi})_{jk} \\
&= \sum_{\pi\in\hat{G}} e^{-t\lambda_\pi} d_\pi \mathrm{tr}(W_{u,\pi}W_{v^*,\pi}) \\
&= \sum_{\pi\in\hat{G}} e^{-t\lambda_\pi} d_\pi \mathrm{tr}(W_{u*v^*,\pi}) \\
&= \sum_{\pi\in\hat{G}} e^{-t\lambda_\pi} d_\pi \sum_{k=1}^{d_\pi} \int_G (u*v^*)(g)(\pi(g^{-1})\psi_k, \psi_k)_{X_\pi} d\mu(g) \\
&= \sum_{\pi\in\hat{G}} e^{-t\lambda_\pi} d_\pi \int_G (u*v^*)(g)\mathrm{tr}(\pi(g^{-1}))\, d\mu(g) \\
&= \sum_{\pi\in\hat{G}} e^{-t\lambda_\pi} d_\pi \int_G (u*v^*)(g)\chi_\pi(g^{-1})\, d\mu(g) \\
&= \sum_{\pi\in\hat{G}} e^{-t\lambda_\pi} d_\pi I,
\end{aligned}$$

where

$$I = \int_G \int_G u(gh^{-1})v^*(h)\chi_\pi(g^{-1})\, d\mu(g)\, d\mu(h).$$

So,

$$\begin{aligned}
I &= \int_G \overline{v(h^{-1})} \left(\int_G u(gh^{-1})\chi_\pi(g^{-1})\, d\mu(g) \right) d\mu(h) \\
&= \int_G \overline{v(h^{-1})} \left(\int_G u(k)\chi_\pi(h^{-1}k^{-1})\, d\mu(k) \right) d\mu(h) \\
&= \int_G \overline{v(h^{-1})}(\chi_\pi * u)(h^{-1})\, d\mu(h) = \int_G \overline{v(h)}(\chi_\pi * u)(h)\, d\mu(h) = (\chi_\pi * u, v)_{L^2(G)}.
\end{aligned}$$

Thus,

$$(e^{-t\Delta_G} u, v)_{L^2(G)} = \sum_{\pi \in \hat{G}} e^{-t\lambda_\pi} d_\pi (\chi_\pi * u, v)_{L^2(G)},$$

and we get

$$e^{-t\Delta_G} u = \sum_{\pi \in \hat{G}} e^{-t\lambda_\pi} d_\pi \chi_\pi * u.$$

The conclusion then is that the heat kernel k_t, $t > 0$, of the Laplacian Δ_G on the compact Lie group G is given by

$$k_t(g, h) = \sum_{\pi \in \hat{G}} e^{-t\lambda_\pi} d_\pi \chi_\pi(gh^{-1}), \quad g, h \in G.$$

The preceding formula for the heat kernel can be found on page 45 of the book [9] by Stein.

9. Green functions

Green functions are the kernels of the integral operators for the inverse. We derive in this section the Green function of the Laplacian Δ_G on a compact Lie group G. By the very definition of a Laplacian, the constant functions on G are eigenfunctions of Δ_G corresponding to the eigenvalue 0. As such, it is desirable to quotient out these constant functions from $L^2(G)$. To do this, we write $L^2(G) = L^2_{\mathbb{C}}(G) \oplus \mathbb{C}$, where $L^2_{\mathbb{C}}(G)$ is the orthogonal complement of $\mathbb{C}$ in $L^2(G)$. Then Δ_G is reduced in the sense that $L^2_{\mathbb{C}}(G)$ and $\mathbb{C}$ are both invariant subspaces with respect to Δ_G. It is then clear that $\Delta_G^{-1} : L^2_{\mathbb{C}}(G) \to L^2_{\mathbb{C}}(G)$ exists and as in the derivation of the heat kernel, we have

$$(\Delta_G^{-1} u)(g) = \sum_{\lambda \neq 0} \frac{d_\pi}{\lambda_\pi} \chi_\pi * u$$

for all $u \in L^2_{\mathbb{C}}(G)$. So, the Green function $\mathcal{G}$ of Δ_G is given by

$$\mathcal{G}(g, h) = \sum_{\lambda_\pi \neq 0} \frac{d_\pi}{\lambda_\pi} \chi_\pi(gh^{-1}), \quad g, h \in G.$$

An alternative way of getting the Green function is to integrate the heat kernel with respect to t from 0 to ∞. We get the same formula for the Green function.

10. Riemann zeta functions

We can now define the Riemann zeta function ζ_G of the Laplacian Δ_G by

$$\zeta_G(s) = \frac{1}{\Gamma(s)} \int_0^\infty \left(\int_G k_t(g,g)\, d\mu(g) - 1 \right) t^s \frac{dt}{t}$$

for all $s \in \mathbb{C}$ such that the integral makes sense. It is the Mellin transform of $\int_G k_t(g,g)\, d\mu(g) - 1$, which is the regularized trace of the heat kernel at time t.

Theorem 10.1. *Let $s \in \mathbb{C}$ be such that $\zeta(s)$ is defined. Then*

$$\zeta(s) = \sum_{\lambda_\pi \neq 0} \frac{d_\pi^2}{\lambda_\pi^s}.$$

Proof. By the definition of the Riemann zeta function, we get

$$\begin{aligned}
\zeta_G(s) &= \frac{1}{\Gamma(s)} \int_0^\infty \left(\int_G \sum_{\pi \in \hat{G}} e^{-t\lambda_\pi} d_\pi^2 d\mu(g) - 1 \right) t^s \frac{dt}{t} \\
&= \frac{1}{\Gamma(s)} \int_0^\infty \left(\sum_{\pi \in \hat{G}} e^{-t\lambda_\pi} d_\pi^2 - 1 \right) t^s \frac{dt}{t} \\
&= \frac{1}{\Gamma(s)} \int_0^\infty \sum_{\lambda_\pi \neq 0} e^{-t\lambda_\pi} d_\pi^2 t^s \frac{dt}{t} \\
&= \frac{1}{\Gamma(s)} \sum_{\lambda_\pi \neq 0} \int_0^\infty e^{-t\lambda_\pi} d_\pi^2 t^s \frac{dt}{t} \\
&= \frac{1}{\Gamma(s)} \sum_{\lambda_\pi \neq 0} d_\pi^2 \int_0^\infty e^{-t\lambda_\pi} t^s \frac{dt}{t} \\
&= \frac{1}{\Gamma(s)} \sum_{\lambda_\pi \neq 0} \frac{d_\pi^2}{\lambda_\pi^s} \int_0^\infty e^{-\delta} \delta^s \frac{d\delta}{\delta} \\
&= \sum_{\lambda \neq 0} \frac{d_\pi^2}{\lambda_\pi^s},
\end{aligned}$$

as asserted. □

11. The unit circle S^1 in $\mathbb{R}^2$

The unit circle S^1 with center at the origin in $\mathbb{R}^2$ is given by

$$S^1 = \{e^{i\theta} : \theta \in \mathbb{R}\}.$$

It is obviously a compact Lie group in which the group law is given by the multiplication of complex numbers. So, the group S^1 can be identified with the group $\mathbb{R}/2\pi\mathbb{Z}$ in which the group law is given by addition modulo $2\pi\mathbb{Z}$. The striking feature to note here is that S^1 is abelian. As such, all its irreducible and unitary representations are one-dimensional. Let us find all of them.

For all $n \in \mathbb{Z}$, let $\chi_n : S^1 \to \mathbb{C}$ be the mapping defined by

$$\chi_n(\theta) = e^{in\theta}, \quad \theta \in \mathbb{R}.$$

Then χ_n is an irreducible and unitary representation of S^1. In fact, deg(χ_n)=1 for all n in $\mathbb{Z}$. To see that every irreducible and unitary representation χ of degree 1 is equal to χ_n for some integer n, we let $\chi(1) = e^{i\alpha}$. Then, using the homogeneity and the continuity of χ, we get

$$\chi(\theta) = \chi(1)^\theta = e^{i\alpha\theta}, \quad \theta \in \mathbb{R}.$$

Thus, for every integer k, we get

$$1 = \chi(0) = \chi(2k\pi) = e^{i\alpha 2k\pi} \Rightarrow \alpha 2k\pi = 2n\pi$$

for some integer n. If we let $k = 1$, then α has to be an integer. Thus, we have proved the following theorem.

Theorem 11.1. $S^1 = \{\chi_n : n \in \mathbb{Z}\}$. *Consequently, we can identify* S^1 *with* $\mathbb{Z}$.

It follows from the Peter-Weyl theorem that $\{\chi_n : n \in \mathbb{Z}\}$ is an orthonormal basis for $L^2(S^1)$.

Let $F \in L^1(S^1)$. Then the Weyl transform $W_{F,\chi_n} : \mathbb{C} \to \mathbb{C}$ associated to the symbol F and the representation χ_n is given by

$$W_{F,\chi_n} = \frac{1}{2\pi} \int_0^{2\pi} F(\theta) e^{-in\theta} d\theta,$$

which is nothing but the nth Fourier coefficient of the function F.

Using polar coordinates, the Laplacian Δ on $\mathbb{R}^2$ is given by

$$\Delta = \frac{\partial^2}{\partial r^2} + \frac{1}{r}\frac{\partial}{\partial r} + \frac{1}{r^2}\frac{\partial^2}{\partial \theta^2}.$$

Hence

$$\Delta_{S^1} = -\frac{\partial^2}{\partial \theta^2}$$

and we have

$$\Delta_{S^1} \chi_n = n^2 \chi_n, \quad n \in \mathbb{Z}.$$

The heat kernel k_t is then given by

$$k_t(\theta) = \frac{1}{2\pi} \sum_{n \in \mathbb{Z}} e^{-n^2 t} e^{in\theta}, \quad t > 0, \theta \in \mathbb{R}.$$

The Green function $\mathcal{G}$ of Δ_{S^1} is given by

$$\mathcal{G}(\theta) = \frac{1}{2\pi} \sum_{n \neq 0} \frac{1}{n^2} e^{in\theta}, \quad \theta \in \mathbb{R}.$$

Remark 11.2. There is an explicit formula for $\mathcal{G}(\theta)$. Indeed,

$$\mathcal{G}(\theta)=\frac{1}{2\pi}\sum_{n=1}^{\infty}\frac{1}{n^2}(e^{in\theta}+e^{-in\theta})=\frac{1}{\pi}\sum_{n=1}^{\infty}\frac{1}{n^2}\cos n\theta=\frac{\pi}{6}-\frac{1}{2}\theta+\frac{1}{4\pi}\theta^2$$

for $0\le\theta\le 2\pi$. See, for instance, the formula (3.8.5) in the book [6] by Hardy and Rogosinski.

By Theorem 10.1, the Riemann zeta function ζ_{S^1} for the Laplacian Δ_{S^1} is given by

$$\zeta_{S^1}(s)=2\sum_{n=1}^{\infty}\frac{1}{n^{2s}}=2\zeta(2s)$$

for all s in $\mathbb{C}$ such that $\zeta(2s)$ is defined, where ζ is the classical Riemann zeta function, which can be found in the book [12] by Titchmarsh among others.

12. The unit sphere S^3 in $\mathbb{R}^4$

In order to see that the unit sphere S^3 with center at the origin in $\mathbb{R}^4$ is a compact Lie group, we use the identification of $\mathbb{R}^4$ with $\mathbb{C}^2$ and the identification

$$\mathbb{R}^4\ni(x,y,\xi,\eta)\mapsto(z,w)\in\mathbb{C}^2,$$

where $z=x+iy$ and $w=\xi+i\eta$.

We let SU(2) be the set defined by

$$\mathrm{SU}(2)=\left\{\begin{pmatrix}a & -\overline{b}\\ b & \overline{a}\end{pmatrix}: a,b\in\mathbb{C},\ |a|^2+|b|^2=1\right\}.$$

It is obviously a group with respect to the multiplication of matrices and is known as the special unitary group. We sometimes denote the matrix $\begin{pmatrix}a & -\overline{b}\\ b & \overline{a}\end{pmatrix}$ by $U_{a,b}$. Matrices in SU(2) are unitary matrices with determinant 1, and hence we get for all a and b in $\mathbb{C}$ with $|a|^2+|b|^2=1$,

$$U_{a,b}^{-1}=U_{a,b}^{*}=U_{\overline{a},-b}.$$

The group action of SU(2) on $\mathbb{C}^2$ is given by

$$\begin{pmatrix}a & -\overline{b}\\ b & \overline{a}\end{pmatrix}\begin{pmatrix}z\\ w\end{pmatrix}=\begin{pmatrix}az-\overline{b}w\\ bz+\overline{a}w\end{pmatrix}$$

or

$$U_{a,b}(z,w)=(az-\overline{b}w,bz+\overline{a}w)$$

for all $\left(\begin{smallmatrix}a & -\overline{b}\\ b & \overline{a}\end{smallmatrix}\right)$ in SU(2) and $\left(\begin{smallmatrix}z\\ w\end{smallmatrix}\right)$ in $\mathbb{C}^2$. Therefore the identification

$$\mathrm{SU}(2)\ni U_{a,b}\leftrightarrow(a,b)=U_{a,b}(1,0)$$

allows us to look at SU(2) as the unit sphere S^3 with center at the origin in $\mathbb{R}^4$, and the identity element in SU(2) is identified with the north pole $(1, 0)$ in $\mathbb{C}^2$. Therefore SU(2) is isomorphic to S^3, which is a group with respect to the multiplication $\circ$ given by

$$\sigma \circ \tau = \omega$$

for all σ and τ in S^3, where $\omega \in S^3$ is such that

$$U_\omega = U_\sigma U_\tau.$$

It is now clear that S^3 is a compact group. Our first task is to identify all the irreducible and unitary representations of S^3. Since S^3 is compact, they have to be finite-dimensional. To see what they are, we let $\mathcal{P}$ be the set of all polynomials in two complex variables. To wit, a polynomial P in $\mathcal{P}$ is given by

$$P(z, w) = \sum_{1 \le j,k \le N} c_{jk} z^j w^k, \quad c_{jk}, z, w \in \mathbb{C}.$$

We can look at $\mathcal{P}$ as a subspace of $L^2(S^3)$, where the surface measure ν on S^3 is chosen to be such that $\nu(S^3) = 1$. For every nonnegative integer m, we let $\mathcal{P}_m$ be the subset of $\mathcal{P}$ consisting of all homogeneous polynomials of degree m. Thus,

$$\mathcal{P}_m = \left\{ \sum_{j=0}^{m} c_j z^j w^{m-j} : c_1, c_2, \ldots, c_m, z, w \in \mathbb{C} \right\}.$$

Since $\mathcal{P}_m$ is a finite-dimensional vector space with dimension d_m, it follows that $\mathcal{P}_m$ is a Hilbert space with respect to the product $(\,,\,)_{L^2(S^3)}$ in $L^2(S^3)$ given by

$$(P, Q)_{L^2(S^3)} = \int_{S^3} P(\sigma)\overline{Q(\sigma)}\, d\nu(\sigma), \quad P, Q \in \mathcal{P}_m.$$

The exact value of d_m can be found on Pages 138-139 of the book [10] by Stein and Weiss. The following result is Proposition 5.34 in the book [4] by Folland.

Proposition 12.1. *The family $\{\mathcal{P}_m : m = 0, 1, 2, \ldots\}$ of Hilbert spaces is mutually orthogonal in $L^2(S^3)$ and for $m = 0, 1, 2, \ldots,$*

$$\left\{ \sqrt{\frac{(m+1)!}{j!(m-j)!}} z^j w^{m-j} : 0 \le j \le m \right\}$$

is an orthonormal basis for $\mathcal{P}_m$.

The group action of SU(2) on $\mathbb{C}^2$ induces a representation π of S^3 on $\mathcal{P}$ given by

$$(\pi(a, b)P)(z, w) = P(U_{a,b}^{-1}(z, w)) = P(\overline{a}z + \overline{b}w, -bz + aw)$$

for all (a, b) in S^3 and (z, w) in $\mathbb{C}^2$.

Proposition 12.2. *For $m = 0, 1, 2, \ldots$, $\mathcal{P}_m$ is an invariant subspace of π.*

Proof. Let $P \in \mathcal{P}_m$. Then for all (a, b) in S^3,

$$(\pi(a,b)P)(z,w) = P(\overline{a}z + \overline{b}w, -bz + aw), \quad z, w \in \mathbb{C}.$$

It is then clear that $\pi(a, b)P$ is in $\mathcal{P}_m$. □

For $m = 0, 1, 2, \ldots$, let π_m be the sub-representation of π on $\mathcal{P}_m$. In other words, π_m is the representation of S^3 on $\mathcal{P}_m$ given by

$$\pi_m(U(a,b)P = \pi(a,b)P$$

for all (a, b) in S^3 and P in $\mathcal{P}_m$.

Proposition 12.3. *For $m = 0, 1, 2, \ldots$, π_m is a unitary representation of S^3 on $\mathcal{P}_m$.*

Proof. It suffices to show that $\pi_m(a, b)$ is unitary for all (a, b) in S^3. But, for all P and Q in $\mathcal{P}_m$, using the fact that the surface measure ν on S^3 is rotation invariant,

$$\begin{aligned}(\pi_m(a,b)P, Q)_{L^2(S^3)} &= \int_{S^3} P(U_{a,b}^{-1}\omega)\overline{Q(\omega)}\, d\nu(\omega) \\ &= \int_{S^3} P(\omega)\overline{Q(U_{a,b}\omega)}\, d\omega \\ &= (P, \pi_m(U_{a,b}^{-1})Q)_{L^2(S^3)}.\end{aligned}$$

Therefore $\pi_m(a, b)^* = \pi_m(a, b)^{-1}$ and we are done. □

The following result is Theorem 5.37 in the book [4] by Folland.

Theorem 12.4. *For $m = 0, 1, 2, \ldots$, π_m is an irreducible representation of S^3 on $\mathcal{P}_m$.*

In fact, it is proved in [4] that up to unitary equivalence, the family $\{\pi_m : m = 0, 1, 2, \ldots\}$ of representations of S^3 on $\mathcal{P}_m$ gives all the irreducible and unitary representations of S^3 on $\mathcal{P}_m$.

Realizing SU(2) as S^3, we see that for $m = 0, 1, 2, \ldots$, π_m can be expressed as

$$(\pi_m(\tau)P)(\sigma) = P(\tau^{-1} \circ \sigma), \quad \sigma, \tau \in S^3,$$

for all P in $\mathcal{P}_m$. Now, let $F \in L^2(S^3)$. Then the Weyl transform $W_{F,\pi_m} : \mathcal{P}_m \to \mathcal{P}_m$ is given by

$$(W_{F,\pi_m}P, Q)_{L^2(S^3)} = \int_{S^3} F(\tau)(\pi_m(\tau)P, Q)_{L^2(S^3)} d\nu(\tau), \quad P, Q \in \mathcal{P}_m.$$

Thus, for all P and Q in $\mathcal{P}_m$,

$$\begin{aligned}(W_{F,\pi_m}P, Q)_{L^2(S^3)} &= \int_{S^3} F(\tau)\left(\int_{S^3} P(\tau^{-1} \circ \sigma)\overline{Q(\sigma)}\right) d\nu(\tau) \\ &= \int_{S^3}\left(\int_{S^3} F(\tau)P(\tau^{-1} \circ \sigma)\, d\nu(\tau)\right)\overline{Q(\sigma)}\, d\nu(\sigma).\end{aligned}$$

Thus, for all P in $\mathcal{P}_m$, $W_{F,\pi_m}P$ is the same as the convolution $F * P$ of F and P.

Now, a basis for the Lie algebra $\mathfrak{su}(2)$ of SU(2) is given by the 2×2 skew-symmetric matrices with zero trace

$$A_1 = \frac{1}{2}\begin{pmatrix} i & 0 \\ 0 & -i \end{pmatrix}, A_2 = \begin{pmatrix} 0 & 1 \\ -1 & 0 \end{pmatrix}, A_3 = \begin{pmatrix} 0 & i \\ i & 0 \end{pmatrix}.$$

For $j = 1, 2, 3$, let $\gamma_j : \mathbb{R} \to \mathrm{SU}(2)$ be a homomorphic curve such that

$$\gamma_j(0) = I \qquad \text{and} \qquad \gamma_j'(0) = A_j,$$

where I is the identity matrix in SU(2). Then the Laplacian $\Delta_{\mathrm{SU}(2)}$ on SU(2) can be identified with the Laplacian Δ_{S^3} on S^3. This identification allows us to use spherical harmonics to construct the heat kernel and Green function for $\Delta_{\mathrm{SU}(2)}$.

Using Theorem 10.1, the Riemann zeta function ζ_{S^3} of the Laplacian Δ_{S^3} is given by

$$\zeta_{S^3}(s) = \sum_{n=1}^{\infty} \frac{d_n^2}{\lambda_n^s} = \frac{1}{4}\sum_{n=1}^{\infty} \frac{(n+1)^2}{n^s(n+2)^{2-s}}$$

for all s in $\mathbb{C}$ such that the series is defined.

13. Spherical harmonics

In this section, we use spherical harmonics in, e.g., the work [1] by Chavel, the book [10] by Stein and Weiss, and the book [11] by Terras, to construct the heat kernel and Green function for the Laplacian $\Delta_{S^{n-1}}$ on the unit sphere S^{n-1} in $\mathbb{R}^n$, where $n > 2$. The Laplacian $\Delta_{S^{n-1}}$ is obtained from the Laplacian $\sum_{j=1}^n \frac{\partial^2}{\partial x_j^2}$ using polar coordinates and is given by

$$\sum_{j=1}^{n} \frac{\partial^2}{\partial x_j^2} = \frac{\partial^2}{\partial r^2} + \frac{n-1}{r}\frac{\partial}{\partial r} - \frac{1}{r^2}\Delta_{S^{n-1}}.$$

A spherical harmonic Y^k of degree k is the restriction to S^{n-1} of a homogeneous and harmonic polynomial of degree k. If we denote by $\mathcal{H}_k$ the set of all spherical harmonics of degree k, then $\mathcal{H}_k$ is a finite-dimensional Hilbert space with respect to the inner product in $L^2(S^{n-1})$ and its dimension d_k is given by

$$d_k = \binom{n+k-1}{k} - \binom{n+k-3}{k-2}.$$

If we let $\{Y_1^k, \ldots, Y_{d_k}^k\}$ be an orthonormal basis for $\mathcal{H}_k$, then it can be proved that $\cup_{k=0}^{\infty}\{Y_1^k, \ldots, Y_{d_k}^k\}$ is an orthonormal basis for $L^2(S^{n-1})$. It can also be proved that for all $Y^k \in \mathcal{H}_k$,

$$\Delta_{S^{n-1}} Y^k = k(k+n-2)Y^k.$$

So, for each nonnegative integer k, $\mathcal{H}_k$ is an eigenspace of $\Delta_{S^{n-1}}$ corresponding to the eigenvalue $k(k+n-2)$. Therefore the heat kernel k_t for $\Delta_{S^{n-1}}$ is given by

$$k_t(\sigma,\tau) = \sum_{k=0}^{\infty} e^{-k(k+n-2)t} \sum_{j=1}^{d_k} Y_j^k(\sigma)\overline{Y_j^k(\tau)}, \quad t>0;\ \sigma,\tau \in S^{n-1}.$$

For $k = 0,1,2,\ldots$, there exists a constant $c_{k,n}$ such that

$$\sum_{j=1}^{d_k} Y_j^k(\sigma)\overline{Y_j^k(\tau)} = c_{k,n} P_k^{(n-2)/2}(\sigma\cdot\tau), \quad \sigma,\tau\in S^{n-1},$$

where P_k^λ is the Gegenbauer polynomial given by the generating function

$$(1-2rt+r^2)^{-\lambda} = \sum_{k=0}^{\infty} P_k^\lambda(t) r^k, \quad |r|<1, |t|\le 1, \lambda>0.$$

For more information about Gegenbauer polynomials, see, for instance, the work [2] by Erdélyi, Magnus, Oberhettinger and Tricomi. In order to compute $c_{k,n}$, we note that

$$\begin{aligned} d_k &= \int_{S^{n-1}} \sum_{j=1}^{d_k} Y_j^k(\sigma)\overline{Y_j^k(\sigma)}\, d\nu(\sigma) \\ &= c_{k,n} \int_{S^{n-1}} P_k^{(n-2)/2}(\sigma\cdot\sigma)\, d\nu(\sigma) \\ &= c_{k,n} \int_{S^{n-1}} P_k^{(n-2)/2}(1)\, d\nu(\sigma) \\ &= c_{k,n}|S^{n-1}|P_k^{(n-2)/2}(1). \end{aligned}$$

Therefore

$$c_{k,n} = \frac{d_k}{|S^{n-1}|P_k^{(n-2)/2}(1)}.$$

So,

$$k_t(\sigma,\tau) = \frac{1}{|S^{n-1}|} \sum_{k=0}^{\infty} e^{-k(k+n-2)t} \frac{d_k}{P_k^{(n-2)/2}(1)} P_k^{(n-2)/2}(\sigma\cdot\tau)$$

for all $t>0$ and all σ and τ in S^{n-1}. The Green function $\mathcal{G}$ for $\Delta_{S^{n-1}}$ is given by

$$\mathcal{G}(\sigma,\tau) = \frac{1}{|S^{n-1}|} \sum_{k=1}^{\infty} \frac{1}{k(k+n-2)} \frac{d_k}{P_k^{(n-2)/2}(1)} P_k^{(n-2)/2}(\sigma\cdot\tau)$$

for all σ and τ in S^{n-1}.

References

[1] I. Chavel, The Laplacian on Riemannian manifolds, in *Spectral Theory and Geometry*, Editors: E.B. Davies and Y. Safarov, Cambridge University Press, 1999, 30–75.

[2] A. Erdélyi, W. Magnus, F. Oberhettinger and F.G. Tricomi, *Higher Transcendental Functions, Vol.* 2, Krieger, 1981.

[3] H.D. Fegan, *An Introduction to Compact Lie Groups*, World Scientific, 1991.

[4] G.B. Folland, *A Course in Abstract Harmonic Analysis*, CRC Press, 1995.

[5] A. Grossmann, J. Morlet and T. Paul, Transforms associated to square integrable group representations I: General results, *J. Math. Phys.* **26** (1985), 2473–2479.

[6] G.H. Hardy and W.W. Rogosinski, *Fourier Series*, Dover, 1999.

[7] L. Hörmander, *The Analysis of Linear Partial Differential Operators III*, Springer-Verlag, 1985.

[8] J.J. Kohn and L. Nirenberg, *An algebra of pseudo-differential operators*, *Comm. Pure Appl. Math.* **18** (1965), 269–305.

[9] E.M. Stein, *Topics in Harmonic Analysis*, Princeton University Press, 1971.

[10] E.M. Stein and G. Weiss, *Introduction to Fourier Analysis on Euclidean Spaces*, Princeton University Press, 1971.

[11] A. Terras, *Harmonic Analysis on Symmetric Spaces and Applications I*, Springer-Verlag, 1985.

[12] E.C. Titchmarsh, *The Theory of the Riemann Zeta-function, Second Edition*, Oxford University Press, 1986.

[13] H. Weyl, *The Theory of Groups and Quantum Mechanics*, Dover, 1950.

[14] M.W. Wong, *Weyl Transforms*, Springer-Verlag, 1998.

[15] M.W. Wong, *Wavelet Transforms and Localization Operators*, Birkhäuser, 2002.

[16] M.W. Wong, A new look at pseudo-differential operators, in *Wavelets and their Applications*, Editors: M. Krishna, R. Radha and S. Thangavelu, Allied Publishers, 2003, 283–290.

[17] M.W. Wong, Weyl transforms and convolution operators on the Heisenberg group, in *Pseudo-differential Operators and Related Topics*, Editors: P. Boggiatto, L. Rodino, J. Toft and M.W. Wong, Birkhäuser, 2006, 115–121.

M.W. Wong
Department of Mathematics and Statistics
York University
4700 Keele Street
Toronto, Ontario M3J 1P3, Canada
e-mail: mwwong@mathstat.yorku.ca

Operator Theory:
Advances and Applications, Vol. 172, 87–105

On the Fourier Analysis of Operators on the Torus

Michael Ruzhansky and Ville Turunen

Abstract. Basic properties of Fourier integral operators on the torus $\mathbb{T}^n = (\mathbb{R}/2\pi\mathbb{Z})^n$ are studied by using the global representations by Fourier series instead of local representations. The results can be applied in studying hyperbolic partial differential equations.

Mathematics Subject Classification (2000). Primary 35L40; Secondary 58J40.

Keywords. Fourier integral operators, torus, pseudo-differential operators.

1. Introduction

In this paper we will discuss the version of the Fourier analysis and pseudo-differential operators on the torus. Using the toroidal Fourier transform we will show several simplifications of the standard theory. We will also discuss the corresponding toroidal version of Fourier integral operators. To distinguish them from those defined using the Euclidean Fourier transform, we will call them *Fourier series operators*. The use of discrete Fourier transform will allow to use global representation of these operators, thus eliminating a number of topological obstructions known in the standard theory. We will prepare the machinery and describe how it can be further used in the calculus of Fourier series operators and applications to hyperbolic partial differential equations. In fact, the form of the required discrete calculus is not a priori clear, for example, the form of the discrete Taylor's theorem best adopted to the calculus. We will develop the corresponding version of the periodic analysis similar in formulations to the standard Euclidean theory.

It was realised already in the 1970s that on the torus, one can study pseudo-differential operators globally using Fourier series expansions, in analogy to Euclidean pseudo-differential calculus. These *periodic pseudo-differential operators* were treated, e.g., by Agranovich [1, 2]. Contributions have been made by many authors, and the following is a non-comprehensive list of the research on the torus:

This first author would like to thank the UK Royal Society for its support. The second author thanks the Academy of Finland and Magnus Ehrnrooth Foundation for their support.

Agranovich, crediting the idea to Volevich, proposed the Fourier series representation of pseudo-differential operators. Later, he proved the equivalence of the Fourier series representation and Hörmander's definition for $(1,0)$-symbol classes; the case of classical pseudo-differential operators on the circle had been treated by Saranen and Wendland [15]; McLean [9] proved the equivalence of these approaches for Hörmander's general (ρ,δ)-classes, by using charts; in [17], the equivalence for the case of $(1,0)$-classes is proven by studying iterated commutators of pseudo-differential operators and smooth vector fields. Elschner [5] and Amosov [3] constructed asymptotic expansions for classical pseudo-differential operators; these results were generalized for (ρ,δ)-classes in [18]. There are plenty of papers considering applications and numerical computation of pseudo-differential equations on torus, e.g., spline approximations by Prössdorf and Schneider [10], physical applications by, e.g., Vainikko and Lifanov [19, 20], and many others.

On the other hand, the use of operators which are discrete in the frequency variable allows one to weaken regularity assumptions on symbols with respect to ξ. Symbols with low regularity in x have been under intensive study for many years, e.g., Kumano-go and Nagase [8], Sugimoto [16], Boulkhemair [4], Garello and Morando [6], and many others. However, in these papers one assumes symbols to be smooth or sufficiently regular in ξ. The discrete approach in this paper will allow us to reduce regularity assumptions with respect to ξ. For example, no regularity with respect to ξ is assumed for L^2 estimates, and for elements of the calculus. Moreover, one can consider scalar hyperbolic equations with C^1 symbols with respect to ξ. For example, this allows to construct parametrices to certain hyperbolic systems with variable multiplicities. Details of such constructions will appear in our forthcoming paper [14].

Let us now fix the notation. Let $\mathcal{S}(\mathbb{R}^n)$ be the Schwartz test function space with its usual topology, and let $\mathcal{S}'(\mathbb{R}^n)$ be its dual, the space of tempered distributions. Let $\mathcal{F}_E : \mathcal{S}(\mathbb{R}^n) \to \mathcal{S}'(\mathbb{R}^n)$ be the Euclidean Fourier transform (hence the subscript $_E$) defined by

$$(\mathcal{F}_E f)(\xi) = \widehat{f}_E(\xi) := \int_{\mathbb{R}^n} f(x)\, \mathrm{e}^{-\mathrm{i}x\cdot\xi}\, \tilde{\mathrm{d}}x,$$

where $\tilde{\mathrm{d}}x = (2\pi)^{-n}\, \mathrm{d}x$. Then $\mathcal{F}_E$ is a bijection and

$$f(x) = \int_{\mathbb{R}^n} \widehat{f}_E(\xi)\, \mathrm{e}^{\mathrm{i}x\cdot\xi}\, \mathrm{d}\xi,$$

and this Fourier transform can be uniquely extended to $\mathcal{F}_E : \mathcal{S}'(\mathbb{R}^n) \to \mathcal{S}'(\mathbb{R}^n)$.

The main symbol class in the sequel consists of Hörmander's (ρ,δ)-symbols of order m: Let $m \in \mathbb{R}$ and $0 \le \delta < \rho \le 1$. For $\xi \in \mathbb{R}^n$ define $\langle\xi\rangle := (1 + \|\xi\|^2)^{1/2}$, where $\|\xi\|^2 := \sum_{j=1}^n |\xi_j|^2$. Then $S^m_{\rho,\delta}(\mathbb{R}^n \times \mathbb{R}^n)$ consists of those functions $\sigma \in C^\infty(\mathbb{R}^n \times \mathbb{R}^n)$ for which

$$\left|\partial_\xi^\alpha \partial_x^\beta \sigma(x,\xi)\right| \le C_{\sigma\alpha\beta m}\, \langle\xi\rangle^{m-\rho|\alpha|+\delta|\beta|} \tag{1.1}$$

for every $x \in \mathbb{R}^n$ and for every $\alpha, \beta \in \mathbb{N}^n$.

Let $\mathbb{T}^n = (\mathbb{R}/2\pi\mathbb{Z})^n$ denote the n-dimensional torus. We may identify $\mathbb{T}^n$ with the hypercube $[0, 2\pi[^n \subset \mathbb{R}^n$ (or $[-\pi, \pi[^n$). Functions on $\mathbb{T}^n$ may be thought as those functions on $\mathbb{R}^n$ that are 2π-periodic in each of the coordinate directions. Let $\mathcal{D}(\mathbb{T}^n)$ be the vector space $C^\infty(\mathbb{T}^n)$ endowed with the usual test function topology, and let $\mathcal{D}'(\mathbb{T}^n)$ be its dual, the space of distributions on $\mathbb{T}^n$. Inclusion $\mathcal{D}(\mathbb{T}^n) \subset \mathcal{D}'(\mathbb{T}^n)$ is interpreted by

$$\phi(\psi) := \int_{\mathbb{T}^n} \phi(x)\ \psi(x)\ \mathrm{d}x,$$

where we identify the measure on torus with the corresponding restriction of the Euclidean measure on the hypercube. Let $\mathcal{S}(\mathbb{Z}^n)$ denote the space of *rapidly decaying functions* $\mathbb{Z}^n \to \mathbb{C}$. Let $\mathcal{F}_T : \mathcal{D}(\mathbb{T}^n) \to \mathcal{S}(\mathbb{Z}^n)$ be the *toroidal Fourier transform* (hence the subscript $_T$) defined by

$$(\mathcal{F}_T f)(\xi) = \widehat{f}_T(\xi) := \int_{\mathbb{T}^n} f(x)\ \mathrm{e}^{-\mathrm{i}x\cdot\xi}\ \tilde{\mathrm{d}}x,$$

where $\tilde{\mathrm{d}}x = (2\pi)^{-n}\ \mathrm{d}x$. Then $\mathcal{F}_T$ is a bijection and

$$f(x) = \sum_{\xi\in\mathbb{Z}^n} \widehat{f}_T(\xi)\ \mathrm{e}^{\mathrm{i}x\cdot\xi}.$$

This Fourier transform is extended uniquely to $\mathcal{F}_T : \mathcal{D}'(\mathbb{T}^n) \to \mathcal{S}'(\mathbb{Z}^n)$. Notice that $\mathcal{S}'(\mathbb{Z}^n)$ consists of those functions $\mathbb{Z}^n \to \mathbb{C}$ growing at infinity at most polynomially. Any continuous linear operator $A : \mathcal{D}(\mathbb{T}^n) \to \mathcal{D}(\mathbb{T}^n)$ can be presented by a formula

$$(Af)(x) = \sum_{\xi\in\mathbb{Z}^n} \sigma_A(x,\xi)\ \widehat{f}(\xi)\ \mathrm{e}^{\mathrm{i}x\cdot\xi},$$

where the unique function $\sigma_A \in C^\infty(\mathbb{T}^n \times \mathbb{Z}^n)$ is called the *symbol of* A:

$$\sigma_A(x,\xi) = \mathrm{e}^{-\mathrm{i}x\cdot\xi} A e_\xi(x),$$

where $e_\xi(x) := \mathrm{e}^{\mathrm{i}x\cdot\xi}$. Notice that when $\widehat{s_A(x)}_T(\xi) = \sigma_A(x,\xi)$, the Schwartz kernel K_A of A satisfies

$$K_A(x,y) = s_A(x)(x-y)$$

in the sense of distributions.

Next, in analogy to the classical differential calculus, we discuss difference calculus, which is needed when dealing with Fourier series operators.

2. Difference calculus

Let $\sigma : \mathbb{Z}^n \to \mathbb{C}$. Let $v_j \in \mathbb{N}^n$, $(v_j)_j = 1$ and $(v_j)_i = 0$ if $i \neq j$. Let us define the partial difference operator $\triangle_{\xi_j}$ by

$$\triangle_{\xi_j}\sigma(\xi) := \sigma(\xi + v_j) - \sigma(\xi),$$

and define

$$\triangle_\xi^\alpha := \triangle_{\xi_1}^{\alpha_1} \cdots \triangle_{\xi_n}^{\alpha_n} \quad \text{for} \quad \alpha = (\alpha_j)_{j=1}^n \in \mathbb{N}^n.$$

Lemma 2.1. *By the binomial theorem,*

$$\triangle_\xi^\alpha \sigma(\xi) = \sum_{\beta\le\alpha} (-1)^{|\alpha-\beta|} \binom{\alpha}{\beta} \sigma(\xi+\beta).$$

By induction, one can show:

Lemma 2.2 (Leibnitz formula for differences). *Let $\phi, \psi : \mathbb{Z}^n \to \mathbb{C}$. Then*

$$\triangle_\xi^\alpha (\phi\psi)(\xi) = \sum_{\beta\le\alpha} \binom{\alpha}{\beta} \left(\triangle_\xi^\beta \phi(\xi)\right)\ \triangle_\xi^{\alpha-\beta} \psi(\xi+\beta).$$

"Integration by parts" has the discrete analogy "summation by parts"

$$\sum_{\xi\in\mathbb{Z}^n} \phi(\xi)\ (\triangle_\xi^\alpha \psi)(\xi) = -\sum_{\xi\in\mathbb{Z}^n} ((\triangle_\xi{}^\alpha)^t \phi)(\xi)\ \psi(\xi),$$

where $((\triangle_{\xi_j})^t\phi)(\xi) = \phi(\xi) - \phi(\xi - v_j)$, provided that the series converge absolutely.

For $\xi \in \mathbb{Z}^n$ and $\gamma \in \mathbb{Z}^n$, let us define

$$\xi^{(\gamma)} = \prod_{j=1}^n \xi_j^{(\gamma_j)},$$

where

$$\xi_j^{(\gamma_j)} := \begin{cases} \prod_{i=0}^{\gamma_j-1} (\xi_j - i), & \gamma_j > 0, \\ 1, & \gamma_j = 0, \\ \prod_{i=\gamma_j+1}^{0} (\xi_j - i)^{-1}, & \gamma_j < 0. \end{cases}$$

Then

$$\triangle_\xi^\alpha \xi^{(\gamma)} = \gamma^{(\alpha)}\ \xi^{(\gamma-\alpha)},$$

in analogy to $\partial_\xi^\alpha \xi^\gamma = \gamma^{(\alpha)} \xi^{\gamma-\alpha}$.

Let us now discuss the discrete version of the Taylor's theorem. For simplicity, let us consider the one-dimensional case first.

Theorem 2.3 (Discrete Taylor's theorem). *For a function $\sigma : \mathbb{Z} \to \mathbb{C}$,*

$$\sigma(\xi+\eta) = \sum_{\alpha=0}^{N-1} \frac{1}{\alpha!}\ (\triangle_\xi^\alpha \sigma)(\xi)\ \eta^{(\alpha)} + R_N(\xi,\eta), \quad (\xi,\eta\in\mathbb{Z},\ N\in\mathbb{N}),$$

where

$$\begin{aligned} |\triangle_\xi^\alpha R_N(\xi,\eta)| &\le \begin{cases} \frac{1}{N!}\eta^{(N)} \max_{0\le\kappa<\eta} \left|\triangle_\xi^{N+\alpha}\sigma(\xi+\kappa)\right|, & \eta \ge N, \\ 0, & 0\le\eta<N, \\ \frac{1}{N!}\left|\eta^{(N)}\right| \max_{\eta\le\kappa<0} \left|\triangle_\xi^{N+\alpha}\sigma(\xi+\kappa)\right|, & \eta<0, \end{cases} \\ &\le \frac{1}{N!}\left|\eta^{(N)}\right| \max_{\kappa\in\{0,\dots,\eta\}} \left|\triangle_\xi^{N+\alpha}\sigma(\xi+\kappa)\right|. \end{aligned}$$

Notice that the estimate above resembles closely the Lagrange form of the error term in the traditional Taylor theorem:

$$\begin{cases} f(x+y) = \sum_{j=0}^{N-1} \frac{1}{j!} f^{(j)}(x) y^j + R_N(x,y), \\ R_N(x,y) = \frac{1}{N!} f^{(N)}(x+\theta) y^N, & \theta \in [\min\{0,y\}, \max\{0,y\}]. \end{cases}$$

Proof. First assume that $\eta \geq 0$. Then, by the binomial formula,

$$\sigma(\xi+\eta) = (I+\triangle_\xi)^\eta \sigma(\xi) = \sum_{\alpha=0}^{\eta} \binom{\eta}{\alpha} \triangle_\xi^\alpha \sigma(\xi) = \sum_{\alpha=0}^{\eta} \frac{1}{\alpha!} \triangle_\xi^\alpha \sigma(\xi)\, \eta^{(\alpha)}. \tag{2.1}$$

Thus $R_N(\xi,\eta) = 0$ for $0 \leq \eta < N$. Therefore

$$\triangle_\eta^\alpha R_N(\xi,\eta)|_{\eta=0} = 0,$$

when $0 \leq \alpha < N$. Now let η be an arbitrary integer. We notice that $\triangle_\eta^N \eta^{(\alpha)} = \alpha^{(N)}\, \eta^{(\alpha-N)} = 0$ for $0 \leq \alpha < N$, so that when we apply $\triangle_\eta^N$, we get

$$\triangle_\eta^N \sigma(\xi+\eta) = \triangle_\eta^N R_N(\xi+\eta).$$

We have hence the Cauchy problem

$$\begin{cases} \triangle_\eta^N R_N(\xi,\eta) = \triangle_\eta^N \sigma(\xi+\eta), \\ \triangle_\eta^\alpha R_N(\xi,\eta)\big|_{\eta=0} = 0, & 0 \leq \alpha \leq N-1. \end{cases}$$

It is enough to prove the estimate for $|R_N(\xi,\eta)|$ (i.e., $\alpha = 0$). Let us define $\sigma(\eta) := \eta^{(N)}/N!$. Then $\triangle_\eta^N \sigma(\eta) = N^{(N)}\eta^{(N-N)}/N! = 1$, and $\triangle_\xi^\alpha \sigma(\xi)|_{\xi=0} = 0$ when $0 \leq \alpha < N$, so by the uniqueness of the solution of the Cauchy problem it has to be

$$\begin{cases} \sum_{\kappa_N=0}^{-1} \sum_{\kappa_{N-1}=1}^{\kappa_N-1} \cdots \sum_{\kappa_1=1}^{\kappa_2-1} 1 = \frac{1}{N!}\eta^{(N)}, & \eta \geq N, \\ \sum_{\kappa_N=m}^{-1} \sum_{\kappa_{N-1}=\kappa_N}^{-1} \cdots \sum_{\kappa_1=\kappa_2}^{-1} 1 = \frac{1}{N!}\left|\eta^{(N)}\right|, & \eta < 0, \end{cases}$$

completing the proof.

Let us now deal with discrete Taylor polynomial-like expansions for a function $f : \mathbb{Z}^n \to \mathbb{C}$. For $b \in \mathbb{N}$, let us denote

$$I_k^b := \sum_{0 \leq k < b} \quad \text{and} \quad I_k^{-b} := -\sum_{-b \leq k < 0} . \tag{2.2}$$

It is useful to think of $I_\xi^\theta \cdots$ as of a discrete version of the one-dimensional integral $\int_0^\theta \cdots \mathrm{d}\xi$. In this discrete context, the difference $\triangle_\xi$ takes the role of the differential operator $\mathrm{d}/\mathrm{d}\xi$.

In the sequel, we adopt the notational conventions

$$I_{k_1}^\theta I_{k_2}^{k_1} \cdots I_{k_\alpha}^{k_{\alpha-1}} 1 = \begin{cases} 1, & \text{if } \alpha = 0, \\ I_k^\theta 1, & \text{if } \alpha = 1, \\ I_{k_1}^\theta I_{k_2}^{k_1} 1, & \text{if } \alpha = 2, \end{cases}$$

and so on.

Lemma 2.4. *If $\theta \in \mathbb{Z}$ and $\alpha \in \mathbb{N}$ then*

$$I_{k_1}^{\theta} I_{k_2}^{k_1} \cdots I_{k_\alpha}^{k_{\alpha-1}} 1 = \frac{1}{\alpha!}\,\theta^{(\alpha)}. \tag{2.3}$$

Proof. The result follows step-by-step from the observations $k^{(0)} \equiv 1$, $\triangle_k k^{(i)} = i\,k^{(i-1)}$ and $I_k^b \triangle_k k^{(i)} = b^{(i)}$.

Remark 2.5. This is like applying a discrete trivial version of the fundamental theorem of calculus: $\int_0^\theta f'(\xi)\,\mathrm{d}\xi = f(\theta) - f(0)$ for smooth enough $f : \mathbb{R} \to \mathbb{C}$ corresponds to $I_\xi^\theta \triangle_\xi f(\xi) = f(\theta) - f(0)$ for $f : \mathbb{Z} \to \mathbb{C}$.

Corollary 2.6. *If $\theta \in \mathbb{Z}^n$ and $\alpha \in \mathbb{N}^n$ then*

$$\prod_{j=1}^{n} I_{k(j,1)}^{\theta} I_{k(j,2)}^{k(j,1)} \cdots I_{k(j,\alpha_j)}^{k(j,\alpha_j-1)} 1 = \frac{1}{\alpha!}\,\theta^{(\alpha)}, \tag{2.4}$$

where $\prod_{j=1}^n I_j$ means $I_1 I_2 \cdots I_n$, where $I_j := I_{k(j,1)}^{\theta_j} I_{k(j,2)}^{k(j,1)} \cdots I_{k(j,\alpha_j)}^{k(j,\alpha_j-1)}$.

We now have

Theorem 2.7. *Let $p : \mathbb{Z}^n \to \mathbb{C}$ and*

$$r_M(\xi,\theta) := p(\xi+\theta) - \sum_{|\alpha|<M} \frac{1}{\alpha!}\,\theta^{(\alpha)} \triangle_\xi^\alpha p(\xi).$$

Then

$$\left|\triangle_\xi^\omega r_M(\xi,\theta)\right| \le c_M \max_{|\alpha|=M,\ \nu\in Q(\theta)} \left|\theta^{(\alpha)}\ \triangle_\xi^{\alpha+\omega} p(\xi+\nu)\right|, \tag{2.5}$$

where $Q(\theta) := \{\nu \in \mathbb{Z}^n :\ \min(0,\theta_j) \le \nu_j \le \max(0,\theta_j)\}$.

Proof. For $0 \ne \alpha \in \mathbb{N}^n$, let us denote $m_\alpha := \min\{j :\ \alpha_j \ne 0\}$. For $\theta \in \mathbb{Z}^n$ and $i \in \{1,\dots,n\}$, let us define $\nu(\theta,i,k) \in \mathbb{Z}^n$ by

$$\nu(\theta,i,k) := (\theta_1,\dots,\theta_{i-1},k,0,\dots,0),$$

i.e.,

$$\nu(\theta,i,k)_j = \begin{cases} \theta_j, & \text{if } 1 \le j < i, \\ k, & \text{if } j = i, \\ 0, & \text{if } i < j \le n. \end{cases}$$

We claim that the remainder can be written in the form

$$r_M(\xi,\theta) = \sum_{|\alpha|=M} r_\alpha(\xi,\theta), \tag{2.6}$$

where for each α,

$$r_\alpha(\xi,\theta) = \prod_{j=1}^{n} I_{k(j,1)}^{\theta_j} I_{k(j,2)}^{k(j,1)} \cdots I_{k(j,\alpha_j)}^{k(j,\alpha_j-1)}\ \triangle_\xi^\alpha p(\xi + \nu(\theta, m_\alpha, k(m_\alpha,\alpha_{m_\alpha}))); \tag{2.7}$$

recall (2.2) and (2.4). The proof of (2.7) is by induction: The first remainder term r_1 is of the claimed form, since

$$r_1(\xi,\theta) = p(\xi+\theta) - p(\xi) = \sum_{i=1}^{n} r_{v_i}(\xi,\theta),$$

where

$$r_{v_i}(\xi,\theta) = I_k^{\theta_i} \triangle_\xi^{v_i} p(\xi + \nu(\theta,i,k));$$

here r_{v_i} is of the form (2.7) for $\alpha = v_i$, $m(\alpha) = i$ and $\alpha_{m_\alpha} = 1$. So suppose that the claim (2.7) is true up to order $|\alpha| = M$. Then

$$\begin{aligned} r_{M+1}(\xi,\theta) &= r_M(\xi,\theta) - \sum_{|\alpha|=M} \frac{1}{\alpha!}\, \theta^{(\alpha)}\, \triangle_\xi^\alpha p(\xi) \\ &= \sum_{|\alpha|=M} \left(r_\alpha(\xi,\theta) - \frac{1}{\alpha!}\, \theta^{(\alpha)}\, \triangle_\xi^\alpha p(\xi) \right) \\ &= \sum_{|\alpha|=M} \prod_{j=1}^{n} I_{k(j,1)}^{\theta_j} I_{k(j,2)}^{k(j,1)} \cdots I_{k(j,\alpha_j)}^{k(j,\alpha_j-1)} \\ &\quad \triangle_\xi^\alpha \left[p(\xi + \nu(\theta, m_\alpha, k(m_\alpha, \alpha_{m_\alpha}))) - p(\xi) \right], \end{aligned}$$

where we used (2.7) and (2.4) to obtain the last equality. Combining this to the observation

$$p(\xi + \nu(\theta, m_\alpha, k)) - p(\xi) = \sum_{i=1}^{m_\alpha} I_\ell^{\nu(\theta,m_\alpha,k)_i}\, \triangle_\zeta^{v_i} p(\xi + \nu(\theta,i,\ell)),$$

we get

$$\begin{aligned} r_{M+1}(\xi,\theta) &= \sum_{|\alpha|=M} \prod_{j=1}^{n} I_{k(j,1)}^{\theta_j} I_{k(j,2)}^{k(j,1)} \cdots I_{k(j,\alpha_j)}^{k(j,\alpha_j-1)} \sum_{i=1}^{m_\alpha} I_{\ell(i)}^{\nu(\theta,m_\alpha,k(m_\alpha,\alpha_{m_\alpha}))_i} \\ &\quad \triangle_\xi^{\alpha+v_i} p(\xi + \nu(\theta,i,\ell(i))) \\ &= \sum_{|\beta|=M+1} \prod_{j=1}^{n} I_{k(j,1)}^{\theta_j} I_{k(j,2)}^{k(j,1)} \cdots I_{k(j,\beta_j)}^{k(j,\beta_j-1)} \\ &\quad \triangle_\xi^\beta p(\xi + \nu(\theta, m_\beta, k(m_\beta, \beta_{m_\beta}))); \end{aligned}$$

the last step here is just simple tedious book-keeping. Thus the induction proof of (2.7) is complete. Finally, let us prove estimate (2.5). By (2.7), we obtain

$$\left| \triangle_\xi^\omega r_M(\xi,\theta) \right| = \left| \sum_{|\alpha|=M} \triangle_\xi^\omega r_\alpha(\xi,\theta) \right|$$

$$= \Bigg| \sum_{|\alpha|=M} \prod_{j=1}^{n} I_{k(j,1)}^{\theta_j} I_{k(j,2)}^{k(j,1)} \cdots I_{k(j,\alpha_j)}^{k(j,\alpha_j-1)}$$

$$\triangle_\xi^{\alpha+\omega} p(\xi + \nu(\theta, m_\alpha, k(m_\alpha, \alpha_{m_\alpha}))) \Bigg|$$

$$\leq \sum_{|\alpha|=M} \frac{1}{\alpha!} \left|\theta^{(\alpha)}\right| \max_{\nu \in Q(\theta)} \left|\triangle_\xi^{\alpha+\omega} p(\xi+\nu)\right|,$$

where in the last step we used (2.4). The proof is complete.

Remark 2.8. If $n \geq 2$, there are many alternative forms for remainders $r_\alpha(\xi, \theta)$. This is due to the fact that there may be many different shortest discrete step-by-step paths in the space $\mathbb{Z}^n$ from ξ to $\xi + \theta$. In the proof above, we chose just one such path, travelling via the points

$$\xi, \quad \xi + \theta_1 v_1, \ldots, \quad \xi + \sum_{i=1}^{j} \theta_i v_i, \ldots, \quad \xi + \theta.$$

But if $n = 1$, there is just one shortest discrete path from $\xi \in \mathbb{Z}$ to $\theta \in \mathbb{Z}$, and in that case

$$r_M(\xi, \theta) = I_{k_1}^{\theta} I_{k_2}^{k_1} \cdots I_{k_M}^{k_{M-1}} \triangle_\xi^M p(\xi + k_M).$$

Notice also that the discrete Taylor theorem presented above implies the following smooth Taylor result:

Corollary 2.9. *Let* $p \in C^\infty(\mathbb{R}^n)$ *and*

$$r_M(\xi, \theta) := p(\xi+\theta) - \sum_{|\alpha|<M} \frac{1}{\alpha!} \theta^\alpha \left(\frac{\partial}{\partial \xi}\right)^\alpha p(\xi).$$

Then

$$\left|\partial_\xi^\omega r_M(\xi, \theta)\right| \leq c_M \max_{|\alpha|=M,\ \nu \in Q_{\mathbb{R}^n}(\theta)} \left|\theta^\alpha\ \partial_\xi^{\alpha+\omega} p(\xi+\nu)\right|, \tag{2.8}$$

where $Q_{\mathbb{R}^n}(\theta) := \{\nu \in \mathbb{R}^n :\ \min(0, \theta_j) \leq \nu_j \leq \max(0, \theta_j)\}$.

Remark 2.10. We see that in the remainder estimates above, the cubes $Q(\theta) \subset \mathbb{Z}^n$ and $Q_{\mathbb{R}^n}(\theta) \subset \mathbb{R}^n$ could be replaced by (discrete, resp. continuous) paths from 0 to θ; especially, $Q_{\mathbb{R}^n}(\theta)$ could be replaced by the straight line from 0 to θ.

3. Pseudo-differential operators on the torus

Let $m \in \mathbb{R}$ and $0 \leq \delta < \rho \leq 1$. Then $S^m_{\rho,\delta}(\mathbb{T}^n \times \mathbb{Z}^n)$ consists of those functions $\sigma \in C^\infty(\mathbb{T}^n \times \mathbb{Z}^n)$ for which

$$\left|\triangle_\xi^\alpha \partial_x^\beta \sigma(x, \xi)\right| \leq C_{\sigma\alpha\beta m}\ \langle\xi\rangle^{m-\rho|\alpha|+\delta|\beta|} \tag{3.1}$$

for every $x \in \mathbb{T}^n$, for every $\alpha, \beta \in \mathbb{N}^n$. If $\sigma_A \in S^m_{\rho,\delta}(\mathbb{T}^n \times \mathbb{Z}^n)$, we denote $A \in \mathrm{Op}S^m_{\rho,\delta}(\mathbb{T}^n \times \mathbb{Z}^n)$. The class $S^m_{\rho,\delta}(\mathbb{T}^n \times \mathbb{T}^n \times \mathbb{Z}^n)$ consists of the functions $a \in C^\infty(\mathbb{T}^n \times \mathbb{T}^n \times \mathbb{Z}^n)$ such that

$$\left|\triangle_\xi^\alpha \partial_x^\beta \partial_y^\gamma a(x,y,\xi)\right| \leq C_{a\alpha\beta\gamma m} \, \langle\xi\rangle^{m-\rho|\alpha|+\delta|\beta+\gamma|} \tag{3.2}$$

for every $x, y \in \mathbb{T}^n$, for every $\alpha, \beta, \gamma \in \mathbb{N}^n$; such a function a is called an *amplitude* of order $m \in \mathbb{R}$ of type (ρ, δ). Formally we may define

$$(\mathrm{Op}(a)f)(x) := \int_{\mathbb{T}^n} f(y) \sum_{\xi\in\mathbb{Z}^n} a(x,y,\xi) \, \mathrm{e}^{\mathrm{i}(x-y)\cdot\xi} \, \mathrm{\tilde{d}}y$$

for $f \in \mathcal{D}(\mathbb{T}^n)$.

Remark 3.1. On $\mathbb{T}^n$, Hörmander's usual (ρ,δ)-symbol class of order $m \in \mathbb{R}$ coincides with the class $\mathrm{Op}S^m_{\rho,\delta}(\mathbb{T}^n \times \mathbb{Z}^n)$ [9].

Lemma 3.2. *Let $a \in S^m_{\rho,\delta}(\mathbb{T}^n \times \mathbb{T}^n \times \mathbb{Z}^n)$, and define*

$$a_\alpha(x,y,\xi) := \left(\mathrm{e}^{\mathrm{i}(y-x)} - 1\right)^\alpha a(x,y,\xi),$$

where $\alpha \in \mathbb{N}^n$. Then $\triangle_\xi^\alpha a \in S^{m-\rho|\alpha|}_{\rho,\delta}(\mathbb{T}^n \times \mathbb{T}^n \times \mathbb{Z}^n)$ and $\mathrm{Op}(a_\alpha) = \mathrm{Op}(\triangle_\xi^\alpha a)$.

Proof. Clearly $\triangle_\xi^\alpha a \in S^{m-\rho|\alpha|}_{\rho,\delta}(\mathbb{T}^n \times \mathbb{T}^n \times \mathbb{Z}^n)$. Now

$$\begin{aligned}
(\mathrm{Op}(a_\alpha)f)(x) &= \int_{\mathbb{T}^n} f(y) \sum_{\xi\in\mathbb{Z}^n} a_\alpha(x,y,\xi) \, \mathrm{e}^{\mathrm{i}(x-y)\cdot\xi} \, \mathrm{\tilde{d}}y \\
&= \int_{\mathbb{T}^n} f(y) \left(\sum_{\xi\in\mathbb{Z}^n} \left(\mathrm{e}^{\mathrm{i}(y-x)} - 1\right)^\alpha a(x,y,\xi) \, \mathrm{e}^{\mathrm{i}(x-y)\cdot\xi} \right) \mathrm{\tilde{d}}y \\
&= \int_{\mathbb{T}^n} f(y) \left(\sum_{\xi\in\mathbb{Z}^n} a(x,y,\xi) \, (\triangle_\xi^\alpha)^t \mathrm{e}^{\mathrm{i}(x-y)\cdot\xi} \right) \mathrm{\tilde{d}}y \\
&= \int_{\mathbb{T}^n} f(y) \left(\sum_{\xi\in\mathbb{Z}^n} \mathrm{e}^{\mathrm{i}(x-y)\cdot\xi} \, \triangle_\xi^\alpha a(x,y,\xi) \right) \mathrm{\tilde{d}}y.
\end{aligned}$$

Thus $\mathrm{Op}(a_\alpha) = \mathrm{Op}(\triangle_\xi^\alpha a)$.

Theorem 3.3. *For every amplitude $a \in S^m_{\rho,\delta}(\mathbb{T}^n \times \mathbb{T}^n \times \mathbb{Z}^n)$ there exists a unique symbol $\sigma \in S^m_{\rho,\delta}(\mathbb{T}^n \times \mathbb{Z}^n)$ satisfying $\mathrm{Op}(a) = \mathrm{Op}(\sigma)$, where*

$$\sigma(x,\xi) \sim \sum_{\alpha\geq 0} \frac{1}{\alpha!} \, \triangle_\xi^\alpha \, \partial_y^{(\alpha)} a(x,y,\xi)|_{y=x}. \tag{3.3}$$

Proof. Essentially the same as in [18].

4. Periodisation

Let us define the *periodisation operator* $p : \mathcal{S}(\mathbb{R}^n) \to \mathcal{D}(\mathbb{T}^n)$ by

$$pu(x) := \mathcal{F}_T^{-1}((\mathcal{F}_E u)|_{\mathbb{Z}^n})(x).$$

Let us describe the extension of this periodisation for some nice-enough classes of distributions. From the proof of the Poisson summation formula it follows that

$$pu(x) = \sum_{\xi\in\mathbb{Z}^n} u(x + 2\pi\xi).$$

This formula makes sense almost everywhere for $u \in L^1(\mathbb{R}^n)$. Indeed, formally

$$\begin{aligned}(pu)(x) &= \sum_{k\in\mathbb{Z}^n} \mathrm{e}^{\mathrm{i}x\cdot k} \int_{\mathbb{R}^n} \mathrm{e}^{-\mathrm{i}y\cdot k}\, u(y)\, \mathrm{d}y = \int_{\mathbb{R}^n} u(y) \left(\sum_{k\in\mathbb{Z}^n} \mathrm{e}^{\mathrm{i}(x-y)\cdot k}\right) \mathrm{d}y \\ &= \int_{\mathbb{R}^n} u(y)\, \delta_{\mathbb{Z}^n}(2\pi(x-y))\, \mathrm{d}y = \sum_{\xi\in\mathbb{Z}^n} u(x+2\pi\xi).\end{aligned}$$

This calculation can be justified in the standard way. Now $pu \in L^1(\mathbb{T}^n)$ with $\|pu\|_{L^1(\mathbb{T}^n)} \leq \|u\|_{L^1(\mathbb{R}^n)}$. Moreover, if $\xi \in \mathbb{Z}^n$ then

$$\widehat{pu}_T(\xi) = \widehat{u}_E(\xi).$$

Clearly $p : L^1(\mathbb{R}^n) \to L^1(\mathbb{T}^n)$ is a surjection. We will also use that

$$(pu)(x) = \sum_{\xi\in\mathbb{Z}^n} \mathrm{e}^{\mathrm{i}x\cdot\xi}\widehat{pu}_T(\xi) = \sum_{\xi\in\mathbb{Z}^n} \mathrm{e}^{\mathrm{i}x\cdot\xi}\widehat{u}_E(\xi).$$

Let us establish basic properties of pseudo-differential operators with respect to periodisation. Let us call symbol $a(x,\xi)$ periodic if the function $x \mapsto a(x,\xi)$ is 2π-periodic. We will use tildes to denote corresponding restricted operators.

Proposition 4.1. *Let $a \in S^m_{\rho,\delta}(\mathbb{R}^n \times \mathbb{R}^n)$ be a periodic symbol. Let $\tilde{a} = a|_{\mathbb{T}^n\times\mathbb{Z}^n}$. Then $p \circ a(X,D) = \tilde{a}(X,D) \circ p$.*

Notice that a does not have to be a symbol, as the same property holds when we define $a(X,D)$ in the usual sense, by even quite irregular amplitude $a(x,\xi)$.

Proof. Notice that $\tilde{a} \in S^m_{\rho,\delta}(\mathbb{T}^n \times \mathbb{Z}^n)$. Let $f \in L^1(\mathbb{R}^n)$. Then we have

$$\begin{aligned}p(a(X,D)f)(x) &= \sum_{k\in\mathbb{Z}^n} a(x+2\pi k, D) f(x+2\pi k) \\ &= \sum_{k\in\mathbb{Z}^n} \int_{\mathbb{R}^n} \mathrm{e}^{\mathrm{i}(x+2\pi k)\cdot\xi}\, a(x+2\pi k,\xi)\, \widehat{f}_E(\xi)\, \mathrm{d}\xi \\ &= \int_{\mathbb{R}^n} \left(\sum_{k\in\mathbb{Z}^n} \mathrm{e}^{\mathrm{i}2\pi k\cdot\xi}\right) \mathrm{e}^{\mathrm{i}x\cdot\xi}\, a(x,\xi)\, \widehat{f}_E(\xi)\, \mathrm{d}\xi \\ &= \int_{\mathbb{R}^n} \mathrm{e}^{\mathrm{i}x\cdot\xi}\, a(x,\xi)\, \widehat{f}_E(\xi)\, \delta_{\mathbb{Z}^n}(\xi)\, \mathrm{d}\xi\end{aligned}$$

$$= \sum_{\xi \in \mathbb{Z}^n} \mathrm{e}^{\mathrm{i}x\cdot\xi}\, a(x,\xi)\, \widehat{f_E}(\xi)$$
$$= \sum_{\xi \in \mathbb{Z}^n} \mathrm{e}^{\mathrm{i}x\cdot\xi}\, a(x,\xi)\, \widehat{pf_T}(\xi)$$
$$= \tilde{a}(X,D)(pf)(x);$$

these calculations are justified in the sense of distributions. The proof is complete.

Since we will not always work with periodic symbols it may be convenient to periodize them. If $a(X,D)$ is a pseudo-differential operator with symbol $a(x,\xi)$, by $(pa)(X,D)$ we will denote a pseudo-differential operator with symbol $(pa)(x,\xi) = \sum_{k\in\mathbb{Z}^n} a(x+2\pi k,\xi)$. This makes sense if, for example, a in integrable in x.

Proposition 4.2. *Let $a \in S^m_{\rho,\delta}(\mathbb{R}^n \times \mathbb{R}^n)$ satisfy $a(x,\xi) = 0$ for all $x \in \mathbb{R}^n \setminus [-\pi,\pi]^n$. Then we have*

$$a(X,D)f = (pa)(X,D)f + Rf,$$

for all f supported in $[-\pi,\pi]^n$. Here $R : \mathcal{S}'(\mathbb{R}^n) \to \mathcal{S}(\mathbb{R}^n)$ is a smoothing pseudo-differential operator.

Proof. By our definition we can write

$$(pa)(X,D)f(x) = \sum_{k\in\mathbb{Z}^n} \int_{\mathbb{R}^n} \mathrm{e}^{\mathrm{i}x\cdot\xi}\, a(x+2\pi k,\xi)\, \widehat{f_E}(\xi)\, \mathrm{d}\xi,$$

and let $Rf = a(X,D)f - (pa)(X,D)f$. The assumption on the support of a implies that for every x there is only one $k \in \mathbb{Z}^n$ for which $a(x+2\pi k,\xi) \neq 0$, so the sum consists of only one term. It follows that $Rf(x) = 0$ for $x \in [-\pi,\pi]^n$. Let now $x \in \mathbb{R}^n \setminus [-\pi,\pi]^n$. Since

$$Rf(x) = -\sum_{k\in\mathbb{Z}^n, k\neq 0} \int_{\mathbb{R}^n} \int_{\mathbb{R}^n} \mathrm{e}^{\mathrm{i}(x-y)\cdot\xi}\, a(x+2\pi k,\xi)\, f(y)\, \mathrm{d}y\, \mathrm{d}\xi$$

is just a single term and $|x-y| > 0$, we can integrate by parts with respect to ξ any number of times. This implies that $R \in \Psi^{-\infty}$ and that Rf decays at infinity faster than any power. The proof is complete since the same argument can be applied to the derivatives of Rf.

Remark 4.3. Note that if f is compactly supported, but not necessarily in the cube $[-\pi,\pi]^n$, sums in the proof may consist of finite number of terms. This means that on $\mathcal{E}'(\mathbb{R}^n)$, modulo a smoothing operator, we can write $a(X,D)$ as a finite sum of operators with periodic symbols. Moreover, the same argument applies if $a(x,\xi)$ is compactly supported in x, but not necessarily in $[-\pi,\pi]^n$.

This proposition allows us to extend formula of Proposition 4.1 to perturbations of periodic symbols. We will use it when $a(x,D)$ is a sum of a constant coefficient operator and an operator with symbol having compact x-support.

Corollary 4.4. *Let $a(X,D)$ be an operator with symbol*

$$a(x,\xi) = a_1(x,\xi) + a_0(x,\xi),$$

where $a_1 \in S^m_{\rho,\delta}(\mathbb{R}^n \times \mathbb{R}^n)$ *is periodic in* x *and* $a_0 \in S^m_{\rho,\delta}(\mathbb{R}^n \times \mathbb{R}^n)$ *has compact* x*-support. Then there is a symbol* $\tilde{b} \in S^m_{\rho,\delta}(\mathbb{T}^n \times \mathbb{Z}^n)$ *such that*

$$p(a(X,D)f) = \tilde{b}(X,D)(pf) + p(Rf), \; f \in \mathcal{E}'(\mathbb{R}^n),$$

where $R : \mathcal{S}'(\mathbb{R}^n) \to \mathcal{S}(\mathbb{R}^n)$. *In particular, if* $\mathrm{supp}(a_0(\cdot,\xi)), \mathrm{supp}(f) \subset [-\pi,\pi]^n$, *we can take* $\tilde{b}(X,D) = \widetilde{a_1}(X,D) + \widetilde{pa_0}(X,D)$.

We assumed that symbols are smooth, but the requirement of the smoothness of $a_1(x,\xi)$ is not necessary similar to Proposition 4.1.

Proof. By Proposition 4.2, $a(X,D) = a_1(X,D) + (pa_0)(X,D) + R$. Since operator $b(X,D) = a_1(X,D) + (pa_0)(X,D)$ has periodic symbol, by Proposition 4.1 we have $p \circ b(X,D) = \tilde{b}(X,D) \circ p = \widetilde{a_1}(X,D) \circ p + \widetilde{pa_0}(X,D) \circ p$. Since $R : \mathcal{S}'(\mathbb{R}^n) \to \mathcal{S}(\mathbb{R}^n)$, we also have $p \circ R : \mathcal{S}'(\mathbb{R}^n) \to \mathcal{D}(\mathbb{T}^n)$. The proof is complete.

5. Conditions for L^2-boundedness

Next we study conditions on a toroidal symbol σ_A that guarantee L^2-boundedness for the corresponding operator $A : \mathcal{D}(\mathbb{T}^n) \to \mathcal{D}(\mathbb{T}^n)$. Notice that $x \mapsto \sigma_A(x,\xi) \in C^\infty(\mathbb{T}^n)$ for every $\xi \in \mathbb{Z}^n$.

Proposition 5.1. *If*

$$\left|\partial_x^\beta \sigma_A(x,\xi)\right| \le C$$

when $|\beta| \le n/2 + 1$ *then* $A \in \mathcal{L}(L^2(\mathbb{T}^n))$.

Proof. Now

$$\begin{aligned} Af(x) &= \sum_{\xi \in \mathbb{Z}^n} \sigma_A(x,\xi)\ \widehat{f_T}(\xi)\ \mathrm{e}^{\mathrm{i}x\cdot\xi} \\ &= \sum_{\xi,\eta \in \mathbb{Z}^n} \widehat{\sigma_A}_T(\eta,\xi)\ \widehat{f_T}(\xi)\ \mathrm{e}^{\mathrm{i}x\cdot(\xi+\eta)} \\ &= \sum_{\omega \in \mathbb{Z}^n} \mathrm{e}^{\mathrm{i}x\cdot\omega} \sum_{\xi \in \mathbb{Z}^n} \widehat{\sigma_A}_T(\omega-\xi,\xi)\ \widehat{f_T}(\xi). \end{aligned}$$

Here $|\widehat{\sigma_A}_T(\eta,\xi)| \le C\ \langle\eta\rangle^{-k}$, so that

$$\begin{aligned} \|Af\|^2_{L^2(\mathbb{T}^n)} &= \int_{\mathbb{T}^n} |Af(x)|^2\ \tilde{\mathrm{d}}x \\ &= \sum_{\omega \in \mathbb{Z}^n} |\widehat{Af}_T(\omega)|^2 \\ &= \sum_{\omega \in \mathbb{Z}^n} \left| \sum_{\xi \in \mathbb{Z}^n} \widehat{\sigma_A}_T(\omega-\xi,\xi)\ \widehat{f_T}(\xi) \right|^2 \end{aligned}$$

$$
\begin{aligned}
&\leq \left(\sup_{\omega\in\mathbb{Z}^n} \sum_{\xi\in\mathbb{Z}^n} |\widehat{\sigma_{A}}_T(\omega-\xi,\xi)|^2 \right) \\
&\quad \left(\sup_{\xi\in\mathbb{Z}^n} \sum_{\omega\in\mathbb{Z}^n} |\widehat{\sigma_{A}}_T(\omega-\xi,\xi)|^2 \right) \sum_{\xi\in\mathbb{Z}^n} \left|\widehat{f_T}(\xi)\right|^2 \\
&\leq C' \, \|f\|^2_{L^2(\mathbb{T}^n)}.
\end{aligned}
$$

Note that no difference conditions for the ξ-variable were needed. In fact, this is related to the following more general result.

Theorem 5.2. *Let*

$$
Tu(x) = \sum_{k\in\mathbb{Z}^n} \mathrm{e}^{\mathrm{i}\phi(x,k)} \; a(x,k) \; \hat{u}(k).
$$

Assume that for every α *for which* $|\alpha| \leq 2n+1$*,*

$$
|\partial_x^\alpha a(x,k)| \leq C, \qquad |\partial_x^\alpha \Delta_k \phi(x,k)| \leq C. \tag{5.1}
$$

Assume also that for every $x \in \mathbb{T}^n$ *and for every* $k, l \in \mathbb{Z}^n$*,*

$$
|\nabla_x \phi(x,k) - \nabla_x \phi(x,l)| \geq C|k-l|. \tag{5.2}
$$

Then $T \in \mathcal{L}(L^2(\mathbb{T}^n))$*.*

Note that condition (5.2) is a discrete version of the usual local graph condition for Fourier integral operators, necessary for the local L^2-boundedness We also note that those conditions roughly correspond to C^1 properties of the phase in ξ. Finally, if ϕ and a are not 2π–periodic in x, operator T is bounded from $L^2(\mathbb{T}^n)$ to $L^2_{loc}(\mathbb{R}^n)$. Theorem 5.2 is the discrete version of the global boundedness theorem in [12].

6. Extending symbols

It is often useful to extend toroidal symbols from $\mathbb{T}^n \times \mathbb{Z}^n$ to $\mathbb{T}^n \times \mathbb{R}^n$. This can be done with a suitable convolution that respects the symbol inequalities. The idea of the following lemma goes probably back to Y. Meyer.

Lemma 6.1. *There exist* $\theta, \phi_\alpha \in \mathcal{S}(\mathbb{R}^n)$ *such that* $\widehat{\theta}_E|_{\mathbb{Z}^n}(\xi) = \delta_{0,\xi}$ *and* $\partial_\xi^\alpha \widehat{\theta}_E(\xi) = (\triangle_\xi^\alpha)^t \phi_\alpha(\xi)$ *for every multi-index* α*.*

Proof. Let us first consider the case $n=1$. Let $\theta = \theta_1 \in C^\infty(\mathbb{R}^1)$ such that

$$
\operatorname{supp}(\theta_1) \subset]-2\pi, 2\pi[, \quad \theta_1(-x) = \theta_1(x), \quad \theta_1(\pi - y) + \theta_1(\pi + y) = 1
$$

for $x \in \mathbb{R}$ and for $0 \leq y \leq \pi$. These assumptions for θ are enough for us, and of course the choice is not unique. In any case, $\widehat{\theta_1}_E \in \mathcal{S}(\mathbb{R}^1)$. If $\xi \in \mathbb{Z}^n$ then

$$\begin{aligned}
\widehat{\theta_1}_E(\xi) &= \int_{\mathbb{R}^1} \theta_1(x)\ \mathrm{e}^{-\mathrm{i}x\cdot\xi}\ \tilde{\mathrm{d}}x \\
&= \int_0^{2\pi} (\theta_1(x-2\pi)+\theta_1(x))\ \mathrm{e}^{-\mathrm{i}x\cdot\xi}\ \tilde{\mathrm{d}}x \\
&= \int_0^{2\pi} \mathrm{e}^{-\mathrm{i}x\cdot\xi}\ \tilde{\mathrm{d}}x \\
&= \delta_{0,\xi}.
\end{aligned}$$

If desired $\phi_\alpha \in \mathcal{S}(\mathbb{R}^n)$ exists, it must satisfy

$$\int_{\mathbb{R}^1} \mathrm{e}^{\mathrm{i}x\cdot\xi}\ \phi^{(\alpha)}(\xi)\ \mathrm{d}\xi = \int_{\mathbb{R}^1} \mathrm{e}^{\mathrm{i}x\cdot\xi}\ (\triangle_\xi^\alpha)^t\phi_\alpha(\xi)\ \mathrm{d}\xi$$

because $\mathcal{F}_E : \mathcal{S}(\mathbb{R}^1) \to \mathcal{S}(\mathbb{R}^1)$ is bijective. Integration by parts yields

$$(-\mathrm{i}x)^\alpha\theta(x) = (1-\mathrm{e}^{\mathrm{i}x})^\alpha(\mathcal{F}_E^{-1}\phi_\alpha)(x).$$

Thus

$$(\mathcal{F}_E^{-1}\phi_\alpha)(x) = \left(\frac{-\mathrm{i}x}{1-\mathrm{e}^{\mathrm{i}x}}\right)^\alpha \theta(x).$$

The general n-dimensional case is reduced to the 1-dimensional case, since $\theta = (x \mapsto \theta_1(x_1)\theta_1(x_2)\cdots\theta_1(x_n)) \in \mathcal{S}(\mathbb{R}^n)$ has desired properties. The following two results can be easily obtained from the discrete Taylor's theorem.

Lemma 6.2. *Let $a : \mathbb{T}^n \times \mathbb{R}^n \to \mathbb{C}$ belong to $a \in S^m_{\rho,\delta}(\mathbb{R}^n \times \mathbb{R}^n)$. Then the restriction $\sigma = a|_{\mathbb{T}^n\times\mathbb{Z}^n} \in S^m_{\rho,\delta}(\mathbb{T}^n \times \mathbb{Z}^n)$.*

Proposition 6.3. *Let $a, b : \mathbb{T}^n \times \mathbb{R}^n \to \mathbb{C}$ such that $a, b \in S^m_{\rho,\delta}(\mathbb{R}^n \times \mathbb{R}^n)$ and $a|_{\mathbb{T}^n\times\mathbb{Z}^n} = b|_{\mathbb{T}^n\times\mathbb{Z}^n}$. Then $a-b$ is smoothing, $a-b \in S^{-\infty}(\mathbb{R}^n \times \mathbb{R}^n)$.*

The main theorem of this paragraph is that we can extend toroidal symbols in a unique smooth way.

Theorem 6.4. *Let $\sigma \in S^m_{\rho,\delta}(\mathbb{T}^n \times \mathbb{Z}^n)$. Then there exists $a \in S^m_{\rho,\delta}(\mathbb{R}^n \times \mathbb{R}^n)$ such that $\sigma = a|_{\mathbb{T}^n\times\mathbb{Z}^n}$; this extended symbol is unique up to smoothing.*

Proof. Uniqueness up to smoothing follows from Proposition 6.3, so the existence is the main issue here. Let $\theta \in \mathcal{S}(\mathbb{R}^n)$ be as in Lemma 6.1. Define $a : \mathbb{R}^n \times \mathbb{R}^n \to \mathbb{C}$ by

$$a(x,\xi) := \sum_{\eta\in\mathbb{Z}^n} \widehat{\theta}_E(\xi-\eta)\ \sigma(x,\eta).$$

It is easy to see that $\sigma = a|_{\mathbb{T}^n \times \mathbb{Z}^n}$. Furthermore,

$$\begin{aligned}
\left|\partial_\xi^\alpha \partial_x^\beta a(x,\xi)\right| &= \left|\sum_{\eta\in\mathbb{Z}^n} \partial_\xi^\alpha \widehat{\theta}_E(\xi-\eta)\ \partial_x^\beta \sigma(x,\eta)\right| \\
&= \left|\sum_{\eta\in\mathbb{Z}^n} (\triangle_\xi^\alpha)^t \phi_\alpha(\xi-\eta)\ \partial_x^\beta \sigma(x,\eta)\right| \\
&= \left|\sum_{\eta\in\mathbb{Z}^n} \phi_\alpha(\xi-\eta)\ \triangle_\eta^\alpha \partial_x^\beta \sigma(x,\eta)\ (-1)^{|\alpha|}\right| \\
&\le \sum_{\eta\in\mathbb{Z}^n} |\phi_\alpha(\xi-\eta)|\ C_{\alpha\beta m}\ \langle\eta\rangle^{m-\rho|\alpha|+\delta|\beta|} \\
&\le C'_{\alpha\beta m}\ \langle\xi\rangle^{m-\rho|\alpha|+\delta|\beta|} \sum_{\eta\in\mathbb{Z}^n} |\phi_\alpha(\eta)|\ \langle\eta\rangle^{|m-\rho|\alpha|+\delta|\beta||} \\
&\le C''_{\alpha\beta m}\ \langle\xi\rangle^{m-\rho|\alpha|+\delta|\beta|}.
\end{aligned}$$

Thus $a \in S^m_{\rho,\delta}(\mathbb{R}^n \times \mathbb{R}^n)$.

7. Fourier series operator calculus

In this section we will describe composition formulae of Fourier series operators with pseudo-differential operators. They are similar to the global composition formulae in [11] and [13] in $\mathbb{R}^n$. However, the situation on the torus is technically much simpler since it does not require the global in space analysis of the corresponding remainders.

Theorem 7.1 (composition TP). *Let $T : \mathcal{D}(\mathbb{T}^n) \to \mathcal{D}'(\mathbb{T}^n)$ be defined by*

$$Tu(x) := \sum_{\xi\in\mathbb{Z}^n} \int_{\mathbb{T}^n} e^{i(\phi(x,\xi)-y\cdot\xi)}\ a(x,y,\xi)\ u(y)\ \tilde{d}y,$$

where the amplitude $a \in C^\infty(\mathbb{T}^n \times \mathbb{T}^n \times \mathbb{Z}^n)$ satisfies

$$\left|\partial_x^\alpha \partial_y^\beta a(x,y,\xi)\right| \le C_{\alpha\beta m}\ \langle\xi\rangle^m$$

for every $x, y \in \mathbb{T}^n$, $\xi \in \mathbb{Z}^n$ and $\alpha, \beta \in \mathbb{N}^n$; no restrictions for ϕ here. Let $p \in S^t(\mathbb{T}^n \times \mathbb{Z}^n)$. Then

$$TPu(x) = \sum_{\xi\in\mathbb{Z}^n} \int_{\mathbb{T}^n} e^{i(\phi(x,\xi)-z\cdot\xi)}\ c(x,z,\xi)\ u(z)\ \tilde{d}z,$$

where

$$c(x,z,\xi) = \sum_{\eta\in\mathbb{Z}^n} \int_{\mathbb{T}^n} e^{i(y-z)\cdot(\eta-\xi)}\ a(x,y,\xi)\ p(y,\eta)\ \tilde{d}y$$

satisfying

$$\left|\partial_x^\alpha \partial_z^\beta c(x,z,\xi)\right| \le C_{\alpha\beta m t}\ \langle\xi\rangle^{m+t}$$

for every $x, z \in \mathbb{T}^n$, $\xi \in \mathbb{Z}^n$ *and* $\alpha, \beta \in \mathbb{N}^n$. *Moreover,*

$$c(x,z,\xi) \sim \sum_{\alpha \in \ge 0} \frac{1}{\alpha!}\ \partial_y^{(\alpha)} \left[a(x,y,\xi)\ \triangle_\xi^\alpha p(y,\xi)\right]|_{y=z}.$$

Composition in the other direction is given by the following theorem.

Theorem 7.2 (composition PT**).** *Let* $T : \mathcal{D}(\mathbb{T}^n) \to \mathcal{D}'(\mathbb{T}^n)$ *such that*

$$Tu(x) := \sum_{\xi \in \mathbb{Z}^n} \int_{\mathbb{T}^n} \mathrm{e}^{\mathrm{i}(\phi(x,\xi) - y\cdot\xi)}\ a(x,y,\xi)\ u(y)\ \tilde{\mathrm{d}}y,$$

where $a \in C^\infty(\mathbb{T}^n \times \mathbb{T}^n \times \mathbb{Z}^n)$ *satisfying*

$$\left|\partial_x^\alpha \partial_y^\beta a(x,y,\xi)\right| \le C_{\alpha\beta m}\ \langle\xi\rangle^m$$

for every $x, y \in \mathbb{T}^n$, $\xi \in \mathbb{Z}^n$ *and* $\alpha, \beta \in \mathbb{N}^n$; *we assume that* $\phi \in C^\infty(\mathbb{T}^n \times \mathbb{Z}^n)$ *satisfies*

$$C^{-1}\ \langle\xi\rangle \le \langle\nabla_x \phi(x,\xi)\rangle \le C\ \langle\xi\rangle$$

for some C *for every* $x \in \mathbb{T}^n$, $\xi \in \mathbb{Z}^n$, *and that*

$$|\partial_x^\alpha \phi(x,\xi)| \le C_\alpha\ \langle\xi\rangle, \quad \left|\partial_x^\alpha \triangle_\xi^\beta \phi(x,\xi)\right| \le C_{\alpha\beta}$$

for every $x \in \mathbb{T}^n$, $\xi \in \mathbb{Z}^n$ *and* $\alpha, \beta \in \mathbb{N}^n \setminus \{0\}$. *Let* $p \in S^t(\mathbb{T}^n \times \mathbb{Z}^n)$. *Then*

$$p(x,D)Tu(x) = \sum_{\xi \in \mathbb{Z}^n} \int_{\mathbb{R}^n} \mathrm{e}^{\mathrm{i}(\phi(x,\xi) - z\cdot\xi)}\ c(x,z,\xi)\ u(z)\ \tilde{\mathrm{d}}z,$$

where

$$\left|\partial_x^\alpha \partial_z^\beta c(x,z,\xi)\right| \le C_{\alpha\beta}\ \langle\xi\rangle^{m+t}$$

for every $x, z \in \mathbb{T}^n$, $\xi \in \mathbb{Z}^n$ *and* $\alpha, \beta \in \mathbb{N}^n$. *Moreover,*

$$c(x,z,\xi) \sim \sum_{\alpha \ge 0} \frac{\mathrm{i}^{-|\alpha|}}{\alpha!}\ \partial_\eta^\alpha p(x,\eta)|_{\eta = \nabla_x \phi(x,\xi)}\ \partial_y^\alpha \left[\mathrm{e}^{\mathrm{i}\Psi(x,y,\xi)} a(y,z,\xi)\right]|_{y=x},$$

(*here we use a smooth extension for the symbol* $p(x,\eta)$) *where*

$$\Psi(x,y,\xi) := \phi(y,\xi) - \phi(x,\xi) + (x-y)\cdot\nabla_x\phi(x,\xi), \tag{7.1}$$

when $x \approx y$.

8. Applications to hyperbolic equations

Let $a(x, D) \in \Psi^m(\mathbb{R}^n)$ (with some properties to be specified). If u depends on x and t, we write

$$\begin{aligned} a(x,D)u(x,t) &= \int_{\mathbb{R}^n} a(x,\xi)\ \widehat{u}_E(\xi,t)\ \mathrm{e}^{\mathrm{i}x\cdot\xi}\ \mathrm{d}\xi \\ &= \int_{\mathbb{R}^n}\int_{\mathbb{R}^n} \mathrm{e}^{\mathrm{i}(x-y)\cdot\xi}\ a(x,\xi)\ u(y,t)\ \tilde{\mathrm{d}}y\ \mathrm{d}\xi. \end{aligned}$$

Let $u(\cdot, t) \in L^1(\mathbb{R}^n)$ $(0 < t < t_0)$ be a solution to the hyperbolic problem

$$\begin{cases} \mathrm{i}\frac{\partial}{\partial t}u(x,t) = a(x,D)u(x,t), \\ u(x,0) = f(x), \end{cases} \tag{8.1}$$

where $f \in L^1(\mathbb{R}^n)$ is compactly supported.

Assume now that $a(X, D) = a_1(X, D) + a_0(X, D)$ where $a_1(x, \xi)$ is periodic and $a_0(x, \xi)$ is compactly supported in x (assume even that $\mathrm{supp}(a_0(\cdot, \xi)) \subset [-\pi, \pi]^n$). Typically, we will want to have $a_1(x, \xi) = a_1(\xi)$ a constant coefficient operator, not necessarily smooth in ξ. Let us also assume that $\mathrm{supp}(f) \subset [-\pi, \pi]^n$.

We will now describe a way to periodise problem (8.1). According to Proposition 4.2, we can replace (8.1) by

$$\begin{cases} \mathrm{i}\frac{\partial}{\partial t}u(x,t) = (a_1(x,D) + (pa_0)(X,D))u(x,t) + Ru(x,t), \\ u(x,0) = f(x), \end{cases} \tag{8.2}$$

where the symbol $a_1 + pa_0$ is periodic and R is a smoothing operator. To study singularities of (8.1), it is sufficient to analyse the Cauchy problem

$$\begin{cases} \mathrm{i}\frac{\partial}{\partial t}v(x,t) = (a_1(x,D) + (pa_0)(X,D))v(x,t), \\ v(x,0) = f(x) \end{cases} \tag{8.3}$$

since by Duhamel's formula $\mathrm{WF}(u - v) = \emptyset$. This problem can be transferred to the torus. Let $w(x, t) = pv(\cdot, t)(x)$. In view of Proposition 4.1 it will solve the Cauchy problem

$$\begin{cases} \mathrm{i}\frac{\partial}{\partial t}w(x,t) = (\widetilde{a_1}(x,D) + \widetilde{pa_0}(X,D))w(x,t), \\ w(x,0) = pf(x). \end{cases} \tag{8.4}$$

Calculus constructed in previous sections provides the solution in the form

$$w(x,t) = \sum_{k\in\mathbb{Z}^n} \mathrm{e}^{\mathrm{i}\phi(t,x,k)} c(t,x,k) \widehat{f}_E(k).$$

Here we note that $\widehat{pf}_T(k) = \widehat{f}_E(k)$. Also, if the symbol $a_1(x, \xi) = a_1(\xi)$ has constant coefficients and a_0 is of order zero, we have $\phi(t, x, k) = x \cdot k + ta_1(k)$. In particular, $\nabla_x\phi(x, k) = k$, so composition formulas for $b(x, D)w$ in previous sections can be applied. Details of this analysis and investigation of the corresponding properties will appear in [14].

References

[1] M.S. Agranovich, Spectral properties of elliptic pseudodifferential operators on a closed curve, (*Russian*) *Funktsional. Anal. i Prilozhen.* **13** (1979), 54–56.

[2] M.S. Agranovich, Elliptic pseudodifferential operators on a closed curve, (*Russian*) *Trudy Moskov. Mat. Obshch.* **47** (1984), 22–67, 246.

[3] B.A. Amosov, On the theory of pseudodifferential operators on the circle, (*Russian*) *Uspekhi Mat. Nauk.* **43** (1988), 169–170; Translation in *Russian Math. Surveys* **43** (1988), 197–198.

[4] A. Boulkhemair, L^2 estimates for pseudodifferential operators, *Ann. Scuola Norm. Sup. Pisa Cl. Sci.* (4) **22** (1995), 155–183.

[5] J. Elschner, *Singular ordinary differential operators and pseudodifferential equations,* Springer-Verlag, Berlin, 1985.

[6] G. Garello and A. Morando, *L^p-boundedness for pseudodifferential operators with non-smooth symbols and applications, Boll. Unione Mat. Ital. Sez. B Artic. Ric. Mat.* (8) **8** (2005), 461–503.

[7] L. Hörmander, *The Analysis of Linear Partial Differential Operators IV,* Springer-Verlag, 1985.

[8] H. Kumano-go and M. Nagase, Pseudo-differential operators with non-regular symbols and applications, *Funkcial. Ekvac.* **21** (1978), 151–192.

[9] W. McLean, Local and global description of periodic pseudodifferential operators, *Math. Nachr.* **150** (1991), 151–161.

[10] S. Prössdorf and R. Schneider, Spline approximation methods for multidimensional periodic pseudodifferential equations, *Integral Equations Operator Theory* **15** (1992), 626–672.

[11] M. Ruzhansky and M. Sugimoto, Global calculus of Fourier integral operators, weighted estimates, and applications to global analysis of hyperbolic equations, in *Pseudo-differential Operators and Related Topics*, Editors: P. Boggiatto, L. Rodino, J. Toft and M.W. Wong, Birkhäuser, 2006, 65–78.

[12] M. Ruzhansky and M. Sugimoto, Global L^2 boundedness theorems for a class of Fourier integral operators, *Comm. Partial Differential Equations,* to appear.

[13] M. Ruzhansky and M. Sugimoto, Weighted L^2 estimates for a class of Fourier integral operators, *Preprint.*

[14] M. Ruzhansky and V. Turunen, Fourier integral operators on the torus, *in preparation.*

[15] J. Saranen and W.L. Wendland, The Fourier series representation of pseudodifferential operators on closed curves, Complex Variables **8** (1987), 55–64.

[16] M. Sugimoto, Pseudo-differential operators on Besov spaces, *Tsukuba J. Math.* **12** (1988), 43–63.

[17] V. Turunen, Commutator characterization of periodic pseudodifferential operators, *Z. Anal. Anw.* **19** (2000), 95–108.

[18] V. Turunen and G. Vainikko, On symbol analysis of periodic pseudodifferential operators, *Z. Anal. Anw.* **17** (1998), 9–22.

[19] G.M. Vainikko and I.K. Lifanov, Generalization and use of the theory of pseudodifferential operators in the modeling of some problems in mechanics, (*Russian*) *Dokl. Akad. Nauk.* **373** (2000), 157–160.

[20] G.M. Vainikko and I.K. Lifanov, The modeling of problems in aerodynamics and wave diffraction and the extension of Cauchy-type integral operators on closed and open curves, (*Russian*) *Differ. Uravn.* **36** (2000), 1184–1195, 1293; Translation in *Diff. Equ.* **36** (2000), 1310–1322.

Michael Ruzhansky
Department of Mathematics
Imperial College London
180 Queen's Gate
London SW7 2AZ, United Kingdom
e-mail: `m.ruzhansky@imperial.ac.uk`

Ville Turunen
Institute of Mathematics
Helsinki University of Technology
P.O. Box 1100
FIN-02015, Finland
e-mail: `ville.turunen@hut.fi`

Operator Theory:
Advances and Applications, Vol. 172, 107–115

Wave Kernels of the Twisted Laplacian

Jingzhi Tie and M.W. Wong

Abstract. We construct the wave kernels of the non-isotropic twisted Laplacian by means of its heat kernel. We then express the wave kernels at the origin in terms of complex Fourier integrals and we exploit the connections of the phase functions with the complex integrals.

Mathematics Subject Classification (2000). Primary 35A22, 35L05, 35S30; Secondary 47G30.

Keywords. Twisted Laplacian, wave kernels, finite propagation speed.

1. Introduction

Let $a = (a_1, a_2, \ldots, a_n)$, where $a_j > 0$, $j = 1, 2, \ldots, n$. Then the non-isotropic twisted Laplacian L_a associated to a on $\mathbb{C}^n = \mathbb{R}^{2n}$ is given by

$$L_a = -\frac{1}{2}\sum_{j=1}^{n}\left\{\left(\frac{\partial}{\partial x_j} + 2ia_jy_j\right)^2 + \left(\frac{\partial}{\partial y_j} - 2ia_jx_j\right)^2\right\}.$$

If $a_j = \frac{1}{2}$, $j = 1, 2, \ldots, n$, then L_a is the isotropic twisted Laplacian.

For $j = 1, 2, \ldots, n$, we let

$$Z_j = \frac{\partial}{\partial z_j} - a_j\overline{z_j} \qquad \text{and} \qquad \overline{Z_j} = \frac{\partial}{\partial \overline{z_j}} + a_jz_j,$$

where

$$\frac{\partial}{\partial z_j} = \frac{1}{2}\left(\frac{\partial}{\partial x_j} - i\frac{\partial}{\partial y_j}\right) \qquad \text{and} \qquad \frac{\partial}{\partial \overline{z_j}} = \frac{1}{2}\left(\frac{\partial}{\partial x_j} + i\frac{\partial}{\partial y_j}\right).$$

Then a straightforward calculation gives

$$L_a = -\sum_{j=1}^{n}(Z_j\overline{Z_j} + \overline{Z_j}Z_j).$$

This research has been supported by the Natural Sciences and Engineering Research Council of Canada.

When $n = 1$, the fundamental solution, also known as the Green function, and the heat kernel of the isotropic twisted Laplacian are derived from scratch in [9, 10, 11]. The underpinnings of these formulas are Weyl transforms, which can be found in [8]. Similar formulas for the non-isotropic twisted Laplacian for $n > 1$ using the heat kernel of the sub-Laplacian on the Heisenberg group can be found in [7]. We recall these formulas in the following theorems.

Theorem 1.1. *The Green function or the fundamental solution G of the non-isotropic twisted Laplacian L_a is given by*

$$G(z,\zeta) = \frac{1}{2\pi^n}\int_0^\infty \prod_{j=1}^n \left\{\frac{a_j}{\sinh(a_j s)}\right\} e^{-\sum_{j=1}^n a_j[|z_j-\zeta_j|^2\coth(a_j s)+\overline{z_j}\zeta_j - z_j\overline{\zeta_j}]}ds$$

for all z and ζ in $\mathbb{C}^n$, which means that

$$L_a\left(\int_{\mathbb{C}^n} G(\cdot,\zeta)f(\zeta)\,d\zeta\right) = f$$

for all Schwartz functions on $\mathbb{C}^n$.

Theorem 1.2. *The heat kernel H_t of the non-isotropic twisted Laplacian L_a is given by*

$$H_t(z,\zeta) = \frac{1}{\pi^n}\left\{\prod_{j=1}^n \frac{a_j}{\sinh(2a_j t)}\right\} e^{-\sum_{j=1}^n a_j[|z_j-\zeta_j|^2\coth(2a_j t)+\overline{z_j}\zeta_j - z_j\overline{\zeta_j}]} \tag{1.1}$$

for all z and ζ in $\mathbb{C}^n$, which means that H_t is the kernel of the strongly continuous one-parameter semigroup e^{-tL_a}, $t > 0$, generated by L_a.

Remark 1.3. The construction of the heat kernel of the sub-Laplacian on the Heisenberg group can be traced back to [2, 4] and expositions can be found in [1, 6]. The heat kernel of the isotropic twisted Laplacian can also be found on page 85 of [6].

Now, we look at the initial value problem for the wave equation governed by the non-isotropic twisted Laplacian, i.e.,

$$\begin{cases} \frac{\partial^2 u}{\partial t^2}(z,t) = (L_a u)(z,t), & z\in\mathbb{C}^n,\ t>0,\\ u(z,0) = f(z), & z\in\mathbb{C}^n,\\ \frac{\partial u}{\partial t}(z,0) = g(z), & z\in\mathbb{C}^n. \end{cases} \tag{1.2}$$

The unique solution u of the preceding initial value problem is given by

$$u(z,t) = (\cos(tL_a^{1/2})f)(z) + (L_a^{-1/2}\sin(tL_a^{1/2})g)(z), \quad z\in\mathbb{C}^n,\ t>0.$$

The aim of this paper is to give explicit formulas for the wave kernels of the initial value problem (1.2), i.e., the kernel W_t of the linear operator $L_a^{-1/2}\sin(tL_a^{1/2})$ and the kernel Ω_t of the linear operator $\cos(tL_a^{1/2})$ for positive time t. Since

$$\cos(tL_a^{1/2}) = \frac{\partial}{\partial t}\{L_a^{-1/2}\sin(tL_a^{1/2})\}, \quad t>0,$$

it is enough to look at the wave kernel W_t of the operator $L_a^{-1/2}\sin(tL_a^{1/2})$ for positive time t.

As a motivation for the explicit formulas for the wave kernel W_t of the non-isotropic twisted Laplacian L_a in this paper, we first give in Section 2 a recall of the finite propagation speed of the wave kernel W_t in [7]. The main result tells us that for all z and ζ in $\mathbb{C}^n$, $W_t(z,\zeta) = 0$ whenever $t < |z-\zeta|$. This phenomenon poses the interesting problem on finding explicit formulas for W_t for all positive numbers t. The formula for the wave kernel W_t, $t > 0$, in terms of the heat kernel is given in Section 3. In Section 4, we express the wave kernel W_t, $t > 0$, at the origin as a complex integral. The contour of the complex integral in Section 4 can be deformed as in [3] to another contour and this is carried out in Section 5.

2. Finite propagation speed

The starting point is the identity

$$e^{-\alpha} = \frac{1}{\sqrt{\pi}} \int_0^\infty e^{-1/u} u^{-3/2} e^{-\alpha^2/4} du,$$

which is valid for all real numbers α. Differentiating with respect to α, we get

$$\alpha^{-1} e^{-\alpha} = \frac{1}{2\sqrt{\pi}} \int_0^\infty e^{-1/u} u^{-1/2} e^{-u\alpha^2/4} du$$

for every real number α. Thus, we can obtain an integral representation of the linear operator $L_a^{-1/2} e^{-tL_a^{1/2}}$ given by

$$L_a^{-1/2} e^{-tL_a^{1/2}} = \frac{1}{\sqrt{\pi}} \int_0^\infty e^{-t^2/(4u)} u^{-1/2} e^{-uL_a} du, \quad t > 0.$$

So, the kernel P_t of the linear operator $L^{-1/2}e^{-tL^{1/2}}$ is given by

$$P_t(z,\zeta) = \frac{1}{\sqrt{\pi}} \int_0^\infty e^{-t^2/(4u)} u^{-1/2} H_u(z,\zeta)\, du$$

for all positive numbers t and all z and ζ in $\mathbb{C}^n$. Since

$$\sin u = \frac{e^{iu} - e^{-iu}}{2i},$$

it follows that a plausible way to get W_t is to complexify the positive time t to imaginary times it and $-it$. Now, for z and ζ in $\mathbb{C}^n$,

$$|H_u(z,\zeta)| = \frac{1}{\pi^n} \left\{ \prod_{j=1}^n \frac{a_j}{\sinh(2a_j u)} \right\} e^{-\sum_{j=1}^n a_j |z_j - \zeta_j|^2}.$$

Since $\sinh u$ behaves like $e^u/2$ and $\coth u$ is bounded for large and positive values of u, it follows that

$$\int_1^\infty e^{-\tau^2/(4u)}u^{-1/2}\left\{\prod_{j=1}^n \frac{a_j}{\sinh(2a_ju)}\right\}e^{-\sum_{j=1}^n a_j|z_j-\zeta_j|^2\coth(2a_ju)}du$$

is absolutely convergent and is an entire function of τ. For small and positive values of u, $\sinh u$ behaves like u and $\coth u$ behaves like $\frac{1}{u}$. Therefore there exists a positive constant C such that

$$\begin{aligned}\int_0^1 &\left| e^{-\tau^2/4u}u^{-1/2}\left\{\prod_{j=1}^n \frac{a_j}{\sinh(2a_ju)}\right\}e^{-\sum_{j=1}^n a_j|z_j-\zeta_j|^2\coth(2a_ju)}\right| du\\ &\le C\int_0^1 \left|e^{-\tau^2/(4u)}\right| u^{-1/2}u^{-n}e^{-\frac{1}{2u}\sum_{j=1}^n|z_j-\zeta_j|^2}du\\ &= C\int_1^\infty u^{n-(3/2)}e^{-\mathrm{Re}\,(\tau^2+2|z-\zeta|^2)u/4}du,\end{aligned}$$

which is convergent provided that

$$\mathrm{Re}\,(\tau^2+2|z-\zeta|^2)>0.$$

Letting $\tau=\pm it$, $t>0$, we see that the integral

$$\int_0^\infty e^{-(\pm it)^2/(4u)}u^{-1/2}H_u(z,\zeta)\,du$$

exists provided that

$$t<\sqrt{2}|z-\zeta|.$$

Thus, the wave kernel W_t is explicitly given by

$$W_t(z,\zeta)=\frac{P_{it}(z,\zeta)-P_{-it}(z,\zeta)}{2i}=0$$

for all z and ζ in $\mathbb{C}^n$ provided that

$$t<\sqrt{2}|z-\zeta|.$$

So, we have the following theorem on the finite propagation speed of the wave kernel.

Theorem 2.1. *For all z and ζ in $\mathbb{C}^n$ and all positive numbers t,*

$$W_t(z,\zeta)=0$$

whenever $t<|z-\zeta|$.

3. The wave kernel W_t

Theorem 3.1. *The wave kernel W_t of the linear operator $L_a^{-1/2}\sin(tL_a^{1/2})$ is given by*

$$W_t(z,\zeta) = \sqrt{\frac{\pi}{2}}\frac{\sqrt{t}}{2\pi i}\int_{c-i\infty}^{c+i\infty} e^{tu/2}H_{t/(2u)}(z,\zeta)u^{-3/2}du, \quad z,\zeta\in\mathbb{C}^n,\, t>0,$$

where c is a positive number, $H_{t/(2u)}$ is obtained by complexifying the time in the heat kernel given by (1.1), and for every nonzero complex number z, $\sqrt{z}$ is understood throughout this paper to be the principal branch of $\sqrt{z}$.

Proof. We begin with the well-known formula for the Bessel function of the first kind and order $\frac{1}{2}$ to the effect that

$$J_{1/2}(z) = \sqrt{\frac{2}{\pi}}\frac{\sin z}{\sqrt{z}} \tag{3.1}$$

for all nonzero complex numbers z. We also need Schläfli's integral for the Bessel function of the first kind and order ν telling us that

$$2\pi i J_\nu(\alpha z) = z^\nu \int_{c-i\infty}^{c+i\infty} e^{\alpha(u-(z^2/u))/2}u^{-\nu-1}du \tag{3.2}$$

for all z in $\mathbb{C}$ such that the formula is defined, where c, α and ν are positive numbers and the principal branch of z^ν is adopted. This can be obtained by a change of variable and a deformation of the contour in the integral (9.12) on page 58 in [5]. Using (3.1) and (3.2), we get for all positive numbers c and α,

$$\frac{\sin(\alpha z)}{z} = \sqrt{\frac{\pi}{2}}\frac{\sqrt{\alpha}}{2\pi i}\int_{c-i\infty}^{c+i\infty} e^{\alpha(u-(z^2/u))/2}u^{-3/2}du$$

for all nonzero complex numbers z. Thus, letting $z = L_a^{1/2}$, we get

$$L_a^{-1/2}\sin(tL_a^{1/2}) = \sqrt{\frac{\pi}{2}}\frac{\sqrt{t}}{2\pi i}\int_{c-i\infty}^{c+i\infty} e^{tu/2}e^{-(t/(2u))L_a}u^{-3/2}du, \quad t>0,$$

which gives the asserted formula. □

4. A complex Fourier integral

Let ν be the function on $(0,\infty)$ defined by

$$\nu(t) = \prod_{j=1}^n \frac{a_j}{\sinh(2a_jt)}, \quad t\in(0,\infty).$$

Let γ be the function on $(0,\infty)\times\mathbb{C}^n$ defined by

$$\gamma(t,z) = \sum_{j=1}^n a_j|z_j|^2\coth(2a_jt), \quad (t,z)\in(0,\infty)\times\mathbb{C}^n.$$

Then for $t > 0$, the heat kernel h_t at the origin is given by

$$h_t(z) = H_t(z,0) = \frac{1}{\pi^n}\nu(t)e^{-\gamma(t,z)}, \quad z \in \mathbb{C}^n.$$

For $t > 0$, we let w_t be the wave kernel at the origin defined by

$$w_t(z) = W_t(z,0), \quad z \in \mathbb{C}^n.$$

Theorem 4.1. *Let c be a positive number. Then for $t > 0$,*

$$w_t(z) = \frac{1}{2\pi^{(1/2)+n}i}\int_{\Gamma_t} e^{(t^2/(4w))-\gamma(w,z)}\nu(w)\frac{dw}{\sqrt{w}}, \quad z \in \mathbb{C}^n,$$

where Γ_t is the circle with center at $(t/(4c),0)$ and radius $t/(4c)$.

Proof. Using Theorem 3.1 and the heat kernel at the origin, we get for all positive numbers t and all z in $\mathbb{C}^n$,

$$w_t(z) = \sqrt{\frac{\pi}{2}}\frac{\sqrt{t}}{2\pi i}\int_{c-i\infty}^{c+i\infty} e^{tu/2}h_{t/(2u)}(z)u^{-3/2}du.$$

Letting $v = 2u/t$, we get for all t in $(0,\infty)$ and z in $\mathbb{C}^n$,

$$w_t(z) = \frac{1}{2\sqrt{\pi}i}\int_{(2c/t)-i\infty}^{(2c/t)+i\infty} e^{t^2v/4}h_{1/v}(z)v^{-3/2}dv.$$

Letting $w = 1/v$, the contour $\operatorname{Re} v = 2c/t$ is transformed into the circle Γ_t with center at $(t/(4c),0)$ and radius $t/(4c)$, and

$$w_t(z) = \frac{1}{2\sqrt{\pi}i}\int_{\Gamma_t} e^{t^2/(4w)}h_w(z)\frac{dw}{\sqrt{w}}, \quad t > 0,\ z \in \mathbb{C}^n,$$

and the proof is complete. □

5. The phase function

The phase function Φ of the wave kernel at the origin given by Theorem 4.1 is the function on $\mathbb{C}\backslash\{0\} \times \mathbb{C}^n \times (0,\infty)$ defined by

$$\Phi(w,z,t) = \frac{t^2}{4w} - \gamma(w,z) = \frac{t^2}{4w} - \sum_{j=1}^{n} a_j|z_j|^2\coth(2a_jw)$$

for all (w,z,t) in $\mathbb{C}\backslash\{0\} \times \mathbb{C}^n \times (0,\infty)$. In this section, we use the method in [3] to deform the contour Γ in Theorem 3.1 to a contour C, which is symmetric with respect to the real axis in the w-plane and is contained in the region

$$\{(w,z,t) \in \mathbb{C}\backslash\{0\} \times \mathbb{C}^n \times (0,\infty) : \operatorname{Re}(\Phi(w,z,t)) \le 0\}.$$

To that end, we begin with the task of finding the set of points (w,z,t) in $\mathbb{C}\backslash\{0\} \times \mathbb{C}^n \times (0,\infty)$ for which

$$\operatorname{Re}(\Phi(w,z,t)) = 0.$$

If we let $w = u + iv$ and use the formula

$$\coth(u+iv) = \frac{\sinh(2u) - i\sin(2v)}{2(\sinh^2 u + \sin^2 v)}$$

for all $w = u + iv$ such that the denominator does not vanish, then for all (w, z, t) in $\mathbb{C}\setminus\{0\} \times \mathbb{C}^n \times (0, \infty)$,

$$\Phi(w,z,t) = \frac{t^2(u-iv)}{4(u^2+v^2)} - \sum_{j=1}^{n} a_j |z_j|^2 \frac{\sinh(4a_j u) - i\sin(4a_j v)}{2[\sinh^2(2a_j u) + \sin^2(2a_j v)]}$$

and hence

$$\operatorname{Re}(\Phi(w,z,t)) = \frac{t^2 u}{4(u^2+v^2)} - \sum_{j=1}^{n} a_j |z_j|^2 \frac{\sinh(4a_j u)}{2[\sinh^2(2a_j u) + \sin^2(2a_j v)]}.$$

For a fixed point $(z, t) \in \mathbb{C}^n \times (0, \infty)$, let $\lambda = (\lambda_1, \lambda_2, \ldots, \lambda_n)$, where

$$\lambda_j = \frac{2a_j|z_j|^2}{t^2}, \quad j = 1, 2, \ldots, n,$$

and let Θ_λ be set of all points w in $\mathbb{C}\setminus\{0\}$ such that

$$\operatorname{Re}(\Phi(w,z,t)) = 0.$$

Then

$$\Theta_\lambda = \left\{ w = u + iv \in \mathbb{C}\setminus\{0\} : \frac{u}{u^2+v^2} = \sum_{j=1}^{n} \frac{\lambda_j \sinh(4a_j u)}{\sinh^2(2a_j u) + \sin^2(2a_j v)} \right\}.$$

It is obvious that Θ_λ is symmetric with respect to the u-axis and the v-axis, and it contains the v-axis. To find the intersection points of Θ_λ with the u-axis, let $v = 0$ in Θ_λ. Then

$$\frac{1}{u} = \sum_{j=1}^{n} \frac{\lambda_j \sinh(4a_j u)}{\sinh^2(2a_j u)}.$$

Thus,

$$\sum_{j=1}^{n} \frac{2\lambda_j u}{\tanh(2a_j u)} = 1$$

and the set of all its nonnegative solutions is denoted by

$$\left\{ u_\lambda \in [0, \infty) : \sum_{j=1}^{n} \frac{2\lambda_j u_\lambda}{\tanh(2a_j u_\lambda)} = 1 \right\}.$$

To find the intersection points of Θ_λ with the v-axis, we see that there is no information by simply letting $u = 0$ in Θ_λ. Instead, we write the equation in Θ_λ as

$$\frac{1}{u^2+v^2} = \sum_{j=1}^{n} \frac{\sinh(4a_j u)}{u} \frac{\lambda_j}{\sinh^2(2a_j u) + \sin^2(2a_j v)}$$

and take the limit as $u \to 0$. This gives

$$\frac{1}{v^2} = \sum_{j=1}^{n} \frac{4a_j\lambda_j}{\sin^2(2a_j v)}.$$

Hence

$$\sum_{j=1}^{n} \frac{\lambda_j}{a_j} \left[\frac{2a_j v}{\sin(2a_j v)}\right]^2 = 1$$

and the set of all its nonnegative solutions is denoted by

$$\left\{ v_\lambda \in [0, \infty) : \sum_{j=1}^{n} \frac{\lambda_j}{a_j} \left[\frac{2a_j v_\lambda}{\sin(2a_j v_\lambda)}\right\} = 1 \right\}.$$

It is clear that the poles of the integrand in w_t are given by the zeros of

$$\sinh(2a_j w) = 0, \quad j = 1, 2, \ldots, n,$$

i.e.,

$$w = \frac{im}{2a_j}\pi, \quad m \in \mathbb{Z}.$$

So, we can enumerate the poles on the positive v-axis as $0 < w_1 < w_2 < \cdots$. To

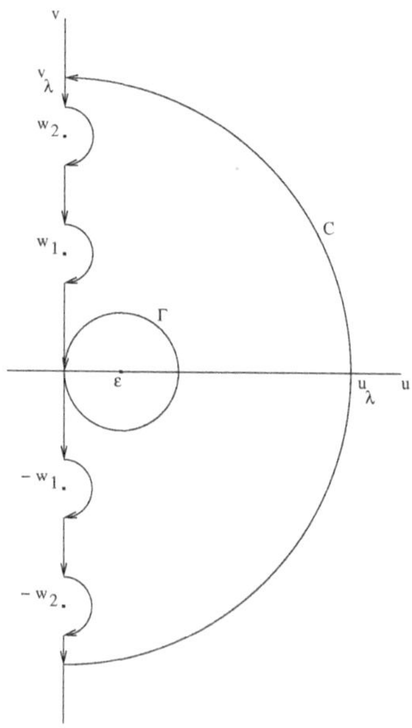

FIGURE 1. The contours Γ and C in the w-plane, where $w = u + iv$

get a contour C, we begin with the point u_λ closest to the origin, and join it with a path having a vertical tangent at (u_λ) to the point v_λ, which is closest to the origin. Then we go down along the v-axis to the origin in such a way that we make a small detour into the right half-plane when we come across a pole. The path is then symmetrized with respect to the u-axis. As an illustration by Figure 1, we assume that two poles w_1 and w_2 are met as we move down to the origin along the v-axis from above.

References

[1] C. Berenstein, D.-C. Chang and J. Tie, *Laguerre Calculus and its Applications on the Heisenberg Group*, AMS/International Press, 2001.

[2] B. Gaveau, Principe de moindre action, propagation de la chaleur et estimées sous elliptiques sur certains groupes nilpotents, *Acta Math.* **139** (1977), 95–153.

[3] P.C. Greiner, D. Holcman and Y. Kannai, Wave kernels related to second order operators, *Duke Math. J.* **114** (2002), 329–386.

[4] A. Hulanicki, The distribution of energy in the Brownian motion in the Gaussian field and analytic-hypoellipticity of certain subelliptic operators on the Heisenberg group, *Studia Math.* **56** (1976), 165–173.

[5] F.W.J. Olver, *Introduction to Asymptotics and Special Functions*, Academic Press, 1974.

[6] S. Thangavelu, *An Introduction to the Uncertainty Principle: Hardy's Theorem on Lie Groups*, Birkhäuser, 2004.

[7] J. Tie, The non-isotropic twisted Laplacian on $\mathbb{C}^n$ and the sub-Laplacian on $\mathbf{H}_n$, *Comm. Partial Differential Equations* **31** (2006), 1047–1069.

[8] M.W. Wong, Weyl Transforms, *Springer-Verlag*, 1998.

[9] M.W. Wong, The heat equation for the Hermite operator on the Heisenberg group, *Hokkaido Math. J.* **34** (2005), 394–404.

[10] M.W. Wong, Weyl transforms, the heat kernel and Green function of a degenerate elliptic operator, *Ann. Global Anal. Geom.* **28** (2005), 271–283.

[11] M.W. Wong, Weyl transforms and a degenerate elliptic partial differential equation, *Proc. Royal Soc. London A* **461** (2005), 3863–3870.

Jingzhi Tie
Department of Mathematics
University of Georgia
Athens, GA 30602-7403, USA
e-mail: `jtie@math.uga.edu`

M.W. Wong
Department of Mathematics and Statistics
York University
4700 Keele Street
Toronto, Ontario M3J 1P3, Canada
e-mail: `mwwong@mathstat.yorku.ca`

Operator Theory:
Advances and Applications, Vol. 172, 117–133

Super-exponential Decay of Solutions to Differential Equations in $\mathbb{R}^d$

Ernesto Buzano

Abstract. We study the exponential decay of solutions to differential equations of hypoelliptic type (see Definition 2.1). In particular we find sufficient conditions on the differential operator A in order for estimates of the kind

$$e^{\varepsilon\langle x\rangle^r} Au \in V \Longrightarrow e^{\varepsilon\langle x\rangle^r} u \in V$$

to hold, for several types of functions and distribution spaces V. For example, V may be $\mathscr{S}$ or $\mathscr{S}'$ or $\mathscr{C}_{\mathrm{b}}^{\infty}$.

Mathematics Subject Classification (2000). Primary 35B05, 35B40; Secondary 35H10, 35S05, 47F05, 47G30.

Keywords. Super-exponential decay, solutions and eigenfunctions of differential operators, pseudo-differential calculus, Λ-hypoellipticity.

1. Introduction

Exponential and super-exponential decay of solutions to differential equations in $\mathbb{R}^d$ have been studied by several people. See for example Agmon [1], Hislop and Sigal [6], Rabinovich [11], Rabier [8, 9] and Rabier and Stuart [10]. We also draw attention on the recent paper of Cappiello, Gramchev and Rodino [3] on super-exponential estimates of solutions to semilinear differential equations with polynomial coefficients.

Here we deal with Sobolev and pointwise super-exponential estimates of solutions to differential operators generalizing the harmonic oscillator. Our results are related to the recent work of Cappiello, Gramchev, Rodino [4, 5] on the exponential decay of solutions to differential equations with elliptic symbol in SG classes and Shubin's classes.

2. Symbol classes

A *weight-function* is a $\mathscr{C}^\infty$-function

$$\Lambda : \mathbb{R}^d \times \mathbb{R}^d \to \mathbb{R}$$

for which

- there exist $0 < \mu_0 \leqslant \mu$ such that [1]

$$\langle x, \xi \rangle^{\mu_0} \lesssim \Lambda(x, \xi) \lesssim \langle x, \xi \rangle^{\mu}, \qquad \text{for all } x,\ \xi \in \mathbb{R}^d; \tag{2.1}$$

- there exists $0 < \sigma \leqslant 1$ such that for all α, $\beta \in \mathbb{N}^d$ we have

$$\left| D_\xi^\alpha D_x^\beta \Lambda(x, \xi) \right| \lesssim \Lambda(x, \xi)^{1-\sigma|\alpha+\beta|}, \qquad \text{for all } x,\ \xi \in \mathbb{R}^d. \tag{2.2}$$

In (2.1) and (2.2) we employ the following notations:

$$\langle x, \xi \rangle = \left(1 + |x|^2 + |\xi|^2\right)^{1/2},$$

$$D_{\xi_j} = -i\frac{\partial}{\partial \xi_j}, \qquad D_{x_j} = -i\frac{\partial}{\partial x_j};$$

moreover

$$f(x) \lesssim g(x), \qquad \text{for all } x \in X,$$

means that there exists a constant $C > 0$ such that

$$f(x) \leqslant C g(x), \qquad \text{for all } x \in X.$$

Given a weight-function Λ, we define the following *symbol-classes*

$$S^m_{\rho,\delta}(\Lambda) = \left\{ a \in \mathscr{C}^\infty(\mathbb{R}^d \times \mathbb{R}^d \to \mathbb{C}) \ :\ \left| D_\xi^\alpha D_x^\beta a \right| \lesssim \Lambda^{m-\rho|\alpha|+\delta|\beta|} \right\},$$

with

$$-\rho \leqslant \delta < \rho \leqslant \sigma.$$

It is clear that

$$\Lambda \in S^1_{\sigma,-\sigma}(\Lambda).$$

When

$$\Lambda(x, \xi) = \langle x, \xi \rangle \quad \text{and} \quad \delta = -\rho,$$

$S^m_{\rho,\delta}(\Lambda)$ reduces to Shubin's symbol class Γ^m_ρ (see [12]).

Under these hypotheses a symbol $a \in S^m_{\rho,\delta}(\Lambda)$ defines a pseudo-differential operator

$$Au(x) = (2\pi)^{-d/2} \int e^{ix\cdot\xi} a(x, \xi) \hat{u}(\xi)\, \mathrm{d}\xi, \tag{2.3}$$

which is continuous on the Schwartz' class $\mathscr{S}$ and extends to a continuous operator (we still denote by A) on temperate distributions $\mathscr{S}'$. In (2.3) $\hat{u}$ is the Fourier Transform

$$\hat{u}(\xi) = (2\pi)^{-d/2} \int e^{-ix\cdot\xi} u(x)\, \mathrm{d}x.$$

---

[1] Actually one can prove that the second assumption in (2.1) is a consequence of assumption (2.2).

Definition 2.1. A symbol $a \in S^m_{\rho,\delta}(\Lambda)$ is Λ-*hypoelliptic* if there exists a constant $R \geqslant 0$ such that:

- for all $\alpha, \beta \in \mathbb{N}^d$ we have

$$\left|D_\xi^\alpha D_x^\beta a(x,\xi)\right| \lesssim |a(x,\xi)|\, \Lambda(x,\xi)^{-\rho|\alpha|+\delta|\beta|}, \qquad \text{for } \max\{|x|, |\xi|\} \geqslant R;$$

- there exists $m_0 \leqslant m$ such that

$$|a(x,\xi)| \gtrsim \Lambda(x,\xi)^{m_0}, \qquad \text{for } \max\{|x|, |\xi|\} \geqslant R.$$

When $m_0 = m$, the symbol is called Λ-*elliptic.*

We denote by $HS^{m,m_0}_{\rho,\delta}(\Lambda)$ the class of all Λ-hypoelliptic symbols belonging to $S^m_{\rho,\delta}(\Lambda)$ and satisfying the above estimates.

An operator with Λ-hypoelliptic symbol is called Λ-hypoelliptic.

In Theorems 2.2 to 2.5 we state some properties of pseudo-differential operators. The proofs are a straightforward modification of the corresponding statements in [2] and [12], and therefore are omitted.

Recall that the *formal adjoint* A^+ of a pseudo-differential operator A is defined by

$$\langle A^+u, \bar{v}\rangle = \langle u, \overline{Av}\rangle, \qquad \text{for all } u, v \in \mathscr{S}.$$

Theorem 2.2. *The formal adjoint and the composition of Λ-hypoelliptic operators are Λ-hypoelliptic. More precisely, if A and B are operators with symbols $a \in HS^{m,m_0}_{\rho,\delta}(\Lambda)$ and $b \in HS^{n,n_0}_{\rho,\delta}(\Lambda)$, we have that A^+ and BA have symbols $a^+ \in HS^{m,m_0}_{\rho,\delta}(\Lambda)$ and $b\#a \in HS^{n+m,n_0+m_0}_{\rho,\delta}(\Lambda)$.*

Theorem 2.3. *A Λ-hypoelliptic operator A with symbol in $HS^{m,m_0}_{\rho,\delta}(\Lambda)$ has a* parametrix, *in the sense that there exist operators Q with symbol in $HS^{-m_0,-m}_{\rho,\delta}(\Lambda)$ and K_1 and K_2 with symbols in $\mathscr{S}$ such that*

$$\begin{cases} QA = I + K_1, \\ AQ = I + K_2. \end{cases}$$

In particular we have the following regularity property

$$Au \in \mathscr{S} \implies u \in \mathscr{S}, \qquad \text{for all } u \in \mathscr{S}'.$$

Theorem 2.4. *Denote by $\bar{A}$ the closure in L^2 of a Λ-hypoelliptic operator A with symbol in $HS^{m,m_0}_{\rho,\delta}(\Lambda)$. Then the* maximal *and the* minimal closed extensions *of A to L^2 coincide, that is*

$$\mathscr{D}(\bar{A}) = \left\{u \in \mathscr{S}' \;:\; Au \in L^2\right\}.$$

Moreover the adjoint A^* is the closure of the formal adjoint A^+ and has domain

$$\mathscr{D}(A^*) = \left\{u \in \mathscr{S}' \;:\; A^+u \in L^2\right\}.$$

In particular $\bar{A}$ is self-adjoint iff A is formally self-adjoint.

Finally, the operator $\bar{A}$ is Fredholm when $m_0 \geqslant 0$.

Theorem 2.5. *Assume that A is a Λ-hypoelliptic operator with symbol $a \in HS^{m,m_0}_{\rho,\delta}$ with $m_0 \geqslant 0$. Then A is a Fredholm operator both on $\mathscr{S}$ and on $\mathscr{S}'$, in the sense that*

$$\ker A = \{u \in \mathscr{S} \ : \ Au = 0\} = \{u \in \mathscr{S}' \ : \ Au = 0\}$$

has finite dimension, and

$$A(\mathscr{S}) = \{u \in \mathscr{S} \ : \ \langle u, \bar{v}\rangle = 0, \text{ for all } v \in \ker A\},$$
$$A(\mathscr{S}') = \{u \in \mathscr{S}' \ : \ \langle u, \bar{v}\rangle = 0, \text{ for all } v \in \ker A\}.$$

In particular A is an isomorphism on $\mathscr{S}$ and $\mathscr{S}'$ iff $\ker A = 0$.

Denote by $\mathscr{C}^\infty_c$ the space of $\mathscr{C}^\infty$-functions with compact support.

Later on we need the following

Theorem 2.6. *Assume that A is a Λ-hypoelliptic formally self-adjoint operator, with symbol $a \in HS^{m,m_0}_{\rho,\delta}(\Lambda)$ and $m_0 \geqslant 0$. Then there exists a real-valued function $\phi \in \mathscr{C}^\infty_c$ such that $A + \phi$ is a pseudo-differential, formally self-adjoint isomorphism on $\mathscr{S}$ and $\mathscr{S}'$ with symbol $a + \phi \in HS^{m,m_0}_{\rho,\delta_+}(\Lambda)$, where $\delta_+ = \max\{\delta, 0\}$.*

Proof. Since $\phi \in S^0_{1,0}(\Lambda)$, we have $a + \phi \in HS^{m,m_0}_{\rho,\delta_+}(\Lambda)$.

Now we prove the existence of ϕ. It suffices to find ϕ such that $A + \phi$ is an isomorphism on $\mathscr{S}$. By Theorem 2.5 it suffices to find ϕ real-valued and such that $A + \phi$ is one-to-one. Thanks to Theorem 2.4, we have that $\bar{A}$ is Fredholm, because $m_0 \geqslant 0$, thus from Lemma 8 of [10] we have that there exists a real-valued function $\phi \in \mathscr{C}^\infty_c$ such that $\bar{A} + \phi$ is one-to-one. This in particular implies that also $A + \phi : \mathscr{S} \to \mathscr{S}$ is one-to-one and the proof is complete. □

3. Basic estimate

Now we state our main result on super-exponential estimates.

Theorem 3.1. *Consider a Λ-hypoelliptic* differential *operator A with symbol $a \in HS^{m,0}_{\rho,\delta}(\Lambda)$.*

Let $s \geqslant \rho\mu_0 > 0$ be such that

$$\langle x\rangle^s \lesssim \Lambda(x,\xi)^\rho, \qquad \text{for all } x,\, \xi \in \mathbb{R}^d, \tag{3.1}$$

where, as usual,

$$\langle x\rangle = \left(1 + |x|^2\right)^{1/2}.$$

Then there exists $\epsilon_0 > 0$ such that for all $0 < \epsilon \leqslant \epsilon_0$ and $0 < r \leqslant 1 + s$ we have

$$e^{\epsilon\langle x\rangle^r} Au \in \mathscr{S}' \implies e^{\epsilon\langle x\rangle^r} u \in \mathscr{S}', \qquad \text{for all } u \in \mathscr{S}'. \tag{3.2}$$

Proof. In order to prove this theorem we need two lemmas.

Lemma 3.2. *Given $\epsilon, r \in \mathbb{R}$ and $\alpha \in \mathbb{N}^d \setminus 0$, we have*

$$D^\alpha e^{-\epsilon\langle x\rangle^r} = e^{-\epsilon\langle x\rangle^r} \sum_{0<k\leqslant|\alpha|} \epsilon^k p_{\alpha,k}(x)\langle x\rangle^{kr-2|\alpha|}, \tag{3.3}$$

where $p_{\alpha,k}$ is a polynomial of degree less than or equal to $2|\alpha| - k$.

Proof. By induction on $|\alpha|$. We have

$$D_j e^{-\epsilon\langle x\rangle^r} = e^{-\epsilon\langle x\rangle^r} i\epsilon r\langle x\rangle^{r-2}x_j,$$

which takes care of the case $|\alpha| = 1$.

Now we differentiate both sides of (3.3):

$$\begin{aligned}
D^\beta e^{-\epsilon\langle x\rangle^r} &= D_j D^\alpha e^{-\epsilon\langle x\rangle^r} \\
&= e^{-\epsilon\langle x\rangle^r} i\epsilon r\langle x\rangle^{r-2}x_j \sum_{0<k\leqslant|\alpha|} \epsilon^k p_{\alpha,k}(x)\langle x\rangle^{kr-2|\alpha|} \\
&\quad + e^{-\epsilon\langle x\rangle^r} \sum_{0<k\leqslant|\alpha|} \epsilon^k \left[D_j p_{\alpha,k}(x)\langle x\rangle^{kr-2|\alpha|} - p_{\alpha,k}(x)i(kr-2|\alpha|)\langle x\rangle^{kr-2|\alpha|-2}x_j\right] \\
&= e^{-\epsilon\langle x\rangle^r} \sum_{0<k\leqslant|\alpha|} \epsilon^{k+1} ix_j p_{\alpha,k}(x)\langle x\rangle^{(k+1)r-2(|\alpha|+1)} \\
&\quad + e^{-\epsilon\langle x\rangle^r} \sum_{0<k\leqslant|\alpha|} \epsilon^k \left[\langle x\rangle^2 D_j p_{\alpha,k}(x) - i(kr-2|\alpha|)x_j p_{\alpha,k}(x)\right]\langle x\rangle^{kr-2(|\alpha|+1)} \\
&= e^{-\epsilon\langle x\rangle^r} \sum_{1<k\leqslant|\beta|} \epsilon^k ix_j p_{\alpha,k-1}(x)\langle x\rangle^{kr-2(|\beta|)} \\
&\quad + e^{-\epsilon\langle x\rangle^r} \sum_{0<k\leqslant|\beta|-1} \epsilon^k \left[\langle x\rangle^2 D_j p_{\alpha,k}(x) - i(kr-2|\alpha|)x_j p_{\alpha,k}(x)\right]\langle x\rangle^{kr-2(|\beta|)} \\
&= e^{-\epsilon\langle x\rangle^r} \sum_{0<k\leqslant|\beta|} \epsilon^k p_{\beta,k}(x)\langle x\rangle^{kr-2(|\beta|)},
\end{aligned}$$

with

$$\begin{aligned}
p_{\beta,1} &= \langle x\rangle^2 D_j p_{\alpha,1}(x) - i(r-2|\alpha|)x_j p_{\alpha,1}(x), \\
p_{\beta,k} &= ix_j p_{\alpha,k-1}(x) + \langle x\rangle^2 D_j p_{\alpha,k}(x) - i(kr-2|\alpha|)x_j p_{\alpha,k}(x),
\end{aligned}$$

for $1 < k < |\beta|$, and

$$p_{\beta,|\beta|} = ix_j p_{\alpha,|\alpha|}(x).$$

Eventually, one verifies that $\deg p_{\beta,k} \leqslant 2|\beta| - k$. □

Lemma 3.3. *Under the hypotheses of Theorem 2.3, the differential operator*

$$A_\epsilon u(x) = e^{\epsilon\langle x\rangle^r} A\left(e^{-\epsilon\langle x\rangle^r}u(x)\right)$$

has symbol $a_\epsilon \in S^m_{\rho,\delta}(\Lambda)$ for all $\epsilon \in \mathbb{R}$ and $0 < r \leqslant 1 + s$.

Moreover, we have

$$a_\epsilon(x,\xi) = a(x,\xi) + \sum_{0<k\leqslant n} \epsilon^k b_k(x,\xi), \tag{3.4}$$

where n is the degree of the polynomial $\xi \mapsto a(x,\xi)$, and for all α, $\beta \in \mathbb{N}^d$

$$\left|D_\xi^\alpha D_x^\beta b_k(x,\xi)\right| \lesssim |a(x,\xi)|\, \Lambda(x,\xi)^{-\rho|\alpha|+\delta|\beta|}, \qquad \text{for all } x,\ \xi \in \mathbb{R}^d. \tag{3.5}$$

In particular there exists $\epsilon_0 > 0$ such that a_ϵ is Λ-hypoelliptic and belongs to $HS^{m,m_0}_{\rho,\delta}(\Lambda)$ for all $|\epsilon| \leqslant \epsilon_0$ and $0 < r \leqslant 1+s$.

Proof. Recall Leibnitz' rule for differential operators: ([7], formula (1.1.10)):

$$A(uv) = \sum_{|\gamma|\leqslant n} \frac{(-1)^{|\gamma|}}{\gamma!} A^{(\gamma)} u D^\gamma v, \tag{3.6}$$

where $A^{(\gamma)}$ is the differential operator with symbol $D_\xi^\gamma a(x,\xi)$.

By using (3.6) one computes

$$A_\epsilon u(x) = e^{\epsilon\langle x\rangle^r} A\left(e^{-\epsilon\langle x\rangle^r} u(x)\right) = e^{\epsilon\langle x\rangle^r} \sum_{|\gamma|\leqslant n} \frac{(-1)^{|\gamma|}}{\gamma!} A^{(\gamma)} u(x) D^\gamma e^{-\epsilon\langle x\rangle^r},$$

and therefore, thanks to Lemma 3.2, we obtain

$$\begin{aligned} a_\epsilon(x,\xi) &= a(x,\xi) + \sum_{0<|\gamma|\leqslant n} \frac{(-1)^{|\gamma|}}{\gamma!} \sum_{0<k\leqslant|\gamma|} \epsilon^k p_{\gamma,k}(x)\langle x\rangle^{kr-2|\gamma|} D_\xi^\gamma a(x,\xi) \\ &= a(x,\xi) + \sum_{0<k\leqslant n} \epsilon^k b_k(x,\xi), \end{aligned}$$

with

$$b_k(x,\xi) = \sum_{k\leqslant|\gamma|\leqslant n} \frac{(-1)^{|\gamma|}}{\gamma!} p_{\gamma,k}(x)\langle x\rangle^{kr-2|\gamma|} D_\xi^\gamma a(x,\xi).$$

In order to prove (3.5), we observe that, for $k \leqslant |\gamma|$ and $0 < r \leqslant 1+s$, thanks to (3.1) for all $\beta \in \mathbb{N}^d$ we have

$$\left|D_x^\eta \left(p_{\gamma,k}(x)\langle x\rangle^{kr-2|\gamma|}\right)\right| \lesssim \langle x\rangle^{k(r-1)-|\eta|} \lesssim \langle x\rangle^{s|\gamma|-|\eta|} \lesssim \Lambda(x,\xi)^{\rho|\gamma|+\delta|\eta|},$$

for all $x, \xi \in \mathbb{R}^d$. It follows that

$$\begin{aligned} &\left|D_\xi^\alpha D_x^\beta b_k(x,\xi)\right| \leqslant \\ &\quad \leqslant \sum_{k\leqslant|\gamma|\leqslant n} \frac{(-1)^{|\gamma|}}{\gamma!} \sum_{\eta\leqslant\beta} \binom{\beta}{\eta} \left|D_x^\eta \left(p_{\gamma,k}(x)\langle x\rangle^{kr-2|\gamma|}\right)\right| \left|D_\xi^{\gamma+\alpha} D_x^{\beta-\eta} a(x,\xi)\right| \\ &\quad \lesssim |a(x,\xi)|\, \Lambda^{-\rho|\alpha|+\delta|\beta|}. \end{aligned}$$

Eventually, from (3.4) and (3.5) we obtain

$$|a_\epsilon(x,\xi)| - |a(x,\xi)| \gtrsim -\epsilon\, |a(x,\xi)|,$$

for $0 < \epsilon \leqslant 1$ and all x, ξ, which implies that a_ϵ is Λ-hypoelliptic when we choose ϵ small enough. □

Now we prove the theorem. Since A^+A is Λ-hypoelliptic with symbol $a^+\#a \in H^{2m,2m_0}_{\rho,\delta}(\Lambda)$, and formally self-adjoint, then from Theorem 2.6 there exists $\phi \in \mathscr{C}_c^\infty$ such that

$$B = A^+A + \phi$$

is an isomorphism on $\mathscr{S}$ and $\mathscr{S}'$.

Moreover one easily verifies that also

$$B_\epsilon = e^{\epsilon\langle x\rangle^r} B e^{-\epsilon\langle x\rangle^r} = (A^+A)_\epsilon + \phi$$

is one-to-one with formal adjoint given by [2]

$$B_{-\epsilon} = e^{-\epsilon\langle x\rangle^r} B e^{\epsilon\langle x\rangle^r}$$

which again is Λ-hypoelliptic and one-to-one. Then Theorem 2.5 implies that B_ϵ is an isomorphism on $\mathscr{S}'$. From Theorem 2.3 we have that B_ϵ^{-1} has symbol in $HS^{-2m_0,-2m}_{\rho,\delta}(\Lambda)$.

From $e^{\epsilon\langle x\rangle^r} Au(x) \in \mathscr{S}'$ we obtain

$$e^{\epsilon\langle x\rangle^r} A^+Au(x) = (A^+)_\epsilon e^{\epsilon\langle x\rangle^r} Au(x) \in \mathscr{S}',$$

and therefore

$$v = B_\epsilon^{-1}\left(e^{\epsilon\langle x\rangle^r} A^+Au + e^{\epsilon\langle x\rangle^r}\phi u\right) \in \mathscr{S}'.$$

On the other hand we have

$$e^{\epsilon\langle x\rangle^r} B e^{-\epsilon\langle x\rangle^r} v = B_\epsilon v = e^{\epsilon\langle x\rangle^r} A^+Au + e^{\epsilon\langle x\rangle^r}\phi u = e^{\epsilon\langle x\rangle^r} Bu,$$

which implies

$$u = e^{-\epsilon\langle x\rangle^r} v,$$

that is

$$e^{\epsilon\langle x\rangle^r} u = v \in \mathscr{S}'.$$

□

4. Pointwise estimates

Proposition 4.1. *Under the hypotheses of Theorem 2.6 we have that there exists $\epsilon_0 > 0$ such that for all $0 < \epsilon \leqslant \epsilon_0$ and $0 < r \leqslant 1 + s$ we have*

$$e^{\epsilon\langle x\rangle^r} Au \in \mathscr{S} \implies e^{\epsilon\langle x\rangle^r} u \in \mathscr{S}, \qquad \text{for all } u \in \mathscr{S}'. \tag{4.1}$$

Proof. From Theorem 3.1 we have that $e^{\epsilon\langle x\rangle^r} u \in \mathscr{S}'$. Since

$$A_\epsilon\left(e^{\epsilon\langle x\rangle^r} u\right) = e^{\epsilon\langle x\rangle^r} Au \in \mathscr{S}$$

and A_ϵ is Λ-hypoelliptic, by Theorem 2.3 it follows that $e^{\epsilon\langle x\rangle^r} u \in \mathscr{S}$. □

[2] As a matter of facts we have

$$\langle B_\epsilon u, \bar{v}\rangle = \langle e^{\epsilon\langle x\rangle^r} B e^{-\epsilon\langle x\rangle^r} u, \bar{v}\rangle = \langle u, \overline{e^{-\epsilon\langle x\rangle^r} B e^{\epsilon\langle x\rangle^r} v}\rangle = \langle u, \overline{B_{-\epsilon} v}\rangle,$$

for all u, $v \in \mathscr{C}_c^\infty$.

Corollary 4.2. *Under the hypotheses of Theorem 2.6, assume that $a \in HS^{m,m_0}_{\rho,\delta}(\Lambda)$ with $m_0 > 0$. Then there exists $\epsilon > 0$ such that*

$$e^{\epsilon \langle x \rangle^{1+s}} u \in \mathscr{S},$$

for all generalized eigenfunctions u of $A : \mathscr{S}' \to \mathscr{S}'$.

Proof. Assume that $u \in \ker(A - \lambda I)^n$, with $n \in \mathbb{Z}^+$. Then the result follows from previous proposition applied to $(A - \lambda I)^n$, which is Λ-hypoelliptic, since $a - \lambda \in HS^{m,m_0}_{\rho,\delta}(\Lambda)$ because $m_0 > 0$. □

The hypothesis that $|a| \gtrsim \Lambda^{m_0}$ with $m_0 > 0$ is essential to estimate the eigenfunctions. Consider for example the ordinary differential operator

$$A = \left(1 + x^2\right)^{-1} D^2 + 1$$

with symbol

$$a(x, \xi) = \frac{\xi^2}{(1 + x^2)} + 1.$$

We have that

$$a \in HS^{2,0}_{1,0}(\Lambda),$$

where

$$\Lambda(x, \xi) = \langle x, \xi \rangle.$$

In fact, for all $|\alpha| + |\beta| > 0$ we have

$$\begin{aligned} \left|D^\alpha_\xi D^\beta_x a(x,\xi)\right| &= D^\alpha_\xi \xi^2 D^\beta_x \left(1 + x^2\right)^{-1} \lesssim \left(1 + x^2 + \xi^2\right)^{1-\alpha/2} \left(1 + x^2\right)^{-1-\beta/2} \\ &\leqslant \frac{\left(1 + x^2 + \xi^2\right)^{1-\alpha/2}}{1 + x^2} = a(x,\xi)\Lambda(x\,\xi)^{-\alpha}, \qquad \text{for all } x,\, \xi \in \mathbb{R}^d. \end{aligned}$$

Then the solutions to $Au = f$ satisfy super-exponential estimate (4.1) with $0 < r \leqslant 2$, but the eigenfunctions corresponding to the eigenvalue $\lambda = 1$ are polynomials of degree 1 and therefore do not decay exponentially.

Denote by $\mathscr{C}^\infty_b$ the space of functions $u \in \mathscr{C}^\infty$, such that $D^\alpha u \in L^\infty$ for all $\alpha \in \mathbb{N}^d$.

Proposition 4.3. *Consider a Λ-hypoelliptic* differential *operator A with symbol $a \in HS^{m,0}_{\rho,0}(\Lambda)$.*

Let $s \geqslant \rho\mu_0 > 0$ be such that

$$\langle x \rangle^s \lesssim \Lambda(x, \xi)^\rho, \qquad \text{for all } x,\, \xi \in \mathbb{R}^d.$$

Then there exists $\epsilon_0 > 0$ such that for all $0 < \epsilon \leqslant \epsilon_0$ and $0 < r \leqslant 1 + s$ we have

$$e^{\epsilon \langle x \rangle^r} Au \in \mathscr{C}^\infty_b \implies e^{\epsilon \langle x \rangle^r} u \in \mathscr{C}^\infty_b, \qquad \text{for all } u \in \mathscr{S}'. \tag{4.2}$$

Proof. First one proves the following simple

Lemma 4.4. *Given a pseudo-differential operator Q with symbol $q \in S^0_{\rho,0}(\Lambda)$, we have that $Q(\mathscr{C}_b^\infty) \subset \mathscr{C}_b^\infty$.* [3]

Proof. Let $\chi : \mathbb{R} \to \mathbb{R}^+$ be a smooth function such that

$$\chi(t) = \begin{cases} 1, & \text{if } |t| \leqslant 1, \\ 0, & \text{if } |t| \geqslant 2 \end{cases}$$

and write

$$\begin{aligned} Qu(x) &= (2\pi)^{-n} \int e^{iy\cdot\xi} q(x,\xi) u(x-y)\,\mathrm{d}y\mathrm{d}\xi \\ &= (2\pi)^{-n} \int e^{iy\cdot\xi} \chi(|y|) q(x,\xi) u(x-y)\,\mathrm{d}y\mathrm{d}\xi \\ &\quad + (2\pi)^{-n} \int e^{iy\cdot\xi} \big(1-\chi(|y|)\big) q(x,\xi) u(x-y)\,\mathrm{d}y\mathrm{d}\xi \end{aligned}$$

From

$$\begin{aligned} &D_x^\alpha \int e^{iy\cdot\xi} \chi(|y|) q(x,\xi) u(x-y)\,\mathrm{d}y\mathrm{d}\xi \\ &\quad = \sum_{\beta \leqslant \alpha} \binom{\alpha}{\beta} \int e^{iy\cdot\xi} \chi(|y|) D_x^\beta q(x,\xi) D_x^{\alpha-\beta} u(x-y)\,\mathrm{d}y\mathrm{d}\xi \\ &\quad = \sum_{\beta \leqslant \alpha} \binom{\alpha}{\beta} \int e^{iy\cdot\xi} \langle\xi\rangle^{-2N} D_x^\beta q(x,\xi)\,(1-\Delta_y)^N \big(\chi(|y|) D_x^{\alpha-\beta} u(x-y)\big)\mathrm{d}y\mathrm{d}\xi, \end{aligned}$$

we obtain that

$$\begin{aligned} &\left| D_x^\alpha \int e^{iy\cdot\xi} \chi(|y|) q(x,\xi) u(x-y)\,\mathrm{d}y\mathrm{d}\xi \right| \lesssim \\ &\quad \leqslant \sum_{\beta \leqslant \alpha} \binom{\alpha}{\beta} \int \langle\xi\rangle^{-2N} \left| D_x^\beta q(x,\xi) \right| \left| (1-\Delta_y)^N \big(\chi(|y|) D_x^{\alpha-\beta} u(x-y)\big) \right| \mathrm{d}y\mathrm{d}\xi \\ &\quad \lesssim \int \langle\xi\rangle^{-2N} d\xi \int\limits_{|y|\leqslant 2} \mathrm{d}y < \infty. \end{aligned}$$

While, from

$$\begin{aligned} &D_x^\alpha \int e^{iy\cdot\xi} \big(1-\chi(|y|)\big) q(x,\xi) u(x-y)\,\mathrm{d}y\mathrm{d}\xi \\ &\quad = \sum_{\beta \leqslant \alpha} \binom{\alpha}{\beta} 2 \int e^{iy\cdot\xi} \big(1-\chi(|y|)\big) D_x^\beta q(x,\xi) D_x^{\alpha-\beta} u(x-y)\,\mathrm{d}y\mathrm{d}\xi \\ &\quad = \sum_{\beta \leqslant \alpha} \binom{\alpha}{\beta} \int e^{iy\cdot\xi} |y|^{-2N} (-\Delta_\xi)^N D_x^\beta q(x,\xi) \big(1-\chi(|y|)\big) D_x^{\alpha-\beta} u(x-y)\mathrm{d}y\mathrm{d}\xi, \end{aligned}$$

[3] Actually Q is continuous on $\mathscr{C}_b^\infty$.

we obtain that

$$\left| D_x^\alpha \int e^{iy\cdot\xi}(1-\chi(|y|))q(x,\xi)u(x-y)\,\mathrm{d}y\mathrm{d}\xi \right| \lesssim$$

$$\leqslant \sum_{\beta\leqslant\alpha} \binom{\alpha}{\beta} \int |y|^{-2N} \left|(-\Delta_\xi)^N D_x^\beta q(x,\xi)\right| \left|(1-\chi(|y|)) D_x^{\alpha-\beta} u(x-y)\right| \mathrm{d}y\mathrm{d}\xi$$

$$\lesssim \int |y|^{-2N} \Lambda(x,\xi)^{-2N\rho} \mathrm{d}y\mathrm{d}\xi \leqslant \int \langle\xi\rangle^{-2N\rho\mu_0} d\xi \int_{|y|\geqslant 2} |y|^{-2N}\, \mathrm{d}y < \infty. \qquad \square$$

Now we prove the proposition. From Lemma 3.3 and Theorem 2.2 we have that for $0 < \epsilon \leqslant \epsilon_0$ the operator $A_\epsilon = e^{\epsilon\langle x\rangle^r} A e^{-\epsilon\langle x\rangle^r}$ has a parametrix Q_ϵ with symbol in $S^0_{\rho,0}(\Lambda)$. This in particular means that $K_\epsilon = I - Q_\epsilon A_\epsilon$ is regularizing. Assume now that $e^{\epsilon\langle x\rangle^r} Au(x) \in \mathscr{C}_b^\infty$. Since $\mathscr{C}_b^\infty \subset \mathscr{S}'$, from the basic estimate we have that $e^{\epsilon\langle x\rangle^r} u(x) \in \mathscr{S}'$. Thus from the above lemma we obtain

$$\begin{aligned} e^{\epsilon\langle x\rangle^r} u(x) &= Q_\epsilon A_\epsilon \left(e^{\epsilon\langle x\rangle^r} u(x)\right) + K_\epsilon \left(e^{\epsilon\langle x\rangle^r} u(x)\right) \\ &= Q_\epsilon \left(e^{\epsilon\langle x\rangle^r} Au(x)\right) + K_\epsilon \left(e^{\epsilon\langle x\rangle^r} u(x)\right) \in \mathscr{C}_b^\infty. \qquad \square \end{aligned}$$

In order to handle the case $\delta > 0$, we introduce the space $\mathscr{C}_t^\infty$ of $\mathscr{C}^\infty$-functions u such that for all $\alpha \in \mathbb{N}^d$ there exists $\kappa_\alpha \in \mathbb{R}$ such that

$$D^\alpha u(x) \lesssim \langle x\rangle^{\kappa_\alpha}, \qquad \text{for all } x \in \mathbb{R}^d.$$

Then we have the following

Proposition 4.5. *Consider a Λ-hypoelliptic* differential *operator A with symbol $a \in HS^{m,0}_{\rho,\delta}(\Lambda)$.*

Let $s \geqslant \rho\mu_0 > 0$ be such that

$$\langle x\rangle^s \lesssim \Lambda(x,\xi)^\rho, \qquad \text{for all } x,\, \xi \in \mathbb{R}^d.$$

Then there exists $\epsilon_0 > 0$ such that for all $0 < \epsilon \leqslant \epsilon_0$ and $0 < r \leqslant 1+s$ we have

$$e^{\epsilon\langle x\rangle^r} Au \in \mathscr{C}_t^\infty \implies e^{\epsilon\langle x\rangle^r} u \in \mathscr{C}_t^\infty, \qquad \text{for all } u \in \mathscr{S}'. \tag{4.3}$$

Proof. The proof is the same as that one of Proposition 4.3, with the following lemma instead of Lemma 4.4:

Lemma 4.6. *Given a pseudo-differential operator Q with symbol $q \in S^0_{\rho,\delta}(\Lambda)$, we have that $Q(\mathscr{C}_t^\infty) \subset \mathscr{C}_t^\infty$.* $\square$

5. Estimates in Sobolev spaces

Now we introduce a scale of Sobolev-like spaces, in order to give further estimates. For all $s \in \mathbb{R}$, we define the operator:

$$W_s u(x) = \int \Lambda(y,\eta)^s \left(\Phi_{y,\eta}, u\right)_{L^2} \Phi_{y,\eta}(x)\, \mathrm{d}y\mathrm{d}\eta,$$

where

$$\Phi_{y,\eta}(x) = \pi^{-n/4} e^{ix\cdot\eta} e^{-\frac{1}{2}|y-x|^2}.$$

W_s is the *anti-Wick quantization of* Λ^s.

The following proposition is a variant of Theorem 8.2 of [2] and therefore its proof is omitted (see [2] and [12] for details on anti-Wick operators).

Proposition 5.1. *W_s is a formally self-adjoint Λ-hypoelliptic pseudo-differential operator with symbol in $HS^{s,s}_{\sigma,-\sigma}(\Lambda)$.*

W_s is an isomorphism on $\mathscr{S}$ and $\mathscr{S}'$; its inverse W_s^{-1} is a Λ-hypoelliptic pseudo-differential operator with symbol in $HS^{-s,-s}_{\sigma,-\sigma}(\Lambda)$.

For each $s \in \mathbb{R}$ we introduce the following *Sobolev-like* spaces:

$$H^s_\Lambda = \left\{u \in \mathscr{S}' \ : \ W_s u \in L^2\right\},$$
$$\|u\|_{H^s_\Lambda} = \|W_s u\|_{L^2}.$$

Observe that W_0 is the identity and $H^0_\Lambda = L^2$. Moreover, when $\Lambda(x,\xi) = \langle x,\xi\rangle$, H^s_Λ reduce to Shubin's space Q^s (see [12]).

The proofs of Propositions 5.2 to 5.5 are standard and therefore are omitted (see [2] and [12]).

Proposition 5.2. *For all $s \in \mathbb{R}$, H^s_Λ is a Hilbert space. The bilinear form*

$$\begin{cases} \mathscr{S} \times \mathscr{S} \to \mathbb{C}, \\ (u,v) \mapsto \langle u,v\rangle = \displaystyle\int uv \,\mathrm{d}x \end{cases}$$

extends to a continuous bilinear mapping

$$H^s_\Lambda \times H^{-s}_\Lambda \to \mathbb{C},$$

which we shall continue to denote by $\langle\cdot,\cdot\rangle$.

The spaces H^s_Λ and H^{-s}_Λ are dual to each other with respect to this pairing.

Proposition 5.3. *We have the compact inclusion $H^s_\Lambda \subset H^t_\Lambda$ whenever $s < t$.*

Moreover

$$\bigcap_{s\in\mathbb{R}} H^s_\Lambda = \mathscr{S}, \qquad \bigcup_{s\in\mathbb{R}} H^s_\Lambda = \mathscr{S}',$$

where the topologies of $\mathscr{S}$ and $\mathscr{S}'$ coincide with the initial and the final one, respectively.

Proposition 5.4. *A pseudo-differential operator A with symbol in $S^m_{\rho,\delta}(\Lambda)$ extends to a continuous operator from H^s_Λ to H^{s-m}_Λ.*

Proposition 5.5. *Given an operator A with Λ-hypoelliptic symbol $a \in HS^{m,m_0}_{\rho,\delta}(\Lambda)$, we have the following regularity property:*

$$Au \in H^s_\Lambda \implies u \in H^{s+m_0}_\Lambda, \qquad \text{for all } u \in \mathscr{S}'.$$

Proposition 5.6. *Consider a Λ-hypoelliptic differential operator A with symbol $a \in HS^{m,m_0}_{\rho,\delta}(\Lambda)$ and $m_0 \geqslant 0$.*

Let $s \geqslant \rho\mu_0 > 0$ be such that

$$\langle x\rangle^s \lesssim \Lambda(x,\xi)^\rho, \qquad \text{for all } x,\, \xi \in \mathbb{R}^d.$$

Then there exists $\epsilon_0 > 0$ such that for all $t \in \mathbb{R}$, $0 < \epsilon \leqslant \epsilon_0$ and $0 < r \leqslant 1+s$ we have

$$e^{\epsilon\langle x\rangle^r} Au \in H^t_\Lambda \implies e^{\epsilon\langle x\rangle^r} u \in H^{t+m_0}_\Lambda, \qquad \text{for all } u \in \mathscr{S}'. \tag{5.1}$$

Proof. It follows from Theorem 3.1 and Proposition 5.5. □

6. Examples

Before considering some examples, we give the following

Definition 6.1. We say that the solutions to a Λ-hypoelliptic differential equation

$$Au = f,$$

with symbol in $HS^{m,m_0}_{\rho,\delta}$, satisfy *super-exponential estimates of type $s > 0$* if they satisfy estimates (3.2), (4.1), (4.3), and (5.1), for $\delta < \rho$ and (4.2) when $\delta \leqslant 0$.

As a first example we consider *operators of Schrödinger type*, that is operators satisfying the assumptions of Proposition 6.3.

Proposition 6.2. *Let $p(\xi)$ be a hypoelliptic polynomial such that*

$$\begin{aligned} \left|D^\alpha_\xi p(\xi)\right| &\lesssim |p(\xi)|^{1-\omega|\alpha|}, &&\text{for } |\xi| \geqslant R, \\ \langle\xi\rangle^{m_0} &\lesssim |p(\xi)| \lesssim \langle\xi\rangle^m, &&\text{for } |\xi| \geqslant R, \end{aligned} \tag{6.1}$$

with $0 < \omega \leqslant 1/m$, $0 < m_0 \leqslant m = \deg p$ and $R \geqslant 0$.

Then for all $l,\, k \in \mathbb{Z}^+$

$$\Lambda(x,\xi) = \left(1 + |p(\xi)|^{2l} + |x|^{2k}\right)^{\frac{1}{2l}} \tag{6.2}$$

is a weight function such that for all $\alpha, \beta \in \mathbb{N}^d$ we have

$$\left|D^\alpha_\xi D^\beta_x \Lambda(x,\xi)\right| \lesssim \Lambda(x,\xi)^{1-\sigma|\alpha+\beta|}, \qquad \text{for all } x, \xi \in \mathbb{R}^d,$$

and

$$\langle x,\xi\rangle^{\mu_0} \lesssim \Lambda(x,\xi) \lesssim \langle x,\xi\rangle^{\mu},$$

for all $x, \xi \in \mathbb{R}^d$, where

$$\sigma = \min\left\{\omega, \frac{l}{k}\right\}. \tag{6.3}$$

$$\mu_0 = \min\left\{m_0, \frac{k}{l}\right\} \quad \text{and} \quad \mu = \max\left\{m, \frac{k}{l}\right\}.$$

Proof. Since $\Lambda = \left(\Lambda^{2l}\right)^{1/(2l)}$, $D_\xi^\alpha |p|^{2l} = 0$ for $|\alpha| > 2lm$ and $D_x^\beta |x|^{2k} = 0$ for $|\beta| > 2k$, it suffices to show that

$$\left|D_\xi^\alpha D_x^\beta \Lambda(x,\xi)^{2l}\right| \lesssim \Lambda(x,\xi)^{2l-\omega|\alpha+\beta|}, \qquad \text{for all } x,\xi \in \mathbb{R}^d,$$

for $0 < |\alpha| \leqslant 2lm$ and $|\beta| = 0$, and for $|\alpha| = 0$ and $0 < |\beta| \leqslant 2k$.

We have

$$\left|D_\xi^\alpha \left(\Lambda(x,\xi)^{2l}\right)\right| = \left|D_\xi^\alpha \left(p(\xi)^l \bar{p}(\xi)^l\right)\right| = \sum_{\beta\leqslant\alpha} \binom{\alpha}{\beta} \left|D_\xi^\beta \left(p(\xi)^l\right)\right| \left|D_\xi^{\alpha-\beta} \left(\bar{p}(\xi)^l\right)\right|$$

$$\lesssim \begin{cases} |p(\xi)|^{2l-\omega|\alpha|} \leqslant \Lambda(x,\xi)^{2l-\omega|\alpha|}, & \text{for } |\xi| \geqslant R, \\ 1 \leqslant \Lambda(x,\xi)^{2l-\omega|\alpha|}, & \text{for } |\xi| < R, \end{cases}$$

and

$$\left|D_x^\beta \left(\Lambda(x,\xi)^{2l}\right)\right| = D_x^\beta \left(|x|^{2k}\right) \lesssim \left(1 + |x|^{2k}\right)^{1-\frac{|\beta|}{2k}}$$

$$\leqslant \Lambda(x,\xi)^{2l-\frac{l}{k}|\beta|}, \qquad \text{for all } x,\xi \in \mathbb{R}^d,$$

because

$$2l - \omega|\alpha| \geqslant 0, \qquad \text{for } |\alpha| \leqslant 2lm,$$

and

$$2l - \frac{l}{k}|\beta| \geqslant 0, \qquad \text{for } |\beta| \leqslant 2k.$$

□

Proposition 6.3. *Consider a differential operator $A = p(D) + q(x)$, where $p(\xi)$ is a hypoelliptic polynomial satisfying the estimates of Proposition 6.2 and $q(x)$ is a $\mathscr{C}^\infty$ function such that*

$$\left|D_x^\beta q(x)\right| \lesssim |q(x)| \langle x\rangle^{\delta|\beta|}, \qquad \text{for } |x| \geqslant R, \tag{6.4}$$

$$\langle x\rangle^{n_0} \lesssim |q(x)| \lesssim \langle x\rangle^{n}, \qquad \text{for } |x| \geqslant R, \tag{6.5}$$

with $0 \leqslant \delta < \omega \leqslant \frac{1}{m}$, $0 < m_0 \leqslant m = \deg p$, $0 < n_0 \leqslant n$, and $R \geqslant 0$.

Assume that

$$|p(\xi)| + |q(x)| \lesssim |p(\xi) + q(x)|, \tag{6.6}$$

for $\max\{|x|, |\xi|\} \geqslant R$, then, if

$$\delta < n_0\omega,$$

we have that

$$p + q \in HS_{\sigma,\delta}^{ln/k,1}(\Lambda)$$

where Λ and σ are defined in (6.2) and (6.3) and $l, k \in \mathbb{Z}^+$ satisfy

$$\frac{k}{l} \leqslant n_0.$$

Remark 6.4. It follows from this proposition that the solutions to the equation

$$p(D)u + qu = f$$

satisfy super-exponential estimates of type

$$s = k\sigma/l = \min\{k\omega/l, 1\} \leqslant \min\{n_0\omega, 1\}.$$

In particular there exists $\epsilon > 0$ such that the eigenfunctions of $p(D) + q$ belong to $e^{-\epsilon\langle x\rangle^{1+k\sigma/l}}\mathscr{S}$.

Proof. Since $D_\xi^\alpha p = 0$ for $|\alpha| > m$, thanks to (6.1), (6.2), (6.4), (6.5) and (6.6) it suffices to observe that

1. $\Lambda(x,\xi) \lesssim |p(\xi)| + \langle x\rangle^{k/l} \lesssim |p(\xi) + q(x)| \lesssim |p(\xi)| + \langle x\rangle^n \lesssim \Lambda(x,\xi)^{ln/k}$,
2. for $0 < |\alpha| \leqslant m$ we have

$$\begin{aligned} D_\xi^\alpha p(\xi) &\lesssim |p(\xi)|^{1-\omega|\alpha|} \lesssim \left(1 + |p(\xi)|^{2l} + |x|^{2k}\right)^{\frac{1-\omega|\alpha|}{2l}} \\ &\lesssim \big(|p(\xi)| + |q(x)|\big)\Lambda(x,\xi)^{-\omega|\alpha|}, \qquad \text{for } |\xi| \geqslant R, \end{aligned}$$

$$\begin{aligned} D_\xi^\alpha p(\xi) &\lesssim 1 \lesssim \left(1 + |p(\xi)|^{2l} + |x|^{2k}\right)^{\frac{1-\omega|\alpha|}{2l}} \\ &\lesssim \big(|p(\xi)| + |q(x)|\big)\Lambda(x,\xi)^{-\omega|\alpha|}, \qquad \text{for } |\xi| \leqslant R \text{ and } |x| \geqslant R, \end{aligned}$$

3. for $|\beta| > 0$ we have

$$\left|D_x^\beta q(x)\right| \lesssim |q(x)|\,\langle x\rangle^{\delta|\beta|} \leqslant \big(|p(\xi)| + |q(x)|\big)\Lambda(x,\xi)^{\delta|\beta|}, \qquad \text{for } |x| \geqslant R$$

and

$$\left|D_x^\beta q(x)\right| \lesssim 1 \leqslant \langle\xi\rangle^{m_0} \lesssim \big(|p(\xi)| + |q(x)|\big)\Lambda(x,\xi)^{\delta|\beta|}, \quad \text{for } |x| \leqslant R \text{ and } |\xi| \geqslant R. \qquad \square$$

For example, the differential operator

$$D_{x_1}^{2m_1} + D_{x_2}^{2m_2} + x_1^{2n_1} + x_2^{2n_2},$$

with $m_1, m_2, n_1, n_2 \in \mathbb{Z}^+$, has symbol

$$p(\xi) + q(x) = \xi_1^{2m_1} + \xi_2^{2m_2} + x_1^{2n_1} + x_2^{2n_2},$$

where for all $\alpha \in \mathbb{N}^2$ we have

$$|D^\alpha p(\xi)| \lesssim |p(\xi)|^{1-\omega|\alpha|}, \qquad \text{for } |\xi| \geqslant 1,$$

with

$$\omega = \frac{1}{2m}, \qquad m = \max\{m_1, m_2\},$$

and for all $\beta \in \mathbb{N}^2$ we have

$$\left|D_x^\beta q(x)\right| \lesssim |q(x)|, \qquad \text{for } |x| \geqslant 1.$$

It follows that

$$p + q \in H_{\sigma,0}^{n/n_0,1}(\Lambda),$$

with

$$\Lambda(x,\xi)=\left(1+\left(\xi_1^{2m_1}+\xi_2^{2m_2}\right)^2+\left(x_1^2+x_2^2\right)^{2n_0}\right)^{1/2},$$

$$n_0=\min\{n_1,n_2\},\qquad n=\max\{n_1,n_2\},$$

$$\sigma=\min\left\{\frac{1}{2m},\frac{1}{2n_0}\right\}.$$

This means that the solutions to the differential equation

$$D_{x_1}^{2m_1}u_1+D_{x_2}^{2m_2}u_2+x_1^{2n_1}+x_2^{2n_2}=f,$$

satisfy super-exponential estimates with

$$s=2n_0\sigma=\min\left\{\frac{n_0}{m},1\right\}.$$

In particular there exists $\epsilon>0$ such that all the eigenfunctions of $p(D)+q(x)$ belong to $e^{-\epsilon\langle x\rangle^{1+s}}\mathscr{S}$.

As a second example we describe the special class of *globally multi-quasi-elliptic differential operators*. We begin with some geometric preliminaries, see [2] for details and proofs.

A convex polyhedron $\mathscr{P}\subset\mathbb{R}^k$ is the convex hull of a finite set of points in $\mathbb{R}^k$. One can show that $\mathscr{P}$ can be obtained as the convex hull of a finite subset $V(\mathscr{P})\subset\mathbb{R}^k$ of convex-linearly independent points, called the *vertices of* $\mathscr{P}$ and univocally determined by $\mathscr{P}$. Moreover there exists a finite set $N(\mathscr{P})=N_0(\mathscr{P})\cup N_1(\mathscr{P})\subset\mathbb{R}^k$ such that

$$|\nu|=1,\qquad\text{for all }\nu\in N_0(\mathscr{P}),$$

and

$$\mathscr{P}=\left\{z\in\mathbb{R}^k\,:\,\nu\cdot z\geqslant 0,\,\forall\,\nu\in N_0(\mathscr{P})\right\}\cap\left\{z\in\mathbb{R}^k\,:\,\nu\cdot z\leqslant 1,\,\forall\,\nu\in N_1(\mathscr{P})\right\}.$$

$N_0(\mathscr{P})$ and $N_1(\mathscr{P})$ are univocally determined by $\mathscr{P}$, if $\mathscr{P}$ has non-empty interior.

Definition 6.5. A *complete polyhedron* is a convex polyhedron with non-empty interior and such that

1. $(0,\dots,0)\in V(\mathscr{P})\subset\mathbb{N}^k$;
2. $N_0(\mathscr{P})=\{(0,\dots,\underbrace{1}_{j\text{-entry}},\dots,0)\in\mathbb{R}^k\,:\,1\leqslant j\leqslant k\}$;
3. $N_1(\mathscr{P})\subset(\mathbb{R}^+)^k$.

Proposition 6.6. *Given a complete polyhedron $\mathscr{P}\subset\mathbb{R}^d\times\mathbb{R}^d$, we have that*

$$\Lambda_{\mathscr{P}}(x,\xi)=\left\{\sum_{(\alpha,\beta)\in V(\mathscr{P})}x^{2\alpha}\xi^{2\beta}\right\}^{1/2}$$

is a weight function such that

$$\langle x,\xi\rangle^{\mu_0}\lesssim\Lambda_{\mathscr{P}}(x,\xi)\lesssim\langle x,\xi\rangle^{\mu},$$

and

$$\left|D_\xi^\alpha D_x^\beta \Lambda_{\mathscr{P}}(x,\xi)\right| \lesssim \Lambda_{\mathscr{P}}(x,\xi)^{1-\sigma|\alpha+\beta|},$$

for all $x, \xi \in \mathbb{R}^d$, *with*

$$\sigma = \inf\left\{\nu_j \ : \ j = 1,\ldots,d \text{ and } \nu \in N_1(\mathscr{P})\right\},$$

and

$$\mu_0 = \min_{(\alpha,\beta)\in V(\mathscr{P})\setminus 0} |\alpha+\beta|,$$

$$\mu = \max_{(\alpha,\beta)\in V(\mathscr{P})\setminus 0} |\alpha+\beta|.$$

Proof. See [2], Chapter I, Proposition 1.1. □

A differential operator A with symbol

$$a(x,\xi) = \sum_{\alpha\in\mathscr{A}} a_\alpha(x)\xi^\alpha$$

is *globally multi-quasi-elliptic* if there exists a complete polyhedron $\mathscr{P}$ such that

$$a \in HS_{\sigma,\delta}^{m,m}(\Lambda_{\mathscr{P}}),$$

with $-\sigma \leqslant \delta < \sigma$.

If the set $N_1(\mathscr{P})$ has only one element ν, the operator is called *globally quasi-elliptic*. If furthermore $\nu_1 = \nu_2 = \cdots = \nu_{2d}$, the operator is called *globally elliptic*.

Proposition 6.7. *The solutions to a globally multi-quasi-elliptic differential equation* $Au = f$ *with symbol* $a \in HS_{\sigma,\delta}^{m,m}(\Lambda_{\mathscr{P}})$, *satisfy super-exponential estimates of type*

$$s = \sigma \min\left\{|\beta| \ : \ (0,\beta) \in V(\mathscr{P}) \setminus 0\right\}.$$

Proof. It follows from Proposition 6.6 and Theorem 3.1. □

References

[1] S. Agmon, *Lectures on Exponential Decay of Solutions of Second-Order Elliptic Equations: Bounds on Eigenfunctions of N-Body Schrödinger Operators*, Princeton University Press, Princeton, NJ, 1982.

[2] P. Boggiatto, E. Buzano and L. Rodino, *Global Hypoellipticity and Spectral Theory*, Akademie Verlag, Berlin, 1996.

[3] M. Cappiello, T. Gramchev and L. Rodino, Super-exponential decay and holomorphic extensions for semilinear equations with polynomial coefficients, *J. Funct. Anal.* **237** (2006), no. 2, 634–654.

[4] M. Cappiello, T. Gramchev and L. Rodino Exponential decay and regularity for SG-elliptic operators with polynomial coefficients, to appear.

[5] M. Cappiello, T. Gramchev and L. Rodino, Gelfand-Shilov spaces, pseudo-differential operators and localization operators, to appear.

[6] P.D. Hislop and I.M. Sigal, *Introduction to Spectral Theory with Applications to Schrödinger Operators*, Springer-Verlag, New York, 1996.

[7] L. Hörmander, *The Analysis of Linear Partial Differential Operators I*, Springer-Verlag, Berlin, 1983.

[8] P.J. Rabier, Fredholm operators, semigroups and the asymptotic and boundary behavior of solutions of PDEs, *J. Diff. Equations* **193** (2003), 460–480.

[9] P.J. Rabier, Asymptotic behavior of the solutions of linear and quasilinear elliptic equations on $\mathbb{R}^N$, *Trans. Amer. Math. Soc.* **356** (2004), 1889–1907.

[10] P.J. Rabier and C.A. Stuart, Fredholm properties of Schrödinger operators in $L^p(\mathbb{R}^N)$, *Differential Integral Equations* **13** (2000), 1429–1444.

[11] V.S. Rabinovich, Exponential estimates for eigenfunctions of Schrödinger operators with rapidly increasing and discontinuous potentials, Complex analysis and dynamical systems, *Contemp. Math.* **364**, Amer. Math. Soc., Providence, RI, 2004, 225–236.

[12] M.A. Shubin, *Pseudodifferential Operators and Spectral Theory*, Springer-Verlag, Berlin, 1987.

Ernesto Buzano
Dipartimento di Matematica
Università di Torino
Via Carlo Alberto 10
I-10123 Torino, Italy
e-mail: ernesto.buzano@unito.it

Operator Theory:
Advances and Applications, Vol. 172, 135–151

Gevrey Local Solvability for Degenerate Parabolic Operators of Higher Order

Alessandro Oliaro and Petar Popivanov

Abstract. In this paper we study the local solvability in Gevrey classes for degenerate parabolic operators of order ≥ 2. We assume that the lower order term vanishes at a suitably smaller rate with respect to the principal part; we then analyze its influence on the behavior of the operator, proving local solvability in Gevrey spaces G^s for small s, and local nonsolvability in G^s for large s.

Mathematics Subject Classification (2000). Primary 35K65.

Keywords. Degenerate parabolic operators, Gevrey classes, local solvability, non local solvability.

1. Introduction

In this paper we study the local solvability of the following degenerate parabolic partial differential operator:

$$P(t, \partial_t, D_x) = \partial_t + at^p D_x^{2k} - bt^q D_x^k, \tag{1.1}$$

where a and b are positive constants, $k \in \mathbb{Z}_+$ and p, q are integers, $p, q \geq 0$. Degenerate parabolic operators have been studied by several authors. The first observation is that they may be locally nonsolvable in C^∞, in fact in 1971 Kannai [8] considered the second order equation

$$\partial_t u + t \sum_{j=1}^{n} \partial_{x_j}^2 u = f, \tag{1.2}$$

proving that for some datum $f \in C^\infty$ it does not admit (locally near the origin) any solution $u \in \mathcal{D}'$. Colombini, Pernazza and Treves [4] studied a class that generalizes (1.2), finding, for the C^∞ local solvability, a necessary and sufficient condition of the kind of the (Ψ) of Nirenberg and Treves for operators of principal type. Results on local nonsolvability in $\mathcal{D}'$ for degenerate parabolic equations can be found in

Popivanov [15], where also the degenerate elliptic and hyperbolic cases are treated. Considering in particular the operator (1.1), it was studied in the simple case $k = 1$ by Rubinstein [18], who proved non local solvability in C^∞ under the condition $p > 2q + 1$; this last requirement is crucial, in fact for $p \leq 2q + 1$, p even, (1.1) is C^∞-locally solvable, as we can deduce by a general result in Helffer [7], who proved C^∞ hypoellipticity for a wide class of equations. In this frame let us mention also Matsuzawa [10], [11] where, for degenerate parabolic second order operators, C^∞ hypoellipticity and correctness of the Cauchy problem is treated; similar equations are studied in Oleĭnik and Radkevič [12], from the point of view of initial-boundary value problems and asymptotic behavior of the solutions. Interesting results on operators of the kind of (1.1) can be found in Popivanov [14], where the case $k = 2$ is analyzed, proving hypoellipticity, positive and negative solvability results in the frame of $\mathcal{D}'$; in the same paper other equations are also treated, in particular necessary and sufficient conditions for C^∞-local solvability are given for a class of two space variable degenerate parabolic operators, for which a subelliptic estimate is also proved.

The results that we have recalled till now, in particular the negative ones, give rise to the questions if we can prove Gevrey G^s-local solvability for (1.1) in the cases when C^∞-local solvability fails, and for which s the operator (1.1) is G^s-locally solvable and for which it is not. For degenerate parabolic equations there are several open problems in Gevrey spaces: in Bove and Tartakoff [1], [2] Gevrey hypoellipticity is studied, but only for degenerate elliptic equations, and the same problem is treated in Rodino [17], for degenerate quasi-elliptic. The parabolic case has been studied for example in Gramchev, Popivanov and Yoshino [6], where positive and negative results about G^s hypoellipticity are showed; in particular it follows from the results in this paper that, in the case $k = 1$, if $p > 2q+1$ then (1.1) is not G^s hypoelliptic for any $s \geq 1$: this leaves open the question if (1.1) is or not G^s-locally solvable, and this is the frame where we work in the present paper. We consider the operator (1.1), for a generic integer k, in the case $p > 2q + 1$, proving its G^s-local solvability for small s and its G^s-local nonsolvability for large s.

In order to precisely state our results let us recall some definitions, starting from Gevrey spaces. Given $s > 1$, an open set $\Omega \subset \mathbb{R}^n$, a relatively compact set $K \subset\subset \Omega$ and a positive constant C, the space $G_0^s(\Omega, K, C)$ is the set of all the functions $f \in C_0^\infty(\Omega)$ with support contained in K and such that the following norm is finite:

$$\|f\|_{s,K,C} := \sup_{\alpha \in \mathbb{Z}_+^n} \left(C^{-|\alpha|} (\alpha!)^{-s} \|\partial^\alpha f\|_{L^\infty(K)} \right). \tag{1.3}$$

Similarly we write $G^s(\Omega, K, C)$, $s \geq 1$, for the space of all the functions $f \in C^\infty(\Omega)$ for which the norm (1.3) of the restriction of f to K is finite. We then define

$$\begin{aligned} G_0^s(\Omega) &= \operatorname*{ind\,lim}_{K \nearrow \Omega\ C \nearrow +\infty} G_0^s(\Omega, K, C), \\ G^s(\Omega) &= \operatorname*{proj\,lim}_{K \nearrow \Omega} \operatorname*{ind\,lim}_{C \nearrow +\infty} G^s(\Omega, K, C); \end{aligned}$$

the space $G_0^s(\Omega)$, $s > 1$, turns out to be the set of all the functions $f \in G^s(\Omega)$ with compact support contained in Ω. We may then define $\mathcal{D}'_s(\Omega)$ and $\mathcal{E}'_s(\Omega)$ as topological duals of $G_0^s(\Omega)$, $G^s(\Omega)$ respectively.

Definition 1.1. *We say that a partial differential operator P with coefficients belonging to $G^s(\Omega)$ is G^s-locally solvable at a point $x_0 \in \Omega$ if there exists a neighborhood U of x_0 such that for every function $f \in G_0^s(U)$ there is a solution $u \in \mathcal{D}'_s(U)$ of the equation $Pu = f$.*

Since for every $t < s$ we have $G^t \subset G^s$ and $\mathcal{D}'_s \subset \mathcal{D}'_t$, we deduce immediately that if P is G^s-locally solvable then it is also G^t-locally solvable for every $t < s$. Therefore, when an operator is not locally solvable in C^∞ the natural question is to investigate its local solvability in Gevrey classes, looking for the largest s up to which it remains G^s-locally solvable. The main result of this paper is the following.

Theorem 1.2. *Let us consider the operator* (1.1), *where we suppose that $p > 2q+1$.*

(i) *If p is even and $\frac{p-2q-1}{p-q} < \frac{1}{k}$, then $P(t, \partial_t, D_x)$ is G^s-locally solvable at $(0,0)$ for*

$$s < \frac{p-q}{k(p-2q-1)}; \tag{1.4}$$

(ii) *if $\frac{p-2q-1}{p-q} < \frac{1}{2k}$ we have that $P(t, \partial_t, D_x)$ is not G^s-locally solvable at $(0,0)$ for*

$$s > 2\frac{p-q}{p-2q-1}. \tag{1.5}$$

(ii)$'$ *if $\frac{1}{2k} \leq \frac{p-2q-1}{p-q} < \frac{1}{k}$, then $P(t, \partial_t, D_x)$ is not G^s-locally solvable at $(0,0)$ for*

$$s > 2\frac{k(p-q)}{(1-k)p-(1-2k)q+k}. \tag{1.6}$$

Regarding the techniques, we shall prove the point (i) by constructing explicitly a solution, following the idea of Trèves [19], cf. also Calvo and Popivanov [3]; the points (ii) and (ii)$'$ are proved by violating the well-known necessary condition of Corli, following a construction that was used in Popivanov [16] in the C^∞ frame and then in Oliaro [13] in Gevrey.

2. Local solvability in Gevrey classes

In this Section we want to prove the local solvability result of Theorem 1.2. We shall proceed as in Tréves [19], by constructing an explicit solution $u(t,x)$ of the equation

$$P(t, \partial_t, D_x)u = f \tag{2.1}$$

for an arbitrary datum $f(t,x) \in G_0^s(\Omega)$, with s satisfying (1.4). At first, a partial Fourier transform of (2.1) gives us

$$\partial_t \hat{u}(\xi, t) + \left(at^p \xi^{2k} - bt^q \xi^k\right) \hat{u}(\xi, t) = \hat{f}(\xi, t), \tag{2.2}$$

where $\hat{u}(\xi,t) = \mathcal{F}_{x\to\xi}\big[u(x,t)\big]$, and similarly $\hat{f}(\xi,t) = \mathcal{F}_{x\to\xi}\big[f(x,t)\big]$. Now writing

$$B(\xi,t,t') = -\int_{t'}^{t} \left(as^p\xi^{2k} - bs^q\xi^k\right) ds$$

we have that the following function is a solution of (2.2):

$$\hat{u}(\xi,t) = \int_{-T}^{t} e^{B(\xi,t,t')}\hat{f}(\xi,t')\,dt', \tag{2.3}$$

where $T > 0$ is fixed. We can then write (formally) a solution of the equation (2.1) in the following way:

$$u(x,t) = \frac{1}{2\pi}\int_{-\infty}^{+\infty} e^{ix\xi}\hat{u}(\xi,t)\,d\xi. \tag{2.4}$$

We shall show that in our hypotheses the function (2.4) is well defined; in particular, we prove now that $\hat{u}(\xi,t)$ is exponentially decreasing at infinity in the ξ-variable.

Lemma 2.1. *For every $t' \in [-T,t]$ we have that*

$$B(\xi,t,t') \leq C|\xi|^{k\frac{p-2q-1}{p-q}}, \tag{2.5}$$

C being a positive constant independent of t, t' and ξ.

Proof. We observe at first that

$$B(\xi,t,t') = g(\xi,t) - g(\xi,t'), \tag{2.6}$$

where

$$g(\xi,r) = -a\frac{r^{p+1}}{p+1}\xi^{2k} + b\frac{r^{q+1}}{q+1}\xi^k.$$

In order to estimate (2.6) we need to analyze the behavior of $g(\xi,r)$, that depends on q and k.

(i) if q is even and k is odd, the function $g(\xi,r)$ has the following form:

In Figure 1 we have set $r_1 = -r_0 = \left[\frac{b(p+1)}{a(q+1)}\xi^{-k}\right]^{\frac{1}{p-q}}$, $r_M = -r_m = \left(\frac{b}{a}\xi^{-k}\right)^{\frac{1}{p-q}}$ and

$$g_M = -g_m = A|\xi|^{k\frac{p-2q-1}{p-q}}, \tag{2.7}$$

where $A = \frac{b}{q+1}\left(\frac{b}{a}\right)^{\frac{q+1}{p-q}} - \frac{a}{p+1}\left(\frac{b}{a}\right)^{\frac{p+1}{p-q}}$ is a positive constant. Now, since $t' \leq t$ and taking into account the particular form of $B(\xi,t,t')$, cf. (2.6), we can prove the following estimates (we assume that $T \geq r_1$, that is the less restrictive case):

- in the case $\xi < 0$, since the function $r \to g(\xi,r)$ is decreasing we have that for every $t \in [-T,T]$ and $t' \in [-T,t]$, $g(\xi,t) \leq g(\xi,t')$, and so by (2.6) we obtain $B(\xi,t,t') \leq 0$; the same arguments show that this last inequality holds when $\xi > 0$, $t' \in [-T,r_0]$, $t \in [-T,0] \cup [r_1,T]$;

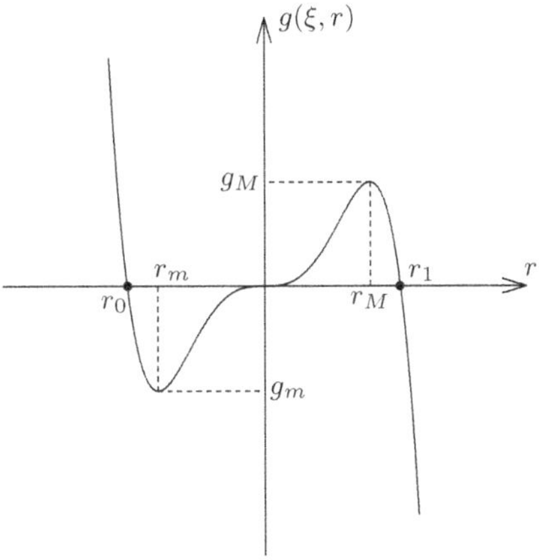

Fig. 1: $g(\xi,r)$ for fixed $\xi>0$

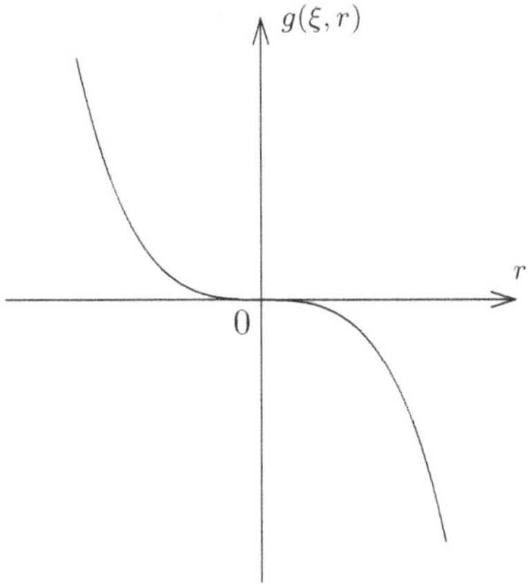

Fig. 2: $g(\xi,r)$ for fixed $\xi<0$

- if $\xi > 0$ and t,t' satisfy $t' \in [-T, r_0]$, $t \in [0, r_1]$, or $t' \in [r_0, 0]$, $t \in [r_1, T]$, or $t, t' \in [0, T]$, $t' \leq t$, we have that $B(\xi, t, t') \leq g_M$, and then by (2.7) the estimate (2.5) is satisfied with $C = A$;
- in the remaining case $t' \in [r_0, 0]$, $t \in [r_0, r_1]$, $t' \leq t$ we have $B(\xi, t, t') \leq g_M - g_m = 2g_M$, which means as in the previous point that (2.5) is satisfied with $C = 2A$;

for q even and k odd the proof of Lemma 2.1 is then complete, with $C = 2A$ in (2.5). Now we want to analyze the other cases:

(ii) if q and k are even, then the function $g(\xi, r)$ behaves as in Fig. 1 for both $\xi > 0$ and $\xi < 0$, and then the same proof as in (i) gives (2.5) with $C = 2A$;

(iii) let us suppose now that q and k are odd. Then one can easily see that in the case $\xi > 0$ we have:

- if $t \leq 0$, since $t' \leq t$ by hypothesis and $r \to g(\xi, r)$ is decreasing for $r \leq 0$ we immediately have by (2.6) that $B(\xi, t, t') \leq 0$;
- in the case $t > 0$, $t' \in [-T, t]$ we obtain $B(\xi, t, t') \leq g_M$, and then by (2.7) we get (2.5) with $C = A$.

If $\xi < 0$ we can proceed in the same way, since the behavior of $g(\xi, r)$ is similar as in the case $\xi > 0$. Summarizing, for both q and k odd we have proved (2.5) with $C = A$;

(iv) the last case that we have to consider is when q is odd and k is even: since in this case the function $g(\xi, r)$ is as in Fig. 3 for both $\xi > 0$ and $\xi < 0$ we can use the same arguments as in (iii). The proof is then complete. □

We can now prove the solvability part of Theorem 1.2.

Proof of Theorem 1.2, (i). We have already written (at least formally) a solution of (2.1) in the form (2.4); now we are interested in giving sense to the formal solution (2.4). We observe at first that, since $f \in G_0^s(\Omega)$ we have that there exist positive constants D and ϵ satisfying

$$\sup_r |\hat{f}(\xi, r)| \le De^{-\epsilon|\xi|^{\frac{1}{s}}}, \tag{2.8}$$

for every $\xi \in \mathbb{R}$. We then have from (2.3), (2.8) and Lemma 2.1 that

$$|\hat{u}(\xi, t)| \le D \int_{-T}^{T} e^{C|\xi|^{k\frac{p-2q-1}{p-q}}} e^{-\epsilon|\xi|^{\frac{1}{s}}} \, dt' = 2DTe^{C|\xi|^{k\frac{p-2q-1}{p-q}}} e^{-\epsilon|\xi|^{\frac{1}{s}}};$$

since by hypothesis $k\frac{p-2q-1}{p-q} < \frac{1}{s}$ we then have that there exist constants $C', \mu > 0$ satisfying

$$|\hat{u}(\xi, t)| \le C'e^{-\mu|\xi|^{\frac{1}{s}}}$$

for all t, and this last estimate shows that (2.4) is a well-defined solution of the linear equation (2.1). □

Remark 2.2. *The same construction as above gives us C^∞ local solvability in the case $p \le 2q + 1$, p even; recall, however, that this result was already known, cf. Helffer* [7], *as we have pointed out in the Introduction.*

3. Non local solvability in Gevrey classes

The aim of this section is to prove Theorem 1.2, (ii)–(ii)′. We recall at first the following known property of the Gevrey seminorms (1.3) (a proof can be found in Corli [5]).

Lemma 3.1. *Let us fix a compact set K and two constants $\eta > \epsilon > 0$. There exists $C > 0$ satisfying*

$$\|fg\|_{s,K,\frac{1}{\eta-\epsilon}} \le C\|f\|_{s,K,\frac{1}{\eta}} \|g\|_{s,K,\frac{1}{\eta}}$$

for all $f, g \in G^s(\Omega, K, \frac{1}{\eta})$.

We shall use the following necessary condition for the local solvability in Gevrey classes, proved by Corli in [5, Proposition 2.2].

Theorem 3.2. *Let us fix $s > 1$ and let P be a linear partial differential operator with coefficients belonging to $G^s(\Omega)$; we suppose that the linear equation $Pu = f$ admits a solution $u \in \mathcal{D}_s'(\Omega)$ for all $f \in G_0^s(\Omega)$. Then for every compact set $K \subset \Omega$, for every $\eta > \epsilon > 0$, there exists a positive constant C such that*

$$\left(\|u\|_{L^\infty(K)}\right)^2 \le C\|u\|_{s,K,\frac{1}{\eta-\epsilon}} \|{}^tPu\|_{s,K,\frac{1}{\eta-\epsilon}} \tag{3.1}$$

for all $u \in G_0^s(\Omega, K, \frac{1}{\eta})$, where ${}^tP$ is the transposed of P.

We are going now to construct a suitable function $u_\lambda(t,x)$ that shall violate the condition (3.1). We follow here an idea developed in Popivanov [16] for the C^∞ case and then applied in Oliaro [13] to the Gevrey frame. Let us consider now the homogeneous equation $P(t,\partial_t,D_x)u=0$; by making a partial Fourier transform with respect to the x-variable we obtain

$$\partial_t\hat{u}(\xi,t)+(at^p\xi^{2k}-bt^q\xi^k)\hat{u}(\xi,t)=0, \tag{3.2}$$

with notations as in (2.2). The equation (3.2) has the following solution:

$$\hat{u}(\xi,t)=e^{g(\xi,t)},$$

where

$$g(\xi,t)=-a\frac{t^{p+1}}{p+1}\xi^{2k}+b\frac{t^{q+1}}{q+1}\xi^k; \tag{3.3}$$

recall that $g(\xi,t)$ has (for fixed $\xi>0$) a positive maximum for

$$t=t_\xi=\left(\frac{b}{a}\right)^{\frac{1}{p-q}}\xi^{-\frac{k}{p-q}}, \tag{3.4}$$

and there exists a constant $A=A(p,q,a,b)>0$ such that

$$g(\xi,t_\xi)=A\xi^{k\frac{p-2q-1}{p-q}}. \tag{3.5}$$

We can then find a constant $e_0>0$ satisfying

$$g(\xi,t)=A\xi^{k\frac{p-2q-1}{p-q}}-e_0(t-t_\xi)^2\xi^{k\frac{p-2q+1}{p-q}}\big(1+o(1)\big) \tag{3.6}$$

with $|o(1)|<\frac{1}{2}$, for $|t-t_\xi|\le\epsilon_0|t_\xi|$, $\epsilon_0>0$ being a sufficiently small constant. Let us fix now

$$s'<2\frac{p-q}{p-2q-1} \tag{3.7}$$

and three cut-off functions $\varphi(x)$, $\psi(\rho)$ and $g_1(t)$ belonging to $G_0^{s'}(\mathbb{R})$ in the following way:

- $\varphi\equiv1$ for $|x|\ll1$, $\varphi\equiv0$ for $|x|>1$ and $0\le\varphi(x)\le1$ for every $x\in\mathbb{R}$;
- $\operatorname{supp}\psi=[1,1+\mu_0]$ for a positive fixed μ_0, $\psi(\rho)>0$ for every $\rho\in(1,1+\mu_0)$ and $\int_{-\infty}^{+\infty}\psi(\rho)\,d\rho=1$;
- there exists a (sufficiently small) $\epsilon_1>0$ such that $g_1\equiv1$ for $|t-1|\le\epsilon_1$, $g_1\equiv0$ for $|t-1|>2\epsilon_1$ and $0\le g_1(t)\le1$ for every $t\in\mathbb{R}$.

We set now, for $\lambda>0$,

$$g_{\lambda\rho}(t,x)=\varphi(x)g_1\Big(\frac{t}{t_{\lambda\rho}}\Big), \tag{3.8}$$

where $t_{\lambda\rho}$ is given by (3.4) with $\lambda\rho$ instead of ξ. Let us consider now the function

$$u_\lambda(t,x)=\int_{-\infty}^{+\infty}e^{ix\lambda\rho}e^{g(\lambda\rho,t)-A(\lambda\rho)^{k\frac{p-2q-1}{p-q}}}\psi(\rho)g_{\lambda\rho}(t,x)\,d\rho, \tag{3.9}$$

$g(\lambda\rho,t)$ being given by (3.3) with $\xi=\lambda\rho$; we shall prove that $u_\lambda(t,x)$ does not satisfy the condition (3.1) for $\lambda\to+\infty$.

Remark 3.3. *The function $u_\lambda(t,x)$ is compactly supported, and*

$$\operatorname{supp} u_\lambda \subset K_\lambda := \{(1-\epsilon_2)t_\lambda \le t \le (1+\epsilon_2)t_\lambda\} \times \{|x| \le 1\}$$

for a suitable $0 < \epsilon_2 < 1$. In the estimate of the quantities appearing in (3.1) *for the function $u_\lambda(t,x)$ we can then limit our attention to $K = K_\lambda$.*

Proposition 3.4 (Lower bound for $\|u_\lambda\|_{L^\infty(K_\lambda)}$). *There exists a positive constant E_0 such that*

$$\|u_\lambda\|_{L^\infty(K_\lambda)} \ge E_0 \lambda^{-k\frac{p-2q+1}{2(p-q)}}, \tag{3.10}$$

for λ sufficiently large.

Proof. We observe at first that, since $\operatorname{supp} g_1\big(\frac{t}{t_{\lambda\rho}}\big) \subset K_1 := \{t \in \mathbb{R} : |t| \le \frac{1}{2}\}$ for $\lambda \gg 0$ and every $\rho \in [1, 1+\mu_0]$, we have

$$\|u_\lambda\|_{L^\infty(K_\lambda)} \ge \|u_\lambda(t,0)\|_{L^\infty(K_1)} \ge \|u_\lambda(t,0)\|_{L^1(K_1)}.$$

Now by applying (3.6) in the expression of $u_\lambda(t,0)$ and by making the change of variables $(t - t_{\lambda\rho})\lambda^{k\frac{p-2q+1}{2(p-q)}} = y$ in the integral we obtain that for λ sufficiently large there exists a positive constant $\tilde{E}_0$ such that

$$\frac{\|u_\lambda(t,0)\|_{L^1(K_1)}}{\lambda^{-k\frac{p-2q+1}{2(p-q)}}} = \int_{\mathbb{R}} \int_1^{1+\mu_0} e^{-e_0 y^2} \psi(\rho) g_1\left(\frac{y}{\lambda^{k\frac{p-2q-1}{2(p-q)}} t_\rho} + 1\right) d\rho\, dy \to \tilde{E}_0$$

for $\lambda \to +\infty$, as we can deduce from the Lebesgue Dominated Convergence Theorem. Then the estimate (3.10) is satisfied for $\lambda \gg 0$ with $E_0 = \frac{\tilde{E}_0}{2}$. □

Now we pass to the estimate of $\|u_\lambda\|_{s,K_\lambda,\frac{1}{\eta-\epsilon}}$; to this aim we need some technical results. The following lemma can be proved in the same way as in Oliaro [13, Lemma 4.4]; the proof is omitted.

Lemma 3.5. *Let us consider the cut-off function* (3.8). *We can find positive constants D and G_0 such that for every $\alpha, \beta \in \mathbb{Z}_+$, $\lambda > 0$, $\rho \in [1, 1+\mu_0]$ and $K \subset \mathbb{R}^2$ compact we have*

$$\left\|(D^\alpha g_1)\Big(\frac{t}{t_{\lambda\rho}}\Big) D^\beta \varphi(x)\right\|_{s,K,\frac{1}{\eta}} \le D^{\alpha+\beta+1}(\alpha!\beta!)^{s'} e^{G_0 \lambda^{\overline{(p-q)(s-s')}^{\,k}}},$$

for arbitrary $s > s'$, where s' is the Gevrey order of the functions $\varphi(\cdot)$ and $g_1(\cdot)$.

Let us recall now some results concerning derivatives of composite functions and Bell Polynomials (here in the simplest case of dimension 1, see for example Mascarello and Rodino [9, Section 5.5] for the n-dimensional case). Given $f, g : \mathbb{R} \to \mathbb{R}$, $f, g \in C^\infty(\mathbb{R})$, we have that for every $\nu \in \mathbb{Z}_+$ the following Faà di Bruno formula holds:

$$\frac{d^\nu}{dy^\nu}(f(g(y))) = \sum_{h=1}^{\nu} f^{(h)}(g(y)) B_{\nu,h}(\{g^{(j)}(y)\}), \tag{3.11}$$

where, given an arbitrary sequence of real numbers $\{y_j\}_{j\in\mathbb{Z}_+}$, the 'Bell Polynomial' $B_{\nu,h}(\{y_j\})$ is given by

$$B_{\nu,h}(\{y_j\}) = \nu! \sum_{\substack{\sum_{j=1}^{\infty} h_j = h \\ \sum_{j=1}^{\infty} jh_j = \nu}} \prod_{j=1}^{\infty} \frac{1}{h_j!}\Big(\frac{1}{j!}\,y_j\Big)^{h_j}, \tag{3.12}$$

for every $\nu, h \in \mathbb{Z}_+$, where h_j are positive integers. The polynomials $B_{\nu,h}(\{y_j\})$ satisfy the following properties:

($\mathcal{B}_1$) For every $z \in \mathbb{R}$ we have

$$\frac{1}{h!}\Big(\sum_{j=1}^{\infty} y_j \frac{z^j}{j!}\Big)^h = \sum_{\nu=h}^{\infty} B_{\nu,h}(\{y_j\})\frac{z^\nu}{\nu!}; \tag{3.13}$$

($\mathcal{B}_2$) the Bell polynomials $B_{\nu,h}(\{y_j\})$ are homogeneous of degree h, with integer non-negative coefficients;

($\mathcal{B}_3$) for $\{y_j\}_{j\in\mathbb{Z}_+}$ and $\{w_j\}_{j\in\mathbb{Z}_+}$ with the property $0 \le y_j \le w_j$ for all j we get

$$B_{\nu,h}(\{y_j\}) \le B_{\nu,h}(\{w_j\}); \tag{3.14}$$

($\mathcal{B}_4$) we have

$$B_{\nu,0}(\{y_j\}) = \begin{cases} 0 & \text{if } \nu > 0 \\ 1 & \text{if } \nu = 0 \end{cases}$$

Remark 3.6. *Let us shortly analyze the case when $\{y_j\}$ in the Bell polynomial (3.12) is a positive constant sequence, namely, $y_j = C$ for every $j \in \mathbb{Z}_+$ and for a constant $C > 0$. From the property ($\mathcal{B}_2$) above we have that $B_{\mu,h}(\{C\}) \ge 0$ for every μ and h. Consider now $B_{\nu,h}(\{C\})$, for $\nu \ge h$; for every $z \in \mathbb{R}$, $z > 0$, from (3.13) we have:*

$$\begin{aligned} &B_{\nu,h}(\{C\}) \\ &= \frac{\nu!}{z^\nu} B_{\nu,h}(\{C\})\frac{z^\nu}{\nu!} \le \frac{\nu!}{z^\nu}\sum_{\mu=h}^{\infty} B_{\mu,h}(\{C\})\frac{z^\mu}{\mu!} = \frac{1}{z^\nu}\frac{\nu!}{h!}\Big(\sum_{j=1}^{\infty} C\frac{z^j}{j!}\Big)^h = \frac{1}{z^\nu}\frac{\nu!}{h!}C^h e^{hz}, \end{aligned}$$

which implies that, by keeping z fixed, we can find positive constants C_1 and C_2 satisfying

$$B_{\nu,h}(\{C\}) \le C_1^\nu C_2^h \frac{\nu!}{h!}.$$

Lemma 3.7. *Let $K \subset \mathbb{R}^2$ be a compact set and fix $C > 0$. There exist positive constants C_1, d_1, d_2 such that*

$$\|e^{ix\lambda\rho + g(\lambda\rho,t)}\|_{s,K,C} \le C_1 e^{m(\lambda\rho) + d_1\lambda^{\frac{1}{s}} + d_2\lambda^{\frac{2k}{s}}}, \tag{3.15}$$

where the function $g(\cdot,t)$ is defined in (3.3) and $m(\lambda\rho) := \sup_{(t,x)\in K} g(\lambda\rho,t)$.

Proof. We observe at first that there exists a constant $\overline{C} \geq 1$ such that $|t| \leq \overline{C}$ for every $(t,x) \in K$; then we have that for every $j \in \mathbb{Z}_+$ and $\lambda \geq 1$

$$|\partial_t^j g(\lambda\rho,t)| \leq ap!\overline{C}^{p+1}(1+\mu_0)^{2k}\lambda^{2k} + bq!\overline{C}^{q+1}(1+\mu_0)^k\lambda^k \leq \tilde{C}\lambda^{2k},$$

where $\tilde{C}$ is a constant independent of λ. We then obtain from (3.11), (3.14) and the property $(\mathcal{B}_2)$ at page 143 that for every $\alpha, \beta \in \mathbb{Z}_+$

$$\begin{aligned}\left|\partial_x^\alpha \partial_t^\beta \left(e^{ix\lambda\rho+g(\lambda\rho,t)}\right)\right| &\leq (\lambda\rho)^\alpha \sum_{h=1}^{\beta} e^{g(\lambda\rho,t)} B_{\beta,h}(\{|\partial_t^j g(\lambda\rho,t)|\}) \\ &\leq (\lambda\rho)^\alpha \sum_{h=1}^{\beta} e^{g(\lambda\rho,t)} B_{\beta,h}(\{\tilde{C}\lambda^{2k}\}) \\ &\leq (1+\mu_0)^\alpha \lambda^\alpha e^{m(\lambda\rho)} \sum_{h=1}^{\beta} \lambda^{2kh} B_{\beta,h}(\{\tilde{C}\}).\end{aligned}$$

We then have

$$\begin{aligned}C^{-\alpha-\beta}(\alpha!\beta!)^{-s} &\sup_{(t,x)\in K} \left|\partial_x^\alpha \partial_t^\beta \left(e^{ix\lambda\rho+g(\lambda\rho,t)}\right)\right| \\ &\leq \frac{\left(C^{-1}(1+\mu_0)\lambda\right)^\alpha}{\alpha!^s} \sum_{h=1}^{\beta} \lambda^{2kh} B_{\beta,h}(\{\tilde{C}\}) \frac{(C^{-1})^\beta}{\beta!^s} e^{m(\lambda\rho)}\end{aligned} \tag{3.16}$$

Now we observe that the property $(\mathcal{B}_2)$ at page 143 guarantees that for $\tilde{C} \geq 1$ we can only have $B_{\beta,h}(\{\tilde{C}\}) = 0$ or $B_{\beta,h}(\{\tilde{C}\}) \geq 1$, which implies that, since we can take $\tilde{C} \geq 1$ without loss of generality,

$$B_{\beta,h}(\{\tilde{C}\}) \leq \left(B_{\beta,h}(\{\tilde{C}\})\right)^s;$$

we then obtain from (3.13) that

$$\begin{aligned}\sum_{h=1}^{\beta} \lambda^{2kh} B_{\beta,h}(\{\tilde{C}\}) \frac{(C^{-1})^\beta}{\beta!^s} &\leq 2^{-\beta} \sum_{h=1}^{\beta} \lambda^{2kh} \left[\sum_{\nu=h}^{\infty} B_{\nu,h}(\{\tilde{C}\}) \frac{\left((2C^{-1})^{1/s}\right)^\nu}{\nu!}\right]^s \\ &= 2^{-\beta} \sum_{h=1}^{\beta} \lambda^{2kh} \left[\frac{1}{h!}\tilde{C}_1^h\right]^s = 2^{-\beta} \sum_{h=1}^{\beta} \left[\frac{\left(\tilde{C}_1 \lambda^{\frac{2k}{s}}\right)^h}{h!}\right]^s \\ &\leq 2^{-\beta} \sum_{h=1}^{\beta} \left(e^{\tilde{C}_1 \lambda^{\frac{2k}{s}}}\right)^s \leq C_1 e^{d_2 \lambda^{\frac{2k}{s}}},\end{aligned} \tag{3.17}$$

where $\tilde{C}_1 = \tilde{C}e^{(2C^{-1})^{1/s}}$, $d_2 = s\tilde{C}_1$ and $C_1 = \sup_{\beta\in\mathbb{Z}_+} 2^{-\beta} \sum_{h=1}^{\beta} 1$ is a constant independent of β. In an analogous way we obtain

$$\frac{\left(C^{-1}(1+\mu_0)\lambda\right)^\alpha}{\alpha!^s} = \left[\frac{\left(\left(C^{-1}(1+\mu_0)\lambda\right)^{1/s}\right)^\alpha}{\alpha!}\right]^s \leq e^{d_1 \lambda^{\frac{1}{s}}}, \tag{3.18}$$

where $d_1 = s\big(C^{-1}(1+\mu_0)\big)^{1/s}$. We finally obtain from (3.16), (3.17) and (3.18) that

$$C^{-\alpha-\beta}(\alpha!\beta!)^{-s} \sup_{(t,x)\in K} \left|\partial_x^\alpha \partial_t^\beta \left(e^{ix\lambda\rho+g(\lambda\rho,t)}\right)\right| \leq e^{m(\lambda\rho)+d_1\lambda^{\frac{1}{s}}+d_2\lambda^{\frac{2k}{s}}},$$

which implies (3.15). □

Proposition 3.8 (Upper bound for $\|u_\lambda\|_{s,K_\lambda,\frac{1}{\eta-\epsilon}}$). *Let $u_\lambda(t,x)$ be the function (3.9). There exist positive constants C, d_1, d_2 and G_0 satisfying*

$$\|u_\lambda\|_{s,K_\lambda,\frac{1}{\eta-\epsilon}} \leq Ce^{d_1\lambda^{\frac{1}{s}}+d_2\lambda^{\frac{2k}{s}}+G_0\lambda^{\frac{k}{(p-q)(s-s')}}},$$

where s' is the Gevrey order of the functions φ, ψ and g_1, cf. page 141.

Proof. At first we have by Lemma 3.1 that

$$\|u_\lambda\|_{s,K_\lambda,\frac{1}{\eta-\epsilon}} \leq$$
$$\leq \int_1^{1+\mu_0} \psi(\rho)e^{-A(\lambda\rho)^{k\frac{p-2q-1}{p-q}}} \left\|e^{ix\lambda\rho+g(\lambda\rho,t)}\right\|_{s,K_\lambda,\frac{1}{\eta}} \|g_{\lambda\rho}(t,x)\|_{s,K_\lambda,\frac{1}{\eta}}\, d\rho;$$

the conclusion then follows from Lemma 3.5 and Lemma 3.7, since now

$$m(\lambda\rho) = \sup_{(t,x)\in K_\lambda} g(\lambda\rho,t) = A(\lambda\rho)^{k\frac{p-2q-1}{p-q}},$$

cf. (3.5). □

Let us pass now to the analysis of $\|{}^tPu_\lambda\|_{s,K_\lambda,\frac{1}{\eta-\epsilon}}$. Since ${}^tP(t,\partial_t,D_x) = P(t,\partial_t,D_x)$ we have by a simple computation that

$${}^tP(t,\partial_t,D_x)\left(e^{ix\lambda\rho+g(\lambda\rho,t)}\right) = 0;$$

we then obtain

$${}^tP(t,\partial_t,D_x)u_\lambda(t,x) = I_1 + I_2 - I_3, \tag{3.19}$$

where

$$I_1 = \int_1^{1+\mu_0} e^{ix\lambda\rho}e^{g(\lambda\rho,t)-A(\lambda\rho)^{k\frac{p-2q-1}{p-q}}}\psi(\rho)\partial_t g_{\lambda\rho}(t,x)\,d\rho, \tag{3.20}$$

$$\begin{aligned} I_2 &= (-i)^j at^p \sum_{j=1}^{2k}\binom{2k}{j}\lambda^{2k-j} \\ &\quad\times \int_1^{1+\mu_0} \rho^{2k-j}e^{ix\lambda\rho}e^{g(\lambda\rho,t)-A(\lambda\rho)^{k\frac{p-2q-1}{p-q}}}\psi(\rho)\partial_x^j g_{\lambda\rho}(t,x)\,d\rho, \end{aligned} \tag{3.21}$$

$$\begin{aligned} I_3 &= (-i)^h bt^q \sum_{h=1}^{k}\binom{k}{h}\lambda^{k-h} \\ &\quad\times \int_1^{1+\mu_0} \rho^{k-h}e^{ix\lambda\rho}e^{g(\lambda\rho,t)-A(\lambda\rho)^{k\frac{p-2q-1}{p-q}}}\psi(\rho)\partial_x^h g_{\lambda\rho}(t,x)\,d\rho. \end{aligned} \tag{3.22}$$

Let us start by analyzing I_1; we have the following result.

Lemma 3.9. *There exist positive constants C_1, L_0, G_0, d_1 and d_2 satisfying*

$$\|I_1\|_{s,K_\lambda,\frac{1}{\eta-\epsilon}} \le C_1\lambda^{\frac{k}{p-q}} e^{-L_0\lambda^{k\frac{p-2q-1}{p-q}} + G_0\lambda^{\frac{k}{(p-q)(s-s')}} + d_1\lambda^{\frac{1}{s}} + d_2\lambda^{\frac{2k}{s}}},$$

s' being as usual the Gevrey order of the cut-offs.

Proof. Observe that in the analysis of the norm of I_1 we can limit our attention to $t \in \operatorname{supp}\big(g_1'\big(\frac{t}{t_{\lambda\rho}}\big)\big)$ and so we can find $\tilde{\epsilon}_1$ and $\tilde{\epsilon}_2$ with $0 < \tilde{\epsilon}_1 < \tilde{\epsilon}_2 < 1$ such that, defining $\tilde{K} = \big\{(t,x) \in \mathbb{R}^2 : \frac{t}{t_\lambda} \in [1-\tilde{\epsilon}_2, 1-\tilde{\epsilon}_1] \cup [1+\tilde{\epsilon}_1, 1+\tilde{\epsilon}_2],\ |x| \le 1\big\}$, we have $\operatorname{supp}\big(\partial_t g_{\lambda\rho}(t,x)\big) \subset \tilde{K}$ for every $\rho \in [1, 1+\mu_0]$; then, from Lemma 3.1,

$$\begin{aligned}\|I_1\|_{s,K_\lambda,\frac{1}{\eta-\epsilon}} &= \|I_1\|_{s,\tilde{K},\frac{1}{\eta-\epsilon}} \\ &\le C\lambda^{\frac{k}{p-q}} \int_1^{1+\mu_0} \|e^{ix\lambda\rho+g(\lambda\rho,t)}\|_{s,\tilde{K},\frac{1}{\eta}} e^{-A(\lambda\rho)^{k\frac{p-2q-1}{p-q}}} \Big\|g_1'\Big(\frac{t}{t_{\lambda\rho}}\Big)\varphi(x)\Big\|_{s,\tilde{K},\frac{1}{\eta}} d\rho.\end{aligned} \tag{3.23}$$

In order to apply Lemma 3.7 to $\|e^{ix\lambda\rho+g(\lambda\rho,t)}\|_{s,\tilde{K},\frac{1}{\eta}}$ we observe that

$$m(\lambda\rho) = \sup_{(t,x)\in\tilde{K}} g(\lambda\rho,t) \le (A-\tilde{L}_0)(\lambda\rho)^{k\frac{p-2q-1}{p-q}}$$

for a positive constant $\tilde{L}_0$, as we can deduce by (3.6). The conclusion then follows from (3.23) and Lemmas 3.5 and 3.7. □

We want now to estimate the norm of I_2 and I_3, cf. (3.21) and (3.22) respectively. We observe at first that we can limit ourselves to consider I_2 and I_3 for $x \in \operatorname{supp}\varphi'$, so, defining $K' = \{(t,x) \in \mathbb{R}^2 : (1-\epsilon_2)t_\lambda \le t \le (1+\epsilon_2)t_\lambda,\ x \in [-1,-\tilde{\epsilon}] \cup [\tilde{\epsilon},1]\}$ for suitable $\tilde{\epsilon}$ we have

$$\|I_j\|_{s,K_\lambda,\frac{1}{\eta-\epsilon}} = \|I_j\|_{s,K',\frac{1}{\eta-\epsilon}}, \tag{3.24}$$

for $j = 2,3$. In order to estimate the norm of I_2 and I_3 we need the following lemma, that we present here without the proof, since it is a trivial modification of [13, Lemmas 4.5 and 4.6].

Lemma 3.10.

(i) *For every $R \in \mathbb{R}$, $s > 1$ and $C > 0$ there exist positive constants d and c satisfying*

$$\|t^R\|_{s,K_\lambda,C} \le d\lambda^{-R\frac{k}{p-q}} e^{c\lambda^{\frac{k}{(p-q)(s-1)}}},$$

for all $\lambda > 0$;

(ii) *we can find a positive constant D such that*

$$\Big\|\frac{1}{(ix)^M}\Big\|_{s,K',C} \le D^{M+1}$$

for every positive integer M.

Remark 3.11. *In Lemma 3.10, (i), we had convenience in making explicit the dependence of the estimate on R, in order to obtain a suitable result for our purposes below. The term $\lambda^{-R\frac{k}{p-q}}$ may become in fact very important when choosing R depending on the compact set K_λ: in the particular case when $R=\lambda$, for example, we get*

$$\|t^\lambda\|_{s,K_\lambda,C} \le b\lambda^{-\lambda\frac{k}{p-q}} e^{c\lambda^{\frac{k}{(p-q)(s-1)}}},$$

which is very nice when $\lambda \to +\infty$, because the term $\lambda^{-\lambda\frac{k}{p-q}}$ is stronger than the exponential $e^{c\lambda^{\frac{k}{(p-q)(s-1)}}}$ and let the expression tend to zero as λ tends to $+\infty$. In the following we do not use Lemma 3.10, (i) exactly for $R=\lambda$, but we use it in (3.32) below, where the corresponding exponent is connected with M that in its turn shall be linked with λ. The term $\lambda^{-R\frac{k}{p-q}}$ is then going to have a crucial importance in the sequel.

We can now estimate the norms of (3.21) and (3.22).

Lemma 3.12. *There exists positive constants $\tilde{C}, c, d_1, d_2$ and G_0 such that for every integer $M>0$ we have*

$$\begin{aligned}&\|I_j\|_{s,K_\lambda,\frac{1}{\eta-\epsilon}}\\ &\quad\le \tilde{C}^{M+1} M!^{s'} \lambda^{2k-M(1-k\frac{p-2q-1}{p-q})} e^{d_1\lambda^{\frac{1}{s}}+d_2\lambda^{\frac{2k}{s}}+G_0\lambda^{\frac{k}{(s-s')(p-q)}}+c\lambda^{\frac{k}{(s-1)(p-q)}}},\end{aligned} \tag{3.25}$$

for $j=2,3$.

Proof. We limit ourselves to prove (3.25) for $j=2$, since the case $j=3$ can be treated in the same way. We observe at first that for all positive integers M we have

$$e^{ix\lambda\rho} = \lambda^{-M}\frac{1}{(ix)^M}\frac{\partial^M}{\partial\rho^M}(e^{ix\lambda\rho}); \tag{3.26}$$

moreover, using (3.6) we have that in the compact K_λ we can write

$$\begin{aligned}&g(\lambda\rho,t)-A(\lambda\rho)^{k\frac{p-2q-1}{p-q}}\\ &\quad= -e_0t^2(\lambda\rho)^{k\frac{p-2q+1}{p-q}} + 2e_0tt_{\lambda\rho}(\lambda\rho)^{k\frac{p-2q+1}{p-q}} - e_0t_{\lambda\rho}^2(\lambda\rho)^{k\frac{p-2q+1}{p-q}}\end{aligned} \tag{3.27}$$

plus lower order terms, that for simplicity we do not write in the following, since they can be treated in the same way as the ones appearing in (3.27). We can now put (3.26) and (3.27) in the expression if I_2; integrating by parts and applying the Leibnitz formula we then get

$$\begin{aligned}I_2 &= \sum \binom{2k}{j}\binom{M}{M_1}\binom{K_1}{M_2}\binom{K_2}{M_3}\binom{K_3}{M_4}\binom{K_4}{M_5}(-1)^{k+M} i^{2k-j} a t^p \lambda^{2k-j}\lambda^{-M}\\ &\times \int_1^{1+\mu_0} \frac{1}{(ix)^M} e^{ix\lambda\rho}\partial_\rho^{M_1}(\rho^{2k-j})\partial_\rho^{M_2}\Big(e^{-e_0t^2(\lambda\rho)^{k\frac{p-2q+1}{p-q}}}\Big)\partial_\rho^{M_3}\Big(e^{2e_0tt_{\lambda\rho}(\lambda\rho)^{k\frac{p-2q+1}{p-q}}}\Big)\\ &\times \partial_\rho^{M_4}\Big(e^{-e_0t_{\lambda\rho}^2(\lambda\rho)^{k\frac{p-2q+1}{p-q}}}\Big)\psi^{(M_5)}(\rho)\partial_\rho^{M_6}\Big(g_1\Big(\frac{t}{t_{\lambda\rho}}\Big)\Big)\varphi^{(j)}(x)\,d\rho\end{aligned}$$

where the sum is performed over $j = 1, \dots, 2k$, $M_5 + M_6 = K_4$, $K_4 + M_4 = K_3$, $K_3 + M_3 = K_2$, $K_2 + M_2 = K_1$, $K_1 + M_1 = M$ and $M_1 \le 2k - j$. We can now write the derivatives appearing in the previous expression via the formula (3.11); applying (3.27) and the property $\mathcal{B}_2$ at page 143 we then obtain

$$\begin{aligned} I_2 = \sum \binom{2k}{j}\binom{M}{M_1}\binom{K_1}{M_2}\binom{K_2}{M_3}\binom{K_3}{M_4}\binom{K_4}{M_5} C(2k-j)\cdots(2k-j-M_1+1) \\ \times t^{p+2h+m+d}\lambda^{2k-j}\lambda^{-M}\lambda^{k\frac{p-2q+1}{p-q}(h+m+r)-\frac{k}{p-q}(m+2r-d)} \int_1^{1+\mu_0} \frac{1}{(ix)^M} e^{ix\lambda\rho+g(\lambda\rho,t)} \\ \times e^{-A(\lambda\rho)^{k\frac{p-2q-1}{p-q}}} \rho^{2k-j-M_1}\psi^{(M_5)}(\rho) g_1^{(d)}\Big(\frac{t}{t_{\lambda\rho}}\Big)\varphi^{(j)}(x) \\ \times B_{M_2,h}\Big(\big\{\partial_\rho^{i_1}\big(\rho^{k\frac{p-2q+1}{p-q}}\big)\big\}_{i_1=1,\dots,M_2}\Big) B_{M_3,m}\Big(\big\{\partial_\rho^{i_2}\big(\rho^{k\frac{p-2q+1}{p-q}-\frac{k}{p-q}}\big)\big\}_{i_2=1,\dots,M_3}\Big) \\ \times B_{M_4,r}\Big(\big\{\partial_\rho^{i_3}\big(\rho^{k\frac{p-2q+1}{p-q}-2\frac{k}{p-q}}\big)\big\}_{i_3=1,\dots,M_4}\Big) B_{M_6,d}\Big(\big\{\partial_\rho^{i_4}\big(\rho^{\frac{k}{p-q}}\big)\big\}_{i_4=1,\dots,M_6}\Big)\, d\rho \end{aligned} \tag{3.28}$$

where $C = (-1)^{k+M+h+r}\, i^{2k-j} a\, e_0^{h+r+m}\, 2^m \big(\frac{b}{a}\big)^{\frac{m+2r-d}{p-q}}$ and the sum is performed over $j = 1, \dots, 2k$, $M_5 + M_6 = K_4$, $K_4 + M_4 = K_3$, $K_3 + M_3 = K_2$, $K_2 + M_2 = K_1$, $K_1 + M_1 = M$, $M_1 \le 2k - j$, $h = 1, \dots, M_2$, $m = 1, \dots, M_3$, $r = 1, \dots, M_4$ and $d = 1, \dots, M_6$; observe that $\sum_{j=1}^6 M_j = M$. We now observe that, due to the inequality $\binom{\alpha}{\beta} \le 2^\alpha$ for all integers α, β with $\alpha \ge \beta$, we have

$$\begin{aligned} \big|(2k-j)\cdots(2k-j-M_1+1)\big| &\le (M_1+4k+1)! \\ &= \binom{M_1+4k+1}{4k+1} M_1!\,(4k+1)! \\ &\le C^{M_1+1} M_1! \end{aligned} \tag{3.29}$$

for a suitable positive constant C (depending on k); similarly we get, using (3.13) and taking into account that $\rho \in [1, 1+\mu_0]$,

$$\begin{aligned} \Big|B_{M_6,d}\Big(\big\{\partial_\rho^{i_4}\big(\rho^{\frac{k}{p-q}}\big)\big\}_{i_4=1,\dots,M_6}\Big)\Big| &\le B_{M_6,d}(\{\tilde{C}^{i_4+1} i_4!\}) \\ &\le \frac{M_6!}{z^{M_6}} \sum_{m_6=d}^{\infty} \Big(B_{m_6,d}(\{\tilde{C}^{i_4+1} i_4!\}) \frac{z^{m_6}}{m_6!}\Big) \\ &= \frac{M_6!}{z^{M_6}} \frac{1}{d!} \Big(\sum_{i_4=1}^{\infty} \tilde{C}^{i_4+1} i_4! \frac{z^{i_4}}{i_4!}\Big)^d; \end{aligned}$$

so choosing $z = \frac{1}{2\tilde{C}}$ we obtain

$$\Big|B_{M_6,d}\Big(\big\{\partial_\rho^{i_4}\big(\rho^{\frac{k}{p-q}}\big)\big\}_{i_4=1,\dots,M_6}\Big)\Big| \le C_1^{M_6+1} M_6! \tag{3.30}$$

and the same kind of estimate holds for all the Bell polynomials appearing in (3.28). By these last considerations, using (3.24), Lemma 3.1, (3.29) and the well-known estimate $\binom{\alpha}{\beta} \le 2^\alpha$ for $0 \le \beta \le \alpha$ integers we then have

$$\begin{aligned}
&\|I_2\|_{s,K_\lambda,\frac{1}{\eta-\epsilon}} \\
&\le \sum C_2^{M+1} \lambda^{2k} \lambda^{-M} M_1!\, M_2!\, M_3!\, M_4!\, M_6!\, \lambda^{k\frac{p-2q+1}{p-q}(h+m+r)-\frac{k}{p-q}(m+2r-d)} \\
&\times \|t^{p+2h+m+d}\|_{s,K',\frac{1}{\eta}} \int_1^{1+\mu_0} \left\| \frac{1}{(ix)^M} \right\|_{s,K',\frac{1}{\eta+\epsilon}} \|e^{ix\lambda\rho+g(\lambda\rho,t)}\|_{s,K',\frac{1}{1+2\epsilon}} \\
&\times e^{-A(\lambda\rho)^{k\frac{p-2q-1}{p-q}}} \rho^{2k} \left|\psi^{(M_5)}(\rho)\right| \left\| g_1^{(d)}\left(\frac{t}{t_{\lambda\rho}}\right)\varphi^{(j)}(x) \right\|_{s,K',\frac{1}{\eta+2\epsilon}} d\rho. \qquad (3.31)
\end{aligned}$$

Now from Lemma 3.10 we have

$$\begin{aligned}
&\lambda^{k\frac{p-2q+1}{p-q}(h+m+r)-\frac{k}{p-q}(m+2r-d)} \|t^{p+2h+m+d}\|_{s,K',\frac{1}{\eta}} \qquad (3.32) \\
&\le d\lambda^{k\frac{p-2q-1}{p-q}(h+m+r)-p\frac{k}{p-q}} e^{c\lambda^{\frac{k}{(p-q)(s-1)}}} \le d\lambda^{Mk\frac{p-2q-1}{p-q}} e^{c\lambda^{\frac{k}{(p-q)(s-1)}}};
\end{aligned}$$

the conclusion then follows from (3.31), (3.32), Lemmas 3.5, 3.7, 3.10 and the fact that $\psi \in G_0^{s'}(\mathbb{R})$. □

Proposition 3.13 (Upper bound for $\|{}^tPu_\lambda\|_{s,K_\lambda,\frac{1}{\eta-\epsilon}}$). *There exist positive constants $C, d_1, d_2, \tilde{G}_0, L_0$ such that for every h satisfying*

$$h > \frac{s'}{1-k\frac{p-2q-1}{p-q}} \qquad (3.33)$$

we have

$$\|{}^tPu_\lambda\|_{s,K_\lambda,\frac{1}{\eta-\epsilon}} \le C\lambda^{2k+\frac{k}{p-q}} e^{d_1\lambda^{\frac{1}{s}}+d_2\lambda^{\frac{2k}{s}}+\tilde{G}_0\lambda^{\frac{k}{(p-q)(s-s')}}} \left\{e^{-L_0\lambda^{k\frac{p-2q-1}{p-q}}} + e^{-s'\lambda^{\frac{1}{h}}}\right\}.$$

Proof. It follows immediately from (3.19), Lemma 3.9 and Lemma 3.12 and the fact that $s' > 1$ that

$$\begin{aligned}
\|{}^tPu_\lambda\|_{s,K_\lambda,\frac{1}{\eta-\epsilon}} &\le C_1\lambda^{\frac{k}{p-q}} e^{-L_0\lambda^{k\frac{p-2q-1}{p-q}}+G_0\lambda^{\frac{k}{(p-q)(s-s')}}+d_1\lambda^{\frac{1}{s}}+d_2\lambda^{\frac{2k}{s}}} \qquad (3.34) \\
&+ 2\tilde{C}^{M+1} M!^{s'} \lambda^{2k-M(1-k\frac{p-2q-1}{p-q})} e^{d_1\lambda^{\frac{1}{s}}+d_2\lambda^{\frac{2k}{s}}+(G_0+c)\lambda^{\frac{k}{(p-q)(s-s')}}};
\end{aligned}$$

now choosing $\lambda = M^h$, with h satisfying (3.33) (observe that $\lambda \to +\infty$ as $M \to +\infty$, due to the hypothesis in Theorem 1.2, (ii)), we have by the Stirling formula that

$$\begin{aligned}
&\tilde{C}^{M+1} M!^{s'} \lambda^{-M(1-k\frac{p-2q-1}{p-q})} \qquad (3.35) \\
&= \tilde{C}^{M+1} M!^{s'} M^{-hM(1-k\frac{p-2q-1}{p-q})} \le \tilde{C}_1 \left(M!\, M^{-M}\right)^{s'} \le \tilde{C}_2 e^{-s'M}.
\end{aligned}$$

The conclusion follows then from (3.34) and (3.35), since $M = \lambda^{1/h}$. □

Now we can pass to the proof of the nonsolvability part of Theorem 1.2.

Proof of Theorem 1.2, (ii)–(ii)′. We suppose ab absurdo that P is G^s-locally solvable at the origin for $s > 2\frac{p-q}{p-2q-1}$ (resp., $s > 2\frac{k(p-q)}{(1-k)p-(1-2k)q+k}$). It follows then from Theorem 3.2 and Propositions 3.4, 3.8 and 3.13 that there exist constants $C', \tilde{d}_1, \tilde{d}_2, \tilde{G}, L_0$ satisfying

$$\lambda^{-k(\frac{p-2q+1}{p-q})} \leq C' \lambda^{2k+\frac{k}{p-q}} e^{\tilde{d}_1 \lambda^{\frac{1}{s}} + \tilde{d}_2 \lambda^{\frac{2k}{s}} + \tilde{G} \lambda^{\frac{k}{(p-q)(s-s')}}} \left\{ e^{-L_0 \lambda^{k \frac{p-2q-1}{p-q}}} + e^{-s' \lambda^{\frac{1}{h}}} \right\} \tag{3.36}$$

for every λ sufficiently large. Let us observe now that in the hypotheses (ii)–(ii)′ we have

$$\max\Big\{\frac{1}{s}, \frac{2k}{s}, \frac{k}{(p-q)(s-s')}\Big\} < \min\Big\{k\,\frac{p-2q-1}{p-q}, \frac{1}{h}\Big\} \tag{3.37}$$

for a suitable choice of s' and h satisfying (3.7) and (3.33) respectively. In fact, choosing $s' < \frac{2p-2q-1}{p-2q-1}$ we have that $\frac{2k}{s} > \frac{k}{(p-q)(s-s')}$; then, since $\frac{2k}{s} > \frac{1}{s}$, the following is equivalent to (3.37):

$$\frac{2k}{s} < \min\Big\{k\,\frac{p-2q-1}{p-q}, \frac{1}{h}\Big\}. \tag{3.38}$$

Now it is enough to prove that

$$\frac{2k}{s} < \min\Big\{k\,\frac{p-2q-1}{p-q}, 1-k\,\frac{p-2q-1}{p-q}\Big\}, \tag{3.39}$$

since if this last inequality holds then for fixed s we can choose $s' > 1$ sufficiently close to 1 and, consequently, h sufficiently close to $\frac{1}{1-k\frac{p-2q-1}{p-q}}$, cf. (3.33) in such a way that (3.38) is satisfied. Now we observe that in the case (ii), $\min\{k\,\frac{p-2q-1}{p-q}, 1-k\,\frac{p-2q-1}{p-q}\} = k\,\frac{p-2q-1}{p-q}$, so (3.39) is true by (1.5), while in the case (ii)′, $\min\{k\,\frac{p-2q-1}{p-q}, 1-k\,\frac{p-2q-1}{p-q}\} = 1-k\,\frac{p-2q-1}{p-q}$, and so (3.39) is a consequence of (1.6). In both cases (3.37) is true, and so (3.36) cannot be satisfied for $\lambda \to +\infty$, for any constant C'. The proof is complete. □

References

[1] A. Bove and D. Tartakoff, Propagation of Gevrey regularity for a class of hypoelliptic equations, *Trans. Amer. Math. Soc.* **348** (1996), 2533–2575.

[2] A. Bove and D. Tartakoff, Optimal non-isotropic Gevrey exponents for sum of squares of vector fields, *Comm. Partial Differential Equations* **22** (1997), 1263–1282.

[3] D. Calvo and P. Popivanov, Solvability in Gevrey classes for second powers of the Mizohata operator, *C. R. Acad. Bulg. Sci.* **57** (2004), 11–18.

[4] F. Colombini, L Pernazza, and F Treves, Solvability and nonsolvability of second-order evolution equations, in *Hyperbolic Problems and Related Topics, Grad. Ser. Anal.*, Int. Press, Somerville, MA, 2003, 111–120.

[5] A. Corli, On local solvability in Gevrey classes of linear partial differential operators with multiple characteristics, *Comm. Partial Differential Equations* **14** (1989), 1–25.

[6] T. Gramchev, P. Popivanov and M. Yoshino, Critical Gevrey index for hypoellipticity of parabolic equations and Newton polygons, *Ann. Mat. Pura Appl.* **170** (1996), 103–131.

[7] B. Helffer, Sur l'hypoellipticité d'une classe d'opérateurs paraboliques dégénérés, in *Sur quelques équations aux dérivées partielles singulières,* Astérisque, **19**, Soc. Math. France, Paris, 1974, 79–105.

[8] Y. Kannai, An unsolvable hypoelliptic differential operator, *Israel J. Math.* **9** (1971), 306–315.

[9] M. Mascarello and L. Rodino, *Partial differential equations with multiple characteristics*, Wiley-Akademie Verlag, Berlin, 1997.

[10] T. Matsusawa, On some degenerate parabolic equations I, *Nagoya Math. J.* **51** (1973), 57–77.

[11] T. Matsusawa, On some degenerate parabolic equations II, *Nagoya Math. J.* **52** (1973), 61–84.

[12] O.A. Oleĭnik and E.V. Radkevič, The method of introducing a parameter for the investigation of evolution equations, *Uspekhi Mat. Nauk* **33**, 5(203) (1978), 7–76, 237.

[13] A. Oliaro, On a Gevrey non-solvable partial differential operator, in *Recent Advances in Operator Theory and its Applications,* Editors: M.A. Kaashoek, C Van der Mee and S. Seatzu, Birkhäuser, 337–356.

[14] P. Popivanov, Local properties of linear pseudodifferential operators with multiple characteristics, *C. R. Acad. Bulg. Sci.* **29**, (1976), 461–464.

[15] P. Popivanov, Local solvability of several classes of Partial Differential Equations, *C. R. Acad. Bulg. Sci.* **48**, (1995), 15–18.

[16] P. Popivanov, On a nonsolvable partial differential operator, *Ann. Univ. Ferrara VII, Sc. Mat.* **49** (2003), 197–208.

[17] L. Rodino, Gevrey hypoellipticity for a class of operators with multiple characteristics, in *Analytic Solutions of Partial Differential Equations (Trento),* Astérisque, **89-90**, Soc. Math. France, Paris, 1981, 249–262.

[18] R. Rubinstein, Examples of nonsolvable partial differential equations, *Trans. Amer. Math. Soc.* **199** (1974), 123–129.

[19] F. Trèves, On the existence and regularity of solutions of linear partial differential equations, in *Proceedings of Symposia in Pure Mathematics,* Vol. XXIII, 33–60.

Alessandro Oliaro
Department of Mathematics
University of Torino
Via Carlo Alberto, 10
I-10123 Torino, Italy
e-mail: `oliaro@dm.unito.it`

Petar Popivanov
Institute of Mathematics and Informatics
Bulgarian Academy of Sciences
Acad. G. Bontchev Str., bl. 8
1113 Sofia , Bulgaria
e-mail: `popivano@math.bas.bg`

Operator Theory:
Advances and Applications, Vol. 172, 153–159

A New Aspect of the L^p-extension Problem for Inhomogeneous Differential Equations

Mitsuru Sugimoto

Abstract. For the differential operator P of order m and the inhomogeneous data $f \in \mathcal{S}'$ on $\mathbf{R}^n$, we say that the L^p-extension of the solution holds if $u \in L^p$, $m \leq n(1-1/p)$, and $Pu = f$ on $\mathbf{R}^n \setminus 0$ imply $Pu = f$ on $\mathbf{R}^n$. In this article, we discuss which kind of inhomogeneous terms $f \in \mathcal{S}'$ admit the L^p-extension of the solution. In previous works, this problem was studied by using classical Bochner's method ([1]) or a new method developed by the author and Uchida ([5], [4]). We consider inhomogeneous terms which are not covered by these results.

Mathematics Subject Classification (2000). Primary 35B60; Secondary 35G05.

Keywords. Extension of weak solution, removable singularity.

1. The L^p-extension problem

First of all, we explain the L^p-extension problem of weak solutions. Let

$$P: \text{ differential operator of order } m \text{ with } C^\infty \text{ coefficient on } \mathbf{R}^n.$$

For a given distribution $f \in \mathcal{D}'(\mathbf{R}^n)$, assume that $u \in L^p(\mathbf{R}^n)$ $(1 < p < \infty)$ is a weak solution of the equation on a punctured space

$$Pu = f \text{ on } \mathbf{R}^n \setminus 0.$$

Our problem is when u is a weak solution of the same equation on the whole space:

$$Pu = f \text{ on } \mathbf{R}^n.$$

Without loss of generality, we may assume that all the derivatives of coefficients of P are bounded and the inhomogeneous term f is a tempered distribution, that is, $f \in \mathcal{S}'(\mathbf{R}^n)$.

The following answer to this problem with the homogeneous case $f = 0$ was given by Bochner [1]:

Theorem 1.1. *If $u \in L^p(\mathbf{R}^n)$, $m \leq n(1-1/p)$, and $Pu = 0$ on $\mathbf{R}^n \setminus 0$, then $Pu = 0$ on $\mathbf{R}^n$.*

The inequality $m \leq n(1-1/p)$ in Theorem 1.1 cannot be removed if we take account of the fundamental solution of an elliptic differential operator P. In fact, if P has analytic coefficients, then

$$u(x) = |x|^{m-n}\{A(x) + B(x)\log|x|\}$$

solves $Pu = \delta$ on $\mathbf{R}^n$ and $Pu = 0$ on $\mathbf{R}^n \setminus 0$ with some $A(x)$, $B(x)$ bounded in a neighborhood of 0 by John [3]. We remark that u belongs to L^p locally in the neighborhood for $m > n(1-1/p)$.

On account of this fact, we say that the L^p-extension of the solution holds for the given inhomogeneous term $f \in \mathcal{S}'$, if $u \in L^p$, $m \leq n(1-1/p)$, and $Pu = f$ on $\mathbf{R}^n \setminus 0$ imply $Pu = f$ on $\mathbf{R}^n$.

In this article, we do not care about the existence of such u and P which solve $Pu = f$ on $\mathbf{R}^n \setminus 0$ for each given $f \in \mathcal{S}'$. But once they exist, $f \in H_p^{-n(1-1/p)}$ is necessary for the L^p-extension, since P maps $u \in L^p$ to $f = Pu \in H_p^{-m} \subset H_p^{-n(1-1/p)}$. On the other hand, it is known that the L^p-extension of the solution always holds for $f \in H_p^{-n(1-1/p)}$ ([4]). The proof of this fact will be also given in Section 2 of this article. Hence we have the following answer to the problem with the inhomogeneous case:

Theorem 1.2. *Let $f \in \mathcal{S}'(\mathbf{R}^n)$. Assume $u \in L^p(\mathbf{R}^n)$, $m \leq n(1-1/p)$, and $Pu = f$ on $\mathbf{R}^n \setminus 0$. Then $Pu = f$ on $\mathbf{R}^n$ if and only if $f \in H_p^{-n(1-1/p)}(\mathbf{R}^n)$.*

Besides this criterion for the inhomogeneous terms, we know more useful ones which can be easily checked. Actually, in a non-linear problem with f replaced by non-linear terms, the following theorem plays an important role ([5]).

Theorem 1.3. *Let $f \in L^1(\mathbf{R}^n)$ (microlocally). If $u \in L^p(\mathbf{R}^n)$, $m \leq n(1-1/p)$, and $Pu = f$ on $\mathbf{R}^n \setminus 0$, then $Pu = f$ on $\mathbf{R}^n$.*

Theorem 1.3 with $f \in L^1$ can be proved by the argument in [1], and its microlocal version by [5]. The typical example of such f is

$$f = h(x) + (x_1 \pm i0)^{-1} \otimes g(x') \in L^1 \quad \text{microlocally},$$

where $x = (x_1, x')$, $x' = (x_2, \ldots, x_n)$, $h(x) \in L^1(\mathbf{R}^n)$, and $g(x') \in \mathcal{S}'(\mathbf{R}^{n-1})$. We remark that $f \in L^1$ (microlocally) dose not always imply $f \in H_p^{-n(1-1/p)}$. But Theorem 1.3 (together with Theorem 1.2) says that we can automatically conclude $f \in H_p^{-n(1-1/p)}$ once we know the existence of $u \in L^p$ and P with $m \leq n(1-1/p)$ satisfying $Pu = f$ on $\mathbf{R}^n \setminus 0$.

2. Two methods

In this article, we will discuss a generalization of Theorem 1.3. In author's previous paper [4], two classes $\mathcal{B}_0$ and $\mathcal{C}_0$ of inhomogeneous terms were introduced, which allow the L^p-extension of the solution. We briefly review these arguments.

First we explain a fundamental idea by Bochner [1] to prove Theorem 1.1:

Let $a(x)$ be a cutoff function of the origin, that is, $a(x) \in C_0^\infty$ and $a(x) = 1$ for x near 0. We use the notation

$$a_\varepsilon(x) = a(x/\varepsilon)$$

for $\varepsilon > 0$. Then, for test functions φ, we have

$$\langle Pu, \varphi\rangle = \langle a_\varepsilon Pu, \varphi\rangle = \langle \tilde{a}_\varepsilon u, {}^tP(a_\varepsilon \varphi)\rangle,$$

where ${}^tP$ is the transpose of P and $\tilde{a}(x)$ is another cutoff function which is equal to 1 on the support of $a(x)$. We remark that

$$\lim_{\varepsilon \searrow 0} \tilde{a}_\varepsilon u \to 0 \text{ in } L^p.$$

Hence, if $\{{}^tP(a_\varepsilon\varphi)\}_{\varepsilon>0}$ is bounded in L^q with $1/p + 1/q = 1$, we get $Pu = 0$, and this can be realized if $m \leq n(1 - 1/p)$.

By the method above, we have easily the following result for inhomogeneous equations:

Definition 2.1. Let $f \in \mathcal{S}'(\mathbf{R}^n)$. We say that $f \in \mathcal{B}_0$ if there exist a strictly decreasing sequence $\{\varepsilon_\nu\}_{\nu=1}^\infty$ of positive numbers and a cutoff function $a(x)$ of the origin such that

$$\lim_{\nu\to\infty} a_{\varepsilon_\nu} f = 0 \text{ in } \mathcal{S}'(\mathbf{R}^n).$$

Theorem 2.2. *Let $f \in \mathcal{B}_0$. If $u \in L^p(\mathbf{R}^n)$, $m \leq n(1-1/p)$, and $Pu = f$ on $\mathbf{R}^n \setminus 0$, then $Pu = f$ on $\mathbf{R}^n$.*

Theorem 2.2 is a generalization of Theorem 1.3 since $L^1 \subset \mathcal{B}_0$ by Lebesgue's convergence theorem.

On the other hand, there is another method to prove Theorem 1.1, found by the author and Uchida [5]:

Assume $Pu = 0$ on $\mathbf{R}^n \setminus 0$. By the structure theorem of distributions with a point support, we can write

$$Pu = Q\delta \text{ on } \mathbf{R}^n$$

where Q is a differential operator with constant coefficients. Then, by the microlocal ellipticity of Q (when $Q \neq 0$) and the mapping property of pseudo-differential operators on the Sobolev space H_p^s, we have

$$\delta = Q^{-1}Pu \in H_p^{-m}$$

(at least in a microlocal sense). It is known that $\delta \in H_p^s$ implies $s < -n(1 - 1/p)$. Then this relation yields $m > n(1 - 1/p)$. Hence $m \leq n(1 - 1/p)$ yields $Q = 0$, hence $Pu = 0$ on $\mathbf{R}^n$. Thus we have obtained Theorem 1.1 again.

We remark that the argument above works for homogeneous equations $Pu = f$ with $f \in H^{-n(1-1/p)}$, which proves the *if* part of Theorem 1.2.

By just following the same argument, we also have a result of another type. Denote by

$$WF(f) = \coprod_{z\in\mathbf{R}^n} WF_z(f)$$

the wave front set of f, which is a subset of the cotangent bundle $T^*\mathbf{R}^n = \coprod_{z\in\mathbf{R}^n} T_z^*\mathbf{R}^n$.

Definition 2.3. Let $f \in \mathcal{S}'(\mathbf{R}^n)$. We say that $f \in \mathcal{C}_0$ if f is microlocally smooth in some direction at the origin, that is, the fiber of the wave front set at 0 has the proper inclusion $WF_0(f) \subsetneq T_0^*\mathbf{R}^n$.

Theorem 2.4. *Let $f \in \mathcal{C}_0$. If $u \in L^p(\mathbf{R}^n)$, $m \leq n(1-1/p)$, and $Pu = f$ on $\mathbf{R}^n \setminus 0$, then $Pu = f$ on $\mathbf{R}^n$.*

In [5], it was shown that $\mathcal{C}_0$ in Theorem 2.4 can be replaced by a larger class $L^1 + \mathcal{C}_0$, which is the microlocal statement of Theorem 1.3. In [4], larger classes than $L^1 + \mathcal{C}_0$ were also discussed.

3. Homogeneous distributions

In Section 2, we have explained two different methods to prove the L^p-extension of the solution. We remark here that the following examples are not integrable, but covered by these method:

$$f_\beta(x) = \mathrm{p.\,v.}\,\frac{1}{x_1} \otimes g(x') \in \mathcal{B}_0,$$

$$f_\gamma^\pm(x) = \frac{1}{x_1 \pm i0} \otimes g(x') \in \mathcal{C}_0,$$

where $x = (x_1, x')$, $x' = (x_2, \ldots, x_n)$, and $g(x') \in \mathcal{S}'(\mathbf{R}^{n-1})$.

If fact, the proof of $f_\beta \in \mathcal{B}_0$ can be reduced to show

$$\left\langle a_\varepsilon(x_1)\,\mathrm{p.\,v.}\,\frac{1}{x_1},\ \varphi(x_1)\right\rangle \to 0 \quad \text{as } \varepsilon \searrow 0$$

for all test function φ of dimension 1. We take a cutoff function $a(x_1)$ of dimension 1 such that $a(-x_1) = a(x_1)$. Then a cancellation property works, and we have

$$\left\langle a_\varepsilon(x_1)\,\mathrm{p.\,v.}\,\frac{1}{x_1},\ \varphi(x_1)\right\rangle = \lim_{\delta\searrow 0}\int_{|x_1|\geq\delta} \frac{(a_\varepsilon\varphi)(x_1)}{x_1}\,dx_1$$

$$= \lim_{\delta\searrow 0}\int_{|x_1|\geq\delta} \frac{a_\varepsilon(x_1)}{x_1}\,dx_1 \cdot \varphi(0) + \int (a_\varepsilon H)(x_1)\,dx_1$$

with H bounded. The first term vanishes since a_ε is an even function, and the second term tends to 0 as $\varepsilon \searrow 0$.

On the other hand, we also have $f_\gamma^\pm \in \mathcal{C}_0$ since

$$WF_0(f_\gamma^\pm) \subset \{0\} \times \{(\xi_1, \xi_2, \ldots, \xi_n); \xi_1 \gtrless 0\} \subsetneq T_0^*\mathbf{R}^n.$$

We remark that we have $f_\beta \notin \mathcal{C}_0$ if we take non-smooth $g(x')$.

On account of these examples, it is natural to ask the question as to whether or not the L^p-extension of the solution holds for the inhomogeneous term

$$f_\mu(x) = \left(\frac{a^+}{x_1 + i0} + \frac{a^-}{x_1 - i0} \right) \otimes g(x'),$$

where $x = (x_1, x')$, $x' = (x_2, \dots, x_n)$, $a^\pm \in \mathbf{C}$, and $g(x') \in \mathcal{S}'(\mathbf{R}^{n-1})$. We remark that

$$\mathrm{p.v.}\, \frac{1}{x_1} \otimes g(x'), \qquad \frac{1}{x_1 \pm i0} \otimes g(x'), \qquad \delta(x_1) \otimes g(x')$$

are typical examples, since

$$\mathrm{p.v.}\, \frac{1}{x_1} = \frac{1}{2}\left(\frac{1}{x_1 + i0} + \frac{1}{x_1 - i0} \right), \qquad \delta(x_1) = \frac{1}{2\pi i}\left(\frac{1}{x_1 - i0} - \frac{1}{x_1 + i0} \right).$$

It is not clear when $f_\mu \in \mathcal{B}_0$, except for the case $a^+ = a^-$. Hence Theorem 2.2 cannot answer this question. Theorem 2.4 neither, except for the case $a^+ = 0$ or $a^- = 0$. In the next section, we will discuss how to deal with this example.

4. A new aspect

By modifying the idea to prove Theorem 2.4, we can provide a new aspect to explain when the L^p-extension of the solution holds for the inhomogeneous term

$$f_\mu(x) = \left(\frac{a^+}{x_1 + i0} + \frac{a^-}{x_1 - i0} \right) \otimes g(x'),$$

as in Section 3. Assume that

$$|\widehat{f_\mu}(\xi)| \to 0 \qquad (|\xi| \to \infty)$$

uniformly in a direction. (See Definition 4.1 below for the precise meaning.) It is equivalent to the same property (of dimension $n-1$) for $g(x')$ since the Fourier transforms of $(x_1 \pm i0)^{-1}$ are just a constant in each direction.

Noticing that, in the direction, $\widehat{f_\mu}(\xi)$ can be regarded as a perturbation to any non-zero polynomial then, let us repeat the argument of the proof of Theorem 2.4. If $Pu = f_\mu$ on $\mathbf{R}^n \setminus 0$, then we have

$$Pu = f_\mu + Q(D)\delta \in H_p^{-m}$$

with a polynomial Q by the structure theorem and the mapping property of P. Furthermore, since $f_\mu = \hat{f}_\mu(D)\delta$, we have

$$\widetilde{Q}(D)\delta \in H_p^{-m}, \qquad \widetilde{Q}(\xi) = Q(\xi) + \hat{f}_\mu(\xi).$$

If the polynomial $Q \neq 0$, then $Q(D)$ is microlocally elliptic in a direction, and if $\hat{f}_\mu(\xi)$ is just a perturbation, the same is true for $\widetilde{Q}(D)$. Then we have $\delta \in H_p^{-m}$ (microlocally), which implies $m > n(1-1/p)$. Hence $m \le n(1-1/p)$ yields $Q = 0$ and we can conclude $Pu = f_\mu$ on $\mathbf{R}^n$.

From this argument, we obtain the following another type of criterion.

Definition 4.1. Let $f \in \mathcal{S}'(\mathbf{R}^n)$. We say that $f \in \mathcal{M}$ if $\hat{f}$ is a function which satisfies

$$|\hat{f}(\xi)| \to 0 \qquad (|\xi| \to \infty)$$

uniformly, that is,

$$\sup_{|\xi|>R} |\hat{f}(\xi)| \to 0 \quad (R \to \infty).$$

We remark that $\mathcal{M}$ is closed under the multiplication of cutoff functions $a(x)$ of the origin. In fact, since

$$\widehat{af}(\xi) = \int \hat{a}(\xi - \eta)\hat{f}(\eta)\, d\eta,$$

we have

$$\begin{aligned}
\sup_{|\xi|>R} \left|\widehat{af}(\xi)\right| &\le \sup_{|\xi|>R} \left(\int_{|\eta|\le R/2} + \int_{|\eta|>R/2} \right) \left|\hat{a}(\xi-\eta)\hat{f}(\eta)\right| d\eta \\
&\le \langle R/2\rangle^{-1} \sup_{\xi\in\mathbf{R}^n} \left|\hat{f}(\xi)\right| \cdot \int \langle\eta\rangle|\hat{a}(\eta)|\, d\eta + \sup_{|\xi|>R/2} \left|\hat{f}(\xi)\right| \cdot \int |\hat{a}(\eta)|\, d\eta \\
&\to 0 \qquad (R \to \infty).
\end{aligned}$$

$\mathcal{M}$ is also closed under the operation of $\chi(D)$, where $\chi(\xi)$ is a bounded function. On account of these arguments, we have the following main result of this article:

Theorem 4.2. *Let $f \in \mathcal{M} + \mathcal{C}_0$. If $u \in L^p(\mathbf{R}^n)$, $m \le n(1-1/p)$, and $Pu = f$ on $\mathbf{R}^n \setminus 0$, then $Pu = f$ on $\mathbf{R}^n$.*

We remark that, by the Riemann-Lebesgue theorem, we have $L^1 \subset \mathcal{M}$. (To justify it, show $\mathcal{S} \subset \mathcal{M}$ and use the fact that $\mathcal{S}$ is dense in L^1.) Hence Theorem 4.2 is also a generalization of Theorem 1.3.

Here is another interesting example:

$$f_\nu = F^{-1}\left[e^{i|\xi|^d} |\xi|^{-nd/2} \psi(|\xi|)\right] \in \mathcal{M},$$

where smooth function $\psi(t)$ is a smooth function of $t \ge 0$, which is equal to 0 for $0 \le t \le 1$ and 1 for $t \ge 2$. Since it is not straightforward to show $f_\nu \in H_p^{-n(1-1/p)}$ except for the case $p = 2$, Theorem 1.2 does not say anything to this example. Theorem 1.3 neither because $f_\nu \notin L^1$ for $0 < d < 1$. In fact, f_ν is of the form

$$f_\nu(x) = K(|x|)|x|^{-n} + O(|x|^\omega)$$

as $|x| \to 0$, where $|K(|x|)|$ is a non-zero constant and $\omega > -n$ (Ishii [2]). The author showed the weak L^p-extension property for it by a different approach ([4]).

As was seen in the above, f_μ is a typical example which belongs to $\mathcal{M}$ if $g(x')$ belongs to $\mathcal{M}$ of dimension $n-1$. More generally, we have

$$\tilde{f}_\mu(x) = g_1(x_1) \otimes g_2(x_2) \cdots \otimes g_n(x_n) \in \mathcal{M},$$

where $x = (x_1, x_2, \ldots, x_n)$, if at least one of $g_j(t)$ $(j = 1, 2, \ldots, n)$ belongs to $\mathcal{M}$ of dimension 1, and all other $g_l(t)$ is a linear combination of $(t \pm i0)^{-1}$. Furthermore, if such $g_j(t) \in L^1(\mathbf{R})$ admits a nice regularity and up to $(k-1)$th derivatives of it are integrable again, then we have a stronger decaying property

$$\sup_{|t|>R} \left|t^{k-1}\widehat{g_j}(t)\right| \to 0 \qquad (R \to \infty)$$

In this case, linear combinations of more general homogeneous distributions

$$(t \pm i0)^{-1}, (t \pm i0)^{-2}, \ldots, (t \pm i0)^{-k}$$

are allowed for all other $g_l(t)$ since their Fourier transforms are polynomial of order up to $k-1$ in each direction, that is,

$$F\left[(t \pm i0)^{-k}\right](\tau) = \begin{cases} c_k|\tau|^{k-1} & \text{for} \quad \pm\tau > 0 \\ 0 & \text{for} \quad \pm\tau < 0 \end{cases}$$

with complex numbers $c_k \in \mathbf{C}$ (for $k = 1, 2, \ldots$).

References

[1] S. Bochner, Weak solutions of linear partial differential equations, *J. Math. Pures Appl.* **35** (1956), 193–202.

[2] H. Ishii, On some Fourier multipliers and partial differential equations, *Math. Japon.* **19** (1974), 139–163.

[3] F. John, The fundamental solution of linear elliptic differential equations with analytic coefficients, *Comm. Pure. Appl. Math.* **3** (1950), 273–304.

[4] M. Sugimoto, A weak extension theorem for inhomogeneous differential equations, *Forum Math* **13** (2001), 323–334.

[5] M. Sugimoto and M. Uchida, A generalization of Bochner's extension theorem and its application, *Ark. Mat.* **38** (2000), 399–409.

Mitsuru Sugimoto
Department of Mathematics
Graduate School of Science
Osaka University
Machikaneyama-cho 1-16
Toyonaka, Osaka 560-0043, Japan
e-mail: sugimoto@math.wani.osaka-u.ac.jp

Operator Theory:
Advances and Applications, Vol. 172, 161–172

Continuity in Quasi-homogeneous Sobolev Spaces for Pseudo-differential Operators with Besov Symbols

Gianluca Garello and Alessandro Morando

Abstract. In this paper a result of continuity for pseudo-differential operators with non-regular symbols on spaces of quasi-homogeneous type is given. More precisely, the symbols $a(x,\xi)$ take their values in a quasi-homogeneous Besov space with respect to the x variable; moreover a finite number of derivatives with respect to the second variable satisfies, in Besov norm, decay estimates of quasi-homogeneous type.

Mathematics Subject Classification (2000). 35S05; 35A17.

Keywords. Pseudo-differential operators, Besov spaces, Sobolev spaces.

1. Introduction

The present paper has to be considered in a series of authors articles about the continuity of pseudo-differential operators with non-regular symbols in the L^p framework, see, e.g., [4], [5], [6]. We consider here the pseudo-differential operators in the classical form:

$$a(x,D)u(x) = (2\pi)^{-n} \int e^{ix\cdot\xi} a(x,\xi)\widehat{u}(\xi)d\xi, \quad u \in \mathcal{S}(\mathbb{R}^n), \tag{1.1}$$

where the symbol $a(x,\xi)$ is a tempered distribution in the Schwartz class $\mathcal{S}'(\mathbb{R}^{2n})$. With standard notations $x\cdot\xi$ denotes the inner product of $x,\xi \in \mathbb{R}^n$, while $\widehat{u} = \mathcal{F}u$ is the Fourier transform of u.

It is well known that the pseudo-differential operators with symbols in the classical zero order Hörmander classes: $S^0_{\rho,\delta}$, $0 \le \delta < \rho \le 1$, defined for any smooth function $a(x,\xi)$ by the estimates:

$$\left|\partial_\xi^\alpha \partial_x^\beta a(x,\xi)\right| \le C_{\alpha,\beta}(1+|\xi|)^{-\rho|\alpha|+\delta|\beta|}, \quad c_{\alpha,\beta} > 0, \quad x,\xi \in \mathbb{R}^n, \tag{1.2}$$

are not in general L^p continuous for $p \neq 2$ unless $\rho = 1$, see, e.g., Fefferman [2].

The authors are supported by F.I.R.B. grant of Italian Government.

In order to obtain L^p continuity when $0 \leq \delta < \rho < 1$ one needs further assumptions. For instance Taylor [[11], Ch. XI] introduces the symbol classes $M_\rho^m \subset S_{\rho,0}^m$, where Marcinkiewicz conditions on the derivatives with respect to ξ are required.

For $M = (m_1, \dots, m_n) \in \mathbb{Z}_+^n$, such that $\min\limits_{1\leq j\leq n} m_j \geq 1$ let us now define the *quasi-homogeneous weight function* in $\mathbb{R}^n$:

$$|\xi|_M := \left(\sum_{j=1}^n \xi_j^{2m_j}\right)^{\frac{1}{2}}, \quad \xi \in \mathbb{R}^n. \tag{1.3}$$

The *quasi-homogeneous* smooth symbol classes S_M^m, $m \in \mathbb{R}$, introduced by Lascar [7], 1977, are defined by the following:

$$\left|\partial_\xi^\alpha \partial_x^\beta a(x,\xi)\right| \leq c_{\alpha,\beta}(1+|\xi|_M)^{m-\frac{1}{M}\cdot\alpha}, \quad c_{\alpha,\beta} > 0, \quad x,\xi \in \mathbb{R}^n, \tag{1.4}$$

where $\frac{1}{M}\cdot\alpha = \sum_{j=1}^n \frac{1}{m_j}\alpha_j$.

It is not difficult to see that $S_M^0 \subset S_{\rho,0}^0$ with $\rho = \dfrac{\min m_j}{\max m_j} < 1$, unless $M = (k, \dots, k)$ for some integer $k \geq 1$; then the related pseudo-differential operators are not expected to be L^p continuous, $p \neq 2$.

In the present paper we deal with a class of pseudo-differential operators whose symbol $a(x,\xi)$ takes its values in a quasi-homogeneous Besov space with respect to x and it satisfies estimates of type (1.4), for a suitable finite number of derivatives with respect to ξ.

Since the weights $|\xi|_M$ display a quasi-homogeneous structure, that is for any $t > 0$:

$$\left|t^{1/M}\xi\right|_M = t|\xi|_M, \quad \text{where} \quad t^{1/M}\xi = \left(t^{1/m_1}\xi_1, \dots, t^{1/m_n}\xi_n\right),$$

we can consider, see the next Proposition 2.3, a dyadic partition of unity on $\mathbb{R}^n$, $\sum_{h=-1}^\infty \varphi_h(\xi) = 1$, with $\varphi_h(\xi)$ supported, for $h \geq 0$, in the *quasi-annulus* of the form $\frac{1}{K} \leq \left|2^{-h/M}\xi\right|_M \leq K$, $K > 1$.

The non-regular symbols $a(x,\xi)$ above considered can then reduce to an expansion of elementary symbols, following the techniques developed by Coifman-Meyer [1], Marschall [9], Taylor [12]. The authors themselves [4] improve such techniques in the more general setup of *multi-quasi-elliptic* pseudo-differential calculus, where any reference to homogeneity is lost.

The zero order pseudo-differential operators considered in this paper satisfy in natural way the Marcinkiewicz-Lizorkin Lemma for Fourier multipliers, thanks to the quasi-homogeneous structure of the symbols; thus their L^p continuity may be proved.

A number of applications of the pseudo-differential operators with non-regular symbols to the non linear partial differential equations are given in the literature of the last twenty years; see, e.g., Taylor [12], [13] and the references given there.

The plan of the paper is the following. In §2, after reviewing the main properties of the quasi-homogeneous weight $|\xi|_M$, we define in the details the symbol classes. We also introduce in this section the quasi-homogeneous dyadic partition of unity and the expansion of $a(x,\xi)$ in *elementary symbols*. A particular stress is given in the proof of the next estimates (2.7), because of their importance in the applications of the Marcinckiewicz Lemma.

In §3 the quasi-homogeneous Besov and Sobolev spaces are introduced and respectively denoted by $B^{r,M}_{\infty,\infty}$, $H^{s,p}_M$, $r,s\in\mathbb{R}$, $1<p<\infty$. The Sobolev spaces are then characterized by means of the above introduced partition of unity and suitable useful properties are derived.

At last in §4 we prove the continuity on the spaces $H^{s,p}_M$, $1<p<\infty$, of the pseudo-differential operators whose symbols belong to the Besov spaces $B^{r,M}_{\infty,\infty}$, with respect to x, for suitable choice of the exponents r,s.

Related to our result let us quote a work of Yamazaki [15], where the L^p continuity is proved for a class of pseudo-differential operators with similar structure, but under different conditions.

2. Symbol classes

In the whole paper $M=(m_1,\dots,m_n)\in\mathbb{Z}^n_+$ shall be a vector with positive integer components, such that $\min\limits_{1\le j\le n} m_j\ge 1$. We will write $1/M:=(1/m_1,\dots,1/m_n)$, and $\langle\xi\rangle_M=\left(1+|\xi|^2_M\right)^{\frac{1}{2}}$.

Proposition 2.1. *For any M we have*

(i) $|\xi+\eta|_M\le C(|\xi|_M+|\eta|_M)$, $\quad C>0$;

(ii) *(quasi-homogeneity) for any $t>0$, $|t^{1/M}\xi|_M=t|\xi|_M$;*

(iii) $\xi^\gamma\partial^{\alpha+\gamma}|\xi|_M\le C\langle\xi\rangle_M^{1-\frac{1}{M}\cdot\alpha}$, *for any $\alpha,\gamma\in\mathbb{Z}^n_+$ and $\xi\neq 0$.*

The proofs of (i), (ii) are quite trivial and (iii) follows with some small changes from the proof of Lemma 2.1 in [3]

Definition 2.2. [symbol classes] For X Banach space and any $r\in\mathbb{R}$, $N\in\mathbb{Z}_+$ we say that the complex-valued measurable function $a(x,\xi)$ on $\mathbb{R}^n_x\times\mathbb{R}^n_\xi$ belongs to the symbol class $XS^r_M(N)$ if for some positive constant C:

$$\left|\partial^\alpha_\xi a(x,\xi)\right|\le C\langle\xi\rangle_M^{r-\frac{1}{M}\cdot\alpha},\quad \text{for any}\quad x,\xi\in\mathbb{R}^n,\quad |\alpha|\le N \tag{2.1}$$

$$\left\|\partial^\alpha_\xi a(\cdot,\xi)\right\|_X\le C\langle\xi\rangle_M^{r-\frac{1}{M}\cdot\alpha},\quad \text{for any}\quad \xi\in\mathbb{R}^n,\quad |\alpha|\le N. \tag{2.2}$$

For some $K>1$ let us consider $\phi(t)\in C_0^\infty([0,+\infty])$ such that $0\le\phi(t)\le 1$, $\phi(t)=1$ for $0\le t\le\frac{1}{2K}$; $\phi(t)=0$, when $t>K$. For $h\ge 0$ we can then introduce the sequence:

$$\varphi_{-1}(\xi) = \phi\left(|\xi|_M\right), \quad \varphi_h(\xi) = \varphi_0\left(2^{-h/M}\xi\right), \tag{2.3}$$

where $\varphi_0(\xi) = \phi\left(\left|2^{-1/M}\xi\right|_M\right) - \phi\left(|\xi|_M\right)$. $\{\varphi_h\}_{h=-1}^{\infty}$ is a quasi-homogeneous dyadic partition of unity.

In fact we have

Proposition 2.3. *For any $\xi \in \mathbb{R}^n$, $\alpha, \gamma \in \mathbb{Z}_+^n$ and some positive constant $C_{\alpha,\gamma,K}$:*

$$\begin{aligned} &\operatorname{supp}\varphi_{-1} \subset C_{-1}^{K,M} := \{\xi \in \mathbb{R}^n \ , \ |\xi|_M \le K\} \\ &\operatorname{supp}\varphi_h \subset C_h^{K,M} := \left\{\xi \in \mathbb{R}^n \ ; \ \tfrac{1}{K}2^{h-1} \le |\xi|_M \le K2^{h+1}\right\}, \ h \ge 0; \end{aligned} \tag{2.4}$$

$$\sum_{h=-1}^{\infty} \varphi_h(\xi) = 1; \tag{2.5}$$

$$\sum_{h=-1}^{\infty} \varphi_h(D)u = u, \quad \text{with convergence in } \mathcal{S}'(\mathbb{R}^n); \tag{2.6}$$

$$\left|\xi^\gamma \partial^{\alpha+\gamma}\varphi_h(\xi)\right| \le C_{\alpha,\gamma,K} 2^{-\left(\frac{1}{M}\cdot\alpha\right)h}, \quad h = -1, 0, \dots. \tag{2.7}$$

Moreover for any fixed $\xi \in \mathbb{R}^n$ the sum in (2.5) reduces to a finite number of terms, independent of the choice of ξ itself.

Proof. It is trivial to prove (2.4). For every fixed $\xi \in \mathbb{R}^n$ we have for any suitably large integer h: $\phi\left(\left|2^{-h/M}\xi\right|_M\right) = 1$; then (2.5), (2.6) follow.

For every integer $h \ge 0$ we compute

$$\left|\xi^\gamma \partial^{\alpha+\gamma}\varphi_h(\xi)\right| = \left|\left(2^{-h/M}\xi\right)^\gamma \left(\partial^{\alpha+\gamma}\varphi_0\right)\left(2^{-h/M}\xi\right)\right| 2^{-h\left(\frac{1}{M}\cdot\alpha\right)} \le C_{\alpha,\gamma,K} 2^{-h\left(\frac{1}{M}\cdot\alpha\right)},$$

where $C_{\alpha,\gamma,K} = \max_{\eta} \left|\eta^\gamma \partial^{\alpha+\gamma}\varphi_0(\eta)\right|$ is independent of h. □

Proposition 2.4. [*Decomposition in elementary symbols.*] *For $N \ge n+1$ let us consider $a(x,\xi) \in XS_M^0(N)$. We then obtain with absolute convergence in $L^\infty(\mathbb{R}^{2n})$:*

$$a(x,\xi) = \sum_{k \in \mathbb{Z}^n} c_k a_k(x,\xi), \tag{2.8}$$

where $\{c_k\}_{k\in\mathbb{Z}^n}$ is a real positive-valued sequence such that $\sum_{k\in\mathbb{Z}^n} c_k < \infty$.

Moreover for some constants $L, C > 0$, $K > 1$ we can find two suitable function sequences $\left\{d_h^k\right\}_{h=-1}^{\infty}$ and $\left\{\psi_h^k\right\}_{h=-1}^{\infty}$ which satisfy, for any $k \in \mathbb{Z}_+^n$, the *following properties:*

$$a_k(x,\xi) = \sum_{h=-1}^{\infty} d_h^k(x)\psi_h^k(\xi); \tag{2.9}$$

$$\|d_h^k\|_X \le L \quad , \quad \|d_h^k\|_{L^\infty} \le L; \tag{2.10}$$

$$\psi_h^k \in C_0^\infty(\mathbb{R}^n) \quad , \quad \operatorname{supp}\psi_h^k \subset C_h^{K,M} \quad ; \tag{2.11}$$

$$\left|\xi^\gamma \partial^{\alpha+\gamma}\psi_h^k(\xi)\right| \le C2^{-\left(\frac{1}{M}\cdot\alpha\right)h}, \quad \text{when } |\alpha+\gamma| \le N-n-1. \tag{2.12}$$

The symbols defined in (2.9) *are called elementary symbols.*

Proof. Let us write $a(x,\xi) = \sum_{h=-1}^{\infty}\varphi_h(\xi)a(x,\xi) = \sum_{h=-1}^{\infty}a_h(x,\xi)$. Then, for any $x \in \mathbb{R}^n$ and $h = -1, 0, \ldots$, $\operatorname{supp} a_h(x,\cdot) \subset C_h^{K,M}$. Moreover when $|\alpha| \le N$ the Leibnitz rule gives:

$$\left|\partial_\xi^\alpha a_h(x,\xi)\right| \le C_\alpha 2^{-\left(\frac{1}{M}\cdot\alpha\right)h}. \tag{2.13}$$

In the same way we obtain

$$\left\|\partial_\xi^\alpha a_h(\cdot,\xi)\right\|_X \le C_\alpha 2^{-\left(\frac{1}{M}\cdot\alpha\right)h}, \quad \xi\in\mathbb{R}^n, \quad |\alpha| \le N. \tag{2.14}$$

By setting now $b_h(x,\xi) = a_h\left(x, 2^{h/M}\xi\right)$ when $h \ge 0$ and $b_{-1}(x,\xi) = a_{-1}(x,\xi)$, it follows for some positive constant c_α independent of h and all $x \in \mathbb{R}^n$:

$$\operatorname{supp} b_h(x,\cdot) \subset C_0^{K,M},\ h \ge 0 \quad , \quad \operatorname{supp} b_{-1}(x,\cdot) \subset C_{-1}^{K,M}; \tag{2.15}$$

$$\left|\partial_\xi^\alpha b_h(x,\xi)\right| \le c_\alpha, \quad \left\|\partial_\xi^\alpha b_h(\cdot,\xi)\right\|_X \le c_\alpha, \quad \xi\in\mathbb{R}^n,\ h = -1, 0, \ldots. \tag{2.16}$$

The functions $b_h(x,\cdot)$ are supported in $[-T,T]^n$, uniformly with respect to x, when $\max_{1\le j\le n}(2K)^{\frac{1}{m_j}} \le T$. We can then construct the $(2T)$-periodic extension of b_h: $B_h(x,\xi) = \sum_{\lambda\in\mathbb{Z}^n} b_h(x,\xi - 2T\lambda)$. For some $K' > K$ such that $\max\limits_{1\le j\le n}(2K')^{\frac{1}{m_j}} \le T$ let us now consider two positive smooth functions $\theta(\xi)$, $\theta_{-1}(\xi)$ supported respectively in the quasi-annulus $C_0^{K',M}$ and in $C_{-1}^{K',M}$, both equal to one in $C_0^{K,M}$ or respectively in $C_{-1}^{K,M}$. Then we plainly obtain

$$b_h(x,\xi) = \theta(\xi)B_h(x,\xi),\ h \ge 0, \quad b_{-1}(x,\xi) = \theta_{-1}(\xi)B_{-1}(x,\xi). \tag{2.17}$$

We are now in the position to apply the classical construction of elementary symbols, introduced by Coifman-Meyer [1]; see also [[5] Proposition 2.9, Lemma 2.11]. By means of the Fourier expansion of $B_h(x,\xi)$ we obtain:

$$a_h(x,\xi) = \sum_{k\in\mathbb{Z}^n}\frac{1}{\left(1+\frac{\pi}{T}|k|\right)^{n+1}}d_h^k(x)\psi_h^k(\xi), \tag{2.18}$$

where $|k| = |k_1| + \cdots + |k_n|$ and

$$d_h^k(x) = (2T)^{-n} \int\limits_{[-T,T]^n} e^{-i\frac{\pi}{T}k\cdot\eta} \mathcal{M}(D_\eta) b_h(x,\eta)\, d\eta, \ h = -1, 0, \ldots; \tag{2.19}$$

$$\begin{aligned} \psi_h^k(\xi) &= \psi^k\left(2^{-h/M}\xi\right), \quad \text{for } h \geq 0, \text{ with} \\ \psi^k(\xi) &= \frac{1}{\left(1 + \frac{\pi}{T}|k|\right)^{N-n-1}} e^{i\frac{\pi}{T}k\cdot\xi}\theta(\xi); \end{aligned} \tag{2.20}$$

$$\psi_{-1}^k(\xi) = \frac{1}{\left(1 + \frac{\pi}{T}|k|\right)^{N-n-1}} e^{i\frac{\pi}{T}k\cdot\xi}\theta_{-1}(\xi). \tag{2.21}$$

$\mathcal{M}(D)$ in (2.19) is a differential operator of integer order N, such that for any $\phi \in C_0^N(\mathbb{R}^n)$ and $m \in \mathbb{Z}^n$: $(1 + |m|)^N \int e^{-im\cdot x}\phi(x)\, dx = \int e^{-im\cdot x}\mathcal{M}(D)\phi(x)\, dx$. For the existence of $\mathcal{M}(D)$ see [[5]; Lemma 2.11].

The functions d_h^k and ψ_h^k in (2.19), (2.20) satisfy the conditions (2.10)–(2.12).

Since $|d_h^k(x)|$, $|\psi_h^k(\xi)|$ are uniformly bounded, using also (2.11) we can assure that the expansion in (2.18) is absolutely convergent in $L^\infty\left(\mathbb{R}_x^n \times \mathbb{R}_\xi^n\right)$. Finally, plugging (2.18) into $a(x,\xi) = \sum_{h=-1}^{+\infty} a_h(x,\xi)$ and setting $a_k(x,\xi) = \sum_{h=-1}^{\infty} d_h^k(x)\psi_h^k(\xi)$ for each $k \in \mathbb{Z}^n$, we obtain:

$$a(x,\xi) = \sum_{k\in\mathbb{Z}^n} \frac{1}{\left(1 + \frac{\pi}{T}|k|\right)^{n+1}} a_k(x,\xi). \tag{2.22}$$

□

3. Quasi-homogeneous Besov and Sobolev spaces

Let us state the Hilbert space version of the classical Marcinkiewicz-Lizorkin Fourier multipliers theorem. For the proof we refer to [[10], §IV.6].

Lemma 3.1. *Let $m_{k,l}(\xi)$ $(k,l = 1,2,\ldots)$ be n times continuously differentiable functions in $A := \mathbb{R}^n \setminus \left\{\prod_{j=1}^n \xi_j = 0\right\}$ and assume that there exists a positive constant B such that*

$$\left|\prod_{j=1}^n \xi_j^{\gamma_j}\right| \left(\sum_{k,l=1}^\infty |\partial^\gamma m_{k,l}(\xi)|^2\right)^{\frac{1}{2}} \leq B, \quad \xi \in A,\ \gamma \in \{0,1\}^n.$$

Then for every $p \in]1,+\infty[$ and some positive constant $c = c_{B,p,n}$,

$$\left\| \left(\sum_{k=1}^\infty |\mathcal{F}^{-1}\sum_{l=1}^\infty m_{k,l}\mathcal{F}f_l|^2\right)^{\frac{1}{2}} \right\|_{L^p} \leq cB \left\| \left(\sum_{l=1}^\infty |f_l|^2\right)^{\frac{1}{2}} \right\|_{L^p} \tag{3.1}$$

holds true for all sequences $\{f_l\}_{l=1}^{\infty} \subset \mathcal{S}(\mathbb{R}^n)$ *such that* $f_l \equiv 0$ *for all but a finite number of indices* $l = 1, 2, \dots$.

Remark 3.2. By a density argument (3.1) extends to the elements of $L^p(\mathbb{R}^n; \ell^2)$, i.e., the space of all the L^p functions taking values in the Banach space ℓ^2.

For any $s \in \mathbb{R}$ we define the *quasi-homogeneous* Besov space $B^{s,M}_{\infty,\infty}$ to be the set of all $u \in \mathcal{S}'(\mathbb{R}^n)$, such that for some quasi homogeneous dyadic partition of unity $\varphi := \{\varphi_h\}_{h=-1}^{\infty}$:

$$\|u\|^{\varphi}_{B^{s,M}_{\infty,\infty}} := \sup_{h=-1,0,\dots} 2^{sh}\|\varphi_h(D)u\|_{L^\infty} < \infty. \tag{3.2}$$

Exploiting the properties of the quasi-homogeneous dyadic decompositions set in Proposition 2.3 it is not difficult to show that different choices of the partition of unity in (3.2) give equivalent norms, noted by $\|\cdot\|_{B^{s,M}_{\infty,\infty}}$; moreover $B^{s,M}_{\infty,\infty}$ realizes to be a Banach space.

For $M = (1, \dots, 1)$ and $s > 0$ $B^{s,M}_{\infty,\infty}$ is the classical Hölder-Zygmund space of order s, deeply studied for example in [14]. We refer also to [[14]; §10.1] for the study of the quasi-homogeneous spaces $B^{s,M}_{p,q}$, with arbitrary $0 < p, q \le +\infty$.

Definition 3.3. For arbitrary $s \in \mathbb{R}$ and $p \in]1, +\infty[$, the quasi-homogeneous Sobolev space $H^{s,p}_M$ is defined as the class of all tempered distributions $u \in \mathcal{S}'(\mathbb{R}^n)$ such that $\langle D\rangle^s_M u \in L^p(\mathbb{R}^n)$.

For every $s \in \mathbb{R}$ and $p \in]1, +\infty[$, $H^{s,p}_M$ is a Banach space with respect to the norm $\|u\|_{H^{s,p}_M} := \|\langle D\rangle^s_M u\|_{L^p}$ and it admits $\mathcal{S}(\mathbb{R}^n)$ as dense subspace. The spaces $H^{s,p}_M$ can be regarded as a particular case of a more general family of weighted Sobolev spaces $H^{s,p}_\Lambda$, defined in terms of an appropriate weight function Λ and deeply studied in [3], [4].

Proposition 3.4. *Let us consider* $s \in \mathbb{R}$ *and* $p \in]1, +\infty[$*; then, for every quasi-homogeneous dyadic partition of unity* $\varphi := \{\varphi_h\}_{h=-1}^{\infty}$, *there exists a constant* $C > 1$ *such that the inequalities*

$$\frac{1}{C}\left\|\left(\sum_{h=-1}^{\infty} 4^{sh}|\varphi_h(D)u|^2\right)^{\frac{1}{2}}\right\|_{L^p} \le \|u\|_{H^{s,p}_M} \le C\left\|\left(\sum_{h=-1}^{\infty} 4^{sh}|\varphi_h(D)u|^2\right)^{\frac{1}{2}}\right\|_{L^p} \tag{3.3}$$

are verified for any $u \in \mathcal{S}'(\mathbb{R}^n)$.

Proposition 3.4 is a straightforward extension of [[14]; Theorem 2.5.6], which provides the characterization of classical Sobolev spaces, $H^{s,p}$, $1 < p < \infty$, by means of dyadic partitions of unity.

Essentially we can achieve the proof by using Lemma 3.1 instead of the Michlin-Hörmander Lemma (see [14], Proposition 2.5.6).

In the same way the following proposition is obtained from [[9] Propositions 1.1, 1.2], where Marschall states the results for classical Sobolev spaces.

Proposition 3.5. *For every $s \in \mathbb{R}$ and $p \in]1,+\infty[$ there exists a positive constant C such that for all sequences $\{u_h\}_{h=-1}^{\infty}$ of distributions $u_h \in \mathcal{S}'(\mathbb{R}^n)$ whose Fourier transform is respectively supported in $\mathcal{C}_h^{K,M}$, $h = -1, 0, \ldots$, we have*

$$\left\| \sum_{h=-1}^{\infty} u_h \right\|_{H_M^{s,p}} \le C \left\| \left(\sum_{h=-1}^{\infty} 4^{sh} |u_h|^2 \right)^{\frac{1}{2}} \right\|_{L^p} . \tag{3.4}$$

Proposition 3.6. *When $s > 0$ and $p \in]1,+\infty[$ there exists a positive constant C such that estimate (3.4) is true for all the sequences $\{u_h\}_{h=-1}^{\infty}$ of distributions $u_h \in \mathcal{S}'(\mathbb{R}^n)$ with $\operatorname{supp} \hat{u}_h \subseteq B_h^{K,M} := \{\xi \in \mathbb{R}^n : |\xi|_M \le K 2^{h+1}\}$, $h = -1, 0, \ldots$.*

Remark 3.7. With the assumptions on the supports of $\hat{u}_h$ respectively made in Propositions 3.5, 3.6, the series $\sum_{h=-1}^{\infty} u_h$ is convergent in $\mathcal{S}'(\mathbb{R}^n)$ as long as the right-hand side of (3.4) is finite.

4. Action on quasi-homogeneous Sobolev spaces

In this section we apply the results of §2 in the particular case $X = B_{\infty,\infty}^{r,M}$ with positive order r. The main result is:

Theorem 4.1. *Let $a(x,\xi)$ belong to the symbol class $B_{\infty,\infty}^{r,M} S_M^m(N)$, with $r > 0$, $m \in \mathbb{R}$ and $N \ge 2n+1$. Then for every $s \in]-r, r[$ and $p \in]1,+\infty[$ the range of the operator $a(x,D) : \mathcal{S}(\mathbb{R}^n) \to \mathcal{S}'(\mathbb{R}^n)$ is contained in $H_M^{s,p}$ and $a(x,D)$ extends to a continuous operator*

$$a(x,D) : H_M^{s+m,p} \to H_M^{s,p}. \tag{4.1}$$

The proof needs some preliminary remarks.

Let us observe that $a(x,\xi) \in B_{\infty,\infty}^{r,M} S_M^m(N)$ yields $a(x,\xi)\langle\xi\rangle_M^{-m} \in B_{\infty,\infty}^{r,M} S_M^0(N)$. Hence, without loss of generality, we are reduced to the case $m = 0$.

Agreeing with Proposition 2.4 the symbols in $B_{\infty,\infty}^{r,M} S_M^0(N)$ reduce to an expansion of $a_k(x,\xi) \in B_{\infty,\infty}^{r,M} S_M^0(N-n-1)$.

Thus we prove at first the result for elementary symbols; from (2.9) it follows

$$a(x,D)u(x) = \sum_{h=-1}^{\infty} d_h(x)\psi_h(D)u(x), \quad u \in \mathcal{S}(\mathbb{R}^n), \tag{4.2}$$

where the expansion is absolutely convergent in $L^\infty(\mathbb{R}^n)$. By means of a given partition of unity $\{\phi_l\}_{l=-1}^{\infty}$ we write (4.2) in the form

$$\begin{aligned} a(x,D)u(x) &= \sum_{h=-1}^{\infty} \sum_{l=-1}^{\infty} d_{h,l}(x) u_h(x) \\ &= T_1^{(N_0)}u(x) + T_2^{(N_0)}u(x) + T_3^{(N_0)}u(x), \end{aligned} \tag{4.3}$$

where $d_{h,l} := \phi_l(D)d_h$, $u_h := \psi_h(D)u$ and the operators $T_j^{(N_0)}$, $j = 1, 2, 3$, are defined by:

$$T_1^{(N_0)}u := \sum_{h=N_0-1}^{\infty} \sum_{l=-1}^{h-N_0} d_{h,l}u_h, \tag{4.4}$$

$$T_2^{(N_0)}u := \sum_{h=-1}^{\infty} \sum_{l=l_h}^{h+N_0-1} d_{h,l}u_h, \quad l_h = \max\{-1; h - N_0 + 1\}, \tag{4.5}$$

$$T_3^{(N_0)}u := \sum_{l=N_0-1}^{\infty} \sum_{h=-1}^{l-N_0} d_{h,l}u_h. \tag{4.6}$$

N_0 is a fixed positive integer to be precised later.

Proposition 4.2. *For every suitably large positive integer N_0, $T_1^{(N_0)}$ extends to a linear continuous operator*

$$T_1^{(N_0)} : H_M^{s,p} \to H_M^{s,p},$$

for all $s \in \mathbb{R}$ and $p \in]1, +\infty[$.

Proof. Set $U_h^{(N_0)}(x) := \sum_{l=-1}^{h-N_0} d_{h,l}(x)u_h(x)$ $(h \geq N_0 - 1)$. Since $(2\pi)^n\widehat{U_h^{(N_0)}} = \sum_{l=-1}^{h-N_0} \widehat{d_{h,l}} * \widehat{u_h}$, we can find a constant $K' > 1$ such that, for all N_0 sufficiently large, $\operatorname{supp} \widehat{U_h^{(N_0)}} \subseteq \mathcal{C}_h^{K',M}$ for $h \geq N_0 - 1$. Proposition 3.5 assures that for any $s \in \mathbb{R}$, $p \in]1, \infty[$ there exists a positive constant C such that

$$\|T_1^{(N_0)}u\|_{H_M^{s,p}} \leq C \left\| \left(\sum_{h=N_0-1}^{\infty} 4^{sh} \left| U_h^{(N_0)} \right|^2 \right)^{\frac{1}{2}} \right\|_{L^p}.$$

Since $d_h \in B_{\infty,\infty}^{r,M}$ with $r > 0$:

$$\left| U_h^{(N_0)}(x) \right| \leq |u_h(x)| \|d_h\|_{B_{\infty,\infty}^{r,M}} \sum_{l=-1}^{h-N_0} 2^{-rl} \leq C\|d_h\|_{B_{\infty,\infty}^{r,M}} |u_h(x)|, \quad x \in \mathbb{R}^n,$$

with a positive constant $C = C_r$. Since $N - n - 1 \geq n$ the previous estimates yield

$$\|T_1^{(N_0)}u\|_{H_M^{s,p}} \leq C \sup_{h=-1,0,\dots} \|d_h\|_{B_{\infty,\infty}^{r,M}} \|u\|_{H_M^{s,p}}, \tag{4.7}$$

with $C = C_{r,s,p}$. This completes the proof □

Proposition 4.3. *For every suitably large positive integer N_0, $T_2^{(N_0)}$ extends to a linear continuous operator*

$$T_2^{(N_0)} : H_M^{s,p} \to H_M^{s+r,p},$$

for all $s > -r$ and $p \in]1, +\infty[$.

Proof. We set $U_h^{(N_0)} := \sum\limits_{l=l_h}^{h+N_0-1} d_{h,l}u_h$ for $h = -1, 0, \ldots$. Arguing as in the proof of Proposition 4.2, we find the existence of a constant $K' > 1$ such that $\operatorname{supp} \widehat{U_h^{(N_0)}} \subseteq B_h^{K',M}$, for all $h = -1, 0, \ldots$ and any N_0 sufficiently large.

For $s > -r$, Proposition 3.6 gives

$$\|T_2^{(N_0)} u\|_{H_M^{s+r,p}} \leq C \left\| \left(\sum_{h=-1}^{\infty} 4^{(s+r)h} \left| U_h^{(N_0)} \right|^2 \right)^{\frac{1}{2}} \right\|_{L^p},$$

with positive constant C depending only on s, r and p. Moreover, for all $h = -1, 0, \ldots$ we get for every $x \in \mathbb{R}^n$ and some positive constant $C = C_{N_0,r}$:

$$\left| U_h^{(N_0)}(x) \right| \leq |u_h(x)| \sum_{l=l_h}^{h+N_0-1} |d_{h,l}(x)| \leq C \|d_h\|_{B^{r,M}_{\infty,\infty}} 2^{-rh} |u_h(x)|. \tag{4.8}$$

By the help of Proposition 3.4 we conclude

$$\|T_2^{(N_0)} u\|_{H_M^{s+r,p}} \leq C \sup_{h=-1,0,\ldots} \|d_h\|_{B^{r,M}_{\infty,\infty}} \|u\|_{H_M^{s,p}}. \qquad \square$$

Proposition 4.4. *For every positive integer N_0 sufficiently large, $T_3^{(N_0)}$ extends to a continuous linear operator*

$$T_3^{(N_0)} : H_M^{s-r+\delta,p} \to H_M^{s,p},$$

for any real s and positive δ such that $s + \delta < r$.

Proof. Let us set $V_l^{(N_0)} := \sum\limits_{h=-1}^{l-N_0} d_{h,l}u_h$, $l \geq N_0 - 1$. Then, arguing as in the proof of Proposition 4.2, we see that for every N_0 sufficiently large $\operatorname{supp} \widehat{V_l^{(N_0)}} \subseteq C_l^{K',M}$, with a suitable $K' > 1$ independent of N_0. Then by Proposition 3.5, for any real s we get

$$\|T_3^{(N_0)} u\|_{H_M^{s,p}} \leq C \left\| \left(\sum_{l=N_0-1}^{\infty} 4^{sl} \left| V_l^{(N_0)} \right|^2 \right)^{\frac{1}{2}} \right\|_{L^p}, \tag{4.9}$$

with positive constant C depending only on s and p. On the other hand, if $s+\delta < r$, by means of the Cauchy-Schwarz inequality, we compute:

$$\begin{aligned}
\left| V_l^{(N_0)}(x) \right| &\leq \sum_{h=-1}^{l-N_0} |d_{h,l}(x)||u_h(x)| \leq \|d_h\|_{B^{r,M}_{\infty,\infty}} 2^{-rl} \sum_{h=-1}^{l-N_0} |u_h(x)| \\
&\leq 2^{-rl} \|d_h\|_{B^{r,M}_{\infty,\infty}} \left(\sum_{h=-1}^{l-N_0} 4^{-(s-r+\delta)h} \right)^{\frac{1}{2}} \left(\sum_{h=-1}^{l-N_0} 4^{(s-r+\delta)h} |u_h(x)|^2 \right)^{\frac{1}{2}} \\
&\leq C 2^{-(s+\delta)(l-N_0)} \|d_h\|_{B^{r,M}_{\infty,\infty}} \left(\sum_{h=-1}^{\infty} 4^{(s-r+\delta)h} |u_h(x)|^2 \right)^{\frac{1}{2}}, \quad x \in \mathbb{R}^n.
\end{aligned} \tag{4.10}$$

The constant C depends only upon r, s, δ, N_0 and p. Then estimates (4.9), (4.10) and Proposition 3.4 give

$$\begin{aligned}\|T_3^{(N_0)}u\|_{H^{s,p}_M} &\le C \sup_{h=-1,0,\dots} \|d_h\|_{B^{r,M}_{\infty,\infty}} \left\| \left(\sum_{l=N_0-1}^{\infty} 4^{-\delta l} \sum_{h=-1}^{\infty} 4^{(s-r+\delta)h} |u_h|^2 \right)^{\frac{1}{2}} \right\|_{L^p} \\ &\le C \sup_{h=-1,0,\dots} \|d_h\|_{B^{r,M}_{\infty,\infty}} \|u\|_{H^{s-r+\delta,p}_M}\end{aligned}$$

where $C = C(r, s, \delta, N_0, p)$. This ends the proof. □

Proof of Theorem 4.1. For any operator carrying the structure in (4.2) let us take the decomposition (4.3) for some N_0 so large that Propositions 4.2, 4.3, 4.4 work; gathering such results we immediately obtain the mapping property (4.1). More precisely, we are able to prove that for every $r > 0$, $s \in]-r, r[$ and $p \in]1, +\infty[$, there exists a positive constant C such that any pseudo-differential operator of type (4.2) satisfies:

$$\|a(\cdot, D)u\|_{H^{s,p}_M} \le C \sup_{h=-1,0,\dots} \|d_h\|_{B^{r,M}_{\infty,\infty}} \|u\|_{H^{s,p}_M}, \quad u \in \mathcal{S}(\mathbb{R}^n). \tag{4.11}$$

For a general symbol $a(x,\xi) \in B^{r,M}_{\infty,\infty} S^0_M(N)$, (2.8) holds for any $a_k(x,\xi) \in B^{r,M}_{\infty,\infty} S^0_M(N-n-1)$ of type (2.9) with $N - n - 1 \ge n$. Then we get

$$\begin{aligned}\|a(., D)u\|_{H^{s,p}_M} &\le \sum_{k\in\mathbb{Z}^n} c_k \|a_k(., D)u\|_{H^{s,p}_M} \\ &\le C \sup_{k,h=-1,0,\dots} \|d^k_h\|_{B^{r,M}_{\infty,\infty}} \|u\|_{H^{s,p}_M}, \quad u \in \mathcal{S}(\mathbb{R}^n),\end{aligned}$$

with positive constant $C = C_{r,s,p}$. Since the sequences $\{d^k_h\}_{h=-1}^{\infty}$ are bounded in $B^{r,M}_{\infty,\infty}$ uniformly with respect to k (cf. Proposition 2.4; (2.10)), the result follows from the density of $\mathcal{S}(\mathbb{R}^n)$ into $H^{s,p}_M$. □

Corollary 4.5. *For every $r > 0$, $s \in]-r, r[$ and $p \in]1, +\infty[$ there exists a positive constant C such that*

$$\|uv\|_{H^{s,p}_M} \le C \|u\|_{H^{s,p}_M} \|v\|_{B^{r,M}_{\infty,\infty}} \tag{4.12}$$

is true for any $u \in \mathcal{S}(\mathbb{R}^n)$ and $v \in B^{r,M}_{\infty,\infty}$.

Proof. Let $\psi = \{\psi_h\}_{h=-1}^{\infty}$ be an arbitrary quasi-homogeneous dyadic partition of unity. For any $u \in \mathcal{S}(\mathbb{R}^n)$ we can write:

$$uv = \sum_{h=-1}^{\infty} v\psi_h(D)u,$$

so that in the Schwartz class $\mathcal{S}(\mathbb{R}^n)$ the multiplication by v can be regarded as a particular pseudo-differential operator of type (4.2), where $d_h = v$ for all $h \ge -1$. In this case (4.11) just reduces to (4.12). □

References

[1] R.R. Coifman and Y. Meyer, Au delà des opérateurs pseudo-differentiels, Astérisque **57**, Soc. Math. France, 1978.

[2] C. Fefferman, L^p bounds for pseudo-differential operators, *Israel J. Math.* **14** (1973), 413–417.

[3] G. Garello and A. Morando, L^p-bounded pseudodifferential operators and regularity for multi-quasi-elliptic equations, *Integral Equations Operator Theory* **51** (2005), 501–517.

[4] G. Garello and A. Morando, L^p boundedness for pseudodifferential operators with nonsmooth symbols and applications, *Boll. Unione Mat. Ital. Sez. B Artic. Ric. Mat.* **8** (2005), 461–503.

[5] G. Garello and A. Morando, Continuity in weighted Sobolev spaces of L^p type for pseudo-differential operators with completely nonsmooth symbols, in *Advances in Pseudo-differential Operators*, Editors: R. Ashino, P. Boggiatto and M.W. Wong, Birkhäuser, 2004, 91–106.

[6] G. Garello and A. Morando, Continuity in weighted Besov spaces for pseudodifferential operators with non-regular symbols, in *Recent Advances in Operator Theory and its Applications,* Editors: M.A. Kaashoek, S. Seatzu and C. van der Mee, Birkhäuser, 2005, 195–216.

[7] R. Lascar, Propagation des singularités des solutions d'équations pseudodifférentielles quasi-homogènes, *Ann. Inst. Fourier Grenoble* **27** (1977), 79–123.

[8] P.I. Lizorkin, (L_p, L_q)-multipliers of Fourier integrals, *Dokl. Akad. Nauk SSSR* **152** (1963), 808–811.

[9] J. Marschall, Pseudo-differential operators with coefficients in Sobolev spaces, *Trans. Amer. Math. Soc.* **307(1)** (1988), 335–361.

[10] E.M. Stein, *Singular Integrals and Differentiability Properties of Functions*, Princeton University Press, 1970.

[11] M.E. Taylor, *Pseudodifferential Operators*, Princeton University Press, 1981.

[12] M.E. Taylor, *Pseudodifferential Operators and Nonlinear PDE*, Birkhäuser, Basel-Boston-Berlin, 1991.

[13] M.E. Taylor, *Tools for PDE: Pseudodifferential Operators, Paradifferential Operators and Layer Potentials*, American Mathematical Society, 2000.

[14] H. Triebel, *Theory of Function Spaces*, Birkhäuser Verlag, Basel, Boston, Stuttgart, 1983.

[15] M. Yamazaki, The L^p-boundedness of pseudodifferential operators with estimates of parabolic type and product type, *J. Math. Soc. Japan* **38(2)** (1986), 199–225.

Gianluca Garello
Dipartimento di Matematica, Università di Torino
Via Carlo Alberto 10, I-10123 Torino, Italy
e-mail: `gianluca.garello@unito.it`

Alessandro Morando
Dipartimento di Matematica, Facoltà di Ingegneria – Università di Brescia
Via Valotti 9, I-25133 Brescia, Italy
e-mail: `morando@dm.unito.it, morando@ing.unibs.it`

Operator Theory:
Advances and Applications, Vol. 172, 173–206

Continuity and Schatten Properties for Pseudo-differential Operators on Modulation Spaces

Joachim Toft

Abstract. Let $M^{p,q}_{(\omega)}$ be the modulation space with parameters p, q and weight function ω. We prove that if $t \in \mathbf{R}$, $p, p_j, q, q_j \in [1, \infty]$, ω_1, ω_2 and ω are appropriate, and $a \in M^{p,q}_{(\omega)}$, then the pseudo-differential operator $a_t(x, D)$ is continuous from $M^{p_1,q_1}_{(\omega_1)}$ to $M^{p_2,q_2}_{(\omega_2)}$. If in addition $p_j = q_j = 2$, then we establish necessary and sufficient conditions on p and q in order to $a_t(x, D)$ should be a Schatten-von Neumann operator of certain order.

Mathematics Subject Classification (2000). Primary 35S05, 47B10, 47B37; Secondary 42B35, 47B35.

Keywords. Pseudo differential operators, Schatten classes, modulation spaces, non-smooth symbols.

0. Introduction

In [25] and [26], Gröchenig and Heil present a method, based on time-frequency analysis when investigating pseudo-differential operators with non-smooth symbols belonging to non-weighted modulation spaces. Here they make suitable Gabor expansions of the symbols, which in some extent reduce the problems in such way that the symbols are translations and modulations of a fix and well-known function. In that end, they are able to make a somewhat detailed study of compactness, and prove embedding properties between Schatten-von Neumann classes of pseudo-differential operators acting on L^2, and modulation spaces.

Furthermore, they prove that any pseudo-differential operator with symbol in the modulation space $M^{\infty,1}$ (denoted by S_w in [41] by Sjöstrand) is continuous on any non-weighted modulation space $M^{p,q}$. Since $L^2 = M^{2,2}$, it follows in particular that such operators are continuous on L^2, a property which was proved by Sjöstrand in [40], where modulation spaces were used as symbol classes for the first time. Furthermore, since S^0_0, the set of functions which are bounded together with all their derivatives, is contained in $M^{\infty,1}$, it follows from these investigations that any pseudo-differential operator with symbol in S^0_0 is continuous on $M^{p,q}$. The latter result was remarked in the L^2-case in [40], and for general p and q, the result is a special case of Theorem 2.1 in [42] by Tachizawa.

Some further improvements and extensions of the results above have been done since [25, 26, 40–42]. In [7], Boulkhemair extend the L^2 continuity to Fourier integral operators with symbols in $M^{\infty,1}$ and phase functions of rather general types. In the independent papers [27] and [48], continuity for pseudo-differential operators with symbol class $M^{p,q}$ acting on modulation spaces, are considered. In [49] these results were further extended in the case of Weyl operators where the symbols belong to weighted modulation spaces. Some further properties concerning embeddings between Schatten-von Neumann classes in the pseudo-differential calculus and modulation spaces can also be found in [48].

Important parts in this context concern modulation spaces, and their properties. These spaces were introduced by Feichtinger in [13] and [15] during the period 1980–1983 as an appropriate family of function and distribution spaces to have in background when discussing certain problems within time-frequency analysis. The basic theory of such spaces were thereafter established and extended by Feichtinger and Gröchenig (see, e.g., [14, 18, 19, 24], and the references therein). Roughly speaking, for an appropriate weight function ω, the modulation space $M^{p,q}_{(\omega)}$ is obtained by imposing a mixed $L^{p,q}_{(\omega)}$-norm on the short-time Fourier transform of a tempered distribution. The non-weighted modulation space $M^{p,q}$ is then obtained by choosing $\omega = 1$. In terms of modulation spaces it is sometimes easy to obtain information concerning growth and decay properties, as well as certain localization and regularity properties for distributions.

In this paper we continue the discussions in [48, 49] concerning continuity for pseudo-differential operators in background of modulation space theory. (Cf. [51].) More precisely, we consider pseudo-differential operators where the corresponding symbols belong to appropriate modulation spaces, and discuss continuity for such operators when acting on modulation spaces. Especially we are concerned with a somewhat detailed study of continuity and compactness for pseudo-differential operators acting between modulation spaces of Hilbert type in terms of Schatten-von Neumann classes. In particular we investigate trace-class and Hilbert-Schmidt properties.

Except for the Hilbert-Schmidt case it is in general a hard task to find complete characterizations of Schatten-von Neumann classes. One is therefore forced to find embeddings between such classes and other spaces which are more convenient. In Section 4 we discuss embeddings between such classes and modulation spaces, and generalize certain results in [26, 45, 48]. In contrast to the latter papers, the situation in Section 4 is more complicated depending on the fact that we consider operators acting on modulation spaces which involve weight functions of general types, instead of operators acting on L^2. In particular, by choosing the involved weight functions in appropriate ways, we may use our results to discuss Schatten-von Neumann properties for pseudo-differential operators acting between weighted Lebesgue spaces and/or Sobolev spaces of Hilbert type.

The general types of modulation spaces which are involved in the continuity investigations cause new problems comparing to [26, 48, 49]. These problems are overcome by using a related Gabor technique as in [26], leading to a convenient way to expand the symbols, and discretization of certain parts of the problems. The requested results are thereafter obtained by using techniques in modulation space theory, which are well known within time-frequency analysis, in combination with certain duality properties for Schatten-von Neumann classes in pseudo-differential calculus, presented in Section 3, and harmonic analysis.

In order to describe our results in more details we recall the definition of modulation spaces. Assume that $\chi \in \mathscr{S}(\mathbf{R}^m) \setminus 0$, $p, q \in [1, \infty]$ and that ω is an appropriate function on $\mathbf{R}^{2m}$, and let $\tau_x\chi(y) = \chi(y-x)$ when $x, y \in \mathbf{R}^m$. (We use the same notation for the usual functions and distribution spaces as in, e.g., [28].) Then the modulation space $M^{p,q}_{(\omega)}(\mathbf{R}^m)$ consists of all $f \in \mathscr{S}'(\mathbf{R}^m)$ such that

$$\begin{aligned} \|f\|_{M^{p,q}_{(\omega)}} &= \|f\|_{M^{p,q,\chi}_{(\omega)}} \\ &\equiv \Big(\int \Big(\int |\mathscr{F}(f\, \tau_x\chi)(\xi)\omega(x,\xi)|^p \, dx \Big)^{q/p} d\xi \Big)^{1/q} < \infty, \end{aligned} \tag{0.1}$$

with the obvious modifications when $p = \infty$ and/or $q = \infty$. Here $\mathscr{F}$ denotes the Fourier transform on $\mathscr{S}'(\mathbf{R}^m)$, which takes the form

$$\mathscr{F}f(\xi) = \widehat{f}(\xi) = (2\pi)^{-m/2} \int f(x) e^{-i\langle x,\xi\rangle} \, dx$$

when $f \in \mathscr{S}(\mathbf{R}^m)$. Moreover, the function $(x,\xi) \mapsto \mathscr{F}(f\,\tau_x\chi)(\xi)$ is called the *short-time Fourier transform* of f with *window function*, or just *window*, χ for f in the literature. (In the literature, the terms coherent state transform and coherent state also occur.)

Next assume that $t \in \mathbf{R}$ is fixed and that $a \in \mathscr{S}(\mathbf{R}^{2m})$. Then the pseudo-differential operator $a_t(x,D)$ is the continuous operator on $\mathscr{S}(\mathbf{R}^m)$ which is defined by the formula

$$\begin{aligned} (a_t(x,D)f)(x) &= (\operatorname{Op}_t(a)f)(x) \\ &= (2\pi)^{-m} \iint a((1-t)x + ty, \xi) f(y) e^{i\langle x-y,\xi\rangle} \, dy d\xi. \end{aligned} \tag{0.2}$$

The definition of $a_t(x,D)$ extends to any $a \in \mathscr{S}'(\mathbf{R}^{2m})$, and then $a_t(x,D)$ is continuous from $\mathscr{S}(\mathbf{R}^m)$ to $\mathscr{S}'(\mathbf{R}^m)$. (See, e.g., [28], or Section 1.) If $t = 1/2$, then $a_t(x,D)$ is equal to the Weyl operator $a^w(x,D)$ for a. If instead $t = 0$, then the standard (Kohn-Nirenberg) representation $a(x,D)$ is obtained.

In Section 4 we discuss continuity for pseudo-differential operators acting on modulation spaces when the operator symbols belong to modulation spaces. Some of these questions were discussed for Weyl operators (i.e., when $t = 1/2$ in (0.2)) already in Section 5 in [49]. In the first part of Section 4 we extend the results in [49] to arbitrary $t \in \mathbf{R}$. In particular we find appropriate conditions on ω, ω_1, ω_2 and p, q, p_j, q_j for $j = 1, 2$, in order for

$$a_t(x, D) : M^{p_1,q_1}_{(\omega_1)} \to M^{p_2,q_2}_{(\omega_2)}$$

to be continuous when $a \in M^{p,q}_{(\omega)}$.

The second part of Section 4 is devoted to the case when the $M^{p_j,q_j}_{(\omega_j)}$ above are Hilbert spaces, i.e., $p_j = q_j = 2$ for $j = 1, 2$. In this case a more detailed compactness study in terms of Schatten-von Neumann properties is performed. These considerations are dependent on some preparations which are made in Section 2 and in Section 3. Recall that an operator T from $M^{2,2}_{(\omega_1)}$ to $M^{2,2}_{(\omega_2)}$ belongs to $\mathscr{I}_p$, the set of Schatten-von Neumann operators of order $p \in [1, \infty]$, if and only if

$$\sup \Big(\sum |(Tf_j, g_j)_{M^2_{(\omega_2)}}|^p \Big)^{1/p} < \infty,$$

where the supremum is taken over all orthonormal sequences (f_j) in $M^{2,2}_{(\omega_1)}$ and (g_j) in $M^{2,2}_{(\omega_2)}$. In particular, this implies that $\mathscr{I}_\infty$ is the set of linear and continuous operators, and that T is compact when $T \in \mathscr{I}_p$ and $p < \infty$. (Cf. [37, 39].) We are then concerned with classification and embedding properties for the set $s_{t,p}(\omega_1, \omega_2)$ which consists of all $a \in \mathscr{S}'(\mathbf{R}^{2m})$ such that $a_t(x, D) \in \mathscr{I}_p$. In Section 4 we prove that

$$M^{p,q_1}_{(\omega)} \subseteq s_{t,p}(\omega_1, \omega_2) \subseteq M^{p,q_2}_{(\omega)}, \tag{0.3}$$

for appropriate choices of q_1, q_2 and ω. In particular, our investigations concern Schatten-von Neumann properties for pseudo-differential operators which map the Sobolev space $H^2_{s_1}$ or the weighted Lebesgue space $L^2_{s_1}$ to $H^2_{s_2}$ or $L^2_{s_2}$, since each one of these spaces agrees with $M^2_{(\omega)}$ if ω is chosen in an appropriate way. Here $H^p_s(\mathbf{R}^m)$ is the Sobolev space of distributions with s derivatives in L^p, i.e., the set of all $f \in \mathscr{S}'$ such that $(1 - \Delta)^{s/2} f \in L^p(\mathbf{R}^m)$.

In the end of Section 4 we also use (0.3) and embedding results in Section 4 in [50] between modulation spaces and Besov spaces, to obtain embedding properties between the $s_{t,p}$-spaces and Besov spaces.

Finally we remark that the results here are applied in [52] in the present volume and in a forthcoming joint paper with Paolo Boggiatto, in order to obtain continuity properties for Toeplitz operators. In particular some improvements of certain results in [9, 48] are presented.

1. Preliminaries

In this section we discuss basic properties for modulation spaces. The proofs are in many cases omitted since they can be found in [5, 13–16, 18–20, 25, 44–47, 49].

We start by recalling some properties of the weight functions which are involved. We say that the function $\omega \in L^\infty_{loc}(\mathbf{R}^m)$ is *v-moderate* for some appropriate positive function $v \in L^\infty_{loc}(\mathbf{R}^m)$, if there is a constant $C > 0$ such that

$$\omega(x_1 + x_2) \le C\omega(x_1)v(x_2), \quad x_1, x_2 \in \mathbf{R}^m. \tag{1.1}$$

The function v is then said to *moderate* ω, and if ω is v-moderate for some polynomial v, then ω is called polynomial moderate. If in addition (1.1) holds for $\omega = v$, then v is said to be a *moderate* or *submultiplicative function.*

As in [49] we let $\mathscr{P}(\mathbf{R}^m)$ denote the cone which consists of all polynomial moderate functions on $\mathbf{R}^m$.

Note that if $\omega \in \mathscr{P}(\mathbf{R}^m)$, then $\omega(x) + \omega(x)^{-1} \le P(x)$, $x \in \mathbf{R}^m$ for some polynomial P on $\mathbf{R}^m$.

Let ω_1 and ω_2 be positive functions. If ω_2/ω_1 is bounded, then we write $\omega_2 \prec \omega_1$. They are called equivalent when $\omega_1 \prec \omega_2 \prec \omega_1$, and then and we write $\omega_1 \sim \omega_2$.

If $\mathscr{H}$ is a Hilbert space, then its scalar product is denoted by $(\,\cdot\,,\,\cdot\,)_{\mathscr{H}}$, or $(\,\cdot\,,\,\cdot\,)$ when there are no confusions about the Hilbert space structure.

The duality between a topological vector space and its dual is denoted by $\langle\,\cdot\,,\,\cdot\,\rangle$. For admissible a and b in $\mathscr{S}'(\mathbf{R}^m)$, we set $(a, b) = \langle a, \overline{b}\rangle$, and it is obvious that $(\,\cdot\,,\,\cdot\,)$ on L^2 is the usual scalar product.

Next assume that $\mathscr{B}_1$ and $\mathscr{B}_2$ are topological spaces. Then $\mathscr{B}_1 \hookrightarrow \mathscr{B}_2$ means that $\mathscr{B}_1$ is continuously embedded in $\mathscr{B}_2$. In the case that $\mathscr{B}_1$ and $\mathscr{B}_2$ are Banach spaces, $\mathscr{B}_1 \hookrightarrow \mathscr{B}_2$ is equivalent to $\mathscr{B}_1 \subseteq \mathscr{B}_2$ and $\|x\|_{\mathscr{B}_2} \le C\|x\|_{\mathscr{B}_1}$, for some constant $C > 0$ which is independent of $x \in \mathscr{B}_1$.

Next let V_1 and V_2 be vector spaces such that $V_1 \oplus V_2 = \mathbf{R}^m$ and $V_2 = V_1^\perp$, and assume that $v_0 \in \mathscr{S}'(V_1)$ and that $v(x_1, x_2) = (v_0 \otimes 1)(x_1, x_2)$, where $x_j \in V_j$ for $j = 1, 2$. Then $v(x_1, x_2)$ is identified with $v_0(x_1)$, and we set $v(x_1, x_2) = v(x_1)$.

Assume that $\omega \in \mathscr{P}(\mathbf{R}^{2m})$, $p, q \in [1, \infty]$, and that $\chi \in \mathscr{S}(\mathbf{R}^m) \setminus 0$. Then recall that the *modulation space* $M^{p,q}_{(\omega)}(\mathbf{R}^m)$ is the set of all $f \in \mathscr{S}'(\mathbf{R}^m)$ such that (0.1) holds. We note that the definition of $M^{p,q}_{(\omega)}(\mathbf{R}^m)$ is independent of the choice of window χ, and that different choices of χ give rise to equivalent norms. (See Proposition 1.2 below.)

If $\omega = 1$, then the notation $M^{p,q}$ is used instead of $M^{p,q}_{(\omega)}$. Moreover we set $M^p_{(\omega)} = M^{p,p}_{(\omega)}$ and $M^p = M^{p,p}$.

Remark 1.1. We are also concerned with the following family of function and distribution spaces which are related to the Wiener amalgam spaces. Assume that $p, q \in [1, \infty]$ and that $\omega \in \mathscr{P}(\mathbf{R}^{2m})$. Then the space $W^{p,q}_{(\omega)}(\mathbf{R}^m)$ consists of all $a \in \mathscr{S}'(\mathbf{R}^m)$ such that

$$\|a\|_{W^{p,q}_{(\omega)}} = \Big(\int \Big(\int |\mathscr{F}(a\,\tau_x\chi)(\xi)\omega(x,\xi)|^p\,d\xi\Big)^{q/p}\,dx\Big)^{1/q}$$

is finite. (Cf. Definition 4 in [20].)

We recall that $W^{p,q}_{(\omega)} = \mathscr{F} M^{p,q}_{(\omega_0)}$ when $\omega_0(x,\xi) = \omega(-\xi,x) \in \mathscr{P}(\mathbf{R}^{2m})$. In fact, let $\check{\chi}(x) = \chi(-x)$ as usual. Then Parseval's formula and a change of the order of integration shows that

$$|\mathscr{F}^{-1}(\widehat{a}\,\tau_\xi\widehat{\chi})(x)| = |\mathscr{F}(a\,\tau_x\check{\chi})(\xi)|, \tag{1.2}$$

and the assertion follows. We refer to [15,20] for more facts about the $W^{p,q}_{(\omega)}$-spaces.

The convention of indicating weight functions with parenthesis is used also in other situations. For example, if $\omega \in \mathscr{P}(\mathbf{R}^m)$, then $L^p_{(\omega)}(\mathbf{R}^m)$ is the set of all measurable functions f on $\mathbf{R}^m$ such that $f\omega \in L^p(\mathbf{R}^m)$, i.e., such that $\|f\|_{L^p_{(\omega)}} \equiv \|f\omega\|_{L^p}$ is finite.

Next we consider the Fourier transform of functions and distributions defined on $\mathbf{R}^{2m}$. By interpreting $\mathbf{R}^{2m}$ as the phase space with dual variables (y,η), we let the phase space Fourier transform, $\widetilde{\mathscr{F}}$, be defined by the formula

$$(\widetilde{\mathscr{F}}f)(y,\eta) = \widehat{f}(y,\eta) \equiv (2\pi)^{-m} \iint f(x,\xi)e^{-i(\langle x,\eta\rangle + \langle y,\xi\rangle)}\,dxd\xi, \tag{1.3}$$

when $f \in L^1(\mathbf{R}^{2m})$, i.e., $(\widetilde{\mathscr{F}}f)(x,\xi) = (\mathscr{F}f)(\xi,x)$. Then $\widetilde{\mathscr{F}}$ is a homeomorphism on $\mathscr{S}(\mathbf{R}^{2m})$ which extends to a homeomorphism on $\mathscr{S}'(\mathbf{R}^{2m})$ and to a unitary map on $L^2(\mathbf{R}^{2m})$, since similar facts hold for $\mathscr{F}$.

We use the notation $\widetilde{M}^{p,q}_{(\omega)}$, $\widetilde{M}^p_{(\omega)}$, $\widetilde{M}^{p,q}$ and $\widetilde{M}^p$ instead of $M^{p,q}_{(\omega)}$, $M^p_{(\omega)}$, $M^{p,q}$ and M^p respectively, when $\widetilde{\mathscr{F}}$ is used instead of $\mathscr{F}$, in the definition of modulation spaces.

The following proposition is a consequence of well-known facts in [15,25]. Here and in what follows, we let p' denote the conjugate exponent of p, i.e., $1/p+1/p' = 1$.

Proposition 1.2. *Assume that $p,q,p_j,q_j \in [1,\infty]$ for $j=1,2$, and that $\omega,\omega_1,\omega_2,v \in \mathscr{P}(\mathbf{R}^{2m})$ are such that ω is v-moderate. Then the following are true:*

(1) *if $\chi \in M^1_{(v)}(\mathbf{R}^m)\setminus 0$, then $f \in M^{p,q}_{(\omega)}(\mathbf{R}^m)$ if and only if (0.1) holds, i.e., $M^{p,q}_{(\omega)}(\mathbf{R}^m)$ is independent of the choice of χ. Moreover, $M^{p,q}_{(\omega)}$ is a Banach space under the norm in (0.1), and different choices of χ give rise to equivalent norms;*

(2) *if $p_1 \le p_2$, $q_1 \le q_2$ and $\omega_2 \prec \omega_1$, then*

$$\mathscr{S}(\mathbf{R}^m) \hookrightarrow M^{p_1,q_1}_{(\omega_1)}(\mathbf{R}^m) \hookrightarrow M^{p_2,q_2}_{(\omega_2)}(\mathbf{R}^m) \hookrightarrow \mathscr{S}'(\mathbf{R}^m);$$

(3) *the L^2 product $(\,\cdot\,,\,\cdot\,)$ from $\mathscr{S}$ extends to a continuous map from $M^{p,q}_{(\omega)}(\mathbf{R}^m) \times M^{p',q'}_{(1/\omega)}(\mathbf{R}^m)$ to $\mathbf{C}$. On the other hand, if $\|a\| = \sup|(a,b)|$, where the supremum is taken over all $b \in M^{p',q'}_{(1/\omega)}(\mathbf{R}^m)$ such that $\|b\|_{M^{p',q'}_{(1/\omega)}} \le 1$, then $\|\cdot\|$ and $\|\cdot\|_{M^{p,q}_{(\omega)}}$ are equivalent norms;*

(4) *if* $p,q<\infty$, *then* $\mathscr{S}(\mathbf{R}^m)$ *is dense in* $M^{p,q}_{(\omega)}(\mathbf{R}^m)$. *The dual space of* $M^{p,q}_{(\omega)}(\mathbf{R}^m)$ *can be identified with* $M^{p',q'}_{(1/\omega)}(\mathbf{R}^m)$, *through the form* $(\,\cdot\,,\,\cdot\,)_{L^2}$. *Moreover,* $\mathscr{S}(\mathbf{R}^m)$ *is weakly dense in* $M^{\infty}_{(\omega)}(\mathbf{R}^m)$.

Proposition 1.2 (1) permits us be rather vague about to the choice of $\chi\in M^1_{(v)}\setminus 0$ in (0.1). For example, if $C>0$ is a constant and Ω is a subset of $\mathscr{S}'$, then $\|a\|_{M^{p,q}_{(\omega)}}\le C$ for every $a\in\Omega$, means that the inequality holds for some choice of $\chi\in M^1_{(v)}\setminus 0$ and every $a\in\Omega$. Evidently, for any other choice of $\chi\in M^1_{(v)}\setminus 0$, a similar inequality is true although C may have to be replaced by a larger constant, if necessary.

Next we discuss weight functions which are common in the applications. For any $s,t\in\mathbf{R}$, set

$$\sigma_t(x)=\langle x\rangle^t,\quad \sigma_{s,t}(x,\xi)=\langle\xi\rangle^s\langle x\rangle^t, \tag{1.4}$$

when $x,\xi\in\mathbf{R}^m$. Then it follows that $\sigma_t\in\mathscr{P}_0(\mathbf{R}^m)$ and $\sigma_{s,t}\in\mathscr{P}_0(\mathbf{R}^{2m})$ for every $s,t\in\mathbf{R}$, and σ_t is $\sigma_{|t|}$-moderate and $\sigma_{s,t}$ is $\sigma_{|s|,|t|}$-moderate. Obviously, $\sigma_s(x,\xi)=(1+|x|^2+|\xi|^2)^{s/2}$, and $\sigma_{s,t}=\sigma_t\otimes\sigma_s$. Moreover, if $\omega\in\mathscr{P}(\mathbf{R}^m)$, then ω is σ_t-moderate provided t is chosen large enough.

For convenience, we use the notations L^p_s, $M^{p,q}_s$ and $M^{p,q}_{s,t}$ instead of $L^p_{(\sigma_s)}$, $M^{p,q}_{(\sigma_s)}$ and $M^{p,q}_{(\sigma_{s,t})}$ respectively.

It is also convenient to let $\mathcal{M}^{p,q}_{(\omega)}(\mathbf{R}^m)$ be the completion of $\mathscr{S}(\mathbf{R}^m)$ under the norm $\|\cdot\|_{M^{p,q}_{(\omega)}}$. Then $\mathcal{M}^{p,q}_{(\omega)}\subseteq M^{p,q}_{(\omega)}$ with equality if and only if $p<\infty$ and $q<\infty$. It follows that most of the properties which are valid for $M^{p,q}_{(\omega)}(\mathbf{R}^m)$, also hold for $\mathcal{M}^{p,q}_{(\omega)}(\mathbf{R}^m)$.

Remark 1.3. Assume that $p,q,q_1,q_2\in[1,\infty]$. Then the following properties for modulation spaces hold:

(1) if $q_1\le\min(p,p')$ and $q_2\ge\max(p,p')$, then $M^{p,q_1}\subseteq L^p\subseteq M^{p,q_2}$. In particular, $M^2=L^2$;

(2) $S^0_0=\cap_{s\in\mathbf{R}}M^{\infty,1}_{s,0}$;

(3) if $\omega\in\mathscr{P}(\mathbf{R}^{2m})$ is such that $\omega(x,\xi)=\omega(x)$, then $M^{p,q}_{(\omega)}(\mathbf{R}^m)\hookrightarrow C(\mathbf{R}^m)$ if and only if $q=1$;

(4) $M^{1,\infty}$ is a convolution algebra which contains all measures on $\mathbf{R}^m$ with bounded mass;

(5) if Ω is a subset of $\mathscr{P}(\mathbf{R}^{2m})$ such that for any polynomial P on $\mathbf{R}^{2m}$, there is an element $\omega\in\Omega$ such that P/ω is bounded, then

$$\mathscr{S}(\mathbf{R}^m)=\cap_{\omega\in\Omega}M^{p,q}_{(\omega)}(\mathbf{R}^m),\quad \mathscr{S}'(\mathbf{R}^m)=\cup_{\omega\in\Omega}M^{p,q}_{(1/\omega)}(\mathbf{R}^m);$$

(6) if $s,t\in\mathbf{R}$ are such that $t\ge 0$, then

$$M^2_{s,0}=H^2_s,\quad M^2_{0,s}=L^2_s,\quad\text{and}\quad M^2_t=L^2_t\cap H^2_t.$$

(See, e.g., [13–15, 17–20, 25, 47–49].)

Next we recall some facts from Chapter 12 and Chapter 13 in [25] concerning extension of the Gabor theory to modulation spaces. Let $(x_j)_{j\in I}$ and $(\xi_k)_{k\in I}$ be lattices in $\mathbf{R}^m$, and assume that $\chi \in \mathscr{S}(\mathbf{R}^m)$ is fixed and satisfies

$$C^{-1} \le \sum_{j\in I} |\chi(x-x_j)|^2 \le C, \quad x \in \mathbf{R}^m, \tag{1.5}$$

for some constant C. If (ξ_k) is sufficiently dense, then it follows from [25] and Section 7.3 in [28] that there exists a function $\psi \in \mathscr{S}(\mathbf{R}^m)$ such that

$$f(x) = \sum_{j,k\in I} c_{j,k}(f) e^{i\langle x,\xi_k\rangle} \chi(x-x_j) \tag{1.6}$$

$$= \sum_{j,k\in I} d_{j,k}(f) e^{i\langle x,\xi_k\rangle} \psi(x-x_j), \tag{1.6$'$}$$

for every $f \in \mathscr{S}'(\mathbf{R}^m)$, where $c_{j,k}(f)$ and $d_{j,k}(f)$ are the "Fourier coefficients" for f, given by the formulas

$$c_{j,k}(f) = c_{j,k} = \mathscr{F}(f\tau_{x_j}\psi)(\xi_k) \quad \text{and} \tag{1.7}$$

$$d_{j,k}(f) = d_{j,k} = \mathscr{F}(f\tau_{x_j}\chi)(\xi_k). \tag{1.7$'$}$$

Here the sums converge in $\mathscr{S}'(\mathbf{R}^m)$.

In order to present some further properties in the case of modulation spaces, it is convenient to consider the following sequence spaces. Assume that $\lambda = (\lambda_{j,k})_{j,k\in I}$ is a (fix) sequence of non-negative numbers, and that $p,q \in [1,\infty]$. Then let $l^{p,q}_{(\lambda)}$ be the Banach space of sequences $(c_{j,k})_{j,k\in I}$ of complex numbers such that

$$\|(c_{j,k})_{j,k\in I}\|_{l^{p,q}_{(\lambda)}} \equiv \Big(\sum_k \Big(\sum_j |c_{j,k}\lambda_{j,k}|^p\Big)^{q/p}\Big)^{1/q}$$

is finite. Also let $\ell^{p,q}_{(\lambda)}$ be the completion of all finite sequences (i.e., all sequences $(c_{j,k})_{j,k\in I}$ such that only a finite numbers of $c_{j,k}$ are non-zero) under the norm $\|\cdot\|_{l^{p,q}_{(\lambda)}}$. Furthermore we set $l^{p,q} = l^{p,q}_{(\lambda)}$ when $\lambda_j = 1$ for every j, $l^p_{(\lambda)} = l^{p,p}_{(\lambda)}$, and similarly for the $\ell^{p,q}_{(\lambda)}$-spaces.

The following proposition shows that essential parts of the Gabor theory, in the context of L^2-spaces, carry over to modulation spaces. This shows in particular that there is a convenient way to discretize modulation spaces. The proof is omitted, since the result follows from Chapter 12 and 13 in [25].

Proposition 1.4. *Let $(x_j)_{j\in I} \subseteq \mathbf{R}^m$ be a lattice, $\omega \in \mathscr{P}(\mathbf{R}^{2m})$, and assume that $\chi \in \mathscr{S}(\mathbf{R}^m)$ satisfies (1.5). Then there is a lattice $(\xi^0_k)_{k\in I_0} \subseteq \mathbf{R}^m$ and a function $\psi \in \mathscr{S}(\mathbf{R}^m)$ such that if $(\xi_k)_{k\in I} \subseteq \mathbf{R}^m$ is a lattice which contains $(\xi^0_k)_{k\in I_0}$, the following are true.*

(1) *if $f \in \mathscr{S}'(\mathbf{R}^m)$, and $c_{j,k}(f)$ and $d_{j,k}(f)$ are given by (1.7) and (1.7)$'$ respectively, then (1.6) and (1.6)$'$ hold;*

(2) *if* $\lambda = (\omega(x_j,\xi_k))_{j,k}$ *and* $f \in \mathscr{S}'(\mathbf{R}^m)$, *then the following conditions are equivalent:*
   (i) $f \in M^{p,q}_{(\omega)}(\mathbf{R}^m)$;
   (ii) $(c_{j,k}(f))_{j,k} \in l^{p,q}_{(\lambda)}$;
   (iii) $(d_{j,k}(f))_{j,k} \in l^{p,q}_{(\lambda)}$.

   Moreover, if the conditions (i)–(iii) *are fulfilled and in addition* $p, q < \infty$, *then the sums in* (1.6) *and* (1.6)$'$ *converge to* f *in* $M^{p,q}_{(\omega)}$. *Furthermore, the norms* $f \mapsto \|(c_{j,k}(f))_{j,k}\|_{l^{p,q}_{(\lambda)}}$, $f \mapsto \|(d_{j,k}(f))_{j,k}\|_{l^{p,q}_{(\lambda)}}$ *and* $\|\cdot\|_{M^{p,q}_{(\omega)}}$ *are equivalent.*

Remark 1.5. We note that the equivalence (i)–(iii) in Proposition 1.4 holds, after $M^{p,q}_{(\omega)}$ and $l^{p,q}_{(\lambda)}$ have been replaced by $\mathcal{M}^{p,q}_{(\omega)}$ and $\ell^{p,q}_{(\lambda)}$ respectively.

Next we discuss (complex) interpolation properties for modulation spaces. Such properties were carefully investigated in [15] for non-weighted modulation spaces, and thereafter extended in several directions in [19], were interpolation properties for coorbit spaces were established. As a consequence of [19] we have the following proposition. In order to be more self-contained, we give a short proof based on Theorem 1.4 and Remark 1.5.

Proposition 1.6. *Assume that* $0 < \theta < 1$ *and that* $p, q, p_j, q_j \in [1, \infty]$ *satisfy*

$$\frac{1}{p} = \frac{1-\theta}{p_1} + \frac{\theta}{p_2} \quad \text{and} \quad \frac{1}{q} = \frac{1-\theta}{q_1} + \frac{\theta}{q_2}.$$

Also assume that $\omega_1, \omega_2 \in \mathscr{P}(\mathbf{R}^{2m})$ *and let* $\omega = \omega_1^{1-\theta}\omega_2^{\theta}$. *Then*

$$(\mathcal{M}^{p_1,q_1}_{(\omega_1)}(\mathbf{R}^m), \mathcal{M}^{p_2,q_2}_{(\omega_2)}(\mathbf{R}^m))_{[\theta]} = \mathcal{M}^{p,q}_{(\omega)}(\mathbf{R}^m).$$

Proof. In view of Remark 1.5 it suffices to prove that

$$(\ell^{p_1,q_1}_{(\lambda^1)}, \ell^{p_2,q_2}_{(\lambda^2)})_{[\theta]} = \ell^{p,q}_{(\lambda)}, \tag{1.8}$$

where $\lambda^i = (\lambda^i_{j,k})_{j,k\ge 0}$ for $i = 1, 2$ and $\lambda = (\lambda_{j,k})_{j,k\ge 0}$ are sequences of positive numbers such that $\lambda_{j,k} = (\lambda^1_{j,k})^{1-\theta}(\lambda^2_{j,k})^{\theta}$. Furthermore, by using the operator $T(z)$ for $0 \le \operatorname{Re} z \le 1$, defined by the formula

$$T(z)\big((c_{j,k})_{j,k\ge 0}\big) = (c_{j,k}(\lambda^1_{j,k})^{1-z}(\lambda^2_{j,k})^{z})_{j,k\ge 0},$$

it follows that we may assume that each $\lambda^i_{j,k}$ is equal to 1. This means that we shall prove $(\ell^{p_1,q_1}, \ell^{p_2,q_2})_{[\theta]} = \ell^{p,q}$. The result is now a consequence of Theorem 5.1.1 and Theorem 5.1.2 in [1]. The proof is complete. □

Next we recall some facts in Chapter XVIII in [28] concerning pseudo-differential operators. Assume that $a \in \mathscr{S}(\mathbf{R}^{2m})$, and that $t \in \mathbf{R}$ is fixed. Then the pseudo-differential operator $a_t(x, D)$ in (0.2) is a linear and continuous operator on

$\mathscr{S}(\mathbf{R}^m)$, as remarked in the introduction. For general $a \in \mathscr{S}'(\mathbf{R}^{2m})$, the pseudo-differential operator $a_t(x,D)$ is defined as the continuous operator from $\mathscr{S}(\mathbf{R}^m)$ to $\mathscr{S}'(\mathbf{R}^m)$ with distribution kernel

$$K_{t,a}(x,y) = (2\pi)^{-m/2}(\mathscr{F}_2^{-1}a)((1-t)x+ty, y-x), \tag{1.9}$$

Here $\mathscr{F}_2 F$ is the partial Fourier transform of $F(x,y) \in \mathscr{S}(\mathbf{R}^{2m})$ with respect to the y-variable. This definition makes sense, since the mappings $\mathscr{F}_2$ and $F(x,y) \mapsto F((1-t)x+ty, y-x)$ are homeomorphisms on $\mathscr{S}'(\mathbf{R}^{2m})$. We also note that this definition of $a_t(x,D)$ agrees with the operator in (0.2) when $a \in \mathscr{S}(\mathbf{R}^{2m})$.

Furthermore, any linear and continuous operator T from $\mathscr{S}(\mathbf{R}^m)$ to $\mathscr{S}'(\mathbf{R}^m)$ has a distribution kernel K in $\mathscr{S}'(\mathbf{R}^{2m})$ in view of kernel theorem of Schwartz. By Fourier's inversion formula we may then find a unique $a \in \mathscr{S}'$ such that (1.9) is fulfilled with $K = K_{t,a}$. Consequently, for every fixed $t \in \mathbf{R}$, there is a one to one correspondence between linear and continuous operators from $\mathscr{S}(\mathbf{R}^m)$ to $\mathscr{S}'(\mathbf{R}^m)$, and $\mathrm{Op}_t(\mathscr{S}'(\mathbf{R}^{2m}))$, the set of all $a_t(x,D)$ such that $a \in \mathscr{S}'(\mathbf{R}^{2m})$.

In particular, if $a \in \mathscr{S}'(\mathbf{R}^{2m})$ and $s,t \in \mathbf{R}$, then there is a unique $b \in \mathscr{S}'(\mathbf{R}^{2m})$ such that $a_s(x,D) = b_t(x,D)$. By straightforward applications of Fourier's inversion formula, it follows that

$$a_s(x,D) = b_t(x,D) \quad \Leftrightarrow \quad b(x,\xi) = e^{i(t-s)\langle D_x, D_\xi\rangle}a(x,\xi). \tag{1.10}$$

(Cf. Section 18.5 in [28].)

We may also express the relations between a and b here above in terms of convolution operators. In fact, by Fourier's inversion formula it follows that if $t \neq 0$ and $\Phi(x,\xi) = \langle x,\xi\rangle/t$, then there is a constant c such that $e^{it\langle D_x,D_\xi\rangle}a = c\,e^{i\Phi} * a$. If instead $t = 0$, then $e^{it\langle D_x,D_\xi\rangle}a = a = \delta * a$. These relations motivate us to consider continuity properties for operators of the form

$$f \mapsto S_\Phi f \equiv (e^{i\Phi} \otimes \delta_{V_2}) * f, \tag{1.11}$$

where δ_{V_2} is the delta function on the vector space $V_2 \subseteq \mathbf{R}^m$ and Φ is a real-valued and non-degenerate quadratic form on $V_1 = V_2^\perp$. The operator S_Φ is essentially a composition of a partial Fourier transform with respect to the variables in V_1 and a non-degenerate matrix $A_\Phi/2$. In particular, Fourier's inversion formula gives that S_Φ is a homeomorphism on $\mathscr{S}(\mathbf{R}^m)$ which extends uniquely to a homeomorphism on $\mathscr{S}'(\mathbf{R}^m)$.

In a similar way as in [45–49], we are especially concerned with continuity properties of S_Φ when acting on modulation spaces. It was proved in [45] that such operators are continuous on any $M^{p,q}$. In [49] the latter result was extended to modulation spaces of the form $M^{p,q}_{(\omega)}$, where $\omega(x,\xi) = \omega(\xi)$. In the following proposition we show that the arguments in [49] can be used in a broader context.

Proposition 1.7. *Assume that* $\chi \in \mathscr{S}(\mathbf{R}^m)$, $\omega \in \mathscr{P}(\mathbf{R}^{2m})$, $p,q \in [1,\infty]$, *and that* $V_1, V_2 \subseteq \mathbf{R}^m$ *are vector spaces such that* $V_2 = V_1^\perp$. *Also assume that* Φ *is a real-valued and non-degenerate quadratic form on* V_1, *and let* $A_\Phi/2$ *be the corresponding* *matrix. If* $\xi = (\xi_1, \xi_2)$ *where* $\xi_j \in V_j$ *for* $j = 1, 2$, *then*

$$\begin{aligned} \|S_\Phi f\|_{M^{p,q,\chi}_{(\omega_\Phi)}} = \|f\|_{M^{p,q,\psi}_{(\omega)}}, \quad &\text{where} \quad f \in \mathscr{S}'(\mathbf{R}^m), \\ \omega_\Phi(x,\xi) = \omega(x - A_\Phi^{-1}\xi_1, \xi) \quad &\text{and} \quad \psi = S_\Phi \chi. \end{aligned} \tag{1.12}$$

In particular, the following are true:

(1) *the map* (1.11) *on* $\mathscr{S}'(\mathbf{R}^m)$ *restricts to a homeomorphism from* $M^{p,q}_{(\omega)}(\mathbf{R}^m)$ *to* $M^{p,q}_{(\omega_\Phi)}(\mathbf{R}^m)$;

(2) *if* $t \in \mathbf{R}$, $\omega_0 \in \mathscr{P}(\mathbf{R}^{2m} \oplus \mathbf{R}^{2m})$, *and*

$$\omega_t(x,\xi,y,\eta) = \omega_0(x - ty, \xi - t\eta, y, \eta),$$

then the map $e^{it\langle D_x, D_\xi\rangle}$ *on* $\mathscr{S}'(\mathbf{R}^{2m})$ *restricts to a homeomorphism from* $\widetilde{M}^{p,q}_{(\omega_0)}(\mathbf{R}^{2m})$ *to* $\widetilde{M}^{p,q}_{(\omega_t)}(\mathbf{R}^{2m})$.

Proof. We only prove the first part of the proposition in the case $V_1 = \mathbf{R}^m$, leaving the modifications to the general case for the reader. Assume that $f \in \mathscr{S}'(\mathbf{R}^m)$ and $\chi \in \mathscr{S}(\mathbf{R}^m)$. It follows easily that the Hilbert adjoint for S_Φ is equal to $S_{-\Phi}$. This gives

$$|\mathscr{F}(S_\Phi f\, \tau_x \chi)(\xi)| = |(S_\Phi f, \tau_x \overline{\chi} e^{i\langle \xi, \cdot\rangle})| = |(f, S_{-\Phi}(\tau_x \overline{\chi} e^{i\langle \xi, \cdot\rangle}))|. \tag{1.13}$$

We have to analyze $S_{-\Phi}(\tau_x \overline{\chi} e^{i\langle \xi, \cdot\rangle})$. By straightforward computations it follows that

$$\begin{aligned} S_{-\Phi}(\tau_x \overline{\chi} e^{i\langle \xi, \cdot\rangle})(y) &= \int e^{-i\Phi(y-z)} \overline{\chi}(z - x) e^{i\langle z, \xi\rangle}\, dz \\ &= e^{i\langle x, \xi\rangle} \int e^{-i\Phi(y-x-z)} \chi(\overline{z}) e^{i\langle z, \xi\rangle}\, dz \\ &= e^{i\langle y, \xi\rangle} e^{i\Psi(x,\xi)} \int e^{-i\Phi(y-(x-A_\Phi^{-1}\xi)-z)} \overline{\chi(z)}\, dz \\ &= e^{i\langle y, \xi\rangle} e^{i\Psi(x,\xi)} \overline{(\tau_{x-A_\Phi^{-1}\xi}(S_\Phi \chi))(y)}, \end{aligned}$$

for some real quadratic form Ψ on $\mathbf{R}^{2m}$. By inserting this into (1.13) we get

$$|\mathscr{F}(S_\Phi f\, \tau_x \chi)(\xi)| = |(f, e^{i\langle\,\cdot\,,\xi\rangle} \overline{(\tau_{x-A_\Phi^{-1}\xi}(S_\Phi \chi))})| = |\mathscr{F}(f\, \tau_{x-A_\Phi^{-1}\xi}\psi)(\xi)|.$$

The assertions (1) and (1.12) now follow by applying the $L^{p,q}_{(\omega_\Phi)}$-norm on the latter equalities, and using Proposition 1.2 (1).

The assertion (2) follows now by replacing $\mathbf{R}^m$ with $\mathbf{R}^{2m}$, and letting $V_1 = \mathbf{R}^{2m}$ and $\Phi(x,\xi) = \langle x, \xi\rangle / t$ when $t \neq 0$, and $V_2 = \mathbf{R}^{2m}$ when $t = 0$. The proof is complete. □

We finish the section by giving some remarks on Wigner distributions and Weyl operators of rank one. The Wigner distribution for $f \in \mathscr{S}(\mathbf{R}^m)$ and $g \in \mathscr{S}(\mathbf{R}^m)$ is defined by the formula

$$W_{f,g}(x,\xi) = (2\pi)^{-m/2} \int f(x - y/2) \overline{g(x + y/2)} e^{i\langle y, \xi\rangle}\, dy. \tag{1.14}$$

For future references we note that the map $(f,g) \mapsto W_{f,g}$ is continuous from $\mathscr{S}(\mathbf{R}^m) \times \mathscr{S}(\mathbf{R}^m)$ to $\mathscr{S}(\mathbf{R}^{2m})$ which extends uniquely to continuous mappings from $\mathscr{S}'(\mathbf{R}^m) \times \mathscr{S}'(\mathbf{R}^m)$ to $\mathscr{S}'(\mathbf{R}^{2m})$, and from $L^2(\mathbf{R}^m) \times L^2(\mathbf{R}^m)$ to $L^2(\mathbf{R}^{2m})$. (See [22, 44, 46].)

Next assume that $a \in \mathscr{S}(\mathbf{R}^{2m})$ and that $f, g \in \mathscr{S}(\mathbf{R}^m)$. Then it follows from (0.2) with $t = 1/2$ and straightforward computations that

$$(a^w(x,D)f, g) = (2\pi)^{-m/2}(a, W_{g,f}), \tag{1.15}$$

In particular, (1.15) and Fourier's inversion formula imply that if $f_1, f_2 \in \mathscr{S}(\mathbf{R}^m)$, then

$$a^w(x,D)f(x) = (2\pi)^{-m/2}(f, f_2)f_1(x) \quad \Leftrightarrow \quad a = W_{f_1,f_2}. \tag{1.16}$$

Consequently, a Weyl operator is a rank one operator if and only if its symbol is a Wigner distribution.

2. Schatten-von Neumann classes for operators acting on Hilbert spaces

In this section we discuss Schatten-von Neumann classes of linear operators from a Hilbert space $\mathscr{H}_1$ to another Hilbert space $\mathscr{H}_2$. Such operator classes were introduced by R. Schatten in [35] in the case $\mathscr{H}_1 = \mathscr{H}_2$. (See also [23, 39].) The general situation when $\mathscr{H}_1$ is not necessarily equal to $\mathscr{H}_2$ have thereafter been discussed in [2, 37]. Here we give a short introduction, based on an argument which essentially reduces the situation to the case $\mathscr{H}_1 = \mathscr{H}_2$.

For any Hilbert space $\mathscr{H}$, we let $\mathrm{ON}(\mathscr{H})$ be the set of orthonormal sequences in $\mathscr{H}$. Assume that $T : \mathscr{H}_1 \to \mathscr{H}_2$ is linear, and that $p \in [1, \infty]$. Then set

$$\|T\|_{\mathscr{I}_p} = \|T\|_{\mathscr{I}_p(\mathscr{H}_1,\mathscr{H}_2)} \equiv \sup \Big(\sum |(Tf_j, g_j)_{\mathscr{H}_2}|^p \Big)^{1/p} \tag{2.1}$$

(with obvious modifications when $p = \infty$). Here the supremum is taken over all $(f_j) \in \mathrm{ON}(\mathscr{H}_1)$ and $(g_j) \in \mathrm{ON}(\mathscr{H}_2)$. Now recall that $\mathscr{I}_p = \mathscr{I}_p(\mathscr{H}_1, \mathscr{H}_2)$, the Schatten-von Neumann class of order p, consists of all linear operators T from $\mathscr{H}_1$ to $\mathscr{H}_2$ such that $\|T\|_{\mathscr{I}_p(\mathscr{H}_1,\mathscr{H}_2)}$ is finite. Obviously, $\mathscr{I}_\infty(\mathscr{H}_1, \mathscr{H}_2)$ consists of all continuous operators from $\mathscr{H}_1$ to $\mathscr{H}_2$. If $\mathscr{H}_1 = \mathscr{H}_2$, then the shorter notation $\mathscr{I}_p(\mathscr{H}_1)$ is used instead of $\mathscr{I}_p(\mathscr{H}_1, \mathscr{H}_2)$. We also let $\mathscr{I}_\sharp(\mathscr{H}_1, \mathscr{H}_2)$ be the set of all linear and compact operators from $\mathscr{H}_1$ to $\mathscr{H}_2$, and equip this space with the norm $\|\cdot\|_{\mathscr{I}_\infty}$ as usual. The spaces $\mathscr{I}_1(\mathscr{H}_1, \mathscr{H}_2)$ and $\mathscr{I}_2(\mathscr{H}_1, \mathscr{H}_2)$ are called the sets of trace-class operators and Hilbert-Schmidt operators respectively. These definitions agree with the old ones when $\mathscr{H}_1 = \mathscr{H}_2$, and in this case the norms $\|\cdot\|_{\mathscr{I}_1}$ and $\|\cdot\|_{\mathscr{I}_2}$ agree with the trace-class norm and Hilbert-Schmidt norm respectively.

Another description of Schatten-von Neumann classes can be obtained in terms of singular numbers. Assume first that T above is compact. Then by the spectral theorem it follows that

$$Tf = \sum_{j=1}^{\infty} \lambda_j (f, g_j)_{\mathscr{H}_1} f_j, \quad f \in \mathscr{H}_1, \tag{2.2}$$

for some sequences $(f_j)_{j=1}^{\infty} \in \mathrm{ON}(\mathscr{H}_2)$ and $(g_j)_{j=1}^{\infty} \in \mathrm{ON}(\mathscr{H}_1)$, and some sequence $\lambda_1 \geq \lambda_2 \geq \cdots \geq 0$. Here the numbers λ_j are called the singular numbers for T, and we use the notation $\sigma_j(T)$ for these numbers, i.e., $\sigma_j(T) = \lambda_j$.

There is a canonical way to extend the definition of singular values to bounded operators, which are not necessarily compact. More precisely, for any closed subspace V of $\mathscr{H}_1$, set

$$\mu_V(T) = \sup_{f \in V,\ \|f\|_{\mathscr{H}_1} \leq 1} \|Tf\|_{\mathscr{H}_2},$$

Then let $\sigma_j(T)$ be defined by the formula

$$\sigma_j(T) = \sigma_{j,\mathscr{H}_1,\mathscr{H}_2}(T) \equiv \inf_{\dim V^{\perp} = j-1} \mu_V(T). \tag{2.3}$$

It is straightforward to verify that $\sigma_j(T)$ agrees with the earlier definition when T is compact. Moreover, $T \in \mathscr{I}_p$ if and only if $(\sigma_j(T)) \in l^p$, and

$$\|T\|_{\mathscr{I}_p} = \|(\sigma_j(T))\|_{l^p}. \tag{2.4}$$

From now on we assume that the involved Hilbert spaces are separable. Then without loss of generality we may in many situations concerning $\mathscr{I}_p(\mathscr{H}_1, \mathscr{H}_2)$ reduce ourself to the case $\mathscr{H}_1 = \mathscr{H}_2$.

In fact, assume that (f_j^0) and (g_j^0) are fixed orthonormal bases for $\mathscr{H}_1$ and $\mathscr{H}_2$ respectively, and let T_0 be the linear map, defined by the formula

$$T_0\Big(\sum \alpha_j f_j^0\Big) = \sum \alpha_j g_j^0.$$

Here $(\alpha_j) \in l^2$ is arbitrary. Then T_0 is an isometric bijection from $\mathscr{H}_1$ to $\mathscr{H}_2$, $T_0^* \circ T_0 = \mathrm{Id}_{\mathscr{H}_1}$ and $T_0 \circ T_0^* = \mathrm{Id}_{\mathscr{H}_2}$. Consequently, $(f_j) \mapsto (T_0 f_j)$ is a bijection from $\mathrm{ON}(\mathscr{H}_1)$ to $\mathrm{ON}(\mathscr{H}_2)$. This in turn implies that $T \mapsto T_0 \circ T$ is an isometric homeomorphism from $\mathscr{I}_p(\mathscr{H}_1)$ to $\mathscr{I}_p(\mathscr{H}_1, \mathscr{H}_2)$, i.e., there is a canonical identification between $\mathscr{I}_p(\mathscr{H}_1)$ and $\mathscr{I}_p(\mathscr{H}_1, \mathscr{H}_2)$.

In Proposition 2.1 till Proposition 2.7 below we list some properties for spaces of the type $\mathscr{I}_p(\mathscr{H}_1, \mathscr{H}_2)$, which are well known in the case $\mathscr{H}_1 = \mathscr{H}_2$. (See [39].) The general case, when $\mathscr{H}_1$ is not necessarily equal to $\mathscr{H}_2$ is now a consequence of the identification here above and the corresponding results in [39]. In the general case, the results can also be found in [2, 23, 37].

Proposition 2.1. *Assume that $p, p_j \in [1, \infty]$ for $1 \leq j \leq 2$ such that $p_1 < p_2 < \infty$. Also assume that $\mathscr{H}_1$ and $\mathscr{H}_2$ are separable Hilbert spaces. Then $\mathscr{I}_p$ is a Banach space,*

$$\mathscr{I}_{p_1}(\mathscr{H}_1, \mathscr{H}_2) \subseteq \mathscr{I}_{p_2}(\mathscr{H}_1, \mathscr{H}_2) \subseteq \mathscr{I}_{\sharp}(\mathscr{H}_1, \mathscr{H}_2) \subseteq \mathscr{I}_{\infty}(\mathscr{H}_1, \mathscr{H}_2),$$

and

$$\|T\|_{\mathscr{I}_\infty} \leq \|T\|_{\mathscr{I}_{p_2}} \leq \|T\|_{\mathscr{I}_{p_1}}, \quad T \in \mathscr{I}_\infty(\mathscr{H}_1, \mathscr{H}_2). \tag{2.5}$$

Moreover, equalities in (2.5) *occur if and only if* T *is a rank one operator, i.e.,* $Tf = (f, g_1)_{\mathscr{H}_1} g_2$ *for some* $g_1 \in \mathscr{H}_1$ *and* $g_2 \in \mathscr{H}_2$*, and then* $\|T\|_{\mathscr{I}_p} = \|g_1\|_{\mathscr{H}_1}\|g_2\|_{\mathscr{H}_2}$ *for every* $p \in [1, \infty]$.

The next result concerns algebraic properties for Schatten-von Neumann classes.

Proposition 2.2. *Assume that* $p, q, r \in [1, \infty]$ *such that* $1/p + 1/q = 1/r$*. Also assume that* $\mathscr{H}_j$ *for* $1 \le j \le 3$ *are separable Hilbert spaces. If* $T_1 \in \mathscr{I}_p(\mathscr{H}_1, \mathscr{H}_2)$ *and* $T_2 \in \mathscr{I}_q(\mathscr{H}_2, \mathscr{H}_3)$*, then* $T = T_2 \circ T_1 \in \mathscr{I}_r(\mathscr{H}_1, \mathscr{H}_3)$*, and*

$$\|T_2 \circ T_1\|_{\mathscr{I}_r} \le \|T_1\|_{\mathscr{I}_p}\|T_2\|_{\mathscr{I}_q}. \tag{2.6}$$

On the other hand, for any $T \in \mathscr{I}_r(\mathscr{H}_1, \mathscr{H}_3)$*, there are operators* $T_1 \in \mathscr{I}_p(\mathscr{H}_1, \mathscr{H}_2)$ *and* $T_2 \in \mathscr{I}_q(\mathscr{H}_2, \mathscr{H}_3)$ *such that* $T = T_2 \circ T_1$ *and equality is attained in* (2.6).

Remark 2.3. Assume that $p \in [1, \infty]$ and that $\mathscr{H}_j$ for $j = 1, \dots, 4$ are Hilbert spaces such that $\mathscr{H}_1 = \mathscr{H}_2$ with equivalent norms, and $\mathscr{H}_3 \hookrightarrow \mathscr{H}_4$. Then it follows from Proposition 2.2 that these embeddings induce the embedding $\mathscr{I}_p(\mathscr{H}_1, \mathscr{H}_3) \hookrightarrow \mathscr{I}_p(\mathscr{H}_2, \mathscr{H}_4)$. This is also a consequence of (2.4), since it follows from the assumptions that there exists a constant C such that if T is linear from $\mathscr{H}_1 = \mathscr{H}_2$ to $\mathscr{H}_3$ and $j \ge 1$, then $\sigma_{j,\mathscr{H}_2,\mathscr{H}_4}(T) \le C\sigma_{j,\mathscr{H}_1,\mathscr{H}_3}(T)$. In particular, if in addition $\mathscr{H}_3 = \mathscr{H}_4$ with equivalent norms, then $\mathscr{I}_p(\mathscr{H}_1, \mathscr{H}_3) = \mathscr{I}_p(\mathscr{H}_2, \mathscr{H}_4)$ with equivalent norms.

We note that $T \in \mathscr{I}_p(\mathscr{H}_1, \mathscr{H}_2)$ if and only if $T^* \in \mathscr{I}_p(\mathscr{H}_2, \mathscr{H}_1)$, in view of (2.1). The next proposition deals with duality properties. Here recall that $p' \in [1, \infty]$ denotes the conjugate exponent for $p \in [1, \infty]$, i.e., $1/p + 1/p' = 1$.

Proposition 2.4. *Assume that* $p \in [1, \infty]$*, and that* $\mathscr{H}_1$ *and* $\mathscr{H}_2$ *are separable Hilbert spaces. Then the form*

$$(T_1, T_2) = (T_1, T_2)_{\mathscr{H}_1,\mathscr{H}_2} \equiv \mathrm{tr}_{\mathscr{H}_1}(T_2^* \circ T_1)$$

on $\mathscr{I}_1(\mathscr{H}_1, \mathscr{H}_2)$ *extends uniquely to a continuous and sesqui-linear form on* $\mathscr{I}_p(\mathscr{H}_1, \mathscr{H}_2) \times \mathscr{I}_{p'}(\mathscr{H}_1, \mathscr{H}_2)$*, and for every* $T_1 \in \mathscr{I}_p$ *and* $T_2 \in \mathscr{I}_{p'}$*, then*

$$(T_1, T_2)_{\mathscr{H}_1,\mathscr{H}_2} = \overline{(T_2, T_1)_{\mathscr{H}_2,\mathscr{H}_1}}$$

$$|(T_1, T_2)_{\mathscr{H}_1,\mathscr{H}_2}| \le \|T_1\|_{\mathscr{I}_p}\|T_2\|_{\mathscr{I}_{p'}}, \qquad \|T_1\|_{\mathscr{I}_p} = \sup |(T_1, S)_{\mathscr{H}_1,\mathscr{H}_2}|,$$

where the supremum is taken over all $S \in \mathscr{I}_{p'}$ *such that* $\|S\|_{\mathscr{I}_{p'}} \le 1$*. If in addition* $p < \infty$*, then the dual space for* $\mathscr{I}_p(\mathscr{H}_1, \mathscr{H}_2)$ *can be identified with* $\mathscr{I}_{p'}(\mathscr{H}_1, \mathscr{H}_2)$ *through this form.*

In view of Proposition 2.4 we note that $\mathscr{I}_2(\mathscr{H}_1, \mathscr{H}_2)$ is a Hilbert space with scalar product $(\cdot, \cdot)_{\mathscr{H}_1,\mathscr{H}_2}$, and that the corresponding norm agrees with the Hilbert-Schmidt norm $\|\cdot\|_{\mathscr{I}_2}$.

The first part of the next proposition follows immediately from Proposition 2.1, since $\mathscr{I}_p \subseteq \mathscr{I}_\sharp$ when $p < \infty$, and $\mathscr{I}_1(\mathscr{H}_1, \mathscr{H}_2)$ contains all operators of finite rank.

Proposition 2.5. *Assume that $p \in [1, \infty)$, and that $\mathscr{H}_1$ and $\mathscr{H}_2$ are separable Hilbert spaces. Then $\mathscr{I}_1(\mathscr{H}_1, \mathscr{H}_2)$ is dense in $\mathscr{I}_\sharp(\mathscr{H}_1, \mathscr{H}_2)$ and in $\mathscr{I}_p(\mathscr{H}_1, \mathscr{H}_2)$. It is dense in $\mathscr{I}_\infty(\mathscr{H}_1, \mathscr{H}_2)$ with respect to the weak* topology.*

The next proposition deals with spectral properties. Here recall that ℓ^∞ is the set of all bounded sequences $(\lambda_j)_{j\ge 1}$ such that $\lambda_j \to \infty$ as j tends to infinity.

Proposition 2.6. *Assume that $T \in \mathscr{I}_\sharp(\mathscr{H}_1, \mathscr{H}_2)$. Then (2.2) holds for some choice of sequences $(f_j)_{j=1}^\infty \in \mathrm{ON}(\mathscr{H}_1)$, $(g_j)_{j=1}^\infty \in \mathrm{ON}(\mathscr{H}_2)$ and $\lambda = (\lambda_j)_{j=1}^\infty \in \ell^\infty$, where the sum on the right-hand side in (2.2) convergences with respect to the operator norm. Moreover, if $1 \le p < \infty$ then $T \in \mathscr{I}_p(\mathscr{H}_1, \mathscr{H}_2)$, if and only if $\lambda \in l^p$, and then $\|T\|_{\mathscr{I}_p} = \|\lambda\|_{l^p}$ and the sum on the right-hand side of (2.2) converges with respect to the norm $\|\cdot\|_{\mathscr{I}_p}$.*

The next result concerns interpolation properties.

Proposition 2.7. *Assume that $p, p_1, p_2 \in [1, \infty]$ and $0 \le \theta \le 1$ such that $1/p = (1-\theta)/p_1 + \theta/p_2$. Assume also that $\mathscr{H}_1$ and $\mathscr{H}_2$ are separable Hilbert spaces. Then the (complex) interpolation space $(\mathscr{I}_{p_1}, \mathscr{I}_{p_2})_{[\theta]}$ is equal to $\mathscr{I}_p$ with equality in norms.*

3. Schatten-von Neumann classes for operators acting on modulation spaces

In this section we present some properties of Schatten-von Neumann classes of linear operators acting from $M^2_{(\omega_1)}(\mathbf{R}^m)$ to $M^2_{(\omega_2)}(\mathbf{R}^m)$, where $\omega_1, \omega_2 \subset \mathscr{P}(\mathbf{R}^{2m})$. We are especially concerned with finding appropriate identifications of the dual of $s_{t,p}(\omega_1, \omega_2)$ when $p < \infty$ (see the introduction or below). In the first part we prove that there is a canonical way to identify the dual of $s_{t,p}(\omega_1, \omega_2)$ with $s_{t,p'}(\omega_1, \omega_2)$. In the second part we use this property to prove that the dual of $s_{t,p}(\omega_1, \omega_2)$ can also be identified with $s_{t,p'}(1/\omega_1, 1/\omega_2)$ through a unique extension of the L^2 product from $\mathscr{S}$.

We start by considering Schatten-von Neumann classes in the context of pseudo-differential calculus. Let $t \in \mathbf{R}$, $p \in [1, \infty]$ and $\omega_1, \omega_2 \in \mathscr{P}(\mathbf{R}^{2m})$ be fixed. From the introduction we recall that $s_{t,p}(\omega_1, \omega_2)$ consists of all $a \in \mathscr{S}'(\mathbf{R}^{2m})$ such that $a_t(x, D) \in \mathscr{I}_p(M^2_{(\omega_1)}, M^2_{(\omega_2)})$. Also let $s_{t,\sharp}(\omega_1, \omega_2)$ be the set of all $a \in \mathscr{S}'(\mathbf{R}^{2m})$ such that $a_t(x, D) \in \mathscr{I}_\sharp(M^2_{(\omega_1)}, M^2_{(\omega_2)})$, We let $s_{t,p}(\omega_1, \omega_2)$ and $s_{t,\sharp}(\omega_1, \omega_2)$ be equipped by the norms

$$\|a\|_{s_{t,p}} = \|a\|_{s_{t,p}(\omega_1,\omega_2)} \equiv \|a_t(x, D)\|_{\mathscr{I}_p(M^2_{(\omega_1)}, M^2_{(\omega_2)})}$$

and $\|\cdot\|_{s_{t,\infty}}$ respectively. Since the Weyl quantization is important in our investigations we also use the notations s^w_p and $s^w_\sharp$ instead of $s_{t,p}$ and $s_{t,\sharp}$ respectively when $t = 1/2$ and $p \in [1, \infty]$. In the case $\omega_1 = \omega_2 = \omega$, then we use the notation $s_{t,p}(\omega)$ and $s^w_p(\omega)$ instead of $s_{t,p}(\omega_1, \omega_2)$ and $s^w_p(\omega_1, \omega_2)$ respectively.

Remark 3.1. Let $\omega_j \in \mathscr{P}(\mathbf{R}^{2m})$ for $j = 1, 2$, $t \in \mathbf{R}$ and $p \in [1, \infty]$. Then recall that $M^2_{(\omega_j)}(\mathbf{R}^m)$ is independent of the choice of windows in $\mathscr{S} \setminus 0$, and that different windows give rise to different norms. Hence Remark 2.3 shows that $s_{t,p}(\omega_1, \omega_2)$ is invariant under the choices of windows and that different windows give rise to equivalent norms.

In order to avoid ambiguity of the $s_{t,p}(\omega_1, \omega_2)$ norm, we assume from now on that the windows for $M^2_{(\omega_1)}$ and $M^2_{(\omega_2)}$ are fixed.

From the fact that each linear and continuous operator from $\mathscr{S}(\mathbf{R}^m)$ to $\mathscr{S}'(\mathbf{R}^m)$ is equal to $a_t(x, D)$, for a unique $a \in \mathscr{S}'(\mathbf{R}^{2m})$, it follows that the map $a \mapsto a_t(x, D)$ is an isometric homeomorphism from $s_{t,p}(\omega_1, \omega_2)$ to $\mathscr{I}_p(M^2_{(\omega_1)}, M^2_{(\omega_2)})$ when $p \in [1, \infty]$, and from $s_{t,\sharp}(\omega_1, \omega_2)$ to $\mathscr{I}_\sharp(M^2_{(\omega_1)}, M^2_{(\omega_2)})$. Consequently, most of the properties which are listed in Section 2 carry over to the $s_{t,p}$-spaces. Hence Proposition 2.1 shows that $s_{t,p}(\omega_1, \omega_2)$ is a scale of Banach spaces which increase with the parameter $p \in [1, \infty]$. Moreover, Proposition 2.4 shows that the norm $\| \cdot \|_{s_{t,2}(\omega_1,\omega_2)}$ in $s_{t,2}(\omega_1, \omega_2)$ induces a scalar product

$$(\cdot, \cdot)_{s_{t,2}} = (\cdot, \cdot)_{s_{t,2}(\omega_1,\omega_2)}$$

and that if $p < \infty$, then the dual for $s_{t,p}(\omega_1, \omega_2)$ can be identified with $s_{t,p'}(\omega_1, \omega_2)$ through a uniquely extension of the $s_{t,2}(\omega_1, \omega_2)$ product from $s_{t,1}(\omega_1, \omega_2)$.

A problem in this context is the somewhat complicated structure of the form $(\cdot, \cdot)_{s_{t,2}(\omega_1,\omega_2)}$, compared to, e.g., the scalar product on L^2, which in general fits pseudo-differential calculus well. In the remaining part of the section we therefore focus on a possible replacement of the form $(\cdot, \cdot)_{s_{t,2}}$ with L^2 when discussing duality. From our investigations it turns out that indeed the dual space of $s_{t,p}(\omega_1, \omega_2)$ may be identified with $s_{t,p'}(1/\omega_1, 1/\omega_2)$, by a unique and continuous extension of the L^2 product from $\mathscr{S}$.

We start by making some preparations. Let $\omega \in \mathscr{P}(\mathbf{R}^{2m})$ be fixed. By Theorem 1.2 it follows that the dual for $M^2_{(\omega)}(\mathbf{R}^m)$ can be identified with $M^2_{(1/\omega)}(\mathbf{R}^m)$ through the scalar product on $L^2(\mathbf{R}^m)$. On the other hand, since $M^2_{(\omega)}$ is a Hilbert space, its dual can also be identified with $M^2_{(\omega)}$ through the scalar product on $M^2_{(\omega)}$. Consequently, there exist unique homeomorphisms

$$\mathcal{T}_\omega : \big(M^2_{(\omega)}\big)^* \to M^2_{(\omega)} \quad \text{and} \quad \mathcal{R}_\omega : M^2_{(\omega)} \to M^2_{(1/\omega)}$$

such that if $\ell \in \big(M^2_{(\omega)}\big)^*$ and $h = \mathcal{T}_\omega \ell \in M^2_{(\omega)}$, then

$$\ell(f) = (f, h)_{M^2_{(\omega)}} = (f, \mathcal{R}_\omega h)_{L^2}, \quad f \in M^2_{(\omega)} \tag{3.1}$$

and

$$C^{-1} \|\mathcal{R}_\omega h\|_{M^2_{(1/\omega)}} \le \|\ell\| = \|h\|_{M^2_{(\omega)}} \le C \|\mathcal{R}_\omega h\|_{M^2_{(1/\omega)}} \tag{3.2}$$

for some constant C which is independent of h. We observe that (3.1) and (3.2) imply that if $(f_j)_{j=1}^\infty \in \mathrm{ON}(M^2_{(\omega)})$ and $g_k = \mathcal{R}_\omega f_k$, then

$$(f_j, g_k)_{L^2} = \delta_{j,k} \quad \text{and} \quad C^{-1} \le \|g_k\|_{M^2_{(1/\omega)}} \le C. \tag{3.3}$$

For convenience we set $\mathrm{ON}_\omega = \mathrm{ON}(M^2_{(\omega)})$, and we let ON^*_ω be the set of all sequences $(g_j)_{j=1}^\infty$ in $M^2_{(1/\omega)}$ such that $g_j = \mathcal{R}_\omega f_j$ for some $(f_j)_{j=1}^\infty \in \mathrm{ON}(M^2_{(\omega)})$. The following characterization of $s_{t,p}(\omega_1,\omega_2)$ follows immediately from (1.15), Proposition 2.6 and the homoeomorphism property of $\mathcal{R}_\omega$.

Proposition 3.2. *Assume that $t \in \mathbf{R}$ and $\omega_1, \omega_2 \in \mathscr{P}(\mathbf{R}^{2m})$, Also assume that $a \in s_{t,\sharp}(\omega_1,\omega_2)$. Then for some $(g_j)_{j=1}^\infty \in \mathrm{ON}^*_{\omega_1}$, $(h_j)_{j=1}^\infty \in \mathrm{ON}_{\omega_2}$ and $\lambda = (\lambda_j)_{j=1}^\infty \in \ell^\infty$ it holds*

$$a = \sum_{j=1}^\infty \lambda_j W_{h_j,g_j} \tag{3.4}$$

(with convergence with respect to the norm $\|\cdot\|_{s_{t,\infty}}$). Moreover, if $1 \le p < \infty$ then $a \in s_{t,p}(\omega_1,\omega_2)$, if and only if $\lambda \in l^p$, and then

$$\|a\|_{s_{t,p}(\omega_1,\omega_2)} = \|\lambda\|_{l^p}$$

for some constant C which is independent of a.

On the other hand, if a is given by (3.4) for some $(g_j)_{j=1}^\infty \in \mathrm{ON}^*_{\omega_1}$, $(h_j)_{j=1}^\infty \in \mathrm{ON}_{\omega_2}$ and $\lambda = (\lambda_j)_{j=1}^\infty \in l^p$, then $a \in s_{t,p}(\omega_1,\omega_2)$.

Next we prove that the dual space of $s^w_1(\omega_1,\omega_2)$ may be identified with $s^w_\infty(1/\omega_1, 1/\omega_2)$ through a continuous extension of the scalar product on $L^2(\mathbf{R}^{2m})$. Assume that $\ell \in s^w_1(\omega_1,\omega_2)^*$. It follows from (1.15) that an arbitrary rank one operator from $M^2_{(\omega_1)}$ to $M^2_{(\omega_2)}$ can be written as

$$Tf = (f,g)_{L^2} h = b^w(x,D)f,$$

where $b = W_{h,g}$, $g \in M^2_{(1/\omega_1)}$ and $h \in M^2_{(\omega_2)}$. Moreover

$$\begin{aligned} C^{-1}\|g\|_{M^2_{(1/\omega_1)}}\|h\|_{M^2_{(\omega_2)}} &< \|b\|_{s^w_1(\omega_1,\omega_2)} \\ &= \|T\|_{\mathscr{I}_1} \le C\|g\|_{M^2_{(1/\omega_1)}}\|h\|_{M^2_{(\omega_2)}}, \end{aligned} \tag{3.5}$$

for some constant C which only depends on ω_1 and ω_2. Hence

$$|\ell(W_{h,g})| \le C\|g\|_{M^2_{(1/\omega_1)}}\|h\|_{M^2_{(\omega_2)}}.$$

In particular it follows that the mappings

$$f_1 \mapsto \overline{\ell(W_{f_2^0,f_1})} \quad \text{and} \quad f_2 \mapsto \ell(W_{f_2,f_1^0})$$

are continuous and linear mappings from $\mathscr{S}(\mathbf{R}^m)$ to $\mathbf{C}$, for every fixed f_1^0 and f_2^0 in $\mathscr{S}(\mathbf{R}^m)$. Hence the Schwartz kernel theorem and its proof (see Theorem 5.2.1 in [28]) show that there is a unique distribution $K \in \mathscr{S}(\mathbf{R}^m \oplus \mathbf{R}^m)$ such that

$$\overline{\ell(W_{f_2,f_1})} = (K, f_2 \otimes \overline{f}_1)$$

for every $f_1, f_2 \in \mathscr{S}(\mathbf{R}^m)$. By letting a be the Weyl symbol for the operator with kernel K, it follows that

$$\overline{\ell(W_{f_2,f_1})} = (a^w(x,D)f_1, f_2) = (a, W_{f_2,f_1})_{L^2}. \tag{3.6}$$

Next we prove that $a \in s_\infty^w(1/\omega_1, 1/\omega_2)$. By the positive linearity in the definition of the trace norm we have

$$\|\ell\| = \sup_{\|b\|_{s_1^w} \le 1} |\ell(b)| = \sup |\ell(b)|$$

where the latter supremum is taken over all $b \in s_1^w(\omega_1, \omega_2)$ such that $\|b\|_{s_1^w} \le 1$ and $b^w(x,D)$ is an operator of rank one. Hence (3.5) and (3.6) give

$$|(a^w(x,D)f_1, f_2)| \le C\|f_1\|_{M^2_{(1/\omega_1)}} \|f_2\|_{M^2_{(\omega_2)}}, \tag{3.7}$$

and it follows from Proposition 1.2 (4) that $a^w(x,D)$ is continuous from $M^2_{(1/\omega_1)}$ to $M^2_{(1/\omega_2)}$, i.e., $a \in s_w^\infty(1/\omega_1, 1/\omega_2)$.

Assume next that $b \in s_1^w(\omega_1, \omega_2)$ is arbitrary. Then $b = \sum \lambda_j W_{h_j, g_j}$ for some $(g_j) \in \mathrm{ON}^*_{\omega_1}$, $(h_j) \in \mathrm{ON}_{\omega_2}$, and $(\lambda_j) \in l^1$. We let $(a,b)_{L^2}$ be defined as

$$(a,b)_{L^2} = \sum \lambda_j (a, W_{h_j,g_j})_{L^2}.$$

This definition makes sense, since (3.2), (3.5) and (3.6) give

$$\sum |\lambda_j||(a, W_{h_j,g_j})| \le C \sum |\lambda_j| = C\|b\|_{s_1^w}$$

for some constant C. Hence $|(a,b)_{L^2}| \le C\|b\|_{s_1^w}$. By combining the latter estimate with (3.5) it follows that $\overline{\ell(b)} = (a,b)_{L^2}$, and that $b \mapsto (a,b)_{L^2}$ is continuous from $s_1^w(\omega_1, \omega_2)$ to $\mathbf{C}$. Consequently, the dual space of $s_1^w(\omega_1,\omega_2)$ can be identified with $s_\infty^w(1/\omega_1, 1/\omega_2)$ through the form $(\cdot,\cdot)_{L^2}$. We also note that the extension of $(\cdot,\cdot)_{L^2}$ to a duality between $s_1^w(\omega_1,\omega_2)$ and $s_\infty^w(1/\omega_1, 1/\omega_2)$ is unique since $\mathscr{S}$ is dense in $M^2_{(\omega_j)}$ for $j = 1, 2$, and that finite rank operators are dense in $\mathscr{I}_1(M^2_{(\omega_1)}, M^2_{(\omega_2)})$. In particular we have proved the following result in the case $p = 1$ and $t = 1/2$.

Theorem 3.3. *Assume that $t \in \mathbf{R}$, $p \in [1, \infty)$ and that $\omega_1, \omega_2 \in \mathscr{P}(\mathbf{R}^{2m})$. Then the scalar product on $L^2(\mathbf{R}^{2m})$ extends uniquely to a duality between $s_{t,p}(\omega_1,\omega_2)$ and $s_{t,p'}(1/\omega_1, 1/\omega_2)$, and the dual space for $s_{t,p}(\omega_1,\omega_2)$ can be identified with $s_{t,p'}(1/\omega_1, 1/\omega_2)$ through this form. Moreover, if $\ell \in s_{t,p}(\omega_1,\omega_2)^*$ and $a \in s_{t,p'}(1/\omega_1,\ 1/\omega_2)$ such that $\overline{\ell(b)} = (a,b)_{L^2}$ when $b \in s_{t,p}(\omega_1,\omega_2)$, then*

$$C^{-1}\|a\|_{s_{t,p'}(1/\omega_1,1/\omega_2)} \le \|\ell\| \le C\|a\|_{s_{t,p'}(1/\omega_1,1/\omega_2)}$$

for some constant C which only depends on ω_1 and ω_2.

Proof. By (1.10) and the fact that $e^{i\langle D_x, D_\xi\rangle}$ is unitary on L^2, it suffices to prove the result in the case $t = 1/2$. Since we have already proved the result in the case $p = 1$, we may assume that $p > 1$. First we prove that $\ell(b) = \overline{(a,b)_{L^2}}$ defines a continuous linear form on $s_p^w(\omega_1,\omega_2)$ for every element $s_{p'}^w(1/\omega_1, 1/\omega_2)$. Since $\mathscr{I}_1$ is dense in $\mathscr{I}_p$ and in $\mathscr{I}_{p'}$, Proposition 3.2 shows that it suffices to prove

$$|(a,b)_{L^2}| \le C\|(\lambda_j)\|_{l^{p'}} \|(\mu_j)\|_{l^p}, \tag{3.8}$$

for some constant C, where

$$a = \sum \lambda_j W_{h_j,g_j} \quad \text{and} \quad b = \sum \mu_j W_{h_j^0, g_j^0}$$

are finite sums. Here $g_j = \mathcal{R}_{1/\omega_1} f_j$ and $g_j^0 = \mathcal{R}_{\omega_1} f_j^0$ for some $(f_j) \in \mathrm{ON}_{1/\omega_1}$, $(f_j^0) \in \mathrm{ON}_{\omega_1}$, $(h_j) \in \mathrm{ON}_{1/\omega_2}$ and $(h_j^0) \in \mathrm{ON}_{\omega_2}$. It is also no restriction to assume that $\|(\lambda_j)\|_{l^{p'}} = \|(\mu_j)\|_{l^p} = 1$,

By straightforward computations we get

$$|(a,b)_{L^2}| = \Big| \sum_{j,k} \lambda_j \overline{\mu_k} (W_{h_j,g_j}, W_{h_k^0,g_k^0})_{L^2} \Big| \le \sum_{j,k} |\lambda_j \mu_k \alpha_{j,k} \beta_{j,k}|,$$

where

$$\alpha_{j,k} = (g_j, g_k^0)_{L^2}, \quad \text{and} \quad \beta_{j,k} = (h_j, h_k^0)_{L^2}.$$

Hence the inequality between arithmetic and geometric mean-values gives

$$|(a,b)_{L^2}| \le \sum_{j,k} \Big(\frac{1}{p'} |\lambda_j|^{p'} + \frac{1}{p} |\mu_k|^p \Big) \frac{1}{2} \Big(|\alpha_{j,k}|^2 + |\beta_{j,k}|^2 \Big). \tag{3.9}$$

By letting $\kappa_j = \mathcal{R}_{1/\omega_2} h_j$ and $\kappa_j^0 = \mathcal{R}_{\omega_2} h_j^0$ it follows from (3.3) that

$$\alpha_{j,k} = (g_j, g_k^0)_{L^2} = (f_j, g_k^0)_{M^2_{(1/\omega_1)}} = (g_j, f_k^0)_{M^2_{(\omega_1)}} = (f_j, f_k^0)_{L^2}, \quad \text{and}$$

$$\beta_{j,k} = (h_j, h_k^0)_{L^2} = (h_j, \kappa_k^0)_{M^2_{(1/\omega_2)}} = (\kappa_j, h_k^0)_{M^2_{(\omega_2)}} = (\kappa_j, \kappa_k^0)_{L^2}.$$

Hence the orthogonality assumptions imply

$$\sum_j |\alpha_{j,k}|^2 \le \|g_k^0\|_{M^2_{(1/\omega_1)}}, \qquad \sum_k |\alpha_{j,k}|^2 \le \|g_j\|_{M^2_{(\omega_1)}},$$

and similarly when g_j, g_k^0 and $\alpha_{j,k}$ are replaced by h_j, h_k^0 and $\beta_{j,k}$ respectively. By combining these estimates with (3.9), it follows that (3.8) holds with

$$C = \sup_{j,k} (\|g_j\|_{M^2_{(\omega_1)}}, \|g_k^0\|_{M^2_{(1/\omega_1)}}, \|\kappa_j\|_{M^2_{(\omega_2)}}, \|\kappa_k^0\|_{M^2_{(1/\omega_2)}}) < \infty.$$

This proves the assertion.

Assume next that $\ell \in s_p^w(\omega_1, \omega_2)^*$. By Proposition 2.4 there is a unique operator $T \in \mathscr{I}_{p'}(M^2_{(\omega_1)}, M^2_{(\omega_2)})$ such that if $\mathscr{H}_1 = M^2_{(\omega_1)}$ and $\mathscr{H}_2 = M^2_{(\omega_2)}$, then

$$\ell(b) = (b^w(x,D), T)_{\mathscr{H}_1, \mathscr{H}_2}$$

holds for any $b \in s_p^w(\omega_1, \omega_2)$. Hence Proposition 2.6 gives

$$Tf = \sum \lambda_j (f, f_j)_{M^2_{(\omega_1)}} h_j, \quad \text{and} \quad b^w(x,D) = \sum \mu_j (f, f_j^0)_{M^2_{(\omega_1)}} h_j^0$$

for some $(f_j), (f_j^0) \in \mathrm{ON}_{\omega_1}$ and $(h_j), (h_j^0) \in \mathrm{ON}_{\omega_2}$, $(\lambda_j) \in l^{p'}$ and $(\mu_j) \in l^p$. This implies that

$$\ell(b) = \sum_{j,k} \overline{\lambda_j} \mu_k (h_k^0, h_j)_{M^2_{(\omega_2)}} (f_j, f_k^0)_{M^2_{(\omega_1)}}.$$

Now let $g_j^0 = \mathcal{R}_{\omega_1} f_j^0$, $\kappa_j = \mathcal{R}_{\omega_2} h_j$ and $a = \sum \lambda_j W_{\kappa_j, f_j}$. Then it follows by straightforward computations that $b = \sum \mu_j W_{h_j^0, g_j^0}$, and that

$$(a,b)_{L^2} = \sum_{j,k} \lambda_j \overline{\mu_k} (W_{\kappa_j,f_j}, W_{h_k^0,g_k^0})_{L^2} = \sum_{j,k} \lambda_j \overline{\mu_k} (\kappa_j, h_k^0)_{L^2} (g_k^0, f_j)_{L^2} = \overline{\ell(b)}.$$

From the first part of the proof it also follows that

$$\|a^w(x,D)\|_{\mathscr{I}_{p'}(M^2_{(1/\omega_1)},M^2_{(1/\omega_2)})} \le C\|(\lambda_j)\|_{l^{p'}} < \infty,$$

for some constant C. The proof is complete. □

Remark 3.4. Theorem 3.3 appeared after fruitful discussions with Paolo Boggiatto.

4. Continuity and Schatten-von Neumann properties for pseudo-differential operators

In this section, we discuss continuity and Schatten-von Neumann properties for pseudo-differential operators, when the operator symbols belong to appropriate classes of modulation spaces. In particular we generalize some of the continuity results in Section 5 in [49]. An important ingredient in these investigations concern continuity properties for the Wigner distributions in context of modulation spaces, as well as the central role for Wigner distributions in the Weyl calculus of pseudo-differential operators.

First assume that $\mathscr{B}_j$ for $j = 1,2,3$ are Frechét spaces such that

$$\mathscr{S}(\mathbf{R}^{2m}) \hookrightarrow \mathscr{B}_1 \hookrightarrow \mathscr{S}'(\mathbf{R}^{2m}), \quad \mathscr{S}(\mathbf{R}^m) \hookrightarrow \mathscr{B}_2, \mathscr{B}_3 \hookrightarrow \mathscr{S}'(\mathbf{R}^m),$$

and that $(a,f,g) \mapsto (a, W_{g,f})$ is well defined and sequently continuous from $\mathscr{B}_1 \times \mathscr{B}_2 \times \mathscr{B}_3$ to $\mathbf{C}$. Then (1.15) is taken as the definition of $a^w(x,D)f$ as an element in $\mathscr{B}_3'$ when $f \in \mathscr{B}_2$, and it follows that $a^w(x,D)$ is a continuous operator from $\mathscr{B}_2$ to $\mathscr{B}_3'$.

Next we discuss continuity properties for pseudo-differential operator, and prove in a moment that if $t \in \mathbf{R}$,

$$1/p_1 - 1/p_2 = 1/q_1 - 1/q_2 = 1 - 1/p - 1/q, \quad q \le p_2, q_2 \le p, \tag{4.1}$$

ω_1, ω_2 and ω are appropriate weight functions and $a \in M^{p,q}_{(\omega)}$, then $a_t(x,D)$ is continuous from $M^{p_1,q_1}_{(\omega_1)}$ to $M^{p_2,q_2}_{(\omega_2)}$. As a first step we consider continuity properties for Wigner distributions in background of modulation space theory.

Proposition 4.1. *Assume that $p_j, q_j, p, q \in [1,\infty]$ are such that $p \le p_j, q_j \le q$, for $j = 1,2$, and that*

$$1/p_1 + 1/p_2 = 1/q_1 + 1/q_2 = 1/p + 1/q. \tag{4.2}$$

holds. Also assume that $\omega_1, \omega_2 \in \mathscr{P}(\mathbf{R}^{2m})$, and that $\omega \in \mathscr{P}(\mathbf{R}^{2m} \oplus \mathbf{R}^{2m})$ satisfy

$$\omega(x,\xi,y,\eta) \le C\omega_1(x - y/2, \xi + \eta/2)\omega_2(x + y/2, \xi - \eta/2). \tag{4.3}$$

Then the map $(f_1,f_2) \mapsto W_{f_1,f_2}$ from $\mathscr{S}'(\mathbf{R}^m) \times \mathscr{S}'(\mathbf{R}^m)$ to $\mathscr{S}'(\mathbf{R}^{2m})$ restricts to a continuous mapping from $M^{p_1,q_1}_{(\omega_1)}(\mathbf{R}^m) \times M^{p_2,q_2}_{(\omega_2)}(\mathbf{R}^m)$ to $\widetilde{M}^{p,q}_{(\omega)}(\mathbf{R}^{2m})$, and for some constant C,

$$\|W_{f_1,f_2}\|_{\widetilde{M}^{p,q}_{(\omega)}} \le C\|f_1\|_{M^{p_1,q_1}_{(\omega_1)}}\|f_2\|_{M^{p_2,q_2}_{(\omega_2)}} \tag{4.4}$$

when $f_1, f_2 \in \mathscr{S}'(\mathbf{R}^m)$.

The result is essentially a restatement of Theorem 5.1 in [49], where a short proof is presented. A detailed proof in the non-weighted case is available in [48]. In order to being more self-contained, we present here a more detailed proof in the general case.

Proof. We only prove the result in the case $p,q<\infty$. The straightforward modifications to the cases when $p=\infty$ or $q=\infty$ are left for the reader. Let $\chi_1,\chi_2\in\mathscr{S}(\mathbf{R}^m)\setminus 0$, and let $\Phi=W_{\chi_2,\chi_1}$. Then it follows from Fourier's inversion formula that if $X=(x,\xi)\in\mathbf{R}^{2m}$ and $Y=(y,\eta)\in\mathbf{R}^{2m}$, then

$$|\widetilde{\mathscr{F}}(W_{f_1,f_2}\tau_X\Phi)(Y)|=F_1(x-y/2,\xi+\eta/2)F_2(x+y/2,\xi-\eta/2)$$

where

$$F_1(X)=|\mathscr{F}(f_1\tau_x\overline{\chi}_1)(\xi)|\quad\text{and}\quad F_2(X)=|\mathscr{F}(f_2\tau_x\overline{\chi}_2)(\xi)|.$$

By applying the $L^{p,q}_{(\omega)}$-norm on the latter equality, and using (4.3) we get

$$\|W_{f_1,f_2}\|_{\widetilde{M}^{p,q}_{(\omega)}}\le C\big(\|G_1*G_2\|_{L^r}\big)^{1/p}=C\Big(\int H(\eta)\,d\eta\Big)^{1/q}$$

for some constant C, where $G_j=(F_j\omega_j)^p$, $r=q/p\ge 1$ and

$$H(\eta)=\int\Big(\iint G_1(y-x,\eta-\xi)G_2(x,\xi)\,dxd\xi\Big)^r\,dy.$$

By Minkowski's inequality we get

$$H(\eta)\le\Big(\int\Big(\int\Big(\int G_1(y-x,\eta-\xi)G_2(x,\xi)\,dx\Big)^r\,dy\Big)^{1/r}\,d\xi\Big)^r.$$

Assume now that $r_j,s_j\in[1,\infty]$ for $j=1,2$ are chosen such that

$$1/r_1+1/r_2=1/s_1+1/s_2=1+1/r.$$

Then Young's inequality gives

$$H(\eta)\le\Big(\int\|G_1(\,\cdot\,,\eta-\xi)\|_{L^{r_1}}\|G_2(\,\cdot\,,\xi)\|_{L^{r_2}}\,d\xi\Big)^r$$

Hence another application of Young's inequality gives

$$\|W_{f_1,f_2}\|_{\widetilde{M}^{p,q}_{(\omega)}}\le C\Big(\int H(\eta)\,d\eta\Big)^{1/q}\le C\big(\|G_1\|_{L^{r_1,s_1}}\|G_2\|_{L^{r_2,s_2}}\big)^{1/p}$$

By letting $p_j=pr_j$ and $q_j=qs_j$, and using the fact that

$$G_j(x,\xi)=|\mathscr{F}(f_j\tau_x\overline{\chi}_j)(\xi)\omega_j(x,\xi)|^p,$$

the last inequality gives (4.4). The proof is complete. □

We shall next apply Proposition 4.1 to pseudo-differential calculus. The following result extends Theorem 14.5.2 in [25], Theorem 7.1 in [27] and Theorem 4.3 in [49].

Theorem 4.2. *Assume that $t \in \mathbf{R}$ and $p, q, p_j, q_j \in [1, \infty]$ for $j = 1, 2$, satisfy (4.1). Also assume that $\omega \in \mathscr{P}(\mathbf{R}^{2m} \oplus \mathbf{R}^{2m})$ and $\omega_1, \omega_2 \in \mathscr{P}(\mathbf{R}^{2m})$ satisfy*

$$\frac{\omega_2(x - ty, \xi + (1-t)\eta)}{\omega_1(x + (1-t)y, \xi - t\eta)} \le C\omega(x, \xi, y, \eta) \tag{4.5}$$

for some constant C. If $a \in \widetilde{M}^{p,q}_{(\omega)}(\mathbf{R}^{2m})$, then $a_t(x, D)$ from $\mathscr{S}(\mathbf{R}^m)$ to $\mathscr{S}'(\mathbf{R}^m)$ extends uniquely to a continuous mapping from $M^{p_1,q_1}_{(\omega_1)}(\mathbf{R}^m)$ to $M^{p_2,q_2}_{(\omega_2)}(\mathbf{R}^m)$.

Moreover, if in addition $a \in \widetilde{\mathcal{M}}^{p,q}_{(\omega)}$, then $a_t(x, D) : M^{p_1,q_1}_{(\omega_1)} \to M^{p_2,q_2}_{(\omega_2)}$ is compact.

Proof. It suffices to prove the theorem in the case $t = 1/2$ in view of Proposition 1.7. The conditions on p_j and q_j implies that

$$p' \le p_1, q_1, p_2', q_2' \le q', \quad 1/p_1 + 1/p_2' = 1/q_1 + 1/q_2' = 1/p' + 1/q'.$$

Hence Proposition 4.1, and (4.5) show that for some constant $C > 0$,

$$\|W_{g,f}\|_{\widetilde{M}^{p',q'}_{(1/\omega)}} \le C\|f\|_{M^{p_1,q_1}_{(\omega_1)}} \|g\|_{M^{p_2',q_2'}_{(1/\omega_2)}}$$

when $f \in M^{p_1,q_1}_{(\omega_1)}(\mathbf{R}^m)$ and $g \in M^{p_2',q_2'}_{(1/\omega_2)}(\mathbf{R}^m)$.

Now assume that $a \in \widetilde{M}^{p,q}_{(\omega)}(\mathbf{R}^{2m})$. Then for some constants $C > 0$ and $C' > 0$ we get

$$\begin{aligned} |(a, W_{g,f})| &\le C\|a\|_{\widetilde{M}^{p,q}_{(\omega)}} \|W_{g,f}\|_{\widetilde{M}^{p',q'}_{(1/\omega)}} \\ &\le C'\|a\|_{\widetilde{M}^{p,q}_{(\omega)}} \|f\|_{M^{p_1,q_1}_{(\omega_1)}} \|g\|_{M^{p_2',q_2'}_{(1/\omega_2)}}. \end{aligned} \tag{4.6}$$

The continuity assertion now follows by combining (4.6) and Proposition 1.2 (3).

It remains to prove that if $a \in \widetilde{\mathcal{M}}^{p,q}_{(\omega)}$, then $a^w(x, D) : M^{p_1,q_1}_{(\omega_1)} \to M^{p_2,q_2}_{(\omega_2)}$ is compact. By straightforward computations it follows that $a^w(x, D)$ is compact when $a \in \mathscr{S}$. For general $a \in \widetilde{\mathcal{M}}^{p,q}_{(\omega)}$, it follows now from the first part of the proof that $a^w(x, D)$ may be approximated in operator norm by compact operators. Hence $a^w(x, D)$ is compact and the proof is complete. □

In Corollary 4.3 below, we prove that Theorem 4.2 may be used to obtain continuity results for pseudo-differential operators, where the corresponding symbols belong to $S_{(\omega)}(\mathbf{R}^{2m})$ for some $\omega \in \mathscr{P}(\mathbf{R}^{2m})$, which consists of all $a \in C^\infty(\mathbf{R}^{2m})$ such that $\partial^\alpha a/\omega$ is bounded for every multi-index α.

Corollary 4.3. *Assume that $t \in \mathbf{R}$, $p, q \in [1, \infty]$, $\omega_1, \omega_2 \in \mathscr{P}(\mathbf{R}^{2m})$,*

$$\omega_0(x, \xi) = \omega_2(x, \xi)/\omega_1(x, \xi) \quad \text{and} \quad \omega(x, \xi, y, \eta) = \omega_0(x, \xi)\sigma_N(y, \eta).$$

where $N \ge 0$ is an integer such that ω_1 and ω_2 are $\sigma_{N/2}$-moderate.

Then the following are true:

(1) *if* $a \in \widetilde{M}^{\infty,1}_{(\omega)}(\mathbf{R}^{2m})$, *then* $a_t(x,D)$ *is continuous from* $M^{p,q}_{(\omega_1)}(\mathbf{R}^m)$ *to* $M^{p,q}_{(\omega_2)}(\mathbf{R}^m)$;

(2) *if* $a \in S_{(\omega_0)}(\mathbf{R}^{2m})$, *then* $a_t(x,D)$ *is continuous from* $M^{p,q}_{(\omega_1)}(\mathbf{R}^m)$ *to* $M^{p,q}_{(\omega_2)}(\mathbf{R}^m)$.

Proof. The assertion (1) is an immediate consequence of Theorem 4.2, and the assertion (2) follows from (1) and the fact that $S_{(\omega_0)} \subseteq M^{\infty,1}_{(\omega)}$ (see [49, 50]). The proof is complete. □

We note that (2) in Corollary 4.3 is deeply related to Theorem 2.1 in [42]. The difference between these two results is that less restrictions are imposed on ω in Corollary 4.3, while less regularity assumptions are imposed in Theorem 2.1 in [42].

In a forthcoming paper we present an alternative proof of (2) in Corollary 4.3, which also holds when $a_t(x,D)$ acts Banach function spaces, a family of functions and distribution spaces which contains the modulation spaces. (See, e.g., [18].)

Remark 4.4. If $p = p_1 = p_2 = \infty$ and $q = q_1 = q_2 = 1$, or $p = q_1 = q_2 = \infty$ and $q = p_1 = p_2 = 1$ in Theorem 4.2, then $\mathscr{S}$ is not dense in any of the involving spaces. However, in spite of these facts, there are no ambiguity in order to define $a_t(x,D)f$ when $f \in M^{p_1,q_1}_{(\omega_1)}$. In fact, by (1.10) and Proposition 1.7 it is no restriction to assume that $t = 1/2$. Then (4.6) shows that $a^w(x,D)f$ makes sense as an element in $\mathscr{S}'$ when $f \in M^{p_1,q_1}_{(\omega_1)}$. By using (4.6) again in combination with Proposition 1.2 (3), it follows that indeed then give that $a^w(x,D)f \in M^{p_2,q_2}_{(\omega_2)}$, and the assertion follows.

Corollary 4.5. *Assume that* $p, q, p_j, q_j \in [1,\infty]$ *for* $j = 1, 2$, *satisfy* $q \le p_2, q_2 \le p$ *and* (4.2)$'$. *Assume also that* $t \in \mathbf{R} \setminus \{0,1\}$, $s_j, t_j \in \mathbf{R}$ *for* $j = 0,1,2$ *such that* $s_1, s_2, t_1, t_2 \ge 0$, *and that* $a \in \widetilde{M}^{p,q}_{(\omega)}(\mathbf{R}^{2m})$, *for some* $\omega \in \mathscr{P}(\mathbf{R}^{2m} \oplus \mathbf{R}^{2m})$. *Then the following are true:*

(1) *if* $|s_0| \le s_1 + s_2$, *and*

$$\omega(x,\xi,y,\eta) = \sigma_{s_1}(x,\xi)\sigma_{s_2}(y,\eta),$$

then $a_t(x,D)$ *is continuous from* $M^{p_1,q_1}_{s_0}(\mathbf{R}^m)$ *to* $M^{p_2,q_2}_{s_0}(\mathbf{R}^m)$;

(2) *if* $|s_0| \le s_1 + s_2$, $|t_0| \le t_1 + t_2$, *and*

$$\omega(x,\xi,y,\eta) = \sigma_{s_1,t_1}(x,\xi)\sigma_{s_2,t_2}(y,\eta),$$

then $a_t(x,D)$ *is continuous from* $M^{p_1,q_1}_{s_0,t_0}(\mathbf{R}^m)$ *to* $M^{p_2,q_2}_{s_0,t_0}(\mathbf{R}^m)$.

Proof. The assertion (1) follows by letting $\omega_1 = \omega_2 = \sigma_{s_0}$ in Theorem 4.2, and using the fact that

$$\begin{aligned}
&\langle (x - ty, \xi + (1-t)\eta/2)\rangle^{-1} \langle (x + (1-t)y, \xi - t\eta/2)\rangle \\
&\qquad \le C_t \min(\langle x\rangle, \langle y\rangle) \cdot \min(\langle \xi\rangle, \langle \eta\rangle)
\end{aligned}$$

for some constant C. The assertion (2) is proved in a similar way, and is left for the reader. The proof is complete. $\square$

In the case $p = q = \infty$ in Theorem 4.2, the converse is also true, i.e., we have the following result.

Theorem 4.6. *Assume that $t \in \mathbf{R}$, $a \in \mathscr{S}'(\mathbf{R}^{2m})$, $\omega \in \mathscr{P}(\mathbf{R}^{2m} \oplus \mathbf{R}^{2m})$, and $\omega_1, \omega_2 \in \mathscr{P}(\mathbf{R}^{2m})$ such that (4.5) holds. Then the operator $a_t(x, D)$ from $\mathscr{S}(\mathbf{R}^m)$ to $\mathscr{S}'(\mathbf{R}^m)$ extends to a continuous mapping from $M^1_{(\omega_1)}(\mathbf{R}^m)$ to $M^\infty_{(\omega_2)}(\mathbf{R}^m)$, if and only if $a \in \widetilde{M}^\infty_{(\omega)}(\mathbf{R}^{2m})$.*

For the proof we need the following two propositions of independent interest, where the first proposition is a slight generalization of Feichtinger-Gröchenig's kernel theorem, named as Schwartz-Gröchenig's kernel theorem in Theorem 4.1 in [49]. (See also Theorem 14.4.1 in [25].)

Proposition 4.7. *Assume that $m = m_1 + m_2$, $\omega_j \in \mathscr{P}(\mathbf{R}^{2m_j})$ for $j = 1, 2$ and $\omega \in \mathscr{P}(\mathbf{R}^m \oplus \mathbf{R}^m)$ such that*

$$\omega(x, y, \xi, \eta) = \omega_2(x, \xi)/\omega_1(y, -\eta). \tag{4.7}$$

Also assume that T is a linear and continuous map from $\mathscr{S}(\mathbf{R}^{m_1})$ to $\mathscr{S}'(\mathbf{R}^{m_2})$. Then T extends to a continuous mapping from $M^1_{(\omega_1)}(\mathbf{R}^{m_1})$ to $M^\infty_{(\omega_2)}(\mathbf{R}^{m_2})$, if and only if it exists an element $K \in M^\infty_{(\omega)}(\mathbf{R}^m)$ such that

$$(Tf)(x) = \langle K(x, \cdot), f \rangle. \tag{4.8}$$

Here the right-hand side in (4.8) should be interpreted as the distribution u, given by the formula $\langle u, g \rangle = \langle K, g \otimes f \rangle$, or alternatively, by the formula $(u, g) = (K, g \otimes \overline{f})$, when $f \in \mathscr{S}(\mathbf{R}^{m_1})$ and $g \in \mathscr{S}(\mathbf{R}^{m_2})$.

Proof. Assume that T extends to continuous map from $M^1_{(\omega_1)}$ to $M^\infty_{(\omega_2)}$. It follows from the kernel theorem of Schwartz that (4.8) holds for some $K \in \mathscr{S}'(\mathbf{R}^m)$. We shall prove that K belongs to $M^\infty_{(\omega)}$.

From the assumptions and Proposition 1.2 (3) it follows that

$$|(K, g \otimes \overline{f})_{L^2}| \le C \|f\|_{M^1_{(\omega_1)}} \|g\|_{M^1_{(1/\omega_2)}} \tag{4.9}$$

holds for some constant C which is independent of $f \in \mathscr{S}(\mathbf{R}^{m_1})$ and $g \in \mathscr{S}(\mathbf{R}^{m_2})$. By letting $\chi = \overline{g} \otimes f$ be fixed, and replacing f and g with

$$f_{y,\eta} = e^{-i\langle \cdot, \eta \rangle} f(\cdot - y) \quad \text{and} \quad g_{x,\xi} = e^{i\langle \cdot, \xi \rangle} f(\cdot - x),$$

it follows that (4.9) takes the form

$$|\mathscr{F}(K \tau_{(x,y)} \chi)(\xi, \eta)| \le C \|f_{y,\eta}\|_{M^1_{(\omega_1)}} \|g_{x,\xi}\|_{M^1_{(1/\omega_2)}}. \tag{4.9$'$}$$

We have to analyze the right-hand side of (4.9)′. If v is chosen such that ω_1 is v-moderate, and $\chi_1 \in \mathscr{S}(\mathbf{R}^{m_1}) \setminus 0$ is a fixed function, then we obtain

$$\begin{aligned}\|f_{y,\eta}\|_{M^1_{(\omega_1)}} &= \iint |\mathscr{F}(f\tau_{z-y}\chi_1)(\zeta+\eta)\omega_1(z,\zeta)|\,dzd\zeta \\ &= \iint |\mathscr{F}(f\tau_z\chi_1)(\zeta)\omega_1(z+y,\zeta-\eta)|\,dzd\zeta \\ &\le C\omega_1(y,-\eta)\|f\|_{M^1_{(v)}} = C'\omega_1(y,-\eta).\end{aligned}$$

In the same way we get

$$\|g_{x,\xi}\|_{M^1_{(1/\omega_2)}} \le C\omega_2(x,\xi)^{-1}.$$

If these estimates are inserted into (4.9)′, we obtain

$$|\mathscr{F}(K\tau_{(x,y)}\chi)(\xi,\eta)\omega(x,y,\xi,\eta)| \le C,$$

for some constant C which is independent of x, y, ξ and η. By taking the supremum of the left-hand side it follows that $\|K\|_{M^\infty_{(\omega)}} < \infty$. Hence $K \in M^\infty_{(\omega)}$, and the necessity follows.

The sufficiency follows by straightforward computations. The details are left for the reader. (See also the proof of Theorem 4.1 in [49].) The proof is complete. □

Proposition 4.8. *Assume that $a \in \mathscr{S}'(\mathbf{R}^{2m})$, and that $K \in \mathscr{S}'(\mathbf{R}^{2m})$ is the distribution kernel for the Weyl operator $a^w(x,D)$. Also assume that $p \in [1,\infty]$, and that $\omega, \omega_0 \in \mathscr{P}(\mathbf{R}^{2m} \oplus \mathbf{R}^{2m})$ are such that*

$$\omega(x,\xi,y,\eta) = \omega_0(x-y/2, x+y/2, \xi+\eta/2, -\xi+\eta/2).$$

Then $a \in \widetilde{M}^p_{(\omega)}(\mathbf{R}^{2m})$ if and only if $K \in M^p_{(\omega_0)}(\mathbf{R}^{2m})$. Moreover, if $\chi \in \mathscr{S}(\mathbf{R}^{2m})$ and

$$\psi(x,y) = \int \overline{\chi((x+y)/2,\xi)}e^{i\langle y-x,\xi\rangle}\,d\xi,$$

then $\|a\|_{\widetilde{M}^{p,\chi}_{(\omega)}} = \|K\|_{M^{p,\psi}_{(\omega_0)}}$.

Proof. By Fourier's inversion formula, it follows by straightforward computations that

$$|\mathscr{F}(K\,\tau_{(x-y/2,x+y/2)}\psi)(\xi+\eta/2,-\xi+\eta/2)| = |\widetilde{\mathscr{F}}(a\,\tau_{(x,\xi)}\chi)(y,\eta)|.$$

The result now follows by applying the $L^p_{(\omega)}$-norm on these expressions. The proof is complete. □

Proof of Theorem 4.6. In the case $t = 1/2$, the result follows immediately by combining Theorem 4.2, Proposition 4.7 and Proposition 4.8. For general t the result is now a consequence of (1.10) and Proposition 1.7. The proof is complete. □

Remark 4.9. Proposition 4.7 was proved in the case that ω_1 and $1/\omega_2$ are moderate functions in [25]. It was proved in the non-weighted case (i.e., $\omega_1 = \omega_2 = 1$) already in [12].

Remark 4.10. Let ω_1, ω_2 and ω be the same as in Proposition 4.7, $f_0 \in M^1_{(\omega_2)}$ and let $g_0 \in (M^\infty_{(\omega_1)})' \setminus M^1_{(1/\omega_1)}$, where the dual space of $M^\infty_{(\omega_1)}$ is identified through an extension of the sesqui-linear form on $\mathscr{S}$. Such g_0 exists, since $M^1_{(1/\omega_1)}$ is a strict subset of $(M^\infty_{(\omega_1)})'$ (see, e.g., [25]).

Then $T_K f(x) \equiv (f, g_0) f_0(x)$ has the distribution kernel $K = f_0 \otimes \overline{g_0}$, and defines an element in $\mathcal{L}(\omega_1, \omega_2)$, the set of all linear and continuous operators from $M^\infty_{(\omega_1)}$ to $M^1_{(\omega_2)}$. Since $K \notin M^1_{(\omega)}$, it follows that the map $K \mapsto T_K$ from $M^1_{(\omega)}$ to $\mathcal{L}(\omega_1, \omega_2)$ is injective but not surjective. (See also [21].) Consequently, the converse of Theorem 4.2 when $p = q = 1$ (the "dual case" of Theorem 4.6) does not hold.

On the other hand, let the operator norm define the topology for $\mathcal{L}(\omega_1, \omega_2)$, and let $\mathcal{L}_0(\omega_1, \omega_2)$ be the completion of the set of all operators with kernels in $\mathscr{S}$, under this norm. Then Theorem 7.4.1 in [21] and its proof show that $M^1_{(\omega)}$ is homeomorphic with $\mathcal{L}_0(\omega_1, \omega_2)$.

Next we discuss embedding properties between Schatten-von Neumann classes in pseudo-differential calculus and modulation spaces. As a first step we characterize Hilbert-Schmidt operators acting on modulation spaces of Hilbert type.

Proposition 4.11. *Assume that $\omega_1, \omega_2 \in \mathscr{P}(\mathbf{R}^{2m})$ and $\omega \in \mathscr{P}(\mathbf{R}^{2m} \oplus \mathbf{R}^{2m})$ are such that (4.7) holds, and that T is a linear and continuous operator from $\mathscr{S}(\mathbf{R}^m)$ to $\mathscr{S}'(\mathbf{R}^m)$ with distribution kernel $K \in \mathscr{S}'(\mathbf{R}^{2m})$. Then $T \in \mathscr{I}_2(M^2_{(\omega_1)}, M^2_{(\omega_2)})$, if and only if $K \in M^2_{(\omega)}(\mathbf{R}^{2m})$, and then*

$$\|T\|_{\mathscr{I}_2} = \|K\|_{M^2_{(\omega)}}. \tag{4.10}$$

Proof. Let $(f_j) \in \mathrm{ON}_{\omega_1}$ and $(h_k) \in \mathrm{ON}_{\omega_2}$ be orthonormal basis for $M^2_{(\omega_1)}(\mathbf{R}^m)$ and $M^2_{(\omega_2)}(\mathbf{R}^m)$ respectively. Then

$$\|T\|^2_{\mathscr{I}_2} = \sum_{j,k} |(Tf_j, h_k)_{M^2_{(\omega_2)}})|^2 = \sum_{j,k} |(K, h_k \otimes \overline{f_j})_{M^2_{(\omega_2)} \otimes L^2}|^2 \tag{4.11}$$

Next we consider the operator $T'_{\omega_0} = I_{M^2_{(\omega_2)}} \otimes \mathcal{R}_{1/\omega_0}$, where $\omega_0(x, \xi) = \omega_1(x, -\xi)$, which acts from $M^2_{(\omega_2)} \otimes M^2_{(1/\omega_0)}$ to $M^2_{(\omega_2)} \otimes M^2_{(\omega_0)}$ (Hilbert tensor products). Then (4.11) gives

$$\begin{aligned}
\|T\|^2_{\mathscr{I}_2} &= \sum_{j,k} |(T'_{\omega_0} K, h_k \otimes \overline{f_j})_{M^2_{(\omega_2)} \otimes M^2_{(\omega_1)}}|^2 \\
&= \|T'_{\omega_0} K\|^2_{M^2_{(\omega_2)} \otimes M^2_{(\omega_0)}} = \|K\|^2_{M^2_{(\omega_2)} \otimes M^2_{(1/\omega_0)}} = \|K\|^2_{M^2_{(\omega)}}
\end{aligned}$$

Hence (4.10) holds, and the proof is complete. □

The following result is now an immediate consequence of Proposition 4.11 and Proposition 4.8 for $p = 2$.

Proposition 4.12. *Assume that $a \in \mathscr{S}'(\mathbf{R}^{2m})$, $\omega_1, \omega_2 \in \mathscr{P}(\mathbf{R}^{2m})$ and that $\omega \in \mathscr{P}(\mathbf{R}^{2m} \oplus \mathbf{R}^{2m})$ are such that equality is attained in (4.5) for $t = 1/2$ and some* *constant C. Then $a^w(x,D) \in \mathscr{I}_2(M^2_{(\omega_1)}, M^2_{(\omega_2)})$, if and only if $a \in \widetilde{M}^2_{(\omega)}(\mathbf{R}^{2m})$. Moreover, for some constant $C > 0$ it holds*

$$C^{-1}\|a\|_{\widetilde{M}^2_{(\omega)}} \le \|a^w(x,D)\|_{\mathscr{I}_2} \le C\|a\|_{\widetilde{M}^2_{(\omega)}}.$$

For general Schatten-von Neumann classes, we have the following generalization of Proposition 4.12.

Theorem 4.13. *Assume that $t \in \mathbf{R}$ and $p, q, p_j, q_j \in [1,\infty]$ for $j = 1, 2$, satisfy*

$$p_1 \le p \le p_2, \quad q_1 \le \min(p,p') \quad \text{and} \quad q_2 \ge \max(p,p'). \tag{4.12}$$

Also assume that $\omega \in \mathscr{P}(\mathbf{R}^{2m} \oplus \mathbf{R}^{2m})$ and $\omega_1, \omega_2 \in \mathscr{P}(\mathbf{R}^{2m})$ are such that equality is attained in (4.5), for some constant C. Then

$$\widetilde{M}^{p_1,q_1}_{(\omega)}(\mathbf{R}^{2m}) \hookrightarrow s_{t,p}(\omega_1,\omega_2) \hookrightarrow \widetilde{M}^{p_2,q_2}_{(\omega)}(\mathbf{R}^{2m}) \tag{4.13}$$

For the proof we need the following lemma.

Lemma 4.14. *Assume that*

$$(x_{j_1})_{j_1\in I_1}, \quad (\xi_{j_2})_{j_2\in I_2}, \quad (y_{k_1})_{k_1\in I_1} \quad \text{and} \quad (\eta_{k_2})_{k_2\in I_2}$$

are lattices in $\mathbf{R}^m$. Also assume that $\varphi(x) = e^{-|x|^2/2}$ where $x \in \mathbf{R}^m$, $\omega_1, \omega_2 \in \mathscr{P}(\mathbf{R}^{2m})$, $(f_l) \in \mathrm{ON}_{\omega_1}$, $(h_l) \in \mathrm{ON}_{\omega_2}$, and set $\kappa_l = \mathcal{R}_{\omega_2} h_l$ and

$$\theta_1(j,k,l) = \mathscr{F}(f_l \tau_{x_{j_1}+y_{k_1}/2}\varphi)(\xi_{j_2} - \eta_{k_2}/2)\omega_1(x_{j_1} + y_{k_1}/2, \xi_{j_2} - \eta_{k_2}/2),$$

$$\theta_2(j,k,l) = \mathscr{F}(\kappa_l \tau_{x_{j_1}-y_{k_1}/2}\varphi)(\xi_{j_2} + \eta_{k_2}/2)/\omega_2(x_{j_1} - y_{k_1}/2, \xi_{j_2} + \eta_{k_2}/2),$$

where $j = (j_1, j_2) \in I_1 \times I_2 \equiv I$, $k = (k_1,k_2) \in I$. Then for some constant C and integer $N \ge 0$ it holds

$$\sum_l |\theta_i(j,k,l)|^2 \le C\|\varphi\|^2_{M^2_N} < \infty, \quad i = 1,2, \tag{4.14}$$

$$\sum_{k\in I} |\theta_1(j,k,l)|^2 \le C\|f_l\|^2_{M^2_{(\omega_1)}} \le C < \infty, \tag{4.15}$$

$$\sum_{k\in I} |\theta_2(j,k,l)|^2 \le C\|h_l\|^2_{M^2_{(\omega_2)}} \le C < \infty. \tag{4.16}$$

Proof. Let $N \ge 0$ be chosen such that ω_1 and ω_2 are σ_N-moderate. Then (4.14) in the case $i = 1$ follows if we prove that

$$\sum_l |\mathscr{F}(f_l\tau_z\varphi)(\zeta)|^2 \le C\|\varphi\|^2_{M^2_N}/\omega_1(z,\zeta)^2$$

for some constant C which is independent of (z,ζ). Since

$$\mathscr{F}(f_l\tau_z\varphi)(\zeta) = (f_l, e^{i\langle\,\cdot\,,\zeta\rangle}\tau_z\varphi)_{L^2} = (f_l, \mathcal{R}_{1/\omega_1}(e^{i\langle\,\cdot\,,\zeta\rangle}\tau_z\varphi))_{M^2_{(\omega_1)}}$$

and $(f_l)\in \mathrm{ON}_{\omega_1}$, we obtain for some $\chi\in \mathscr S\setminus 0$ that

$$
\begin{aligned}
\sum_l |\mathscr F(f_l\tau_z\varphi)(\zeta)|^2 &= \sum_l |(f_l,\mathcal R_{1/\omega_1}(e^{i\langle\,\cdot\,,\zeta\rangle}\tau_z\varphi))_{M^2_{(\omega_1)}}|^2\\
&\le \|\mathcal R_{1/\omega_1}(e^{i\langle\,\cdot\,,\zeta\rangle}\tau_z\varphi)\|^2_{M^2_{(\omega_1)}} \le C\|e^{i\langle\,\cdot\,,\zeta\rangle}\tau_z\varphi\|^2_{M^2_{(1/\omega_1)}}\\
&= C\iint |\mathscr F(e^{i\langle\,\cdot\,,\zeta\rangle}\tau_z\varphi\tau_x\chi)(\xi)/\omega_1(x,\xi)|^2\,dxd\xi\\
&= C\iint |\mathscr F(\varphi\tau_{x-z}\chi)(\xi-\zeta)/\omega_1(x,\xi)|^2\,dxd\xi\\
&= C\iint |\mathscr F(\varphi\tau_x\chi)(\xi)/\omega_1(x+z,\xi+\zeta)|^2\,dxd\xi\\
&\le C'\iint |\mathscr F(\varphi\tau_x\chi)(\xi)\sigma_N(x,\xi)|^2\,dxd\xi/\omega_1(z,\zeta)^2\\
&= C'\|\varphi\|^2_{M^2_N}/\omega_1(z,\zeta)^2.
\end{aligned}
$$

This proves the assertion.

The case $i=2$ in (4.14) follows from similar arguments together with the fact that

$$
\mathscr F(\kappa_l\tau_z\varphi)(\zeta) = (\kappa_l,e^{i\langle\,\cdot\,,\zeta\rangle}\tau_z\varphi)_{L^2} = (h_l,e^{i\langle\,\cdot\,,\zeta\rangle}\tau_z\varphi)_{M^2_{(\omega_2)}}.
$$

Next we prove (4.15). For some lattices $(z_{k_1})_{k_1\in I_1}$ and $(\zeta_{k_2})_{k_2\in I_2}$ we have

$$
\sum_k |\theta_1(j,k,l)|^2 \le \sum_{k_1,k_2} |\mathscr F(f_l\tau_{z_{k_1}}\varphi)(\zeta_{k_2})\omega_1(z_{k_1},\zeta_{k_2})|^2 \le C\|f_l\|^2_{M^2_{(\omega_1)}}
$$

for some constant C, where the last inequality follows from Proposition 1.4. This proves (4.15). By replacing the lattices (z_{k_1}) and (ζ_{k_2}) with other ones, if necessary, we obtain

$$
\begin{aligned}
\sum_k |\theta_2(j,k,l)|^2 &\le \sum_{k_1,k_2} |\mathscr F(\kappa_l\tau_{z_{k_1}}\varphi)(\zeta_{k_2})\omega_2(z_{k_1},\zeta_{k_2})|^2\\
&\le C\|\kappa_l\|^2_{M^2_{(1/\omega_2)}} \le C'\|h_l\|^2_{M^2_{(\omega_2)}}
\end{aligned}
$$

for some constants C and C', and (4.16) follows. The proof is complete. □

Proof of Theorem 4.13. We use the same notations as in Lemma 4.14. In view of Proposition 1.7 it is no restriction to assume that $t=1/2$, and that equalities are attained in (4.12). Then the result is an immediate consequence of Proposition 4.12 in the case $p=q=2$. Next we consider the case $q=1$. Let $X_j=(x_{j_1},\xi_{j_2})$ and $Y_k=(y_{k_1},\eta_{k_2})$. Then it follows that $(X_j)_{j\in I}$ and $(Y_k)_{k\in I}$ are lattices in $\mathbf R^{2m}$. Assume that $a\in \widetilde{M}^{p,1}_{(\omega)}$. By Proposition 1.4 it follows that if the lattices here above are chosen sufficiently dense, then

$$
a(X) = \sum_{j,k\in I} c_{j,k}e^{i\langle X,Y_k\rangle}(\tau_{X_j}W_{\varphi,\varphi})(X),
$$

for some sequence $(c_{j,k})_{j,k\in I}$ which satisfies

$$C^{-1}\sum_{j\in I}\Big(\sum_{k\in I}\lambda_{j,k}^p\Big)^{1/p} \le C\|a\|_{\widetilde{M}^{p,1}_{(\omega)}} \le C\sum_{j\in I}\Big(\sum_{k\in I}\lambda_{j,k}^p\Big)^{1/p}, \tag{4.17}$$
$$\lambda_{j,k} = |c_{j,k}\omega(x_{j_1},\xi_{j_2}y_{k_1},\eta_{k_2})|.$$

Here $X=(x,\xi)\in\mathbf{R}^{2m}$ and $\langle X,Y_k\rangle = \langle x,\eta_{k_2}\rangle + \langle y_{k_1},\xi\rangle$. Hence if $(f_l)\in \mathrm{ON}_{\omega_1}$ and $(\kappa_l)\in \mathrm{ON}^*_{\omega_2}$, then

$$|(a^w(x,D)f_l,\kappa_l)| \le \sum_{j,k}|c_{j,k}(e^{i\langle\cdot,Y_k\rangle}\tau_{X_j}W_{\varphi,\varphi},W_{\kappa_l,f_l})|.$$

By straightforward computations it follows that

$$e^{i\langle X,Y_k\rangle}(\tau_{X_j}W_{\varphi,\varphi})(X) = e^{i\Phi(Y_k,X_j)}W_{\varphi^1_{j,k},\varphi^2_{j,k}},$$

where Φ is a real-valued quadratic form on $\mathbf{R}^{2m}\oplus\mathbf{R}^{2m}$, and

$$\varphi^1_{j,k}(x) = e^{i\langle x,\xi_{j_2}+\eta_{k_2}/2\rangle}\varphi(x-x_{j_1}+y_{k_1}/2),$$
$$\varphi^2_{j,k}(x) = e^{i\langle x,\xi_{j_2}-\eta_{k_2}/2\rangle}\varphi(x-x_{j_1}-y_{k_1}/2).$$

This gives

$$\begin{aligned}|(e^{i\langle\cdot,Y_k\rangle}\tau_{X_j}W_{\varphi,\varphi},W_{\kappa_l,f_l})| &= |(W_{\varphi^1_{j,k},\varphi^2_{j,k}},W_{\kappa_l,f_l})| = |(\varphi^1_{j,k},\kappa_l)(\varphi^2_{j,k},f_l)|\\ &= |\theta_1(j,k,l)\theta_2(j,k,l)|\frac{\omega_2(x_{j_1}-y_{k_1}/2,\xi_{j_2}+\eta_j/2)}{\omega_1(x_{j_1}+y_{k_1}/2,\xi_{j_2}-\eta_{k_2}/2)}\\ &\le C|\theta_1(j,k,l)\theta_2(j,k,l)|\omega(x_{j_1},\xi_{j_2},y_{k_1},\eta_{k_2}),\end{aligned}$$

for some constant C. Hence

$$|c_{j,k}(e^{i\langle\cdot,Y_k\rangle}\tau_{X_j}W_{\varphi,\varphi},W_{\kappa_l,f_l})| \le C\lambda_{j,k}|\theta_1(j,k,l)\theta_2(j,k,l)|.$$

From these inequalities we obtain

$$\Big(\sum_l|(a^w(x,D)f_l,\kappa_l)|^p\Big)^{1/p} \le C\sum_{j,k,l}\lambda_{j,k}|\theta_1(j,k,l)\theta_2(j,k,l)| \le C(J_1+J_2)/2,$$

where

$$J_i = \Big(\sum_l\Big(\sum_{j,k}\lambda_{j,k}|(\theta_i(j,k,l)|^2\Big)^p\Big)^{1/p},\quad j=1,2.$$

We have to estimate J_1 and J_2. By Minkowski's and Hölder's inequalities we get

$$J_1 \le C\sum_j \Big(\sum_l \Big(\sum_k \lambda_{j,k}|\theta_1(j,k,l)|^2\Big)^p\Big)^{1/p}$$
$$= C\sum_j \Big(\sum_l \Big(\sum_k (\lambda_{j,k}|\theta_1(j,k,l)|^{2/p})|\theta_1(j,k,l)|^{2/p'}\Big)^p\Big)^{1/p}$$
$$\le C\sum_j \Big(\sum_l \Big(\sum_k \lambda^p_{j,k}|\theta_1(j,k,l)|^2\Big)\Big(\sum_k |\theta_1(j,k,l)|^2\Big)^{p/p'}\Big)^{1/p}.$$

Now (4.14)–(4.17) give

$$J_1 \le C_1\sum_j \Big(\Big(\sum_{k,l} \lambda^p_{j,k}|\theta_1(j,k,l)|^2\Big)\sup_l\Big(\sum_k |\theta_1(j,k,l)|^2\Big)^{p/p'}\Big)^{1/p}$$
$$\le C_2\Big(\sup_l \|f_l\|^{2/p'}_{M^2_{(\omega_1)}}\Big)\sum_j \Big(\sum_k \lambda^p_{j,k}\Big(\sum_l |\theta_1(j,k,l)|^2\Big)\Big)^{1/p}$$
$$\le C_3\|\varphi\|^{2/p}_{M^2_N}\sum_{j\in I}\Big(\sum_{k\in I}\lambda^p_{j,k}\Big)^{1/p} \le C_4\|a\|_{\widetilde{M}^{p,1}_{(\omega)}},$$

for some constants $C_1,\dots,C_4$. In the same way we get $J_2 \le C\|a\|_{\widetilde{M}^{p,1}_{(\omega)}}$ for some constant C. Hence it follows from these estimates that

$$\Big(\sum_l |(a^w(x,D)f_l,\kappa_l)|^p\Big)^{1/p} \le C\|a\|_{M^{p,1}_{(\omega)}}.$$

For some constant which is independent of the choice of sequences (f_l) and (κ_l). The result now follows by taking the supremum of the left-hand side with respect to all sequences (f_l) and (κ_l).

The first embedding in (4.13) now follows by interpolation of the case $q = 1$ and the case $p = q = 2$, using Proposition 1.6 and Proposition 2.7. The second embedding in (4.13) now follows from the first embedding and duality, using Theorem 3.3. The proof is complete. □

Next we present some consequences of Theorem 4.13 when the weight functions ω_1 and ω_2 are the same and satisfy certain properties which are common in the applications. It is for example common that the moderate function $v \in \mathscr{P}(\mathbf{R}^{2m})$ satisfies the symmetry condition

$$v(x,\xi) = v(-x,\xi) = v(x,-\xi) = v(-x,-\xi), \tag{4.18}$$

for every $(x,\xi) \in \mathbf{R}^{2m}$.

Corollary 4.15. *Assume that $t \in \mathbf{R}$, $p, q, p_j, q_j \in [1,\infty]$ for $j = 1,2$ are the same as in Theorem 4.13, and that $\omega_0 \in \mathscr{P}(\mathbf{R}^{2m})$ is v-moderate for some $v \in \mathscr{P}(\mathbf{R}^{2m})$ which satisfies (4.18). Also let $v_0(x,\xi,y,\eta) = v(y,\eta)$. Then*

$$\widetilde{M}^{p_1,q_1}_{(v_0)} \subseteq s_{t,p}(\omega_0) \subseteq \widetilde{M}^{p_2,q_2}_{(1/v_0)}.$$

Proof. It follows from the assumptions that

$$C^{-1}v(y,\eta)^{-1} \le \frac{\omega_0(x-ty,\xi+(1-t)\eta)}{\omega_0(x+(1-t)y,\xi-t\eta)} \le Cv(y,\eta).$$

The result is therefore a consequence of Proposition 1.2 (2) and Theorem 4.13. The proof is complete. □

Example 4.16. Assume that $t \in \mathbf{R}$ and that $p, q, p_j, q_j \in [1,\infty]$ for $j = 1,2$ are the same as in Theorem 4.13. Then it follows from the Corollary 4.15 that the following are true:

(1) if $s \in \mathbf{R}$, then

$$\widetilde{M}^{p_1,q_1}_{|s|,0} \subseteq s_{t,p}(\sigma_s) \subseteq \widetilde{M}^{p_2,q_2}_{-|s|,0};$$

(2) if $s_1, s_2 \in \mathbf{R}$ and $v(x,\xi,y,\eta) = \sigma_{|s_1|,|s_2|}(y,\eta)$, then

$$\widetilde{M}^{p_1,q_1}_{(v)} \subseteq s_{t,p}(\sigma_{s_1,s_2}) \subseteq \widetilde{M}^{p_2,q_2}_{(1/v)}.$$

Remark 4.17. By using embedding properties in [50] between modulation spaces and Besov spaces, the embeddings in Theorem 4.13, Corollary 4.15 and Example 4.16 give rise to embeddings between the $s_{t,p}$ spaces and Besov spaces. More precisely, assume that $t \in \mathbf{R}$, $p, q, q_j \in [1,\infty]$ for $j = 1,2$, ω_0 and v are the same as in Corollary 4.15, and let $\vartheta(p) = |1-2/p|$ and $s, s_1, s_2 \in \mathbf{R}$. Then Theorem 4.4 and Remark 4.6 in [50] and Corollary 4.15 and Example 4.16 give:

$$B^{p,q_1}_{(\sigma_{2m\vartheta(p)}v)}(\mathbf{R}^{2m}) \subseteq s_{t,p}(\omega_0) \quad \subseteq B^{p,q_2}_{(\sigma_{-2m\vartheta(p)}/v)}(\mathbf{R}^{2m}),$$

$$B^{p,q_1}_{2m\vartheta(p)+|s|}(\mathbf{R}^{2m}) \subseteq s_{t,p}(\sigma_s) \quad \subseteq B^{p,q_2}_{-2m\vartheta(p)-|s|}(\mathbf{R}^{2m}),$$

$$B^{p,q_1}_{m\vartheta(p)+|s_2|,m\vartheta(p)+|s_1|}(\mathbf{R}^{2m}) \subseteq s_{t,p}(\sigma_{s_1,s_2}) \subseteq B^{p,q_2}_{-m\vartheta(p)-|s_2|,-m\vartheta(p)-|s_1|}(\mathbf{R}^{2m})$$

(See [6,8] and Section 4 in [50] for strict definitions of the involved Besov spaces.)

Remark 4.18. Assume that $\omega_1 = \omega_2 = 1$ in $s_{t,p} = s_{t,p}(\omega_1,\omega_2)$. Then it is proved in Corollary 3.5 in [48] that

$$B^{p,\min(p,p')}_{2m|1-2/p|} \subseteq M^{p,\min(p,p')} \subseteq s_{t,p} \subseteq M^{p,\max(p,p')} \subseteq B^{p,\max(p,p')}_{-2m|1-2/p|}. \tag{4.19}$$

These relations can also be obtained by combining Theorem 4.4 and Remark 4.6 in [50] or Theorem 3.1 in [48], with Example 4.16 in the case $s = 0$, or $s_1 = s_2 = 0$.

Note here that the embeddings

$$B^{p,p}_{2m|1-2/p|} \subseteq s^w_p, \quad 1 \le p \le 2 \quad \text{and} \quad s^w_p \subseteq B^{p,p}_{-2m|1-2/p|}, \quad 2 \le p \le \infty,$$

which are consequences of (4.19), can not be improved. On the other hand, in the remaining cases, it was proved in [46] that the embeddings

$$s^w_p \subseteq B^{p,p'}_{-m|1-2/p|}, \quad 1 \le p \le 2 \quad \text{and} \quad B^{p,p'}_{m|1-2/p|} \subseteq s^w_p, \quad 2 \le p \le \infty$$

are sharp, which lead to strict improvements of (4.19) when considering embeddings between the s^w_p-spaces and Besov spaces. Note here that if $p = \infty$, then the latter embedding was proved already in [8]. Furthermore, it was proved in [6] that $s_{0,\infty} \subseteq B^{\infty,1}_m$.

Acknowledgements

I am grateful to A. Holst for a careful reading and valuable comments, leading to improvement of the language and the content. I also thank P. Boggiatto, N. Kruglyak and I. Asekritova for fruitful discussions.

References

[1] J. Bergh and J. Löfström, *Interpolation Spaces, An Introduction*, Springer-Verlag, Berlin, Heidelberg, New York, 1976.

[2] M.S. Birman and M.Z. Solomyak, Estimates for the singular numbers of integral operators (Russian), *Uspehi Mat. Nauk.* **32**, (1977), 17–84.

[3] P. Boggiatto, Localization operators with L^p symbols on modulation spaces, in *Advances in Pseudo-differential Operators,* Editors: R. Ashino, P. Boggiatto and M.W. Wong, Birkhäuser, Basel, 2004, 149–163.

[4] P. Boggiatto, E. Cordero and K. Gröchenig, Generalized anti-Wick operators with symbols in distributional Sobolev spaces, *Integral Equations Operator Theory* **48** (2004), 427–442.

[5] P. Boggiatto and J. Toft, Embeddings and compactness for generalized Sobolev-Shubin spaces and modulation spaces, *Applic. Anal.* **84** (2005), 269–282.

[6] A. Boulkhemair, L^2 estimates for pseudodifferential operators, *Ann. Scuola Norm. Sup. Pisa Cl. Sci.* **22** (1995), 155–183.

[7] A. Boulkhemair, Remarks on a Wiener type pseudodifferential algebra and Fourier integral operators, *Math. Res. Letters* **4** (1997), 53–67.

[8] A. Boulkhemair, L^2 estimates for Weyl quantization, *J. Funct. Anal.* **165** (1999), 173–204.

[9] E. Cordero and K. Gröchenig, Time-frequency analysis of localization operators, *J. Funct. Anal.* **205** (2003), 107–131.

[10] W. Czaja and Z. Rzeszotnik, Pseudodifferential operators and Gabor frames: spectral asymptotics, *Math. Nachr.* **233/234** (2002), 77–88.

[11] M. Dimassi and J. Sjöstrand *Spectral Asymptotics in the Semi-Classical Limit*, Cambridge University Press, Cambridge, New York, Melbourne, Madrid, 1999.

[12] H.G. Feichtinger Un espace de Banach de distributions tempérés sur les groupes localement compacts abéliens (French), *C. R. Acad. Sci. Paris Sér. A-B 290* **17** (1980), A791–A794.

[13] H.G. Feichtinger *Banach spaces of distributions of Wiener's type and interpolation,* in *Proc. Conf. Oberwolfach, Functional Analysis and Approximation, August 1980,* Editors: P. Butzer, B. Sz.-Nagy and E. Görlich, Int. Ser. Num. Math. **69** Birkhäuser Verlag, Basel, Boston, Stuttgart, 1981, 153–165.

[14] H.G. Feichtinger, Banach convolution algebras of Wiener's type, in *Proc. Functions, Series, Operators in Budapest*, Colloquia Math. Soc. J. Bolyai, North Holland Publ. Co., Amsterdam, Oxford, New York, 1980.

[15] H.G. Feichtinger, Modulation spaces on locally compact abelian groups, in *Wavelets and their Applications*, Editors: M. Krishna, R. Radha and S. Thangavelu, Allied Publishers, 2003, 99–140.

[16] H.G. Feichtinger, Atomic characterizations of modulation spaces through Gabor-type representations, *Rocky Mountain J. Math.* **19** (1989), 113–126.

[17] H.G. Feichtinger and P. Gröbner, Banach spaces of distributions defined by decomposition methods, I, *Math. Nachr.* **123** (1985), 97–120.

[18] H.G. Feichtinger and K.H. Gröchenig, Banach spaces related to integrable group representations and their atomic decompositions, I, *J. Funct. Anal.* **86** (1989), 307–340.

[19] H.G. Feichtinger and K.H. Gröchenig, Banach spaces related to integrable group representations and their atomic decompositions, II, *Monatsh. Math.* **108** (1989), 129–148.

[20] H.G. Feichtinger and K. Gröchenig, Gabor frames and time-frequency analysis of distributions, *J. Funct. Anal.* **146** (1997), 464–495.

[21] H.G. Feichtinger and W. Kozek, Operator quantization on lca groups, in *Gabor Analysis and Algorithms, Theory and Applications*, Editors: H.G. Feichtinger and T. Strohmer, Birkhäuser, Boston, Basel, Berlin, 1998.

[22] G.B. Folland, *Harmonic Analysis in Phase Space*, Princeton University Press, Princeton, 1989.

[23] I.C. Gohberg and M.G. Krein, *Introduction to the Theory of Linear Non-Selfadjoint Operators in Hilbert Space (Russian)*, Izdat. Nauka, Moscow, 1965.

[24] K. Gröchenig, Describing functions: atomic decompositions versus frames, *Monatsh. Math.* **112** (1991), 1–42.

[25] K. Gröchenig, *Foundations of Time-Frequency Analysis*, Birkhäuser, Boston, 2001.

[26] K. Gröchenig and C. Heil, Modulation spaces and pseudo-differential operators, *Integral Equations Operator Theory* **34** (1999), 439–457.

[27] K. Gröchenig and C. Heil, Modulation spaces as symbol classes for pseudodifferential operators, in *Wavelets and their Applications*, Editors: M. Krishna, R. Radha and S. Thangavelu, Allied Publishers, 2003, 151–170.

[28] L. Hörmander, *The Analysis of Linear Partial Differential Operators I, III*, Springer-Verlag, Berlin, Heidelberg, New York, Tokyo, 1983, 1985.

[29] C. Heil, J. Ramanathan and P. Topiwala, Singular values of compact pseudodifferential operators, *J. Funct. Anal.* **150** (1997), 426–452.

[30] D. Labate, Pseudodifferential operators on modulation spaces, *J. Math. Anal. Appl.* **262** (2001), 242–255.

[31] S. Pilipović and N. Teofanov, Wilson bases and ultramodulation spaces, *Math. Nachr.* **242** (2002), 179–196.

[32] S. Pilipović and N. Teofanov, On a symbol class of elliptic pseudodifferential operators, *Bull. Acad. Serb. Sci. Arts* **27** (2002), 57–68.

[33] M. Reed and B. Simon, *Methods of Modern Mathematical Physics*, Academic Press, London, New York, 1979.

[34] R. Rochberg and K. Tachizawa, Pseudo-differential operators, Gabor frames and local trigonometric bases, in *Gabor Analysis and Algorithms*, Editors: H.G. Feichtinger and T. Strohmer, Birkhäuser, Boston, 1998, 171–192.

[35] R. Schatten, *Norm Ideals of Completely Continuous Operators*, Springer, Berlin, 1960.

[36] B.W. Schulze, *Boundary Value Problems and Singular Pseudo-differential Operators*, Wiley, Chichester, 1998.
[37] B.W. Schulze and N.N. Tarkhanov, Pseudodifferential operators with operator-valued symbols, in *Israel Math. Conf. Proc.* **16**, 2003.
[38] M.A. Shubin, *Pseudodifferential Operators and Spectral Theory*, Springer-Verlag, Berlin, 1987.
[39] B. Simon, *Trace Ideals and their Applications*, Cambridge University Press, Cambridge, London, New York, Melbourne, 1979.
[40] J. Sjöstrand, An algebra of pseudodifferential operators, *Math. Res. Letters* **1** (1994), 185–192.
[41] J. Sjöstrand, Wiener type algebras of pseudodifferential operators, in *Séminaire Equations aux Dérivées Partielles, Ecole Polytechnique,* 1994/1995, *Exposé n° IV.*
[42] K. Tachizawa, The boundedness of pseudo-differential operators on modulation spaces, *Math. Nachr.* **168** (1994), 263–277.
[43] N. Teofanov, *Ultramodulation Spaces and Pseudodifferential Operators*, Endowment Andrejević, Beograd, 2003.
[44] J. Toft, *Continuity and Positivity Problems in Pseudo-differential Calculus*, Thesis, Department of Mathematics, University of Lund, Lund, 1996.
[45] J. Toft, Subalgebras to a Wiener type algebra of pseudo-differential operators, *Ann. Inst. Fourier (Grenoble)* **51** (2001), 1347–1383.
[46] J. Toft, Continuity properties for non-commutative convolution algebras with applications in pseudo-differential calculus, *Bull. Sci. Math.* **126** (2002), 115–142.
[47] J. Toft, Modulation spaces and pseudo-differential operators, *Research Report*, Blekinge Institute of Technology, Karlskrona, 2002.
[48] J. Toft, Continuity properties for modulation spaces with applications to pseudo-differential calculus, I, *J. Funct. Anal.* **207** (2004), 399–429.
[49] J. Toft, Continuity properties for modulation spaces with applications to pseudo-differential calculus, II, *Ann. Global Anal. Geom.* **26** (2004), 73–106.
[50] J. Toft, Convolution and embeddings for weighted modulation spaces, in *Advances in Pseudo-differential Operators,* Editors: R. Ashino, P. Boggiatto and M.W. Wong, Birkhäuser, Basel 2004, 165–186.
[51] J. Toft, Continuity and Schatten-von Neumann properties for pseudo-differential operators on modulation spaces, *Research Report*, Växjö University, Växjö, 2005.
[52] J. Toft, Continuity and Schatten properties for Toeplitz operators on modulation spaces, in *Modern Trends in Pseudo-differential Operators,* Editors: J. Toft, M.W. Wong and H. Zhu, Birkhäuser, Basel, this Volume, 313–328.
[53] H. Triebel, Modulation spaces on the Euclidean n-space, *Z. Anal. Anwendungen* **2** (1983), 443–457.
[54] M.W. Wong, *Weyl Transforms*, Springer-Verlag, 1998.

Joachim Toft
Department of Mathematics and Systems Engineering
Växjö University, Sweden
e-mail: `joachim.toft@vxu.se`

Operator Theory:
Advances and Applications, Vol. 172, 207–233

Algebras of Pseudo-differential Operators with Discontinuous Symbols

Yu.I. Karlovich

Abstract. Using the boundedness of the maximal singular integral operator related to the Carleson-Hunt theorem we prove the boundedness and study the compactness of pseudo-differential operators $a(x, D)$ with bounded measurable $V(\mathbb{R})$-valued symbols $a(x, \cdot)$ on the Lebesgue spaces $L^p(\mathbb{R})$ with $1 < p < \infty$, where $V(\mathbb{R})$ is the Banach algebra of all functions of bounded total variation on $\mathbb{R}$. Replacement of absolutely continuous functions of bounded total variation by arbitrary functions of bounded total variation allows us to study pseudo-differential operators with symbols admitting discontinuities of the first kind with respect to the spatial and dual variables. Appearance of discontinuous symbols leads to non-commutative algebras of Fredholm symbols. Three different Banach algebras of pseudo-differential operators with discontinuous symbols acting on the spaces $L^p(\mathbb{R})$ are studied. We construct Fredholm symbol calculi for these algebras and establish Fredholm criteria for the operators in these algebras in terms of their Fredholm symbols. For the operators in the first algebra we also obtain an index formula. An application to the Haseman boundary value problem is given.

Mathematics Subject Classification (2000). Primary 47G30, 47L15; Secondary 47A53, 47G10, 30E25.

Keywords. Maximal singular integral operator, function of bounded total variation, slow oscillation, pseudo-differential operator, Lebesgue space, boundedness, compactness, algebra, Fredholm symbols, Fredholmness, index, Haseman problem.

1. Introduction

The study of pseudo-differential operators with non-regular symbols, motivated by an extension of the sphere of their applications, is one of the natural directions in the development of the theory of pseudo-differential operators (see, for example,

Partially supported by CONACYT (México) Project No. 49992.

[10], [11], [40], [41]). There are several calculi for pseudo-differential operators with non-smooth symbols, which were elaborated by J. Marschall [33]–[35], W. Hoh [21], [22], and by N. Jacob and A.G. Tokarev [24]. Recently a new calculus was worked out in [27] for the Banach algebra of pseudo-differential operators with slowly oscillating $V_0(\mathbb{R})$-valued symbols on the Lebesgue spaces $L^p(\mathbb{R})$ with $1 < p < \infty$, where $V_0(\mathbb{R})$ is the Banach algebra of absolutely continuous functions of bounded total variation on $\mathbb{R}$ (also see [28] for an application to singular integral operators).

The present paper continues the investigations of [27]. Applying the integral analogue [30] of the famous Carleson-Hunt theorem (see [9], [23] and [25]) we study the boundedness and compactness of pseudo-differential operators $a(x, D)$ with bounded measurable $V(\mathbb{R})$-valued symbols $a(x, \cdot)$ on the Lebesgue spaces $L^p(\mathbb{R})$ with $1 < p < \infty$, where $V(\mathbb{R})$ is the Banach algebra of all functions of bounded total variation on $\mathbb{R}$. Replacement of absolutely continuous functions of bounded total variation by arbitrary functions of bounded total variation essentially extends the class of pseudo-differential operators in comparison with [27]. In particular, this allows one to study pseudo-differential operators with symbols admitting discontinuities of the first kind with respect to the spatial and dual variables. Appearance of discontinuous symbols leads to non-commutative algebras of Fredholm symbols, which completely changes the strategy of the Fredholm study for corresponding pseudo-differential operators because we cannot approximate initial symbols by infinitely differentiable ones. As usual, Fredholm symbols depend only on the behavior of initial symbols $a(x, \lambda)$ in the neighborhood of infinity but the presence of discontinuities results in an appearance of an additional variable μ in the Fredholm symbols $\mathcal{A}(x, \lambda, \mu)$ associated with the initial symbols $a(x, \lambda)$, where μ runs through a set $\mathcal{L}_p$ depending on $p \in (1, \infty)$.

Given $p \in (1, \infty)$, let $\mathcal{B} := \mathcal{B}(L^p(\mathbb{R}))$ denote the Banach algebra of all bounded linear operators acting on the Banach space $L^p(\mathbb{R})$ and let $\mathcal{K} := \mathcal{K}(L^p(\mathbb{R}))$ be the closed two-sided ideal of all compact operators in $\mathcal{B}$. As is well known (see, for example, [7], [18], [5]), an operator $A \in \mathcal{B}$ is said to be *Fredholm*, if its image is closed and the spaces $\ker A$ and $\ker A^*$ are finite-dimensional. In that case $\operatorname{Ind} A = \dim \ker A - \dim \ker A^*$ is referred to as the *index* of A.

Three different Banach algebras ($\mathfrak{A}$, $\widetilde{\mathfrak{A}}$ and $\widehat{\mathfrak{A}}$) of pseudo-differential operators with discontinuous symbols acting on the space $L^p(\mathbb{R})$ are studied in the present paper. We construct Fredholm symbol calculi for these algebras and establish Fredholm criteria for the operators in these algebras in terms of their Fredholm symbols. For the operators $A \in \mathfrak{A}$ we also obtain an index formula. The results obtained for the algebra $\widetilde{\mathfrak{A}}$ are applied to studying the Haseman boundary value problem on $\mathbb{C} \setminus \overline{\mathbb{R}}_+$ (see [1], [26], [32], and the references therein).

The paper is organized as follows. In Section 2 we introduce necessary classes of functions and symbols of pseudo-differential operators. In Sections 3 and 4 we study, respectively, the boundedness and compactness of pseudo-differential operators with symbols in $L^\infty(\mathbb{R}, V(\mathbb{R}))$ on the Lebesgue spaces $L^p(\mathbb{R})$ where $1 < p < \infty$. In Section 5, following [14, Section 7], [13], [37, Section 15] and [29, Section 6], we expose the Fredholm theory for the Banach subalgebra $\mathfrak{D}$ of $\mathcal{B}(L^p(\mathbb{R}))$ generated by convolution type operators $cW_b := c\mathcal{F}^{-1}b\mathcal{F}$ where $\mathcal{F}$ is the Fourier transform, c and b are piecewise continuous functions on $\mathbb{R}$, and b is also a Fourier multiplier on $L^p(\mathbb{R})$. In Section 6 we study the Banach algebra $\mathfrak{A}$ of pseudo-differential operators with piecewise continuous $V(\mathbb{R})$-valued symbols on the spaces $L^p(\mathbb{R})$ by reduction to the algebra $\mathfrak{D}$. In Section 7 we study the Banach algebra $\widetilde{\mathfrak{A}}$ generated by pseudo-differential operators with slowly oscillating (in the sense of [12]) $V_0(\mathbb{R})$-valued symbols and piecewise continuous $V_0(\mathbb{R})$-valued symbols. In Section 8 we study the Banach algebra $\widehat{\mathfrak{A}}$ generated by pseudo-differential operators with slowly oscillating (in the sense of [36]) $V_0(\mathbb{R})$-valued symbols and piecewise continuous $V(\mathbb{R})$-valued symbols. In Section 9 we study a singular integral operator N with a shift which is associated with the Haseman boundary value problem on $\mathbb{C} \setminus \mathbb{R}_+$. We reduce the operator N to a Mellin pseudo-differential operator related to an operator in the algebra $\widetilde{\mathfrak{A}}$. Finally, applying results of Section 7, we establish a Fredholm criterion for the operator N provided that the shift has a slowly oscillating derivative in contrast to preceding papers on the Haseman problem.

2. Classes of symbols

As usual, let $C_0^\infty(\mathbb{R})$ be the set of all functions in $C^\infty(\mathbb{R})$ of compact support.

Functions of bounded total variation. Let $a : \mathbb{R} \to \mathbb{C}$ be a function of bounded total variation $V(a)$ on $\mathbb{R}$ where

$$V(a) := \sup\left\{\sum_{k=1}^{n} \left|a(x_k) - a(x_{k-1})\right| : -\infty < x_0 < x_1 < \cdots < x_n < +\infty,\ n \in \mathbb{N}\right\}.$$

Hence (see, for example, [20, Chapter 9]), at every point $x \in \dot{\mathbb{R}} := \mathbb{R} \cup \{\infty\}$ there exist finite one-sided limits $a(x \pm 0) = \lim\limits_{t \to x^\pm} a(t)$, where $a(\pm\infty) = a(\infty \mp 0)$, the set of discontinuities is at most countable, and thus a is piecewise continuous on $\overline{\mathbb{R}} := [-\infty, +\infty]$. Without loss of generality (see the proof of Theorem 3.1 below) we will assume that functions of bounded total variation are continuous from the left at every discontinuity point $x \in \dot{\mathbb{R}}$. The set $V(\mathbb{R})$ of all continuous from the left functions of bounded total variation on $\mathbb{R}$ is a unital non-separable Banach algebra with the norm $\|a\|_V := \|a\|_{L^\infty(\mathbb{R})} + V(a)$.

Absolutely continuous functions of bounded total variation. Let $V_0(\mathbb{R})$ be the unital Banach subalgebra of $V(\mathbb{R})$ consisting of all absolutely continuous functions of bounded total variation on $\mathbb{R}$. Hence (see [20, Chapter 9]), every function $a \in V_0(\mathbb{R})$ is continuous on $\overline{\mathbb{R}} = [-\infty, +\infty]$, and

$$a' \in L^1(\mathbb{R}), \quad a(x) = \int_{-\infty}^{x} a'(y)dy + a(-\infty) \ \text{ for all } x \in \overline{\mathbb{R}}, \quad V(a) = \int_{\mathbb{R}} |a'(y)|dy.$$

Fourier multipliers. Let $\mathcal{M}_p$ denote the Banach algebra of Fourier multipliers on $L^p(\mathbb{R})$ for $1 < p < \infty$. From Stechkin's inequality $\|b\|_{\mathcal{M}_p} \le \big\|S_{\mathbb{R}}\big\|_{\mathcal{B}(L^p(\mathbb{R}))}\|b\|_V$ (see, for example, [13, Theorem 2.11]) it follows that $V(\mathbb{R}) \subset \mathcal{M}_p$ for all $p \in (1,\infty)$. Let $PC_p(\overline{\mathbb{R}})$ and $C_p(\overline{\mathbb{R}})$ be the closures in $\mathcal{M}_p$ of the sets of all piecewise constant functions on $\mathbb{R}$ with at most finite sets of jumps and, respectively, of all continuous functions of bounded total variation on $\mathbb{R}$. Clearly, $PC_p(\overline{\mathbb{R}})$ and $C_p(\overline{\mathbb{R}})$ are Banach subalgebras of $\mathcal{M}_p$. Moreover, $PC_p(\overline{\mathbb{R}})$ is also the closure of $V(\mathbb{R})$ in $\mathcal{M}_p$ (see [13, Remark 2.12]), and, by [39, Lemma 1.1], $C_p(\overline{\mathbb{R}})$ is the closure of $V_0(\mathbb{R})$ in $\mathcal{M}_p$. Hence, $C_p(\dot{\mathbb{R}}) := PC_p(\overline{\mathbb{R}}) \cap C(\dot{\mathbb{R}})$ is the closure in $\mathcal{M}_p$ of the set $V_0(\mathbb{R}) \cap C(\dot{\mathbb{R}})$.

Slowly oscillating functions. Let $C_b(\mathbb{R}) := C(\mathbb{R}) \cap L^\infty(\mathbb{R})$ be the C^*-algebra of all bounded continuous functions $a : \mathbb{R} \to \mathbb{C}$. For a continuous function $a : \mathbb{R} \to \mathbb{C}$, let

$$cm_x(a) := \max\Big\{\big|a(x+h) - a(x)\big| : \ h \in \mathbb{R},\ |h| \le 1\Big\} \tag{2.1}$$

be the local oscillation of a at a point $x \in \mathbb{R}$. According to [12, p. 122], a function $a \in C_b(\mathbb{R})$ is called *slowly oscillating at* ∞ if $\lim\limits_{|x|\to\infty} cm_x(a) = 0$. Clearly, the set SO of all functions in $C_b(\mathbb{R})$ which are slowly oscillating at ∞ is a C^*-subalgebra of $C_b(\mathbb{R})$. Following [36] we also consider the C^*-subalgebra $SO^\diamond$ of $C_b(\mathbb{R})$ which consists of functions that *slowly oscillate at* ∞ in another sense. Namely,

$$SO^\diamond := \Big\{a \in C_b(\mathbb{R}) : \ \lim_{x\to+\infty} \sup_{t,\tau\in[-2x,-x]\cup[x,2x]} \big|a(t) - a(\tau)\big| = 0\Big\}.$$

Obviously, $SO^\diamond \subset SO$.

Let $M(\mathcal{A})$ stand for the maximal ideal space of a unital C^*-algebra $\mathcal{A}$. Identifying the points $t \in \overline{\mathbb{R}}$ with the evaluation functionals $t(f) = f(t)$ for $f \in C(\overline{\mathbb{R}})$, where $f(\pm\infty) = \lim\limits_{x\to\pm\infty} f(x)$, we obtain $M(C(\overline{\mathbb{R}})) = \overline{\mathbb{R}}$. Consider the fibers

$$M_{\pm\infty}(SO) := \big\{\xi \in M(SO) : \xi|_{C(\overline{\mathbb{R}})} = \pm\infty\big\}$$

of the maximal ideal space $M(SO)$ over the points $t = \pm\infty$. Analogously, the fiber $M_\infty(SO^\diamond)$ of the maximal ideal space $M(SO^\diamond)$ at the point $t = \infty$ is defined by

$$M_\infty(SO^\diamond) := \big\{\xi \in M(SO^\diamond) : \xi|_{C(\dot{\mathbb{R}})} = \infty\big\}.$$

According to [27, Proposition 2.4], [4, Proposition 5] and [6, Proposition 4.1],

$$M_{-\infty}(SO) \cup M_{+\infty}(SO) = (\mathrm{clos}_{SO^*}\mathbb{R}) \setminus \mathbb{R}, \quad M_\infty(SO^\diamond) = (\mathrm{clos}_{(SO^\diamond)^*}\mathbb{R}) \setminus \mathbb{R},$$

where $\mathrm{clos}_{SO^*}\mathbb{R}$ is the weak-star closure of $\mathbb{R}$ in SO^*, the dual space of SO, and $\mathrm{clos}_{(SO^\diamond)^*}\mathbb{R}$ is the weak-star closure of $\mathbb{R}$ in $(SO^\diamond)^*$, the dual space of $SO^\diamond$.

Bounded symbols. In what follows we denote by $L^\infty(\mathbb{R}, V(\mathbb{R}))$ the set of all functions $a : \mathbb{R}\times\overline{\mathbb{R}} \to \mathbb{C}$ such that $\widehat{a} : x \mapsto a(x,\cdot)$ is a bounded measurable $V(\mathbb{R})$-valued function on $\mathbb{R}$. Since the Banach space $V(\mathbb{R})$ is non-separable, the measurability of $\widehat{a}$ means that the map $\widehat{a} : \mathbb{R} \to V(\mathbb{R})$ possesses the Luzin property: for any compact $K \subset \mathbb{R}$ and any $\delta > 0$ there is a compact $K_\delta \subset K$ such that the Lebesgue measure of the set $K \setminus K_\delta$ is less than δ and the function $\widehat{a} : K_\delta \to V(\mathbb{R})$ is continuous (see, for example, [38, Chapter IV, § 4]). This implies that the functions $x \mapsto a(x, \lambda \pm 0)$ for all $\lambda \in \dot{\mathbb{R}}$ and the function $x \mapsto \|a(x,\cdot)\|_V$ are measurable on $\mathbb{R}$ as well. Note that the limits $a(x, \lambda \pm 0)$ exist for all $\lambda \in \dot{\mathbb{R}}$ and for almost all $x \in \mathbb{R}$, and $a(x,\lambda) = a(x, \lambda - 0)$ for all $\lambda \in \mathbb{R}$ and $a(x, \pm\infty) = \lim\limits_{\lambda \to \pm\infty} a(x,\lambda)$. Therefore, the functions $a(\cdot, \lambda \pm 0)$ for every $\lambda \in \dot{\mathbb{R}}$ and the function $x \mapsto \|a(x,\cdot)\|_V$, where

$$\|a(x,\cdot)\|_V := \|a(x,\cdot)\|_{L^\infty(\mathbb{R})} + V\big(a(x,\cdot)\big), \tag{2.2}$$

belong to $L^\infty(\mathbb{R})$. Clearly, $L^\infty(\mathbb{R}, V(\mathbb{R}))$ is a unital Banach algebra with the norm

$$\|a\|_{L^\infty(\mathbb{R},V(\mathbb{R}))} := \operatorname*{ess\,sup}_{x \in \mathbb{R}} \|a(x,\cdot)\|_V.$$

Piecewise continuous symbols. Let $PC(\overline{\mathbb{R}}, V(\mathbb{R}))$ stand for the set of all functions $a : \overline{\mathbb{R}} \times \overline{\mathbb{R}} \to \mathbb{C}$ such that $\widehat{a} : x \mapsto a(x,\cdot)$ is a piecewise continuous $V(\mathbb{R})$-valued function on $\overline{\mathbb{R}}$. If $a \in PC(\overline{\mathbb{R}}, V(\mathbb{R}))$, then the functions $x \mapsto a(x, \lambda \pm 0)$ for all $\lambda \in \dot{\mathbb{R}}$ and the function $x \mapsto \|a(x,\cdot)\|_V$ given by (2.2) belong to $PC(\overline{\mathbb{R}})$. Clearly, $PC(\overline{\mathbb{R}}, V(\mathbb{R})) \subset PC(\overline{\mathbb{R}} \times \overline{\mathbb{R}})$ is a unital Banach subalgebra of $L^\infty(\mathbb{R}, V(\mathbb{R}))$.

Slowly oscillating symbols. Let $C_b(\mathbb{R}, V_0(\mathbb{R}))$ denote the Banach algebra of all bounded continuous $V_0(\mathbb{R})$-valued functions on $\mathbb{R}$. Following [27], for a function $a \in C_b(\mathbb{R}, V_0(\mathbb{R}))$ by analogy with (2.1), we put

$$cm_x^V(a) := \max\Big\{ \|a(x+h,\cdot) - a(x,\cdot)\|_V : \ h \in \mathbb{R}, \ |h| \le 1 \Big\}.$$

Let $SO(\mathbb{R}, V_0(\mathbb{R}))$ be the Banach algebra of all bounded continuous $V_0(\mathbb{R})$-valued functions on $\mathbb{R}$ that slowly oscillate at ∞, that is, satisfy the condition

$$\lim_{|x| \to \infty} cm_x^V(a) = 0.$$

Clearly, every $V_0(\mathbb{R})$-valued function $x \mapsto a(x,\cdot)$ in $SO(\mathbb{R}, V_0(\mathbb{R}))$ is uniformly continuous on $\mathbb{R}$.

We also introduce the Banach subalgebra $SO^\diamond(\mathbb{R}, V_0(\mathbb{R}))$ of $C_b(\mathbb{R}, V_0(\mathbb{R}))$ consisting of all bounded continuous $V_0(\mathbb{R})$-valued functions on $\mathbb{R}$ that slowly oscillate at ∞ in the sense of [36]. Thus, $SO^\diamond(\mathbb{R}, V_0(\mathbb{R}))$ is the set

$$\Big\{ f \in C_b(\mathbb{R}, V_0(\mathbb{R})) : \lim_{x \to +\infty} \sup_{t,\tau \in [-2x,-x] \cup [x,2x]} \|a(t,\cdot) - a(\tau,\cdot)\|_V = 0 \Big\}.$$

Let $\widetilde{\mathcal{E}}$ denote the subset of all functions $a : \mathbb{R} \times \overline{\mathbb{R}} \to \mathbb{C}$ in $SO(\mathbb{R}, V_0(\mathbb{R}))$ that satisfy the conditions

$$\lim_{M \to -\infty} \sup_{x \in \mathbb{R}} V_{-\infty}^{M}\big(a(x,\cdot)\big) = \lim_{M \to +\infty} \sup_{x \in \mathbb{R}} V_M^{+\infty}\big(a(x,\cdot)\big) = 0 \tag{2.3}$$

and

$$\lim_{|h| \to 0} \sup_{x \in \mathbb{R}} \|a(x,\cdot) - a^h(x,\cdot)\|_V = 0 \tag{2.4}$$

where $V_b^c(f)$ is the variation of f on an interval (b,c), and $a^h(x,\lambda) := a(x,\lambda+h)$ for all $(x,\lambda) \in \mathbb{R}\times\mathbb{R}$. Analogously, let $\widehat{\mathcal{E}}$ be the subset of all functions $a : \mathbb{R}\times\overline{\mathbb{R}} \to \mathbb{C}$ in $SO^\diamond(\mathbb{R}, V_0(\mathbb{R}))$ that satisfy the conditions (2.3) and (2.4). Obviously, $\widetilde{\mathcal{E}}$ and $\widehat{\mathcal{E}}$ are Banach subalgebras of the Banach algebra $C_b(\mathbb{R}, V_0(\mathbb{R}))$, and $\widehat{\mathcal{E}} \subset \widetilde{\mathcal{E}}$.

By [27, Lemmas 2.7 and 2.9], for every $a(x,\lambda) \in \widetilde{\mathcal{E}}$ and every $\xi \in M_{\pm\infty}(SO)$, there exist a sequence $x_n \to \pm\infty$ and a function $a(\xi,\cdot) \in V_0(\mathbb{R})$ such that

$$\lim_{n\to\infty} \big\|a(x_n,\cdot) - a(\xi,\cdot)\big\|_V = 0; \tag{2.5}$$

and, conversely, every sequence $y_n \to \pm\infty$ contains a subsequence $x_n \to \pm\infty$ such that (2.5) holds for some $\xi \in M_{\pm\infty}(SO)$. Obviously, the functions $a(x,\lambda) \in \widehat{\mathcal{E}}$ also possess such the property for every $\xi \in M_\infty(SO^\diamond)$ (with $x_n \to \infty$ and $y_n \to \infty$) because $\widehat{\mathcal{E}} \subset \widetilde{\mathcal{E}}$.

Discontinuous slowly oscillating symbols. Finally, we consider discontinuous slowly oscillating symbols that belong to the subsets $\widetilde{\Omega} := PC(\overline{\mathbb{R}}, V_0(\mathbb{R})) \cup \widetilde{\mathcal{E}}$ and $\widehat{\Omega} := PC(\overline{\mathbb{R}}, V(\mathbb{R})) \cup \widehat{\mathcal{E}}$ of $L^\infty(\mathbb{R}, V(\mathbb{R}))$.

3. Boundedness of pseudo-differential operators on Lebesgue spaces

For $-\infty < a < b < +\infty$, we consider the *partial sum operators* $S_{(a,b)}$ given by

$$(S_{(a,b)}f)(x) = \frac{1}{2\pi}\int_a^b \widehat{f}(\lambda)e^{ix\lambda}d\lambda, \quad \text{for } x \in \mathbb{R}, \tag{3.1}$$

where

$$\widehat{f}(\lambda) := (\mathcal{F}f)(\lambda) := \int_{\mathbb{R}} f(y)e^{-i\lambda y}dy, \quad \text{for } \lambda \in \mathbb{R},$$

is the Fourier transform of f. Since $S_{(a,b)} = \mathcal{F}^{-1}\chi_{(a,b)}\mathcal{F}$ where $\chi_{(a,b)}$ is the characteristic function of the interval (a,b), we infer from Stechkin's inequality that the operators $S_{(a,b)}$ are bounded on all the spaces $L^p(\mathbb{R})$ with $1 < p < \infty$.

According to the famous Carleson-Hunt theorem on almost everywhere convergence (more precisely, by its integral analog for $L^p(\mathbb{R})$, see [30]), the maximal singular integral operator S_* given by

$$(S_*f)(x) = \sup_{-\infty<a<b<+\infty} \big|(S_{(a,b)}f)(x)\big|, \quad \text{for } x \in \mathbb{R},$$

is also bounded on every space $L^p(\mathbb{R})$ with $1 < p < \infty$ (also see [15, Chapter 2, Section 2.2]). In particular, for every $f \in L^p(\mathbb{R})$ and for almost every $x \in \mathbb{R}$,

$$\sup_{\lambda\in\mathbb{R}} \big|(S_{(0,\lambda)}f)(x)\big| \le (S_*f)(x) < \infty. \tag{3.2}$$

Modifying the proofs of [15, Chapter 3, Section 3.3] and [27, Theorem 3.1], we obtain the following result.

Theorem 3.1. *If $a \in L^\infty(\mathbb{R}, V(\mathbb{R}))$, then the pseudo-differential operator $a(x,D)$ defined for functions $u \in C_0^\infty(\mathbb{R})$ by the iterated integral*

$$\big[a(x,D)u\big](x) = \frac{1}{2\pi}\int_{\mathbb{R}} d\lambda \int_{\mathbb{R}} a(x,\lambda)e^{i(x-y)\lambda}u(y)dy, \quad \text{for } x \in \mathbb{R}, \tag{3.3}$$

extends to a bounded linear operator on every Lebesgue space $L^p(\mathbb{R})$ with $p \in (1,\infty)$, and

$$\big\|a(x,D)\big\|_{\mathcal{B}(L^p(\mathbb{R}))} \le 2\,\big\|a\big\|_{L^\infty(\mathbb{R},V(\mathbb{R}))}\big\|S_*\big\|_{\mathcal{B}(L^p(\mathbb{R}))}. \tag{3.4}$$

Proof. Fix $u \in C_0^\infty(\mathbb{R})$. Due to (3.3),

$$\big[a(x,D)u\big](x) = \frac{1}{2\pi}\int_{\mathbb{R}} a(x,\lambda)\widehat{u}(\lambda)e^{i\lambda x}d\lambda, \quad \text{for } x \in \mathbb{R}, \tag{3.5}$$

where the integral on the right of (3.5) converges a.e. on $\mathbb{R}$ because the function $\lambda \mapsto a(x,\lambda)$ is bounded on $\mathbb{R}$ for almost all $x \in \mathbb{R}$ and $\widehat{u}$ is a rapidly decreasing function. Since for every $x \in \mathbb{R}$ the function $\lambda \mapsto \big(S_{(0,\lambda)}u\big)(x)$ given by (3.1) is continuous on $\mathbb{R}$ and for almost every $x \in \mathbb{R}$ the function $\lambda \mapsto a(x,\lambda)$ is of bounded total variation on $\mathbb{R}$, we conclude that the Riemann-Stieltjes integral

$$\int_{-M}^{M} \big(S_{(0,\lambda)}u\big)(x)\,da(x,\lambda)$$

with the parameter $x \in \mathbb{R}$ also converges a.e. on $\mathbb{R}$ for every $M > 0$. Therefore, it follows from the formula of integration by parts for the Riemann-Stieltjes integrals that

$$\frac{1}{2\pi}\int_{-M}^{M} a(x,\lambda)\widehat{u}(\lambda)e^{i\lambda x}d\lambda = -\int_{-M}^{M} \big(S_{(0,\lambda)}u\big)(x)\,da(x,\lambda)$$
$$+\Big[a(x,M)\big(S_{(0,M)}u\big)(x) + a(x,-M)\big(S_{(-M,0)}u\big)(x)\Big]$$

for almost all $x \in \mathbb{R}$. Therefore, taking into account (3.2) and (2.2), for almost every $x \in \mathbb{R}$ and every $M > 0$ such that the function $\lambda \mapsto a(x,\lambda)$ is continuous at the points $\lambda = \pm M$, we obtain

$$\left|\frac{1}{2\pi}\int_{-M}^{M} a(x,\lambda)\widehat{u}(\lambda)e^{i\lambda x}d\lambda\right| \le \left|\int_{-M}^{M} \big(S_{(0,\lambda)}u\big)(x)\,da(x,\lambda)\right|$$
$$+ \big|a(x,M)\big|\big|\big(S_{(0,M)}u\big)(x)\big| + \big|a(x,-M)\big|\big|\big(S_{(-M,0)}u\big)(x)\big|$$
$$\le \big(S_*u\big)(x)\Big(V\big(a(x,\cdot)\big) + 2\big\|a(x,\cdot)\big\|_{L^\infty(\mathbb{R})}\Big) \le 2\big(S_*u\big)(x)\big\|a(x,\cdot)\big\|_V.$$

The latter estimate and (3.5) imply that

$$\big|\big[a(x,D)u\big](x)\big| \le 2\big(S_*u\big)(x)\big\|a(x,\cdot)\big\|_V \quad \text{a.e. on } \mathbb{R}.$$

Consequently, in view of the boundedness of the maximal singular operator S_* on the spaces $L^p(\mathbb{R})$ for $p \in (1,\infty)$, we obtain

$$\big\|a(x,D)u\big\|_{L^p(\mathbb{R})} \le 2\,\big\|a\big\|_{L^\infty(\mathbb{R},V(\mathbb{R}))}\big\|S_*\big\|_{\mathcal{B}(L^p(\mathbb{R}))}\big\|u\big\|_{L^p(\mathbb{R})},$$

which implies (3.4). □

4. Compactness of pseudo-differential operators on Lebesgue spaces

Here, by analogy with [27, Theorem 4.1], we get sufficient conditions for compactness of pseudo-differential operators $\sigma(x,D)$ with symbols $\sigma(x,\lambda)\in L^\infty(\mathbb{R},V(\mathbb{R}))$.

Theorem 4.1. *The pseudo-differential operator $\sigma(x,D)$ with a discontinuous symbol $\sigma(x,\lambda)\in L^\infty(\mathbb{R},V(\mathbb{R}))$ is compact on every Lebesgue space $L^p(\mathbb{R})$ $(1<p<\infty)$ if*

$$\sigma(x,\pm\infty)=0 \ \ for\ almost\ all\ \ x\in\mathbb{R}, \qquad \lim_{|x|\to\infty} V\big(\sigma(x,\cdot)\big)=0, \tag{4.1}$$

$$\lim_{M\to+\infty}\ \operatorname*{ess\,sup}_{|x|\le N}\Big(V_{-\infty}^{-M}\big(\sigma(x,\cdot)\big)+V_M^{+\infty}\big(\sigma(x,\cdot)\big)\Big)=0 \ \ for\ every\ \ N>0. \tag{4.2}$$

Proof. Since $\sigma(x,\lambda)\in L^\infty(\mathbb{R},V(\mathbb{R}))$, the operator $\sigma(x,D)$ is bounded on all the spaces $L^p(\mathbb{R})$ with $1<p<\infty$ by virtue of Theorem 3.1. Following the proof of [27, Theorem 4.1] we consider the sequence of pseudo-differential operators $\{\sigma_{k,n}(x,D)\}_{k,n\in\mathbb{N}}$ with the symbols

$$\sigma_{k,n}(x,\lambda)=\psi_k(x)\sigma(x,\lambda)\psi_n(\lambda) \tag{4.3}$$

where $\psi_n(x)=1$ for $|x|<n$, $\psi_n(x)=\psi(|x|-n)$ for $|x|\in[n,n+1]$, $\psi_n(x)=0$ for $|x|>n+1$, and $\psi(x)=(x-1)^2(2x+1)$. Obviously, $\psi_n\in C_0^1(\mathbb{R})$ because

$$\psi:[0,1]\to[1,0],\ \ \psi'(0)=\psi'(1)=0,\ \ \text{and}\ \ \psi'(x)<0\ \ \text{for}\ \ x\in(0,1). \tag{4.4}$$

Since $\sigma(x,\cdot)\in V(\mathbb{R})$ for almost all $x\in\mathbb{R}$, we conclude from the first equality in (4.1) that for almost all $x\in\mathbb{R}$ and for all $M>0$,

$$|\sigma(x,\lambda)|\le V_{-\infty}^{-M}\big(\sigma(x,\cdot)\big)\ \text{ if }\ \lambda\le -M,\ \ |\sigma(x,\lambda)|\le V_M^{+\infty}\big(\sigma(x,\cdot)\big)\ \text{ if }\ \lambda\ge M.$$

Therefore, taking into account the facts that according to (4.4),

$$\|\psi_n\|_{L^\infty(\mathbb{R})}=1,\ V_n^{n+1}(\psi_n)=V_{-n-1}^{-n}(\psi_n)=\int_0^1|\psi'(x)|dx=|\psi(1)-\psi(0)|=1, \tag{4.5}$$

we obtain

$$\begin{aligned} V\big(\sigma_{k,n}(x,\cdot)\big)&=|\psi_k(x)|\,V\big(\sigma(x,\cdot)\psi_n(\cdot)\big)\\ &\le V_{-n-1}^{n+1}\big(\sigma(x,\cdot)\psi_n(\cdot)\big)\le\|\psi_n\|_{L^\infty[-n-1,n+1]}V_{-n-1}^{n+1}\big(\sigma(x,\cdot)\big)\\ &+\|\sigma(x,\cdot)\|_{L^\infty[-n-1,-n]}V_{-n-1}^{-n}(\psi_n)+\|\sigma(x,\cdot)\|_{L^\infty[n,n+1]}V_n^{n+1}(\psi_n)\\ &\le V\big(\sigma(x,\cdot)\big)+V_{-\infty}^{-n}\big(\sigma(x,\cdot)\big)+V_n^{+\infty}\big(\sigma(x,\cdot)\big)\le 2V\big(\sigma(x,\cdot)\big), \end{aligned} \tag{4.6}$$

$$\|\sigma_{k,n}(x,\cdot)\|_V\le\|\sigma_{k,n}(x,\cdot)\|_{L^\infty(\mathbb{R})}+V\big(\sigma_{k,n}(x,\cdot)\big)\le 2\|\sigma(x,\cdot)\|_V<\infty. \tag{4.7}$$

Clearly, (4.7) means that $\sigma_{k,n}(x,\lambda)\in L^\infty(\mathbb{R},V(\mathbb{R}))$. Hence, by Theorem 3.1, the operators $\sigma_{k,n}(x,D)$ are bounded on all the spaces $L^p(\mathbb{R})$ $(1<p<\infty)$.

Analogously to (4.6), making use of the definition of ψ_n and (4.5), we obtain

$$\begin{aligned} &V\big(\sigma(x,\cdot)[1-\psi_n(\cdot)]\big) = V_{-\infty}^{-n}\big(\sigma(x,\cdot)[1-\psi_n(\cdot)]\big) + V_n^{+\infty}\big(\sigma(x,\cdot)[1-\psi_n(\cdot)]\big) \\ &\le \big\|\sigma(x,\cdot)\big\|_{L^\infty(-\infty,-n]} + \big\|\sigma(x,\cdot)\big\|_{L^\infty[n,+\infty)} + V_{-\infty}^{-n}\big(\sigma(x,\cdot)\big) + V_n^{+\infty}\big(\sigma(x,\cdot)\big) \\ &\le 2\Big(V_{-\infty}^{-n}\big(\sigma(x,\cdot)\big) + V_n^{+\infty}\big(\sigma(x,\cdot)\big)\Big). \end{aligned} \tag{4.8}$$

Since $\sigma(x,\pm\infty) - \sigma_{k,n}(x,\pm\infty) = 0$, applying (4.8) we get

$$\begin{aligned} &\operatorname*{ess\,sup}_{x\in\mathbb{R}} \big\|\sigma(x,\cdot) - \sigma_{k,n}(x,\cdot)\big\|_V \le 2 \operatorname*{ess\,sup}_{x\in\mathbb{R}} V\big(\sigma(x,\cdot) - \sigma_{k,n}(x,\cdot)\big) \\ &\le 2 \operatorname*{ess\,sup}_{x\in\mathbb{R}} \Big(\big|1-\psi_k(x)\big| V\big(\sigma(x,\cdot)\big)\Big) + 2 \operatorname*{ess\,sup}_{x\in\mathbb{R}} \Big(\psi_k(x) V\big(\sigma(x,\cdot)[1-\psi_n(\cdot)]\big)\Big) \\ &\le 2 \operatorname*{ess\,sup}_{|x|\ge k} V\big(\sigma(x,\cdot)\big) + 4 \operatorname*{ess\,sup}_{|x|\le k+1} \Big(V_{-\infty}^{-n}\big(\sigma(x,\cdot)\big) + V_n^{+\infty}\big(\sigma(x,\cdot)\big)\Big). \end{aligned} \tag{4.9}$$

Fix $\varepsilon > 0$. By (4.1), there exists a $k_0 > 0$ such that

$$\operatorname*{ess\,sup}_{|x|\ge k} V\big(\sigma(x,\cdot)\big) < \varepsilon \quad \text{for all} \quad k > k_0. \tag{4.10}$$

According to (4.2), for every $k > k_0$ there is an $n_k \in \mathbb{N}$ such that

$$\operatorname*{ess\,sup}_{|x|\le k+1} \Big(V_{-\infty}^{-n_k}\big(\sigma(x,\cdot)\big) + V_{n_k}^{+\infty}\big(\sigma(x,\cdot)\big)\Big) < k^{-1}. \tag{4.11}$$

From (4.9), (4.10) and (4.11) it follows that $\lim\limits_{k\to\infty} \big\|\sigma - \sigma_{k,n_k}\big\|_{L^\infty(\mathbb{R},V(\mathbb{R}))} = 0$. This implies according to (3.4) that

$$\lim_{k\to\infty} \big\|\sigma(x,D) - \sigma_{k,n_k}(x,D)\big\|_{\mathcal{B}(L^p(\mathbb{R}))} = 0. \tag{4.12}$$

Obviously, by (4.3), all the operators $\sigma_{k,n}(x,D)$ are compact on the space $L^2(\mathbb{R})$. By the Krasnoselskii theorem on interpolation of compactness (see [31, Theorem 3.10]), every operator $\sigma_{k,n}(x,D)$, being bounded on all the spaces $L^p(\mathbb{R})$ for $1 < p < \infty$ and compact on the space $L^2(\mathbb{R})$, is compact on all $L^p(\mathbb{R})$ $(1 < p < \infty)$.

Finally, from (4.12) it follows that the operator $\sigma(x,D)$, being the uniform limit of compact operators $\sigma_{k,n_k}(x,D)$, is itself compact. □

Corollary 4.2. *If $\sigma(x,\lambda) \in PC(\overline{\mathbb{R}}, V(\mathbb{R}))$ and (4.1) holds, then the pseudo-differential operator $\sigma(x,D)$ is compact on every Lebesgue space $L^p(\mathbb{R})$ with $1 < p < \infty$.*

Proof. Since $\sigma(x,\lambda) \in PC(\overline{\mathbb{R}}, V(\mathbb{R}))$ and for every $M \in \mathbb{R}$,

$$\big|V_M^{+\infty}\big(\sigma(x,\cdot)\big) - V_M^{+\infty}\big(\sigma(y,\cdot)\big)\big| \le V_M^{+\infty}\big(\sigma(x,\cdot) - \sigma(y,\cdot)\big) \le V\big(\sigma(x,\cdot) - \sigma(y,\cdot)\big),$$

the function $x \mapsto V_M^{+\infty}\big(\sigma(x,\cdot)\big)$ and, analogously, the function $x \mapsto V_{-\infty}^{M}\big(\sigma(x,\cdot)\big)$ are piecewise continuous on $\overline{\mathbb{R}}$. Furthermore, for every $x \in \mathbb{R}$,

$$\lim_{\lambda\to-\infty} \sigma(x\pm 0,\lambda) = \sigma(x\pm 0,-\infty), \qquad \lim_{\lambda\to+\infty} \sigma(x\pm 0,\lambda) = \sigma(x\pm 0,+\infty),$$

whence

$$\lim_{M\to+\infty} \Big(V_{-\infty}^{-M}\big(\sigma(x\pm 0,\cdot)\big) + V_M^{+\infty}\big(\sigma(x\pm 0,\cdot)\big)\Big) = 0 \quad \text{for all } x \in \mathbb{R}. \tag{4.13}$$

As the function $x \mapsto \big(V_{-\infty}^{-M} + V_{M}^{+\infty}\big)\big(\sigma(x,\cdot)\big)$ is piecewise continuous on every segment $[-N,N] \subset \mathbb{R}$, choosing an appropriate finite covering of this segment and taking into account (4.13) it is easily seen that (4.2) is fulfilled automatically. It remains to apply Theorem 4.1. □

5. Algebra $\mathfrak{D}$ of convolution type operators

Recall the symbol calculus and the Fredholm criterion for the Banach algebra

$$\mathfrak{D} := \operatorname{alg}\big\{cW_b : c \in PC(\overline{\mathbb{R}}),\ b \in PC_p(\overline{\mathbb{R}})\big\} \subset \mathcal{B}(L^p(\mathbb{R}))$$

where $cW_b = c\mathcal{F}^{-1}b\mathcal{F}$ (see [14, Section 7], [13], and also [37, Section 15], [8], [3]).

Let $\mathcal{K}$ be the closed two-sided ideal of all compact operators in the Banach algebra $\mathcal{B} = \mathcal{B}(L^p(\mathbb{R}))$. By analogy with [17] one can show that $\mathcal{K} \subset \mathfrak{D}$. As is known, the Fredholmness of operators $A \in \mathfrak{D}$ is equivalent to the invertibility of the cosets $A^\pi := A + \mathcal{K}$ in the Calkin algebra $\mathcal{B}^\pi := \mathcal{B}/\mathcal{K}$.

Given $1 < p < \infty$, we consider the set

$$M := \big(\mathbb{R} \times \{\infty\} \times \mathcal{L}_p\big) \cup \big(\{\infty\} \times \mathbb{R} \times \mathcal{L}_p\big) \cup \big(\{\infty\} \times \{\infty\} \times \{0,1\}\big) \tag{5.1}$$

equipped with the discrete topology, where

$$\mathcal{L}_p := \Big\{\mu = 2^{-1}\big(1 + \coth(\pi x + \pi i/p)\big) : x \in \overline{\mathbb{R}}\Big\}$$

is a circular arc with the endpoints 0 and 1 if $p \neq 2$, and the line segment joining these points if $p = 2$.

Following [3, Subsection 2.4], for $A = cW_b$ (with $c \in PC(\overline{\mathbb{R}})$ and $b \in PC_p(\overline{\mathbb{R}})$) and for $(x,\lambda,\mu) \in M$, we define the matrix function

$$\mathcal{A}(x,\lambda,\mu) =$$
$$\begin{bmatrix} c(x+0)\big(b(\lambda-0)\mu + b(\lambda+0)(1-\mu)\big) & c(x+0)\big(b(\lambda-0) - b(\lambda+0)\big)\nu(\mu) \\ c(x-0)\big(b(\lambda-0) - b(\lambda+0)\big)\nu(\mu) & c(x-0)\big(b(\lambda-0)(1-\mu) + b(\lambda+0)\mu\big) \end{bmatrix}, \tag{5.2}$$

where, by convention, $a(\infty \pm 0) = a(\mp\infty)$, $b(\infty \pm 0) = b(\mp\infty)$, and $\nu(\mu)$ is the continuous branch of the function $\sqrt{\mu(1-\mu)}$ in $\mathbb{C} \setminus \big((-\infty,0) \cup (1,+\infty)\big)$ which satisfies the condition $\nu(1/2) = 1/2$.

Let $BC(M,\mathbb{C}^{2\times 2})$ stand for the set of bounded continuous functions of M into $\mathbb{C}^{2\times 2}$. From [14, Section 7], [13], [37, Section 15] and [8] we obtain the following.

Theorem 5.1. *Let $1 < p < \infty$ and let M be given by* (5.1).

(i) *The map Ψ associating with the operator $A = cW_b$ ($c \in PC(\overline{\mathbb{R}})$, $b \in PC_p(\overline{\mathbb{R}})$) the matrix function* (5.2) *extends to a Banach algebra homomorphism $\Psi : \mathfrak{D} \to BC(M,\mathbb{C}^{2\times 2})$ whose kernel contains $\mathcal{K}$.*

(ii) *The quotient algebra $\mathfrak{D}^\pi = \mathfrak{D}/\mathcal{K}$ is inverse closed in the Calkin algebra $\mathcal{B}^\pi$.*

(iii) *An operator $A \in \mathfrak{D}$ is Fredholm on the space $L^p(\mathbb{R})$ if and only if*

$$[\det \Psi(A)](x,\lambda,\mu) \neq 0 \quad \textit{for all} \quad (x,\lambda,\mu) \in M.$$

Now we equip the set M given by (5.1) with the following topology. Let the neighborhood base for $(x,\lambda,\mu) \in M$ consist of the sets $U(x,\lambda,\mu)$ of the form

$$U(x,\infty,\mu) = \begin{cases} \{(x,\infty)\} \times V_\mu & \text{if} \quad x \in \mathbb{R},\ \mu \in \mathcal{L}_p \setminus \{0,1\}; \\ \big(\{(x,\infty)\} \times V_0\big) \cup \big((x-\varepsilon,x) \times \{\infty\} \times \mathcal{L}_p\big) & \text{if} \quad x \in \mathbb{R},\ \mu = 0; \\ \big(\{(x,\infty)\} \times V_1\big) \cup \big((x,x+\varepsilon) \times \{\infty\} \times \mathcal{L}_p\big) & \text{if} \quad x \in \mathbb{R},\ \mu = 1; \end{cases}$$

$$U(\infty,\lambda,\mu) = \begin{cases} \{(\infty,\lambda)\} \times V_\mu & \text{if} \quad \lambda \in \mathbb{R},\ \mu \in \mathcal{L}_p \setminus \{0,1\}; \\ \big(\{(\infty,\lambda)\} \times V_0\big) \cup \big(\{\infty\} \times (\lambda,\lambda+\varepsilon) \times \mathcal{L}_p\big) & \text{if} \quad \lambda \in \mathbb{R},\ \mu = 0; \\ \big(\{(\infty,\lambda)\} \times V_1\big) \cup \big(\{\infty\} \times (\lambda-\varepsilon,\lambda) \times \mathcal{L}_p\big) & \text{if} \quad \lambda \in \mathbb{R},\ \mu = 1; \end{cases}$$

$$U(\infty,\infty,0) = \{(\infty,\infty,0)\} \times \big((c,+\infty) \times \{\infty\} \times \mathcal{L}_p\big) \cup \big(\{\infty\} \times (-\infty,d) \times \mathcal{L}_p\big),$$
$$U(\infty,\infty,1) = \{(\infty,\infty,1)\} \times \big((-\infty,d) \times \{\infty\} \times \mathcal{L}_p\big) \cup \big(\{\infty\} \times (c,+\infty) \times \mathcal{L}_p\big),$$

where $\varepsilon > 0$, $c, d \in \mathbb{R}$, and $V_\mu := \mathcal{L}_p \cap \{\nu \in \mathbb{C} : |\nu - \mu| < \varepsilon\}$. Then M becomes a compact Hausdorff space.

Let $\mathcal{A}(x,\lambda,\mu) := [\Psi(A)](x,\lambda,\mu)$ for $A \in \mathfrak{A}$. With the Fredholm symbol $\mathcal{A}(x,\lambda,\mu) = \big(a_{ij}(x,\lambda,\mu)\big)_{i,j=1}^2$ of any Fredholm operator $A \in \mathfrak{D}$ we associate the function

$$f_A(x,\lambda,\mu) := \frac{\det \mathcal{A}(x,\lambda,\mu)\, a_{22}(x,\infty,0)}{a_{11}(x,\infty,0)\, a_{22}(x,\lambda,0)\, a_{22}(x,\lambda,1)}, \quad \text{for } (x,\lambda,\mu) \in M. \tag{5.3}$$

By analogy with [29, Section 6] one can prove that the function (5.3) is continuous on M and separated from zero.

Let us show that $f_A(M)$ is the graph of a bounded, closed and continuous curve $A^\#$ in the complex plane. For $x \in \dot{\mathbb{R}}$, define $A_1(x) := f_A(x,\infty,0)$. It follows from (5.2) that the limits $A_1(x \pm 0)$ exist at each point $x \in \dot{\mathbb{R}}$, and

$$A_1(x-0) = f_A(x,\infty,0), \quad A_1(x+0) = f_A(x,\infty,1).$$

Thus, these limits are finite, non-zero, and $A_1(x-0) = A_1(x)$. Let J_1 denote the set of points $x \in \mathbb{R}$ at which the functions $f_A(x,\infty,\cdot)$ are not constant on $\mathcal{L}_p$. It is clear that the set J_1 is at most countable. For every $x \in J_1$ we join the one-sided limits $A_1(x \pm 0)$ by the continuous curve $A^\#_{x,\infty} := \{f_A(x,\infty,\mu) : \mu \in \mathcal{L}_p\}$ oriented from $A_1(x-0) = f_A(x,\infty,0)$ to $A_1(x+0) = f_A(x,\infty,1)$. Therefore, we obtain the open continuous oriented curve

$$A_1^\# := \{f_A(\infty,\infty,1)\} \cup A_1(\mathbb{R}) \cup \Big(\bigcup\nolimits_{x \in J_1} A^\#_{x,\infty}\Big) \cup \{f_A(\infty,\infty,0)\}$$

with the starting point $A_1(-\infty) = f_A(\infty,\infty,1)$ and the terminating point $A_1(+\infty) = f_A(\infty,\infty,0)$. Moreover, $0 \notin A_1^\#$.

Analogously, for $\lambda \in \mathbb{R}$, we define the function $A_2(\lambda) := f_A(\infty,\lambda,1)$. It follows from (5.2) that the limits $A_2(\lambda \pm 0)$ also exist at each point $\lambda \in \dot{\mathbb{R}}$, and

$$A_2(\lambda-0) = f_A(\infty,\lambda,1), \quad A_2(\lambda+0) = f_A(\infty,\lambda,0).$$

These limits also are finite, non-zero, and $A_2(\lambda-0) = A_2(\lambda)$. Let J_2 denote the at most countable set of points $\lambda \in \mathbb{R}$ at which the functions $f_A(\infty,\lambda,\cdot)$ are not constant on $\mathcal{L}_p$. For each $\lambda \in J_2$ we join the one-sided limits $A_2(\lambda \pm 0)$ of the function A_2 at the point λ by the continuous curve $A^{\#}_{\infty,\lambda} := \{f_A(\infty, \lambda, \mu) : \mu \in \mathcal{L}_p\}$ oriented from $A_2(\lambda - 0) = f_A(\infty, \lambda, 1)$ to $A_2(\lambda + 0) = f_A(\infty, \lambda, 0)$. Therefore, we obtain the open continuous oriented curve

$$A^{\#}_2 := \{f_A(\infty, \infty, 0)\} \cup A_2(\mathbb{R}) \cup \Big(\bigcup\nolimits_{\lambda \in J_2} A^{\#}_{\infty,\lambda}\Big) \cup \{f_A(\infty, \infty, 1)\}$$

with the starting point $A_2(-\infty) = f_A(\infty, \infty, 0)$ and the terminating point $A_2(+\infty) = f_A(\infty, \infty, 1)$, where again $0 \notin A^{\#}_2$.

As $A_2(+\infty) = A_1(-\infty)$ and $A_2(-\infty) = A_1(+\infty)$, the curve $A^{\#} := A^{\#}_1 \cup A^{\#}_2$ is closed. Thus, for every Fredholm operator $A \in \mathfrak{D}$, we get the bounded, closed, continuous and oriented curve $A^{\#} \subset \mathbb{C}$. Obviously, $A^{\#} = f_A(M)$ and $0 \notin A^{\#}$.

Finally, from [14, Section 7], [13] and [29, Section 6] we deduce the following.

Theorem 5.2. *If an operator $A \in \mathfrak{D}$ is Fredholm on the space $L^p(\mathbb{R})$ $(1 < p < \infty)$, then*

$$\operatorname{Ind} A = -\operatorname{wind} A^{\#} \tag{5.4}$$

where wind $A^{\#}$ *denotes the winding number of the curve $A^{\#}$ about the origin.*

6. Algebra $\mathfrak{A}$ of pseudo-differential operators

Let $\mathfrak{A} = \mathfrak{A}_p$ denote the minimal Banach subalgebra of $\mathcal{B}(L^p(\mathbb{R}))$ that contains all pseudo-differential operators $a(x, D)$ with symbols $a \in PC(\overline{\mathbb{R}}, V(\mathbb{R}))$. Repeating literally the proof of [27, Lemma 10.1] we obtain the following.

Lemma 6.1. *For every $p \in (1, \infty)$, the Banach algebra $\mathfrak{A}$ contains all compact operators acting on the space $L^p(\mathbb{R})$.*

Let us show that pseudo-differential operators $A \in \mathfrak{A}$ are related to convolution type operators in the Banach algebra $\mathfrak{D}$. To this end with every function $a(x, \lambda) \in PC(\overline{\mathbb{R}}, V(\mathbb{R}))$ we associate the function

$$\widetilde{a}(x, \lambda) = a(-\infty, \lambda)\zeta_-(x) + a(+\infty, \lambda)\zeta_+(x) + c_-(x)\zeta_-(\lambda) + c_+(x)\zeta_+(\lambda) \tag{6.1}$$

where $x, \lambda \in [-\infty, +\infty]$ and

$$c_\pm(x) = a(x, \pm\infty) - a(-\infty, \pm\infty)\zeta_-(x) - a(+\infty, \pm\infty)\zeta_+(x), \tag{6.2}$$

$$\zeta_\pm(\lambda) = (1 \pm \tanh \lambda)/2. \tag{6.3}$$

Since $c_\pm(-\infty) = c_\pm(+\infty) = 0$, it is easily seen that

$$\widetilde{a}(x, \pm\infty) = a(x, \pm\infty), \quad \widetilde{a}(\pm\infty, \lambda) = a(\pm\infty, \lambda) \quad \text{for all } x, \lambda \in [-\infty, +\infty]. \tag{6.4}$$

By (6.1)–(6.3), the function $\widetilde{a}(x, \lambda)$ also is in the Banach algebra $PC(\overline{\mathbb{R}}, V(\mathbb{R}))$. Hence, by Theorem 3.1, the pseudo-differential operator $\widetilde{a}(x, D)$ is bounded on each Lebesgue space $L^p(\mathbb{R})$ with $1 < p < \infty$.

On the other hand, the operator $\widetilde{a}(x, D)$ is represented in the form

$$\widetilde{a}(x, D) = \zeta_- W_{a(-\infty,\cdot)} + \zeta_+ W_{a(+\infty,\cdot)} + c_- W_{\zeta_-} + c_+ W_{\zeta_+},$$

that is, $\widetilde{a}(x,D)$ is a convolution type operator that belongs to the Banach subalgebra $\mathfrak{D}$ of $\mathcal{B}(L^p(\mathbb{R}))$ generated by all operators of the form $cW_b = c\mathcal{F}^{-1}b\mathcal{F}$ with $c \in PC(\overline{\mathbb{R}})$ and $b \in PC_p(\overline{\mathbb{R}})$.

Lemma 6.2. *If $a \in PC(\overline{\mathbb{R}}, V(\mathbb{R}))$, then the pseudo-differential operator $a(x,D) - \widetilde{a}(x,D)$ is compact on every Lebesgue space $L^p(\mathbb{R})$ with $1 < p < \infty$.*

Proof. Since $a(x,\lambda) - \widetilde{a}(x,\lambda) = 0$ for all $(x,\lambda) \in \partial(\overline{\mathbb{R}} \times \overline{\mathbb{R}})$, we infer that the function $\sigma(x,\lambda) = a(x,\lambda) - \widetilde{a}(x,\lambda)$ belongs to $PC(\overline{\mathbb{R}}, V(\mathbb{R}))$ and satisfies (4.1). Hence, by Corollary 4.2, the operator $\sigma(x,D) = a(x,D) - \widetilde{a}(x,D)$ is compact on every Lebesgue space $L^p(\mathbb{R})$ with $1 < p < \infty$. □

Hence the quotient Banach algebras $\mathfrak{A}^\pi = \mathfrak{A}/\mathcal{K}$ and $\mathfrak{D}^\pi = \mathfrak{D}/\mathcal{K}$ coincide, and therefore the Banach algebra $\mathfrak{A}^\pi$ is not commutative.

By (6.4) and (5.2), to every pseudo-differential operator $A = a(x,D)$ with symbol $a(x,\lambda) \in PC(\overline{\mathbb{R}}, V(\mathbb{R}))$ we assign the Fredholm symbol

$$\mathcal{A}(x,\lambda,\mu) = \begin{bmatrix} a(x+0,\lambda-0)\mu + a(x+0,\lambda+0)(1-\mu) & \big[a(x+0,\lambda-0) - a(x+0,\lambda+0)\big]\nu(\mu) \\ \big[a(x-0,\lambda-0) - a(x-0,\lambda+0)\big]\nu(\mu) & a(x-0,\lambda-0)(1-\mu) + a(x-0,\lambda+0)\mu \end{bmatrix} \tag{6.5}$$

where $(x,\lambda,\mu) \in M$. Then Theorems 5.1 and 5.2 immediately give the following.

Theorem 6.3. *Let $1 < p < \infty$ and let M be given by (5.1).*

(i) *The map $\Psi : A \mapsto \mathcal{A}(x,\lambda,\mu)$ associating matrix functions of the form (6.5) to pseudo-differential operators $A = a(x,D)$ with symbols $a(x,\lambda) \in PC(\overline{\mathbb{R}}, V(\mathbb{R}))$ extends to a Banach algebra homomorphism $\Psi : \mathfrak{A} \to BC(M, \mathbb{C}^{2\times 2})$ whose kernel contains $\mathcal{K}$.*

(ii) *The quotient algebra $\mathfrak{A}^\pi$ is inverse closed in the Calkin algebra $\mathcal{B}^\pi$.*

(iii) *An operator $A \in \mathfrak{A}$ is Fredholm on the space $L^p(\mathbb{R})$ if and only if*

$$\det \mathcal{A}(x,\lambda,\mu) \neq 0 \quad \textit{for all} \;\; (t,x,\mu) \in M,$$

where $\mathcal{A}(x,\lambda,\mu) = [\Psi(A)](x,\lambda,\mu)$ is the Fredholm symbol of the operator A.

(iv) *If $A \in \mathfrak{A}$ is Fredholm on the space $L^p(\mathbb{R})$, then its index is calculated by (5.4).*

Thus, the algebra $\Psi(\mathfrak{A})$ of Fredholm symbols for $A \in \mathfrak{A}$ is non-commutative.

7. Algebra $\widetilde{\mathfrak{A}}$ of pseudo-differential operators

Let now $a(x,\lambda) \in \widetilde{\Omega}$ where $\widetilde{\Omega} = PC(\overline{\mathbb{R}}, V_0(\mathbb{R})) \cup \widetilde{\mathcal{E}}$ is a subset of $L^\infty(\mathbb{R}, V_0(\mathbb{R}))$. Combining results of [27] and [14] we obtain the following.

Lemma 7.1. *If $c \in SO$, $b \in C_p(\dot{\mathbb{R}})$, and $a(x,\lambda) \in \widetilde{\Omega}$, then the commutator $cW_b\, a(x,D) - a(x,D)\, cW_b$ is compact on every Lebesgue space $L^p(\mathbb{R})$ $(1 < p < \infty)$.*

Proof. Fix $c \in SO$, $b \in C_p(\dot{\mathbb{R}})$, and $p \in (1,\infty)$. Since $C_p(\dot{\mathbb{R}})$ is the closure of $V_0(\mathbb{R}) \cap C(\dot{\mathbb{R}})$ in $\mathcal{M}_p$, we conclude that every operator cW_b with $c \in SO$ and $b \in C_p(\dot{\mathbb{R}})$ can be approximated in $\mathcal{B}(L^p(\mathbb{R}))$ by pseudo-differential operators with symbols in $\widetilde{\mathcal{E}}$. Then, by [27, Corollary 8.4], the commutator $cW_b\, a(x,D) - a(x,D)\, cW_b$ is compact on the space $L^p(\mathbb{R})$ for every $a(x,\lambda) \in \widetilde{\mathcal{E}}$. If $a \in PC(\overline{\mathbb{R}}, V_0(\mathbb{R}))$, then the functions $a(\pm\infty,\cdot)$ and $\zeta_\pm$ belong to $\widetilde{\mathcal{E}}$. Therefore, by (6.1) and by [27, Corollary 8.4], the commutator $c\,\widetilde{a}(x,D) - \widetilde{a}(x,D)\, cI$ is compact on the space $L^p(\mathbb{R})$. By [14, Lemma 7.3] and by (6.1) and (6.2), the commutator $W_b\,\widetilde{a}(x,D) - \widetilde{a}(x,D)\, W_b$ is also compact on $L^p(\mathbb{R})$. As a result, the commutator $cW_b\,\widetilde{a}(x,D) - \widetilde{a}(x,D)\, cW_b$ is compact on $L^p(\mathbb{R})$ too. Finally, applying Lemma 6.2, we deduce the compactness of the commutator $cW_b\, a(x,D) - a(x,D)\, cW_b$ for $a \in PC(\overline{\mathbb{R}}, V_0(\mathbb{R}))$. □

By [27, Lemma 10.1], the Banach subalgebra $\widetilde{\mathfrak{A}} = \widetilde{\mathfrak{A}}_p$ of $\mathcal{B}(L^p(\mathbb{R}))$ generated by all pseudo-differential operators $a(x,D)$ with symbols $a(x,\lambda) \in \widetilde{\Omega}$ contains all compact operators in $\mathcal{B}(L^p(\mathbb{R}))$. Consider the quotient Banach algebra $\widetilde{\mathfrak{A}}^\pi := \widetilde{\mathfrak{A}}/\mathcal{K}$. Let $\widetilde{\mathcal{Z}}$ be the Banach subalgebra of $\mathcal{B}(L^p(\mathbb{R}))$ generated by all operators cW_b, where $c \in SO$ and $b \in C_p(\dot{\mathbb{R}})$, and by all compact operators on $L^p(\mathbb{R})$. In virtue of Lemma 7.1, the quotient Banach algebra $\widetilde{\mathcal{Z}}^\pi := \widetilde{\mathcal{Z}}/\mathcal{K}$ is a central subalgebra of $\widetilde{\mathfrak{A}}^\pi$. From [27, Sections 10–12] it follows that the maximal ideal space of $\widetilde{\mathcal{Z}}^\pi$ can be identified with the set

$$M(\widetilde{\mathcal{Z}}^\pi) := (M_{-\infty}(SO) \times \dot{\mathbb{R}}) \cup (\mathbb{R} \times \{\infty\}) \cup (M_{+\infty}(SO) \times \dot{\mathbb{R}}), \tag{7.1}$$

where $M_{\pm\infty}(SO)$ are the fibers of $M(SO)$ at $\pm\infty$. Let $\widetilde{\Lambda}$ be the Banach algebra generated by all operators in $\mathcal{B}(L^p(\mathbb{R}))$ that commute with every operator $cW_b \in \widetilde{\mathcal{Z}}$ to within compact operators (the elements of $\widetilde{\Lambda}$ are called operators of local type with respect to $\widetilde{\mathcal{Z}}$). Then $\widetilde{\mathfrak{A}}^\pi$ is a Banach subalgebra of $\widetilde{\Lambda}^\pi := \widetilde{\Lambda}/\mathcal{K}$. Let $\widetilde{J}^\pi_{\xi,\eta}$ be the closed two-sided ideal of $\widetilde{\Lambda}^\pi$ generated by the maximal ideal of $\widetilde{\mathcal{Z}}^\pi$ identified with the point $(\xi,\eta) \in M(\widetilde{\mathcal{Z}}^\pi)$. Applying the Allan-Douglas local principle (see, for example, [7, Theorem 1.34]), we obtain the following criterion.

Theorem 7.2. *An operator $A \in \widetilde{\mathfrak{A}}$ is Fredholm on the space $L^p(\mathbb{R})$ with $1 < p < \infty$ if and only if for every point $(\xi,\eta) \in M(\widetilde{\mathcal{Z}}^\pi)$ the coset $\widetilde{A}^\pi_{\xi,\eta} := A^\pi + \widetilde{J}^\pi_{\xi,\eta}$ is invertible in the Banach algebra $\widetilde{\Lambda}^\pi_{\xi,\eta} := \widetilde{\Lambda}^\pi / \widetilde{J}^\pi_{\xi,\eta}$.*

Given $y \in \mathbb{R}$, let χ^-_y and χ^+_y denote the characteristic functions of $(-\infty, y)$ and $(y,+\infty)$, respectively. Put $\chi_\pm := \chi^\pm_0$. One can easily obtain the following.

Lemma 7.3. *Let $a(x,\lambda) \in \widetilde{\Omega}$. Then*

$$\big[a(x,D) - a(\xi-0,-\infty)\chi^-_\xi W_{\chi_-} - a(\xi+0,-\infty)\chi^+_\xi W_{\chi_-}$$
$$-a(\xi-0,+\infty)\chi^-_\xi W_{\chi_+} - a(\xi+0,+\infty)\chi^+_\xi W_{\chi_+}\big]^\pi \in \widetilde{J}^\pi_{\xi,\eta} \quad \text{if} \quad (\xi,\eta) \in \mathbb{R} \times \{\infty\},$$
$$\big[a(x,D) - a(\xi,\eta) I\big]^\pi \in \widetilde{J}^\pi_{\xi,\eta} \quad \text{if} \quad \xi \in M_{\pm\infty}(SO), \quad \eta \in \mathbb{R},$$
$$\big[a(x,D) - a(\xi,-\infty)\zeta_\pm W_{\zeta_-} - a(\xi,+\infty)\zeta_\pm W_{\zeta_+}\big]^\pi \in \widetilde{J}^\pi_{\xi,\eta} \quad \text{if } \xi \in M_{\pm\infty}(SO),\ \eta = \infty.$$

Lemma 7.3 immediately implies the following.

Lemma 7.4. *For every* $(\xi,\eta) \in M(\widetilde{\mathcal{Z}}^\pi)$, *the mapping* $\widetilde{\delta}_{\xi,\eta} : A \mapsto \widetilde{A}^\pi_{\xi,\eta}$, *defined on the generators* $a(x,D)$ *of the Banach algebra* $\widetilde{\mathfrak{A}}$ *by the rule*

$$\widetilde{\delta}_{\xi,\eta}[a(x,D)] :=$$

$$\begin{cases} \big[a(\xi-0,-\infty)\chi_\xi^- W_{\chi_-} + a(\xi+0,-\infty)\chi_\xi^+ W_{\chi_-} \\ + a(\xi-0,+\infty)\chi_\xi^- W_{\chi_+} + a(\xi+0,+\infty)\chi_\xi^+ W_{\chi_+}\big]^\pi_{\xi,\eta}, & (\xi,\eta) \in \mathbb{R}\times\{\infty\}, \\ \big[a(\xi,\eta)I\big]^\pi_{\xi,\eta}, \qquad (\xi,\eta) \in M_{\pm\infty}(SO)\times\mathbb{R}, \\ \big[a(\xi,-\infty)\zeta_\pm W_{\zeta_-} + a(\xi,+\infty)\zeta_+ W_{\zeta_+}\big]^\pi_{\xi,\eta}, & (\xi,\eta) \in M_{\pm\infty}(SO)\times\{\infty\}, \end{cases}$$

extends to a Banach algebra homomorphism $\widetilde{\delta}_{\xi,\eta} : \widetilde{\mathfrak{A}} \to \widetilde{\Lambda}^\pi_{\xi,\eta}$, *and*

$$\sup_{(\xi,\eta)\in M(\widetilde{\mathcal{Z}}^\pi)} \big\|\widetilde{\delta}_{\xi,\eta}(A)\big\|_{\widetilde{\Lambda}^\pi_{\xi,\eta}} \le \|A^\pi\| := \inf_{K\in\mathcal{K}} \|A+K\| \quad \text{for all} \quad A \in \widetilde{\mathfrak{A}}.$$

Thus, $\widetilde{\delta}_{\xi,\eta}(\widetilde{\mathfrak{A}}) \subset \{\widetilde{A}^\pi_{\xi,\eta} = A^\pi + \widetilde{J}^\pi_{\xi,\eta} : A \in \mathfrak{D}\} \subset \widetilde{\Lambda}^\pi_{\xi,\eta}$ for every $(\xi,\eta) \in M(\widetilde{\mathcal{Z}}^\pi)$.

To study the invertibility of the cosets $\widetilde{A}^\pi_{\xi,\eta}$ in the local algebras $\widetilde{\Lambda}^\pi_{\xi,\eta}$ in the case $(\xi,\eta) \in \mathbb{R}\times\{\infty\}$ we need the following special version of the two projections theorem (see [16], [19] and [5]).

Theorem 7.5. *Let* $\mathcal{A}$ *be a Banach algebra with identity* e, *let* r *and* q *be idempotents in* $\mathcal{A}$, *and let* $\mathcal{R} = \operatorname{alg}(e,r,q)$ *denote the smallest closed subalgebra of* $\mathcal{A}$ *which contains* e,r,q. *Let* $x = e-(r-q)^2$ *and suppose that* 0 *and* 1 *are not isolated points of the spectrum* $sp_{\mathcal{A}}x$ *of* x *in* $\mathcal{A}$ *and that* $sp_{\mathcal{A}}x = sp_{\mathcal{R}}x$. *Then* $\mathcal{R}$ *is inverse closed in* $\mathcal{A}$ *and*

(i) *for every* $\mu \in sp_{\mathcal{A}}x$, *the map* $\Upsilon_\mu : \{e,r,q\} \to \mathbb{C}^{2\times 2}$ *given by*

$$\Upsilon_\mu(e) = \begin{bmatrix} 1 & 0 \\ 0 & 1 \end{bmatrix}, \Upsilon_\mu(r) = \begin{bmatrix} 1 & 0 \\ 0 & 0 \end{bmatrix}, \quad \Upsilon_\mu(q) = \begin{bmatrix} \mu & \nu(\mu) \\ \nu(\mu) & 1-\mu \end{bmatrix}$$

extends to a Banach algebra homomorphism $\Upsilon_\mu : \mathcal{R} \to \mathbb{C}^{2\times 2}$, *where* $\nu(\mu) := \sqrt{\mu(1-\mu)}$ *means an arbitrary value of the square root;*

(ii) *an element* $a \in \mathcal{R}$ *is invertible in the algebra* $\mathcal{A}$ *(equivalently, in* $\mathcal{R}$*) if and only if* $\det \Upsilon_\mu(a) \ne 0$ *for all* $\mu \in sp_{\mathcal{A}}x$.

By analogy with (5.1) and (7.1) we introduce the sets

$$\begin{aligned} M_{\widetilde{\mathfrak{A}}} &:= (M_{-\infty}(SO)\times\overline{\mathbb{R}}) \cup (\mathbb{R}\times\{\infty\}\times\mathcal{L}_p) \cup (M_{+\infty}(SO)\times\overline{\mathbb{R}}), \\ \widetilde{M} &:= (M_{-\infty}(SO)\times\overline{\mathbb{R}}) \cup (\mathbb{R}\times\{\pm\infty\}) \cup (M_{+\infty}(SO)\times\overline{\mathbb{R}}). \end{aligned}$$

Applying Lemma 7.4 and Theorem 7.5 we infer the following (cf. Theorem 5.1).

Lemma 7.6. *For every* $(x,\lambda,\mu) \in \mathbb{R}\times\{\infty\}\times\mathcal{L}_p$, *the map* $A \mapsto \mathcal{A}(x,\lambda,\mu)$ *given for pseudo-differential operators* $a(x,D)$ *with symbols* $a(x,\lambda) \in \widetilde{\Omega}$ *by (6.5) extends to a Banach algebra homomorphism of* $\widetilde{\mathfrak{A}}$ *into* $\mathbb{C}^{2\times 2}$. *For every* $(\xi,\lambda) \in \widetilde{M}$, *the map* $A \mapsto \mathcal{A}(\xi,\lambda)$ *given for pseudo-differential operators* $a(x,D)$ *with symbols*

$a(x,\lambda) \in \widetilde{\Omega}$ by $\mathcal{A}(\xi,\lambda) := a(\xi,\lambda)$ extends to a Banach algebra homomorphism of $\widetilde{\mathfrak{A}}$ onto $\mathbb{C}$.

Theorem 7.7. *A pseudo-differential operator $A \in \widetilde{\mathfrak{A}}$ is Fredholm on the space $L^p(\mathbb{R})$ $(1 < p < \infty)$ if and only if*

$$\min_{\mu\in\mathcal{L}_p} |\det \mathcal{A}(x,\infty,\mu)| > 0 \quad \textit{for every} \quad x \in \mathbb{R}, \tag{7.2}$$

$$\liminf_{x\to-\infty} \min_{\lambda\in\overline{\mathbb{R}}} |\mathcal{A}(x\pm 0,\lambda)| > 0, \quad \liminf_{x\to+\infty} \min_{\lambda\in\overline{\mathbb{R}}} |\mathcal{A}(x\pm 0,\lambda)| > 0. \tag{7.3}$$

Proof. From Lemmas 7.4 and 7.6 it follows that for $(\xi,\eta) \in M_{\pm\infty}(SO) \times \dot{\mathbb{R}}$ the cosets $\widetilde{A}^\pi_{\xi,\eta}$ are invertible in the quotient algebras $\widetilde{\Lambda}^\pi_{\xi,\eta}$ if and only if $\mathcal{A}(\xi,\eta) \neq 0$ in case $\eta \in \mathbb{R}$ and $\mathcal{A}(\xi,\pm\infty) \neq 0$ in case $\eta = \infty$. On the other hand, if $\xi \in \mathbb{R}$ and $\eta = \infty$, then setting $r := \big[\chi^+_\xi I\big]^\pi_{\xi,\infty}$ and $q := \big[W_{\chi_+}\big]^\pi_{\xi,\infty}$ in Theorem 7.5, we conclude from [8, Subsection 4.6] (also see [3, Corollary 5.3]) that the spectra of the element $e-(r-q)^2$ in the Banach algebras $\mathcal{R} = \operatorname{alg}(e,r,q)$ and $\widetilde{\Lambda}^\pi_{\xi,\infty}$ coincide with $\mathcal{L}_p$. Therefore, Theorem 7.5 together with Lemmas 7.4 and 7.6 immediately imply that for $(\xi,\eta) \in \mathbb{R} \times \{\infty\}$ the cosets $\widetilde{A}^\pi_{\xi,\eta}$ are invertible in the quotient algebras $\widetilde{\Lambda}^\pi_{\xi,\eta}$ if and only if $\det \mathcal{A}(\xi,\infty,\mu) \neq 0$ for every $(\xi,\mu) \in \mathbb{R} \times \mathcal{L}_p$. Applying Theorem 7.2 we obtain the Fredholm criterion for the operators $A \in \widetilde{\mathfrak{A}}$ in the form

$$\begin{aligned} \det \mathcal{A}(x,\infty,\mu) \neq 0 \quad &\text{for every} \quad (x,\mu) \in \mathbb{R} \times \mathcal{L}_p, \\ \mathcal{A}(\xi,\lambda) \neq 0 \quad &\text{for every} \quad (\xi,\lambda) \in M_{\pm\infty}(SO) \times \overline{\mathbb{R}}, \end{aligned}$$

or, equivalently, in the form

$$\inf_{x\in\mathbb{R}} \min_{\mu\in\mathcal{L}_p} |\det \mathcal{A}(x,\infty,\mu)| > 0, \qquad \inf_{\xi\in M_{\pm\infty}(SO)} \min_{\lambda\in\overline{\mathbb{R}}} |\mathcal{A}(\xi,\lambda)| > 0. \tag{7.4}$$

Since the property (2.5) holds for $\mathcal{A}(\xi,\lambda)$ too, we infer from [27, Lemma 11.9] that the inequalities (7.4) are equivalent to (7.2)–(7.3). □

Thus, the algebra of pairs $\big((x,\mu) \mapsto \mathcal{A}(x,\infty,\mu), (\xi,\lambda) \mapsto \mathcal{A}(\xi,\lambda)\big)$, where $(x,\mu) \in \mathbb{R} \times \mathcal{L}_p$ and $(\xi,\lambda) \in M_{\pm\infty}(SO) \times \overline{\mathbb{R}}$, is a non-commutative algebra of Fredholm symbols for the operators $A \in \widetilde{\mathfrak{A}}$.

8. Algebra $\widehat{\mathfrak{A}}$ of pseudo-differential operators

Let $a(x,\lambda) \in \widehat{\Omega}$ where $\widehat{\Omega} = PC(\overline{\mathbb{R}}, V(\mathbb{R})) \cup \widehat{\mathcal{E}}$ is a subset of $L^\infty(\mathbb{R}, V(\mathbb{R}))$. Combining results of [27], [2] and [14] we obtain the following.

Lemma 8.1. *If $c \in SO^\diamond$, $b \in C_p(\dot{\mathbb{R}})$, and $a(x,\lambda) \in \widehat{\Omega}$, then the commutator $cW_b\, a(x,D) - a(x,D)\, cW_b$ is compact on every Lebesgue space $L^p(\mathbb{R})$ $(1 < p < \infty)$.*

Proof. Fix $c \in SO^\diamond$, $b \in C_p(\dot{\mathbb{R}})$, and $p \in (1,\infty)$. Since $SO^\diamond \subset SO$ and since $C_p(\dot{\mathbb{R}})$ is the closure of $V_0(\mathbb{R}) \cap C(\dot{\mathbb{R}})$ in $\mathcal{M}_p$, we conclude that every operator cW_b with $c \in SO^\diamond$ and $b \in C_p(\dot{\mathbb{R}})$ can be approximated in $\mathcal{B}(L^p(\mathbb{R}))$ by pseudo-differential operators with symbols in $\widetilde{\mathcal{E}}$. Therefore, taking into account the inclusion $\widehat{\mathcal{E}} \subset \widetilde{\mathcal{E}}$ we infer from [27, Corollary 8.4] that the commutator $cW_b\, a(x,D) - a(x,D)\, cW_b$ is compact on the space $L^p(\mathbb{R})$ for every $a(x,\lambda) \in \widehat{\mathcal{E}}$. If $a \in PC(\overline{\mathbb{R}}, V(\mathbb{R}))$, then the function $\widetilde{a}$ given by (6.1) also belongs to $PC(\overline{\mathbb{R}}, V(\mathbb{R}))$. Therefore, by (6.1) and by [2, Theorem 4.2], the commutator $c\,\widetilde{a}(x,D) - \widetilde{a}(x,D)\, cI$ is compact on the space $L^p(\mathbb{R})$. By [14, Lemma 7.3] and by (6.1) and (6.2), the commutator $W_b\, \widetilde{a}(x,D) - \widetilde{a}(x,D)\, W_b$ is also compact on the space $L^p(\mathbb{R})$. Thus, the commutator $cW_b\, \widetilde{a}(x,D) - \widetilde{a}(x,D)\, cW_b$ is compact on the space $L^p(\mathbb{R})$. Applying Lemma 6.2, we obtain the compactness of the commutator $cW_b\, a(x,D) - a(x,D)\, cW_b$ for $a \in PC(\overline{\mathbb{R}}, V(\mathbb{R}))$, which completes the proof. □

By [27, Lemma 10.1], the Banach subalgebra $\widehat{\mathfrak{A}} = \widehat{\mathfrak{A}}_p$ of $\mathcal{B}(L^p(\mathbb{R}))$ generated by all pseudo-differential operators $a(x,D)$ with symbols $a(x,\lambda) \in \widehat{\Omega}$ contains all compact operators in $\mathcal{B}(L^p(\mathbb{R}))$. Consider the quotient Banach algebra $\widehat{\mathfrak{A}}^\pi := \widehat{\mathfrak{A}}/\mathcal{K}$. Let $\widehat{\mathcal{Z}}$ be the Banach subalgebra of $\mathcal{B}(L^p(\mathbb{R}))$ generated by all operators cW_b, where $c \in SO^\diamond$ and $b \in C_p(\dot{\mathbb{R}})$, and by all compact operators in $\mathcal{B}(L^p(\mathbb{R}))$. In virtue of Lemma 8.1, the quotient Banach algebra $\widehat{\mathcal{Z}}^\pi := \widehat{\mathcal{Z}}/\mathcal{K}$ is a central subalgebra of $\widehat{\mathfrak{A}}^\pi$. From [3, Section 6] it follows that the maximal ideal space of $\widehat{\mathcal{Z}}^\pi$ can be identified with the set

$$M(\widehat{\mathcal{Z}}^\pi) := (M_\infty(SO^\diamond) \times \dot{\mathbb{R}}) \cup (\mathbb{R} \times \{\infty\}),$$

where $M_\infty(SO^\diamond)$ is the fiber of $M(SO^\diamond)$ at ∞. Let $\widehat{\Lambda}$ be the Banach algebra generated by all operators in $\mathcal{B}(L^p(\mathbb{R}))$ that commute with every operator $cW_b \in \widehat{\mathcal{Z}}$ to within compact operators (the elements of $\widehat{\Lambda}$ are called operators of local type with respect to $\widehat{\mathcal{Z}}$). Then $\widehat{\mathfrak{A}}^\pi$ is a Banach subalgebra of $\widehat{\Lambda}^\pi := \widehat{\Lambda}/\mathcal{K}$. Let $\widehat{J}^\pi_{\xi,\eta}$ be the closed two-sided ideal of $\widehat{\Lambda}^\pi$ generated by the maximal ideal $(\xi,\eta) \in M(\widehat{\mathcal{Z}}^\pi)$. Applying again the Allan-Douglas local principle, we obtain the following criterion.

Theorem 8.2. *An operator $A \in \widehat{\mathfrak{A}}$ is Fredholm on the space $L^p(\mathbb{R})$ with $1 < p < \infty$ if and only if for every $(\xi,\eta) \in M(\widehat{\mathcal{Z}})$ the coset $\widehat{A}^\pi_{\xi,\eta} := A^\pi + \widehat{J}^\pi_{\xi,\eta}$ is invertible in the Banach algebra $\widehat{\Lambda}^\pi_{\xi,\eta} := \widehat{\Lambda}^\pi / \widehat{J}^\pi_{\xi,\eta}$.*

For $(\xi,\eta) \in M_\infty(SO^\diamond) \times \dot{\mathbb{R}}$, we set

$$a(\xi \pm 0, \eta \pm 0) := \begin{cases} a(\mp\infty, \eta \pm 0) & \text{if } a(x,\lambda) \in PC(\overline{\mathbb{R}}, V(\mathbb{R})), \\ a(\xi, \eta \pm 0) & \text{if } a(x,\lambda) \in \widehat{\mathcal{E}}. \end{cases}$$

By analogy with Lemma 7.3 one can easily obtain the following.

Lemma 8.3. *Let $a(x,\lambda) \in \widehat{\Omega}$. Then*

$\big[a(x,D) - a(\xi - 0, -\infty)\chi_\xi^- W_{\chi_-} - a(\xi + 0, -\infty)\chi_\xi^+ W_{\chi_-} - a(\xi - 0, +\infty)\chi_\xi^- W_{\chi_+}$
$-a(\xi + 0, +\infty)\chi_\xi^+ W_{\chi_+}\big]^\pi \in \widehat{J}^\pi_{\xi,\eta}$ *if* $(\xi,\eta) \in \mathbb{R} \times \{\infty\}$,
$\big[a(x,D) - a(\xi - 0, \eta + 0)\chi_+ W_{\chi_\eta^+} - a(\xi + 0, \eta + 0)\chi_- W_{\chi_\eta^+} - a(\xi - 0, \eta - 0)\chi_+ W_{\chi_\eta^-}$
$-a(\xi + 0, \eta - 0)\chi_- W_{\chi_\eta^-}\big]^\pi \in \widehat{J}^\pi_{\xi,\eta}$ *if* $(\xi,\eta) \in M_\infty(SO^\diamond) \times \mathbb{R}$,

$\big[a(x,D) - a(\xi-0,-\infty)\chi_+ W_{\chi_-} - a(\xi+0,-\infty)\chi_- W_{\chi_-} - a(\xi-0,+\infty)\chi_+ W_{\chi_+}$
$-a(\xi+0,+\infty)\chi_- W_{\chi_+}\big]^\pi \in \widehat{J}^\pi_{\xi,\eta}$ *if* $(\xi,\eta) \in M_\infty(SO^\diamond) \times \{\infty\}$.

For $\xi \in M_\infty(SO^\diamond)$ and $\eta = \infty$, we put $\chi_\xi^\pm = \chi_\eta^\pm = \chi_0^\mp$. Then we immediately infer from Lemma 8.3 the following assertion.

Lemma 8.4. *For every* $(\xi,\eta) \in M(\widehat{\mathcal{Z}}^\pi) = \big(M_\infty(SO^\diamond) \times \dot{\mathbb{R}}\big) \cup (\mathbb{R} \times \{\infty\})$, *the mapping* $\widehat{\delta}_{\xi,\eta} : A \mapsto \widehat{A}^\pi_{\xi,\eta}$ *defined on the generators* $a(x,D)$ *of the Banach algebra* $\widehat{\mathfrak{A}}$ *by the rule*

$$\begin{aligned}\widehat{\delta}_{\xi,\eta}[a(x,D)] \;&:=\; \big[a(\xi-0,\eta+0)\chi_\xi^- W_{\chi_\eta^+} + a(\xi+0,\eta+0)\chi_\xi^+ W_{\chi_\eta^+}\\ &+\; a(\xi-0,\eta-0)\chi_\xi^- W_{\chi_\eta^-} + a(\xi+0,\eta-0)\chi_\xi^+ W_{\chi_\eta^-}\big]^\pi_{\xi,\eta},\end{aligned}$$

extends to a Banach algebra homomorphism $\widehat{\delta}_{\xi,\eta} : \widehat{\mathfrak{A}} \to \widehat{\Lambda}^\pi_{\xi,\eta}$, *and*

$$\sup_{(\xi,\eta)\in M(\widehat{\mathcal{Z}}^\pi)} \big\|\widehat{\delta}_{\xi,\eta}(A)\big\|_{\widehat{\Lambda}^\pi_{\xi,\eta}} \le \|A^\pi\| := \inf_{K\in\mathcal{K}} \|A+K\| \quad \textit{for all } A \in \widehat{\mathfrak{A}}.$$

Theorem 8.2, Lemma 8.4 and Theorem 7.5 immediately imply the following.

Theorem 8.5. *Let* $1 < p < \infty$ *and let* $(\xi,\lambda,\mu) \in \widehat{M}$ *where the set*

$$\widehat{M} := \big(\mathbb{R} \times \{\infty\} \times \mathcal{L}_p\big) \cup \big(M_\infty(SO^\diamond) \times \mathbb{R} \times \mathcal{L}_p\big) \cup \big(M_\infty(SO^\diamond) \times \{\infty\} \times \{0,1\}\big)$$

is equipped with the discrete topology. The map $\widehat{\Psi} : A \mapsto \widehat{\mathcal{A}}(\xi,\lambda,\mu)$ *given for pseudo-differential operators* $A = a(x,D) \in \widehat{\mathfrak{A}}$ *with symbols* $a(x,\lambda) \in \widehat{\Omega}$ *by*

$$\widehat{\mathcal{A}}(\xi,\lambda,\mu) =$$
$$\begin{bmatrix} a(\xi+0,\lambda-0)\mu + a(\xi+0,\lambda+0)(1-\mu) & \big[a(\xi+0,\lambda-0) - a(\xi+0,\lambda+0)\big]\nu(\mu) \\ \big[a(\xi-0,\lambda-0) - a(\xi-0,\lambda+0)\big]\nu(\mu) & a(\xi-0,\lambda-0)(1-\mu) + a(\xi-0,\lambda+0)\mu \end{bmatrix},$$

extends to a Banach algebra homomorphism $\widehat{\Psi} : \widehat{\mathfrak{A}} \to BC(\widehat{M}, \mathbb{C}^{2\times 2})$ *whose kernel contains* $\mathcal{K}$. *An operator* $A \in \widehat{\mathfrak{A}}$ *is Fredholm on the space* $L_p(\mathbb{R})$ *if and only if*

$$[\det \widehat{\Psi}(A)](\xi,\lambda,\mu) \ne 0 \quad \textit{for all } (\xi,\lambda,\mu) \in \widehat{M}.$$

Thus, we again obtain the non-commutative algebra $\widehat{\Psi}(\widehat{\mathfrak{A}})$ of Fredholm symbols $\widehat{\mathcal{A}}$ possessing the property: the Fredholmness of any pseudo-differential operator $A \in \widehat{\mathfrak{A}}$ on the space $L^p(\mathbb{R})$ is equivalent to the invertibility of $\widehat{\mathcal{A}}$ on $\widehat{M}$.

9. An application to the Haseman boundary value problem

Let $L^p(\mathbb{R}_+, w)$ be the weighted Lebesgue space over $\mathbb{R}_+ := (0,\infty)$ with the norm $\|f\|_{L^p(\mathbb{R}_+,w)} = \|wf\|_{L^p(\mathbb{R}_+)}$ where $1 < p < \infty$ and $w \in A_p(\mathbb{R}_+)$, that is, w is a nonnegative weight on $\mathbb{R}_+$ that satisfies the Muckenhoupt condition

$$\sup_{\varepsilon>0}\sup_{t\in\mathbb{R}_+} \frac{1}{\varepsilon}\bigg(\int_{\mathbb{R}_+(t,\varepsilon)} w(\tau)^p d\tau\bigg)^{1/p}\bigg(\int_{\mathbb{R}_+(t,\varepsilon)} w(\tau)^{-q} d\tau\bigg)^{1/q} < \infty \qquad (9.1)$$

where $\mathbb{R}_+(t,\varepsilon) := \mathbb{R}_+ \cap (t-\varepsilon, t+\varepsilon)$ and $1/p + 1/q = 1$. Then (see, for example, [5]) the Cauchy singular integral operator $S_{\mathbb{R}_+}$ given by

$$(S_{\mathbb{R}_+}\varphi)(t) = \lim_{\varepsilon\to 0} \frac{1}{\pi i} \int_{\mathbb{R}_+\setminus(t-\varepsilon,t+\varepsilon)} \frac{\varphi(\tau)}{\tau - t}\, d\tau, \quad \text{for } t \in \mathbb{R}_+,$$

is bounded on the space $L^p(\mathbb{R}_+, w)$. Consider an orientation-preserving diffeomorphism α of $\mathbb{R}_+$ onto itself with $\ln \alpha' \in C_b(\mathbb{R}_+)$. Assume that

$$(w \circ \alpha)/w,\ w/(w \circ \alpha) \in L^\infty(\mathbb{R}_+). \tag{9.2}$$

Then the shift operator $B_\alpha : \varphi \mapsto \varphi \circ \alpha$ is also bounded on the space $\mathcal{B}(L^p(\mathbb{R}_+, w))$. Obviously, B_α is invertible and $(B_\alpha)^{-1} = B_{\alpha^{-1}} \in \mathcal{B}(L^p(\mathbb{R}_+, w))$.

Let us apply the results of Section 7 to studying the Haseman boundary value problem (see, for example, [1], [32]): Find a function Φ analytic in $\mathbb{C} \setminus \overline{\mathbb{R}}_+$, represented by the Cauchy type integral over $\mathbb{R}_+$ with a density $\varphi \in L^p(\mathbb{R}_+, w)$, where $1 < p < \infty$ and $w \in A_p(\mathbb{R}_+)$, and satisfying the boundary condition

$$\Phi^+(\alpha(t)) = G(t)\Phi^-(t) + g(t), \quad \text{for } t \in \mathbb{R}_+, \tag{9.3}$$

where $\Phi^\pm(t)$ are angular boundary values of Φ on $\mathbb{R}_+$, $G \in L^\infty(\mathbb{R}_+)$, $g \in L^p(\mathbb{R}_+, w)$. By the Sokhotski-Plemelj formulas $\Phi^\pm = \pm P_\pm \varphi$, with the Haseman problem (9.3) we can associate the equivalent singular integral operator with a shift

$$N = B_\alpha P_+ + GP_- \tag{9.4}$$

where $P_\pm = \frac{1}{2}(I \pm S_{\mathbb{R}_+})$, I is the identity operator and $S_{\mathbb{R}_+}$ is the Cauchy singular integral operator. Clearly, $N \in \mathcal{B}(L^p(\mathbb{R}_+, w))$.

Let us study the Fredholmness of the operator (9.4) on the space $L^p(\mathbb{R}_+, w)$. First, following [6], we introduce the slowly oscillating data of the operator (9.4). A function $a \in C_b(\mathbb{R}_+)$ is called slowly oscillating at the points 0 and ∞ if the function $x \mapsto a(e^x)$ is in SO, that is, if

$$\lim_{r\to s} \max\left\{|a(x) - a(y)| : x, y \in [r, 2r]\right\} = 0, \quad \text{for } s \in \{0, \infty\}.$$

Let $SO(\mathbb{R}_+)$ stand for the C^*-algebra of all functions in $C_b(\mathbb{R}_+)$ which are slowly oscillating at 0 and ∞.

We call a function $w : \mathbb{R}_+ \to \mathbb{R}_+$ a slowly oscillating weight (at 0 and ∞) if

$$w(r) = e^{v(r)}, \quad \text{for } r \in \mathbb{R}_+, \tag{9.5}$$

where v is a real-valued function in $C^3(\mathbb{R}_+)$ and the function $r \mapsto rv'(r)$ is in $SO(\mathbb{R}_+)$. One can show (see, for example, [5, Theorem 2.36] and [29, Section 5]) that the weight (9.5) satisfies the Muckenhoupt condition (9.1) if and only if

$$\begin{aligned} -1/p &< \liminf_{r\to 0} rv'(r) \le \limsup_{r\to 0} rv'(r) < 1/q, \\ -1/p &< \liminf_{r\to\infty} rv'(r) \le \limsup_{r\to\infty} rv'(r) < 1/q. \end{aligned} \tag{9.6}$$

Let $w \in A_p^{SO}(\mathbb{R}_+)$ denote the set of all slowly oscillating weights on $\mathbb{R}_+$ that satisfy (9.6).

We call an orientation-preserving diffeomorphism $\alpha : \mathbb{R}_+ \to \mathbb{R}_+$ a slowly oscillating shift (at 0 and ∞) if

$$\alpha(r) = re^{\omega(r)}, \quad \text{for } r \in \mathbb{R}_+, \tag{9.7}$$

where ω is a real-valued function in $C^3(\mathbb{R}_+)$ and the functions ω and $r \mapsto r\omega'(r)$ are in $SO(\mathbb{R}_+)$. Let α be a slowly oscillating shift. Hence $\alpha' > 0$ on $\mathbb{R}_+$, and since $\alpha'(r) = (1 + r\omega'(r))e^{\omega(r)}$ for all $r \in \mathbb{R}_+$, we conclude that $\ln \alpha' \in C_b(\mathbb{R}_+)$, $\alpha' \in SO(\mathbb{R}_+)$, and

$$\inf_{r \in \mathbb{R}_+} (1 + r\omega'(r)) > 0. \tag{9.8}$$

Moreover, because $r\omega'(r)$ is slowly oscillating at 0 and ∞, the slow oscillation of ω at 0 and ∞ is equivalent to the property:

$$\lim_{r \to 0} r\omega'(r) = 0, \quad \lim_{r \to \infty} r\omega'(r) = 0. \tag{9.9}$$

Finally, let G belong to the C^*-algebra $PSO(\mathbb{R}_+) \subset L^\infty(\mathbb{R}_+)$ generated by all functions in $SO(\mathbb{R}_+)$ and by all piecewise continuous functions on $\overline{\mathbb{R}}_+ := [0, +\infty]$.

Under these conditions singular integral operators in the Banach subalgebra $\mathfrak{S} := \operatorname{alg}\{cI, S_{\mathbb{R}_+} : c \in PSO(\mathbb{R}_+)\}$ of $\mathcal{B}(L^p(\mathbb{R}_+, w))$ can be reduced to Mellin pseudo-differential operators $OP(b) = E\, a(x, D)E^{-1}$ acting on the Lebesgue space $L^p(\mathbb{R}_+, d\mu)$ with measure $d\mu(r) = dr/r$, where $a(x, D)$ are pseudo-differential operators in the algebra $\widetilde{\mathfrak{A}}$ studied in Section 7, $b(r, \lambda) = a(\ln r, \lambda)$ for $(r, \lambda) \in \mathbb{R}_+ \times \overline{\mathbb{R}}$, and E is the following isometric isomorphism:

$$E : L^p(\mathbb{R}) \to L^p(\mathbb{R}_+, d\mu), \quad (Ef)(x) = f(\ln x), \quad \text{for } x \in \mathbb{R}_+. \tag{9.10}$$

Hence, Theorem 3.1 implies the following.

Theorem 9.1. *If $b \in L^\infty(\mathbb{R}_+, V(\mathbb{R}))$, then the Mellin pseudo-differential operator $OP(b)$ defined for functions $u \in C_0^\infty(\mathbb{R}_+)$ by the iterated integral*

$$\big[OP(b)f\big](r) = \frac{1}{2\pi} \int_{\mathbb{R}} d\lambda \int_{\mathbb{R}_+} b(r, \lambda) \left(\frac{r}{\varrho}\right)^{i\lambda} f(\varrho)\, \frac{d\varrho}{\varrho}, \quad \textit{for } r \in \mathbb{R}_+,$$

extends to a bounded linear operator on every space $L^p(\mathbb{R}_+, d\mu)$ with $p \in (1, \infty)$.

Consider the operator R given by

$$(Rf)(t) = \frac{1}{\pi i} \int_{\mathbb{R}_+} \frac{f(\tau)}{\tau + t}\, d\tau, \quad \text{for } t \in \mathbb{R}_+.$$

By [29, Corollary 4.2], the operator R with fixed singularities at 0 and ∞ is bounded on every space $L^p(\mathbb{R}_+, w)$ with $1 < p < \infty$ and $w \in A_p(\mathbb{R}_+)$. Put $m_a(r, \varrho) := (a(r) - a(\varrho))/(\ln r - \ln \varrho)$ for $a \in C(\mathbb{R}_+)$ and consider the map

$$\Gamma : \mathcal{B}(L^p(\mathbb{R}_+, w)) \to \mathcal{B}(L^p(\mathbb{R}_+, d\mu)), \quad A \mapsto YAY^{-1}, \tag{9.11}$$

where Y is the isometric isomorphism of $L^p(\mathbb{R}_+, w)$ onto $L^p(\mathbb{R}_+, d\mu)$ given by

$$(Yf)(r) = e^{v(r)} r^{1/p} f(r), \quad \text{for } r \in \mathbb{R}_+. \tag{9.12}$$

Theorem 9.2. *If $1 < p < \infty$, $w \in A_p^{SO}(\mathbb{R}_+)$, α is a slowly oscillating shift on $\mathbb{R}_+$, (9.2) is fulfilled and*

$$0 < \inf_{r,\varrho\in\mathbb{R}_+} \big(1/p + m_v(r,\varrho)\big) \le \sup_{r,\varrho\in\mathbb{R}_+} \big(1/p + m_v(r,\varrho)\big) < 1,$$

$$0 < \inf_{r,\varrho\in\mathbb{R}_+} \frac{1/p + m_v(r,\varrho)}{1 + m_\omega(r,\varrho)} \le \sup_{r,\varrho\in\mathbb{R}_+} \frac{1/p + m_v(r,\varrho)}{1 + m_\omega(r,\varrho)} < 1,$$

then for the operators $S_{\mathbb{R}_+}$, $B_\alpha S_{\mathbb{R}_+} B_\alpha^{-1}$, R, $B_\alpha R \in \mathcal{B}(L^p(\mathbb{R}_+, w))$ we have

$$\Gamma(S_{\mathbb{R}_+}) = OP(\sigma) + K_0, \quad \Gamma(B_\alpha S_{\mathbb{R}_+} B_\alpha^{-1}) = OP(\sigma_\alpha) + K_\alpha, \tag{9.13}$$

$$\Gamma(R) = OP(\nu_0) + \widetilde{K}_0, \quad \Gamma(B_\alpha R) = OP(\nu_\alpha) + \widetilde{K}_\alpha, \tag{9.14}$$

where $K_0, K_\alpha, \widetilde{K}_0, \widetilde{K}_\alpha$ are compact operators on the space $L^p(\mathbb{R}_+, d\mu)$ and, for $(r,\lambda) \in \mathbb{R}_+ \times \mathbb{R}$,

$$\sigma(r,\lambda) := \coth\big(\pi\lambda + \pi i(1/p + rv'(r))\big), \tag{9.15}$$

$$\sigma_\alpha(r,\lambda) := \coth\left(\pi\frac{\lambda + i(1/p + rv'(r))}{1 + r\omega'(r)}\right), \tag{9.16}$$

$$\nu_0(r,\lambda) := 1/\sinh\big(\pi\lambda + \pi i(1/p + rv'(r))\big), \tag{9.17}$$

$$\nu_\alpha(r,\lambda) := \frac{\exp\big(i\lambda\omega(r) - (1/p + rv'(r))\omega(r)\big)}{\sinh\big(\pi\lambda + \pi i(1/p + rv'(r))\big)}. \tag{9.18}$$

Proof. Relations (9.13) follow from [28, Theorems 8.3 and 6.3]. A straightforward computation on the basis of (9.7), (9.11) and (9.12) shows that for $r > 0$,

$$[\Gamma(B_\alpha R)f](r) = \frac{1}{\pi i}\int_{\mathbb{R}_+} \frac{(r/\varrho)^{1/p + m_v(r,\varrho)}}{1 + e^{\omega(r)}(r/\varrho)} f(\varrho)\frac{d\varrho}{\varrho} =: I_0. \tag{9.19}$$

Further, taking into account the Mellin transform identity

$$\frac{1}{\pi i}\frac{x^\eta}{1 + \kappa x} = \frac{1}{2\pi}\int_{\mathbb{R}} \frac{\exp(i(\lambda + i\eta)\ln\kappa)}{\sinh(\pi(\lambda + i\eta))} x^{i\lambda}\, d\lambda \quad (x > 0,\ 0 < \eta < 1,\ \kappa > 0)$$

and setting $x = r/\varrho$, $\eta = 1/p + m_v(r,\varrho)$ and $\kappa = e^{\omega(r)}$, we get

$$I_0 = \frac{1}{2\pi}\int_{\mathbb{R}} d\lambda \int_{\mathbb{R}_+} \frac{\exp\big(i\lambda\omega(r) - (1/p + m_v(r,\varrho))\omega(r)\big)}{\sinh\big(\pi\lambda + \pi i(1/p + m_v(r,\varrho))\big)} \left(\frac{r}{\varrho}\right)^{i\lambda} f(\varrho)\frac{d\varrho}{\varrho}. \tag{9.20}$$

Replacing $m_v(r,\varrho)$ by $rv'(r)$ in (9.20), we obtain $[OP(\nu_\alpha)f](r)$, with ν_α given by (9.18). Finally, using (9.19), (9.20) and [28, Theorem 6.3], we infer by analogy with [28, Theorem 8.3] that the operator $\Gamma(B_\alpha R) - OP(\nu_\alpha)$ is compact on the space $L^p(\mathbb{R}_+, d\mu)$. This implies for $B_\alpha = I$ and $\omega = 0$ that the operator $\Gamma(R) - OP(\nu_0)$ with ν_0 given by (9.17) is also compact. Thus, relations (9.14) hold too. □

Consider the isomeric shift operator $U_\alpha := c_\alpha B_\alpha \in \mathcal{B}(L^p(\mathbb{R}_+, w))$ where

$$c_\alpha(r) := \big[(\alpha')^{1/p}(w\circ\alpha)/w\big](r) = (1 + r\omega'(r))^{1/p} e^{\omega(r)/p} e^{v(re^{\omega(r)}) - v(r)}. \tag{9.21}$$

Obviously, c_α is invertible in $C_b(\mathbb{R}_+)$ in view of (9.8), and $(U_\alpha)^{-1} = U_{\alpha^{-1}}$.

Lemma 9.3. *If all the conditions of Theorem 9.2 are fulfilled, then the operator* $U_\alpha S_{\mathbb{R}_+} U_\alpha^{-1} - S_{\mathbb{R}_+}$ *is compact on the space* $L^p(\mathbb{R}_+, w)$.

Proof. As the functions $r \mapsto rv'(r)$ and ω are in $C_b(\mathbb{R}_+)$ and

$$\left|v\big(re^{\omega(r)}\big) - v(r)\right| = \left|\int_r^{re^{\omega(r)}} v'(t)\,dt\right| \le \sup_{t\in\mathbb{R}_+} |tv'(t)|\,\omega(r),$$

we deduce that the function $\eta(r) := v\big(re^{\omega(r)}\big) - v(r)$ also is in $C_b(\mathbb{R}_+)$. Moreover, η is slowly oscillating at 0 and ∞ because the function $r \mapsto rv'(r)$ is in $SO(\mathbb{R}_+)$ and therefore, due to (9.9),

$$\lim_{r\to s} r\eta'(r) = \lim_{r\to s} \left[(1 + r\omega'(r))re^{\omega(r)}v'\big(re^{\omega(r)}\big) - rv'(r)\right] = 0 \quad \text{for } s \in \{0, \infty\}.$$

Hence $(w \circ \alpha)/w \in SO(\mathbb{R}_+)$. Since the function $\alpha' : r \mapsto (1 + r\omega'(r))e^{\omega(r)}$ is in $SO(\mathbb{R}_+)$ too, we conclude from (9.21) that $c_\alpha \in SO(\mathbb{R}_+)$.

It follows from (9.15), (9.16) and (9.9) that $\sigma_\alpha - \sigma \in C(\overline{\mathbb{R}}_+, V_0(\mathbb{R}))$ and $\sigma_\alpha - \sigma = 0$ on $\partial(\mathbb{R}_+ \times \mathbb{R})$. Hence, passing to pseudo-differential operators on $\mathbb{R}$ and using Corollary 4.2, we infer that the operators $OP(\sigma_\alpha - \sigma)$ and $\Gamma(B_\alpha S_{\mathbb{R}_+} B_\alpha^{-1} - S_{\mathbb{R}_+})$ are compact on the space $L^p(\mathbb{R}_+, d\mu)$. Analogously, [27, Corollary 8.2] implies that the operator $\Gamma(c_\alpha S_{\mathbb{R}_+} - S_{\mathbb{R}_+} c_\alpha I)$ is also compact on $L^p(\mathbb{R}_+, d\mu)$. Hence the operator $\Gamma(U_\alpha S_{\mathbb{R}_+} U_\alpha^{-1} - S_{\mathbb{R}_+})$ is compact, which implies the assertion of the lemma. □

Theorem 9.4. *Under the conditions of Theorem 9.2, the operator* $T_\alpha := U_\alpha P_+ + P_-$ *is Fredholm on the space* $L^p(\mathbb{R}_+, w)$ *and* $\operatorname{Ind} T_\alpha = 0$.

Proof. Since $-4P_+P_- = S_{\mathbb{R}_+}^2 - I = R^2$ and, by Lemma 9.3,

$$\begin{aligned} (U_\alpha P_+ + P_-)(U_\alpha^{-1} P_+ + P_-) &= I + (U_\alpha + U_\alpha^{-1} - 2I)P_+P_- + K_1, \\ (U_\alpha^{-1} P_+ + P_-)(U_\alpha P_+ + P_-) &= I + (U_\alpha + U_\alpha^{-1} - 2I)P_+P_- + K_2, \end{aligned} \tag{9.22}$$

where K_1, K_2 are compact operators, we infer from Theorem 9.2 that

$$\Gamma(I + (U_\alpha + U_\alpha^{-1} - 2I)P_+P_-) = OP(\beta_\alpha) + K \tag{9.23}$$

where K is a compact operator in $\mathcal{B}(L^p(\mathbb{R}_+, d\mu))$ and

$$\beta_\alpha := 1 - 4^{-1}(c_\alpha \nu_\alpha \nu_0 + c_{\alpha^{-1}} \nu_{\alpha^{-1}} \nu_0 - 2\nu_0^2). \tag{9.24}$$

Taking into account the relations

$$c_\alpha(r)\nu_\alpha(r, \lambda) - e^{i\lambda\omega(r)}\nu_0(r, \lambda) = \frac{(1 + r\omega'(r))^{1/p} e^{v(re^{\omega(r)}) - v(r) - rv'(r)\omega(r)} - 1}{e^{-i\lambda\omega(r)} \sinh\big(\pi\lambda + \pi i(1/p + rv'(r))\big)},$$

$$\lim_{r\to s} \left[v\big(re^{\omega(r)}\big) - v(r) - rv'(r)\omega(r)\right] = \lim_{r\to s} \int_r^{re^{\omega(r)}} \left[tv'(t) - rv'(r)\right] dt/t = 0,$$

where $s \in \{0, \infty\}$, we deduce that

$$c_\alpha(r)\nu_\alpha(r, \lambda) - e^{i\lambda\omega(r)}\nu_0(r, \lambda) = 0 \quad \text{for all } (r, \lambda) \in \partial(\mathbb{R}_+ \times \mathbb{R}).$$

Analogously,

$$c_{\alpha^{-1}}(r)\nu_{\alpha^{-1}}(r, \lambda) - e^{-i\lambda\omega(r)}\nu_0(r.\lambda) = 0 \quad \text{for all } (r, \lambda) \in \partial(\mathbb{R}_+ \times \mathbb{R}).$$

Hence, taking into account (9.24) and setting

$$\gamma_\alpha(r,\lambda)=1-\frac{\exp(i\lambda\omega(r))+\exp(-i\lambda\omega(r))-2}{4\sinh^2\big(\pi\lambda+\pi i(1/p+rv'(r))\big)},\quad \text{for } (r,\lambda)\in\mathbb{R}_+\times\mathbb{R}, \tag{9.25}$$

we conclude that $\beta_\alpha-\gamma_\alpha\in C(\overline{\mathbb{R}}_+,V_0(\mathbb{R}))$ and $\beta_\alpha-\gamma_\alpha=0$ on $\partial(\mathbb{R}_+\times\mathbb{R})$. Therefore, Corollary 4.2 implies that the operator $OP(\beta_\alpha-\gamma_\alpha)$ is compact. Thus, by (9.23),

$$\Gamma(I+(U_\alpha+U_\alpha^{-1}-2I)P_+P_-)=OP(\gamma_\alpha)+K' \tag{9.26}$$

where K' is a compact operator on $L^p(\mathbb{R}_+,d\mu)$.

One can prove that the function $(x,\lambda)\mapsto\gamma_\alpha(e^x,\lambda)$ belongs to the Banach algebra $\widetilde{\mathcal{E}}\subset SO(\mathbb{R},V_0(\mathbb{R}))$ defined in Section 2. Then passing to pseudo-differential operators on $\mathbb{R}$ we infer from [27, Theorems 12.2 and 12.5] that the operator $OP(\gamma_\alpha)$ is Fredholm on the space $L^p(\mathbb{R}_+,d\mu)$ if and only if

$$\inf_{r\in\mathbb{R}_+}|\gamma_\alpha(r,\pm\infty)|>0,\quad \liminf_{r\to s}\min_{\lambda\in\overline{\mathbb{R}}}|\gamma_\alpha(r,\lambda)|>0\quad \text{for } s\in\{0,\infty\}, \tag{9.27}$$

and in the case of Fredholmness

$$\operatorname{Ind}OP(\gamma_\alpha)=\lim_{m\to+\infty}\frac{1}{2\pi}\Big\{\arg\gamma_\alpha(r,\lambda)\Big\}_{(r,\lambda)\in\partial\Pi_m} \tag{9.28}$$

where $\Pi_m=[1/m,m]\times\overline{\mathbb{R}}$ and $\big\{\arg\gamma_\alpha(r,\lambda)\big\}_{(r,\lambda)\in\partial\Pi_m}$ denotes the increment of $\arg\gamma_\alpha(r,\lambda)$ when the point (r,λ) traces the boundary $\partial\Pi_m$ of Π_m counter-clockwise. Clearly, conditions (9.27) are fulfilled because, according to (9.25),

$$\gamma_\alpha(r,\lambda)=\frac{\exp\big(-2\pi\lambda-2\pi i(1/p+rv'(r))\big)}{4\sinh^2\big(\pi\lambda+\pi i(1/p+rv'(r))\big)}\,f_+(r,\lambda)\,f_-(r,\lambda)$$

and $\gamma_\alpha(r,\pm\infty)=1$ for all $r\in\mathbb{R}_+$, where the functions

$$f_\pm(r,\lambda):=\exp(\pm i\lambda\omega(r))-\exp\big(2\pi\lambda+2\pi i(1/p+rv'(r))\big)$$

are separated from 0 for all $(r,\lambda)\in\mathbb{R}_+\times\overline{\mathbb{R}}$. Furthermore, since the function

$$\widetilde{\gamma}_\alpha(r,\lambda,\theta)=1-\frac{\exp(i\lambda\theta\omega(r))+\exp(-i\lambda\theta\omega(r))-2}{4\sinh^2\big(\pi\lambda+\pi i(1/p+rv'(r))\big)}$$

is continuous and separated from 0 for all $(r,\lambda,\theta)\in\mathbb{R}_+\times\overline{\mathbb{R}}\times[0,1]$, we conclude that $\big\{\arg\widetilde{\gamma}_\alpha(r,\lambda,\theta)\big\}_{(r,\lambda)\in\partial\Pi_m}$ does not depend on $\theta\in[0,1]$. Consequently,

$$\frac{1}{2\pi}\Big\{\arg\gamma_\alpha(r,\lambda)\Big\}_{(r,\lambda)\in\partial\Pi_m}=\frac{1}{2\pi}\Big\{\arg\widetilde{\gamma}_\alpha(r,\lambda,0)\Big\}_{(r,\lambda)\in\partial\Pi_m}=0,$$

and therefore, by (9.28), $\operatorname{Ind}OP(\gamma_\alpha)=0$. Hence, applying (9.26) and (9.22), we deduce that the operators $T_\alpha=U_\alpha P_++P_-$ and $T_{\alpha^{-1}}=U_\alpha^{-1}P_++P_-$ are Fredholm on the space $L^p(\mathbb{R}_+,w)$ and

$$\operatorname{Ind}T_\alpha+\operatorname{Ind}T_{\alpha^{-1}}=0. \tag{9.29}$$

It remains to observe that U_α is invertible and, with $Cf = \overline{f}$,

$$C(U_\alpha P_+ + P_-)C = U_\alpha P_- + P_+ = U_\alpha(U_\alpha^{-1}P_+ + P_-).$$

Hence $\operatorname{Ind} T_\alpha = \operatorname{Ind} T_{\alpha^{-1}}$, which implies due to (9.29) that $\operatorname{Ind} T_\alpha = 0$. □

Theorem 9.5. *Let $G \in PSO(\mathbb{R}_+)$ and all conditions of Theorem 9.2 hold. Then the operator $N = B_\alpha P_+ + GP_-$ is Fredholm on the space $L^p(\mathbb{R}_+, w)$ if and only if*

$$\min_{\mu\in\mathcal{L}_p} \big|G(r+0)(1-\mu) + G(r-0)\mu\big| > 0 \quad \text{for every } r \in \mathbb{R}_+, \tag{9.30}$$

$$\liminf_{r\to s} \min_{\lambda\in\overline{\mathbb{R}}} \big|e^{i\omega(r)(\lambda+i(1/p+rv'(r)))}\mathcal{P}_+(r,\lambda) + G(r)\mathcal{P}_-(r,\lambda)\big| > 0, \tag{9.31}$$

where $s \in \{0,\infty\}$ and $\mathcal{P}_\pm(r,\lambda) := 2^{-1}\big[1 \pm \coth\big(\pi\lambda + \pi i(1/p + rv'(r))\big)\big]$.

Proof. By analogy with (9.22), we obtain

$$(B_\alpha P_+ + GP_-)(U_\alpha^{-1}P_+ + P_-) = c_\alpha^{-1}N_\alpha + K, \tag{9.32}$$

where K is a compact operator on $L^p(\mathbb{R}_+, w)$,

$$N_\alpha := P_+ + G_\alpha P_- + 4^{-1}(U_\alpha + G_\alpha U_\alpha^{-1} - (1+G_\alpha)I)P_+P_-, \tag{9.33}$$

and $G_\alpha := c_\alpha G$. Hence, by (9.32) and Theorem 9.4, the operators N and N_α are Fredholm on the space $L^p(\mathbb{R}_+, w)$ only simultaneously, and in the case of Fredholmness $\operatorname{Ind} N = \operatorname{Ind} N_\alpha$. It is easily seen that

$$\Gamma(N_\alpha) = OP(\zeta_\alpha) + K', \tag{9.34}$$

where K' is a compact operator on the space $L^p(\mathbb{R}_+, d\mu)$ and according to (9.33),

$$\zeta_\alpha := (\mathcal{P}_+ + G_\alpha\mathcal{P}_-) - 4^{-1}(c_\alpha\nu_\alpha\nu_0 + G_\alpha c_{\alpha^{-1}}\nu_{\alpha^{-1}}\nu_0 - (1+G_\alpha)\nu_0^2).$$

By analogy with (9.25), we define the function

$$\begin{aligned}\phi_\alpha(r,\lambda) &:= \big(\mathcal{P}_+(r,\lambda) + \widetilde{G}_\alpha(r)\mathcal{P}_-(r,\lambda)\big)\\ &\quad - \frac{\exp(i\lambda\omega(r)) + \widetilde{G}_\alpha(r)\exp(-i\lambda\omega(r)) - (1+\widetilde{G}_\alpha(r))}{4\sinh^2\big(\pi\lambda + \pi i(1/p + rv'(r))\big)}\end{aligned}$$

for $(r,\lambda) \in \mathbb{R}_+ \times \mathbb{R}$, where $\widetilde{G}_\alpha(r) := e^{(1/p+rv'(r))\omega(r)}G(r)$ and $G_\alpha(r) - \widetilde{G}_\alpha(r) \to 0$ as $r \to s \in \{0,\infty\}$. Then $\zeta_\alpha - \phi_\alpha\delta_\alpha \in PC(\overline{\mathbb{R}}_+, V_0(\mathbb{R}))$ and $\zeta_\alpha - \phi_\alpha\delta_\alpha = 0$ on $\partial(\mathbb{R}_+ \times \mathbb{R})$, where $\delta_\alpha := \mathcal{P}_+ + (G_\alpha/\widetilde{G}_\alpha)\mathcal{P}_- \in C(\overline{\mathbb{R}}_+, V_0(\mathbb{R}))$. Therefore, from Corollary 4.2 it follows that the operator $OP(\zeta_\alpha - \phi_\alpha\delta_\alpha)$ is compact. Lemmas 6.2 and 7.1 imply that the operator $OP(\phi_\alpha\delta_\alpha) - OP(\phi_\alpha)OP(\delta_\alpha)$ also is compact. As $\delta_\alpha \neq 0$ on $\partial(\mathbb{R}_+ \times \mathbb{R})$, the operator $OP(\delta_\alpha)$ is Fredholm on the space $L^p(\mathbb{R}_+, d\mu)$ by Theorem 5.1. Hence, taking into account (9.34) we infer that the operator N is Fredholm on the space $L^p(\mathbb{R}_+, w)$ if and only if the Mellin pseudo-differential operator $OP(\phi_\alpha)$ is Fredholm on the space $L^p(\mathbb{R}_+, d\mu)$, and in the case of Fredholmness

$$\operatorname{Ind} N = \operatorname{Ind} OP(\phi_\alpha) + \operatorname{Ind} OP(\delta_\alpha). \tag{9.35}$$

One can show that the function $a(x,\lambda) := \phi_\alpha(e^x, \lambda)$ is the symbol of a pseudo-differential operator belonging to the algebra $\widetilde{\mathfrak{A}}$ (see Section 7). Therefore, passing to pseudo-differential operators $a(x,D) = E^{-1}OP(\phi_\alpha)E \in \widetilde{\mathfrak{A}}$, where E is given by (9.10), and applying Theorem 7.7 we infer that the Mellin pseudo-differential operator $OP(\phi_\alpha)$ is Fredholm on the space $L^p(\mathbb{R}_+, d\mu)$ if and only if

$$\min_{\mu\in\mathcal{L}_p} |\det \Phi_\alpha(r,\infty,\mu)| > 0 \quad \text{for every } r \in \mathbb{R}_+, \tag{9.36}$$

$$\liminf_{r\to 0} \min_{\lambda\in\overline{\mathbb{R}}} |\phi_\alpha(r\pm 0,\lambda)| > 0, \quad \liminf_{r\to\infty} \min_{\lambda\in\overline{\mathbb{R}}} |\phi_\alpha(r\pm 0,\lambda)| > 0, \tag{9.37}$$

where for $r \in \mathbb{R}_+$ and $\lambda \in \mathcal{L}_p$,

$$\Phi_\alpha(r,\infty,\mu) := \begin{bmatrix} \phi_\alpha^-(r+0)\mu + \phi_\alpha^+(r+0)(1-\mu) & \big[\phi_\alpha^-(r+0) - \phi_\alpha^+(r+0)\big]\nu(\mu) \\ \big[\phi_\alpha^-(r-0) - \phi_\alpha^+(r-0)\big]\nu(\mu) & \phi_\alpha^-(r-0)(1-\mu) + \phi_\alpha^+(r-0)\mu \end{bmatrix},$$

with $\phi_\alpha^-(r) := \phi_\alpha(r,+\infty) = 1$ and $\phi_\alpha^+(r) := \phi_\alpha(r,-\infty) = \widetilde{G}_\alpha(r)$. Since

$$\begin{aligned} \det \Phi_\alpha(r,\infty,\mu) &= \widetilde{G}_\alpha(r+0)(1-\mu) + \widetilde{G}_\alpha(r-0)\mu \\ &= e^{(1/p+rv'(r))\omega(r)}\big(G(r+0)(1-\mu) + G(r-0)\mu\big), \end{aligned}$$

we infer that condition (9.36) is equivalent to (9.30).

Finally, since $\widetilde{G}_\alpha(r) := e^{(1/p+rv'(r))\omega(r)}G(r)$, it follows from the equality

$$\phi_\alpha(r,\lambda) = \big[e^{i\lambda\omega(r)}\mathcal{P}_+(r,\lambda) + \widetilde{G}_\alpha(r)\mathcal{P}_-(r,\lambda)\big]\big[e^{-i\lambda\omega(r)}\mathcal{P}_+(r,\lambda) + \mathcal{P}_-(r,\lambda)\big]$$

that inequalities (9.37) are equivalent to the relations (9.31). □

If G has a finite set of discontinuous, then combining (9.35), Theorem 6.3 and [27, Theorem 12.5] one can also calculate $\operatorname{Ind} N$.

References

[1] A.V. Aĭzenshtat, Yu.I. Karlovich and G.S. Litvinchuk, The method of conformal gluing for the Haseman boundary value problem on an open contour, *Complex Variables* **28** (1996), 313–346.

[2] M.A. Bastos, A. Bravo and Yu.I. Karlovich, Convolution type operators with symbols generated by slowly oscillating and piecewise continuous matrix functions, in *Operator Theoretical Methods and Applications to Mathematical Physics, The Erhard Meister Memorial Volume*, Editors: I. Gohberg at al., Birkhäuser, 2004, 151–174.

[3] M.A. Bastos, A. Bravo and Yu.I. Karlovich, Symbol calculus and Fredholmness for a Banach algebra of convolution type operators with slowly oscillating and piecewise continuous data, *Math. Nachr.* **269–270** (2004), 11–38.

[4] M.A. Bastos, Yu.I. Karlovich and B. Silbermann, Toeplitz operators with symbols generated by slowly oscillating and semi-almost periodic matrix functions, *Proc. London Math. Soc.* **89** (2004), 697–737.

[5] A. Böttcher and Yu.I. Karlovich, *Carleson Curves, Muckenhoupt Weights, and Toeplitz Operators,* Birkhäuser, Basel, 1997.

[6] A. Böttcher, Yu.I. Karlovich and V.S. Rabinovich, The method of limit operators for one-dimensional singular integrals with slowly oscillating data, *J. Operator Theory* **43** (2000), 171–198.

[7] A. Böttcher and B. Silbermann, *Analysis of Toeplitz Operators,* Akademie-Verlag, Berlin, 1989 and Springer, Berlin, 1990.

[8] A. Böttcher and I.M. Spitkovsky, Pseudodifferential operators with heavy spectrum, *Integral Equations Operator Theory* **19** (1994), 251–269.

[9] L. Carleson, On convergence and growth of partial sums of Fourier series, *Acta Math.* **116** (1966), 135–157.

[10] R.R. Coifman and Y. Meyer, *Au delà des opérateurs pseudodifférentiels,* Astérisque **57** (1978), 1–184.

[11] H.O. Cordes, On compactness of commutators of multiplications and convolutions, and boundedness of pseudodifferential operators, *J. Funct. Anal.* **18** (1975), 115–131.

[12] H.O. Cordes, *Elliptic Pseudo-differential Operators – An Abstract Theory,* Springer, Berlin, 1979.

[13] R.V. Duduchava, Integral equations of convolution type with discontinuous coefficients, *Soobshch. Akad. Nauk Gruz. SSR* **92** (1978), 281–284 [Russian].

[14] R.V. Duduchava, *Integral Equations with Fixed Singularities,* Teubner, Leipzig, 1979.

[15] E.M. Dynkin, Methods of the theory of singular integrals: Hilbert transform and Calderón-Zygmund theory, in *Commutative Harmonic Analysis I: General Surveys, Classical Results*, Editors: V. P. Khavin and N.K. Nikol'skiy, Encyclopedia of Mathematical Sciences **15**, Springer, Berlin, 1991; Russian original, VINITI, Moscow, 1987.

[16] T. Finck, S. Roch and B. Silbermann, Two projection theorems and symbol calculus for operators with massive local spectra, *Math. Nachr.* **162** (1993), 167–185.

[17] I. Gohberg and N. Krupnik, Singular integral operators with piecewise continuous coefficients and their symbols, *Math. USSR - Izv.* **5** (1971), 955–979 [Russian].

[18] I. Gohberg and N. Krupnik, *One-Dimensional Linear Singular Integral Equations,* Vols. 1 and 2, Birkhäuser, Basel, 1992; Russian original, Shtiintsa, Kishinev, 1973.

[19] I. Gohberg and N. Krupnik, Extension theorems for invertibility symbols in Banach algebras, *Integral Equations Operator Theory* **15** (1992), 991–1010.

[20] N.B. Haaser and J.A. Sullivan, *Real Analysis,* Dover Publications, New York, 1991.

[21] W. Hoh, A symbolic calculus for pseudo-differential operators generating Feller semigroups, *Osaka J. Math.* **35** (1998), 789–820.

[22] W. Hoh, Perturbations of pseudodifferential operators with negative definite symbol, *Appl. Math. Optimization* **45** (2002), 269–281.

[23] R.A. Hunt, On the convergence of Fourier series, in *Orthogonal Expansions and Their Continuous Analogues*, Editor: D. Haimo, Southern Illinois Univ. Press, Carbondale, 1968, 235–255.

[24] N. Jacob and A.G. Tokarev, A parameter-dependent symbolic calculus for pseudodifferential operators with negative-definite symbols, *J. London Math. Soc.* **70** (2004), 780–796.

[25] O.G. Jørsboe and L. Melbro, *The Carleson–Hunt Theorem on Fourier Series,* Springer, Berlin, 1982.

[26] Yu.I. Karlovich, On the Haseman problem, *Demonstratio Mathematica* **26** (1993), No. 3–4, 581–595.

[27] Yu.I. Karlovich, An algebra of pseudodifferential operators with slowly oscillating symbols, *Proc. London Math. Soc.* **92** (2006), 713–761.

[28] Yu.I. Karlovich, Pseudodifferential operators with compound slowly oscillating symbols, in: Oper. Theory: Adv. Appl., Vol. 171, 2006, 189–224, to appear.

[29] Yu.I. Karlovich and E. Ramírez de Arellano, Singular integral operators with fixed singularities on weighted Lebesgue spaces, *Integral Equations Operator Theory* **48** (2004), 331–363.

[30] C.E. Kenig and P.A. Tomas, Maximal operators defined by Fourier multipliers, *Studia Math.* **68** (1980), 79–83.

[31] M.A. Krasnoselskii, P.P. Zabreiko, E.I. Pustylnik and P.E. Sobolevskii, *Integral Operators in Spaces of Summable Functions,* Noordhoff I.P., Leyden, 1976; Russian original, Nauka, Moscow, 1966.

[32] G.S. Litvinchuk, *Solvability Theory of Boundary Value Problems and Singular Integral Equations with Shift,* Kluwer, Dordrecht, 2000.

[33] J. Marschall, Pseudo-differential operators with nonregular symbols of the class $S^m_{\rho\delta}$, *Comm. Partial Differential Equations* **12** (1987), 921–965.

[34] J. Marschall, On the boundedness and compactness of nonregular pseudo-differential operators, *Math. Nachr.* **175** (1995), 231–262.

[35] J. Marschall, Nonregular pseudo-differential operators, *Z. Anal. Anwendungen* **15** (1996), 109–148.

[36] S.C. Power, Fredholm Toeplitz operators and slow oscillation, *Canad. J. Math.* **32** (1980), 1058–1071.

[37] S. Roch and B. Silbermann, *Algebras of Convolution Operators and Their Image in the Calkin Algebra,* Report R-Math-05/90, Karl-Weierstrass-Inst. f. Math., Berlin, 1990.

[38] L. Schwartz, *Analyse Mathématique*, Vol. **1**, Hermann, 1967.

[39] I.B. Simonenko and Chin Ngok Min, *Local Method in the Theory of One-Dimensional Singular Integral Equations with Piecewise Continuous Coefficients, Noetherity,* University Press, Rostov on Don, 1986 [Russian].

[40] E.M. Stein, *Harmonic Analysis: Real-Variable Methods, Orthogonality, and Oscillatory Integrals,* Princeton Univ. Press, Princeton, NJ, 1993.

[41] M.E. Taylor, *Tools for PDE. Pseudodifferential Operators, Paradifferential Operators, and Layer Potentials,* American Mathematical Society, Providence, RI, 2000.

Yu.I. Karlovich
Facultad de Ciencias
Universidad Autónoma del Estado de Morelos
Av. Universidad 1001, Col. Chamilpa,
C.P. 62209 Cuernavaca, Morelos, México
e-mail: karlovich@buzon.uaem.mx

Operator Theory:
Advances and Applications, Vol. 172, 235–249

A Class of Quadratic Time-frequency Representations Based on the Short-time Fourier Transform

Paolo Boggiatto, Giuseppe De Donno and Alessandro Oliaro

Abstract. Motivated by problems in signal analysis, we define a class of time-frequency representations which is based on the short-time Fourier transform and depends on two fixed windows. We show that this class can be viewed as a link between the classical Rihaczek representation and the spectrogram. Correspondingly we formulate for this class a suitable general form of the uncertainty principle which have, as limit case, the uncertainty principles for the Rihaczek representation and for the spectrogram. We finally consider the questions of marginal distributions. We compute them in terms of convolutions with the windows and prove simple conditions for which average and standard deviation of the distributions in our class coincide with that of their marginals.

Mathematics Subject Classification (2000). Primary 47G30; Secondary 42B10, 47G10.

Keywords. Time-frequency representations, short-time Fourier transform, uncertainty principles, Rihaczek distribution.

1. Time-frequency representations

The total energy of a signal $f(x)$, ($x \in \mathbb{R}^d$), is commonly interpreted in physics and in signal analysis as its L^2 norm $\|f\|_{L^2}$. Accordingly, the positive function $|f(x)|^2$ represents the density of the energy of the signal with respect to the time (although x is the "time" variable we shall assume here $x \in \mathbb{R}^d$ for generality and also in view of possible other interpretations). As the Fourier transform $\hat{f}(\omega)$ gives the frequencies contained in the signal, then $|\hat{f}(\omega)|^2$ represents the energy density with respect to the frequency variable ω and the equality $\|f\|_{L^2} = \|\hat{f}\|_{L^2}$, stated by the Plancherel theorem, is interpreted, in this context, as the natural assertion that the total energy of a signal does not change if we switch from the time representation

$f(x)$ to the frequency representation $\hat{f}(\omega)$. Actually the sentence "the signal $f(x)$", which is of common use and we shall also adopt, is somehow misleading because $f(x)$ and $\hat{f}(\omega)$ should be viewed as two different, and in some sense complementary, representations of the same physical phenomenon constituted by the signal. In the L^2 frame, which corresponds to the space of finite energy signals, the fact that the Fourier transform is a bijection means that each of the two representations contain in reality all information about the signal. However in each representation some information are contained in an explicit and others in a more implicit form and this can make the latter difficult to be recognized. For example it is clear that from the time representation $f(x)$ of a signal the support supp f immediately indicates when the signal takes place in time, however the frequencies contained in the signal are not to be detected directly but given only through the "form" of the graph of the function $f(x)$. In a quite symmetrical way the function $\hat{f}(\omega)$ contains explicit information about the frequencies of the signal but the information about when these frequencies are present in the signal is hidden in the "form" of the function $\hat{f}(\omega)$. For example the Fourier transform $e^{2\pi i a\omega}\hat{f}(\omega)$ of the delayed signal $f(x-a)$ differs from the transform $\hat{f}(\omega)$ of the original signal $f(x)$ only in the "shape" factor $e^{2\pi i a\omega}$. A clear exposition on these subjects can be found for example in [13].

One fundamental problem of time-frequency analysis is to find suitable representations for the energy density of a signal simultaneously with respect to time and frequency. Of course this question is equivalent to the definition of a $2d$-dimensional distribution density $Q(f)(x,\omega)$ depending on the variables $x \in \mathbb{R}^d$ and $\omega \in \mathbb{R}^d$. However, differently to what happens with usual distributions in statistics and probability theory, here the uncertainty principle put intrinsic limitations to the natural properties one would expect from a $2d$-dimensional distribution. The basic problem is that "instantaneous frequency" has no physical meaning, and therefore, no matter which time-frequency representation $Q(f)$ is used, its point values do not have any univocal physical interpretation.

More precisely, let $Q(f)(x,\omega)$ be a time-frequency distribution of the energy of a signal f. Then some desirable features are expressed by the following conditions:

- Positivity: $Q(f)(x,\omega)$ should be real and positive.
- No spreading effect:
  If supp $f \subset I$ for an interval $I \subset \mathbb{R}^d$ then $\Pi_x \mathrm{supp}\, Q(f) \subset I$
  (where Π_x is the orthogonal projection from $\mathbb{R}^d_x \times \mathbb{R}^d_\omega$ to $\mathbb{R}^d_x$); (1)
  analogously, if supp $\hat{f} \subset J$ then $\Pi_\omega \mathrm{supp}\, Q(f) \subset J$.
- Marginal distributions:
  $\int_{R^d} Q(f)(x,\omega)dx = |\hat{f}(\omega)|^2; \quad \int_{R^d} Q(f)(x,\omega)d\omega = |f(x)|^2.$

It turns out that these conditions are actually incompatible with the uncertainty principle and they can therefore be satisfied only with a certain degree of approximation.

Many different representations have been defined in the literature with the aim of approaching as near as possible an ideal representation (see [5], [8], [9],

[12], [13]). A general frame for the study of these representations is given by the so-called *Cohen class*, see for reference [3], [4].

Three of the most known representations are the *spectrogram*, the *Rihaczek* and the *Wigner* representations.

For what concerns the spectrogram, the idea, originally due to D. Gabor, is to focus the Fourier transform on small intervals in time and to analyze the frequencies present in these intervals. This is performed by multiplying the signal $f(t)$ by a cut-off, or *window*, function $\phi(t)$ that can be translated by a parameter x along the time axis, before taking the Fourier transform. In this way we are lead to the definition of the *Gabor transform* or *short-time Fourier transform* (briefly *STFT*) of a signal $f(t)$:

$$V_\phi f(x,\omega) = \int_{\mathbb{R}^d} e^{-2\pi it\omega}\overline{\phi(t-x)}f(t)dt. \tag{2}$$

It indicates the frequencies ω contained in the signal in a neighborhood of the time x. The conjugation on the window appears just for mathematical convenience so that $V_\phi f(x,\omega) = (f, \phi_{x,\omega})_{L^2}$, where $\phi_{x,\omega}(t) = e^{2\pi i\omega t}\phi(t-x)$.
The quadratic representation corresponding to the STFT is called *spectrogram* (see [4], [10]) and is defined as

$$Sp_\phi(f)(x,\omega) = |V_\phi f(x,\omega)|^2. \tag{3}$$

It is of course a positive distribution but it does not satisfies the marginals and has a spreading effect depending on the support of the window ϕ.

The Rihaczek quadratic representation is essentially defined as the product of the signal $f(x)$ with its Fourier transform $\hat{f}(\omega)$, more precisely it is the distribution

$$R(f)(x,\omega) = e^{-2\pi ix\omega}f(x)\overline{\hat{f}(\omega)}. \tag{4}$$

Though its elementary definition it has reasonable physical motivations and was widely used in the time-frequency analysis of signals (see [4], [11]). As one can immediately verify, it satisfies the marginals and has no spreading effect, however it is evidently not positive.

The third very popular representation is the Wigner distribution (see [18]) defined as

$$Wig(f)(x,\omega) = \int_{\mathbb{R}^d} e^{-2\pi ix\omega}f(x+t/2)\overline{f(x-t/2)}dt. \tag{5}$$

As the Rihaczek representation, it is not positive but it has no spreading effect and it satisfies the marginals (see [4], [10], [12]).

The Wigner distribution and the STFT are the basic tools for the theory of Weyl and localization operators respectively, about these wide fields there exists a huge literature, see for example [2], [16], [17], [19], [15], [20].

We shall be addressed in this paper essentially to the first two above-mentioned representations; related questions about the Wigner representation will be treated in a further paper. We shall establish a link between the spectrogram and the Rihaczek representation by defining in the next section a class of time-frequency representations $Q(f)(x,\omega) = Q_{\phi,\psi}(f)(x,\omega)$ which depend on two window functions or distributions ϕ, ψ. In a suitable general functional setting for windows and signals, we shall show that a sort of path between the spectrogram and the Rihaczek representation is allowed within this class, permitting to pass from one representation to the other by a continuous variation of the windows. Of course the degree of approximation by which the conditions (1) are satisfied will also change accordingly. As we explain in Section 2, the motivation of our definition is not only a mathematical one but it has been suggested by the attempt to reduce the spreading effect of the spectrogram, result which can be easily confirmed by computer simulations.

In Section 3 we treat the question of the uncertainty principle. We formulate a simple uncertainty principle for the representation $Q_{\phi,\psi}$ in the L^p frame and show that it links different form of uncertainty principles for the Rihaczek and spectrogram representations. Finally in Section 4 we consider the problem of the marginal distributions. We compute the marginals of the representation $Q_{\phi,\psi}$ in terms of convolution with the product of the windows and deduce from this that the case where our representation reduces to the Rihaczek is the only case where both marginal conditions in (1) are satisfied. Further we give in the general case simple conditions on the windows in order that average and standard deviation of the marginals of $Q_{\phi,\psi}$ coincide with that of the "expected" marginal distributions $|f(x)|^2$ and $|\hat{f}(\omega)|^2$.

2. The representation $Q_{\phi,\psi}$

Let $q : E \times E \to F$ be a sesquilinear form from the cartesian product of some functional space E to another space F. Then, as mentioned in [10], there are at least two natural ways of obtaining a quadratic representation Q on E. One can consider $Q(f) = q(f,f)$ or, for fixed ϕ, one can set $Q(f) = |q(f,\phi)|^2$.

The first of these methods is followed in order to pass from the "cross" Rihaczek sesquilinear form $R(f,g)(x,\omega) = e^{-2\pi i x\omega} f(x)\overline{\hat{g}(\omega)}$ to the (quadratic) Rihaczek representation (4), as well as from the "cross" Wigner distribution $Wig(f,g)(x,\omega) = \int_{\mathbb{R}^d} e^{-2\pi i x\omega} f(x+t/2)\overline{g(x-t/2)}dt$ to the quadratic Wigner representation (5).

The second is used to pass from the short-time Fourier transform $V_\phi f(x,\omega)$, which is sesquilinear in the couple (f,ϕ), to the spectrogram $Sp_\phi(f)(x,\omega) = |V_\phi f(x,\omega)|^2$.

We propose here a third way of defining a quadratic representation. Let q be a given sesquilinear form, then, for fixed ϕ and ψ, we set

$$Q_{\phi,\psi}(f)(x,\omega) = q(f,\phi)\overline{q(f,\psi)}. \tag{6}$$

We observe that, in the case $\phi = \psi$ our definition coincides obviously with the second method mentioned above. On the other hand it can also be viewed as an application of the first method to the sesquilinear form $\tilde{q}(f,g) = q(f,\phi)\overline{q(g,\psi)}$.

In order to justify our definition we shall now focus on the case where $q(f, \phi) = V_\phi f$ and introduce some considerations about the spreading effect of the spectrogram. As we mentioned in the previous section, the presence of the window ϕ is responsible for this effect. More precisely, as a consequence of the uncertainty principle, if the support of the window ϕ is well concentrated around the origin then we have a little spreading effect in time, i.e., a good "localization" in time, but a considerable spreading effect in the frequency, i.e., a bad "localization" in frequency. Vice versa, if the support of ϕ is "widespread" then the localization is bad in time but good in frequency. This suggests that instead of the spectrogram $Sp_\phi(f) = |V_\phi f|^2 = V_\phi f \overline{V_\phi f}$, we can fix a second window ψ and consider the form

$$Q_{\phi,\psi}(f) = V_\phi f \overline{V_\psi f}. \tag{7}$$

The advantage is that, if ϕ is very concentrated around the origin and ψ very spread on the plane (i.e., $\hat{\psi}$ is very concentrated), then $V_\phi f$ will show a good localization in time and $V_\psi f$ a good localization in frequency. Due to the reciprocal cut-off effect in the product, $V_\phi f \overline{V_\psi f}$ will have both good localizations. The drawback is of course the lost of positivity of the distribution. In the limit case where ϕ tends to δ and ψ to 1 then we shall have no spreading effect at all, i.e., the third condition in (1) will be exactly satisfied. We shall prove that this case will coincide with that of the Rihaczek representation.

The better behavior with regard to the support property of the representation $Q_{\phi,\psi}$ compared with the spectrogram can be made precise in many different ways. For example we can remark that $V_\phi f = (e^{-2\pi i\omega(\cdot)} f) * \overline{g}$ and therefore, from the support property of the convolution, $\Pi_x(\text{supp } V_\phi f) \subset \overline{\text{supp } \phi + \text{supp } f}$. From the fundamental equality of time-frequency analysis $V_\phi f(x, \omega) = e^{-2\pi i x\omega} V_{\hat{\phi}} \hat{f}(\omega, -x)$ we also have $\Pi_\omega(\text{supp } V_\phi f) \subset \overline{\text{supp } \hat{\phi} + \text{supp } \hat{f}}$, so we have proved the following property.

Proposition 2.1. *Let $B_0^{\delta_j}$ indicate balls in $\mathbb{R}^d$ of radius $\delta_j > 0$, $j = 1, 2$.*

i) *If* $\text{supp } \phi \subset B_0^{\delta_1}$ *then* $\Pi_x(\text{supp } Sp_\phi(f)) \subset \text{supp } f + B_0^{\delta_1}$;
ii) *If* $\text{supp } \hat{\phi} \subset B_0^{\delta_2}$ *then* $\Pi_\omega(\text{supp } Sp_\phi(f)) \subset \text{supp } \hat{f} + B_0^{\delta_2}$.

Of course hypothesis i) and ii) of this proposition cannot be both satisfied, which is one possible form of expressing that for the spectrogram good localization in time and in frequency are complementary.

An immediate consequence of this same proposition is that, on the contrary, for the representation $Q_{\phi,\psi} = V_\phi f \overline{V_\psi f}$ we can have arbitrary good localization both in time and frequency, more precisely:

Proposition 2.2. *Let $B_0^{\delta_j}$ indicate balls in $\mathbb{R}^d$ of radius $\delta_j > 0$, $j = 1, 2$. Suppose that* $\text{supp } \phi \subset B_0^{\delta_1}$ *and* $\text{supp } \hat{\psi} \subset B_0^{\delta_2}$, *then*

i) $\Pi_x(\text{supp } Q_{\phi,\psi}(f)) \subset \text{supp } f + B_0^{\delta_1}$;
ii) $\Pi_\omega(\text{supp } Q_{\phi,\psi}(f)) \subset \text{supp } \hat{f} + B_0^{\delta_2}$.

We present next two possible functional settings for the representation $Q_{\phi,\psi}$. We start by recalling some properties of the STFT. Let $f \otimes \phi$ be the (tensor) product $f \otimes \phi\,(x,t) = f(x)\phi(t)$, τ_a the asymmetric coordinate transform $\tau_a F(x,t) = F(t, t-x)$, and $\mathcal{F}_2$ the partial Fourier transform $\mathcal{F}_2 F(x,\omega) = \int_{\mathbb{R}^d} e^{-2\pi i t\omega} F(x,t)\,dt$ of a function F on $\mathbb{R}^{2d}$. Then an easy computation gives (see [7]):

Lemma 2.3. *If $f, \phi \in \mathrm{L}^2(\mathbb{R}^d)$, then*

$$V_\phi f(x,\omega) = \mathcal{F}_2 \tau_a (f \otimes \overline{\phi}).$$

Lemma 2.3 is used to extend the STFT to a map from $\mathcal{S}'(\mathbb{R}^d) \times \mathcal{S}'(\mathbb{R}^d)$ to $\mathcal{S}'(\mathbb{R}^d)$. Namely, if $f, \phi \in \mathcal{S}'(\mathbb{R}^d)$, then $f \otimes \overline{\phi} \in \mathcal{S}'(\mathbb{R}^{2d})$, with $(f \otimes \overline{\phi}, \chi) = (f_x, \overline{(\overline{\phi}_\omega, \chi)})$, $\chi \in \mathcal{S}(\mathbb{R}^{2d})$, as we consider distributions as anti-linear functionals. Further both operators τ_a and $\mathcal{F}_2$ are isomorphisms on $\mathcal{S}'(\mathbb{R}^{2d})$. Thus $V_\phi f$ is a well-defined tempered distribution, whenever $f, \phi \in \mathcal{S}'(\mathbb{R}^d)$.
Following these general lines $Q_{\phi,\psi}(f,g) = V_\phi f \overline{V_\psi g}$ would result as a product of two tempered distributions. To overcome problems of definition we introduce then some restrictions.

Definiton 2.4. *Let $\mathcal{B}^\infty(\mathbb{R}^d)$ be the space of smooth bounded functions together with all its derivatives, i.e., the space of C^∞ functions h such that for every multi-index $\alpha = (\alpha_1, \ldots, \alpha_d) \in \mathbb{N}^d$ there exists a constant M_α with $|\partial_x^\alpha h(x)| \leq M_\alpha$ on $\mathbb{R}^d$.*

By the differentiation under the integral and the Riemann-Lebesgue theorem one immediately obtain the following result.

Lemma 2.5. *If $\psi \in \mathcal{B}^\infty(\mathbb{R}^d)$ and $g \in \mathcal{S}(\mathbb{R}^d)$, then $V_\psi g \in \mathcal{B}^\infty(\mathbb{R}^{2d})$.*

Of course the role of g and ψ in Lemma 2.5 can be exchanged.

We observe now that the form $q_{\phi,\psi}(f,g) = V_\phi f \overline{V_\psi g}$ makes sense whenever $V_\phi f$ is a tempered distribution and $V_\psi g \in \mathcal{B}^\infty(\mathbb{R}^d)$. From Lemma 2.5 this is the case for example of the next proposition.

Proposition 2.6. *Let $f, \phi \in \mathcal{S}'(\mathbb{R}^d)$ and $g \in \mathcal{S}(\mathbb{R}^d)$, $\psi \in \mathcal{B}^\infty(\mathbb{R}^d)$ (or vice versa $g \in \mathcal{B}^\infty(\mathbb{R}^d)$, $\psi \in \mathcal{S}(\mathbb{R}^d)$). Then $q_{\phi,\psi}(f,g)$ is a well-defined tempered distribution in $\mathcal{S}'(\mathbb{R}^{2d})$.*

Better regularity for the time-frequency representation $q_{\phi,\psi}$ can easily be obtained for example in the following situation.

Proposition 2.7. *Let $f, g \in \mathcal{S}(\mathbb{R}^d)$ and $\phi, \psi \in \mathcal{S}'(\mathbb{R}^d)$ (or vice versa $f, g \in \mathcal{S}'(\mathbb{R}^d)$, and $\phi, \psi \in \mathcal{S}(\mathbb{R}^d)$), then $q_{\phi,\psi}(f,g) \in C^\infty(\mathbb{R}^{2d})$.*

As a second functional setting for the representation (7) we consider L^p spaces.
We shall need the following boundedness result about the STFT, see [14], [1].

Proposition 2.8. *Let us fix q and p satisfying $q \geq 2$ and $q' \leq p \leq q$, where q' is the conjugate of q, i.e., $\frac{1}{q} + \frac{1}{q'} = 1$. Then $V : L^{p'}(\mathbb{R}^d) \times L^p(\mathbb{R}^d) \to L^q(\mathbb{R}^{2d})$ is continuous, in particular*

$$\|V_\phi f\|_{L^q} \leq \|f\|_{L^{p'}} \|\phi\|_{L^p}. \tag{8}$$

We can now expresses the behavior of (7) in the context of L^p spaces by the following proposition.

Theorem 2.9. *Let $q \in [1,\infty]$, $q_j \in [2,\infty]$, $p_j \in [1,\infty]$, $j = 1,2$ satisfy $\frac{1}{q_1} + \frac{1}{q_2} = \frac{1}{q}$, and $q'_j \leq p_j \leq q_j$, where q'_j is the conjugate of q_j, i.e., $\frac{1}{q_j} + \frac{1}{q'_j} = 1$. Then*

$$q : (f, \phi, g, \psi) \in L^{p'_1}(\mathbb{R}^d) \times L^{p_1}(\mathbb{R}^d) \times L^{p'_2}(\mathbb{R}^d) \times L^{p_2}(\mathbb{R}^d) \to q_{\phi,\psi}(f, g) \in L^q(\mathbb{R}^{2d}) \tag{9}$$

is continuous, in particular

$$\|q_{\phi,\psi}(f, g)\|_q \leq \|f\|_{L^{p'_1}} \|\phi\|_{L^{p_1}} \|g\|_{L^{p'_2}} \|\psi\|_{L^{p_2}}. \tag{10}$$

Proof. We begin by remarking that $q = \frac{q_1 q_2}{q_1+q_2} \geq 1$ for $q_j \geq 2$, $j = 1,2$, which means that (9) makes sense under our hypothesis. From Proposition 2.8 we have that $V_\varphi h : L^{p'}(\mathbb{R}^d) \times L^p(\mathbb{R}^d) \to L^q(\mathbb{R}^{2d})$, with $q \geq 2$ and $q' \leq p \leq q$, is continuous, and in particular $\|V_\varphi h\|_{L^q} \leq \|h\|_{L^{p'}} \|\varphi\|_{L^p}$ holds. So we obtain that

$$q = V_\phi f \overline{V_\psi g} : \quad L^{p'_1}(\mathbb{R}^d) \times L^{p_1}(\mathbb{R}^d) \times L^{p'_2}(\mathbb{R}^d) \times L^{p_2}(\mathbb{R}^d) \quad \to \quad L^{q_1}(\mathbb{R}^{2d}) \cdot L^{q_2}(\mathbb{R}^{2d})$$

with

$$\|V_\phi f\|_{L^{q_1}} \leq \|f\|_{L^{p'_1}} \|\phi\|_{L^{p_1}}, \quad \|\overline{V_\psi g}\|_{L^{q_2}} \leq \|g\|_{L^{p'_2}} \|\psi\|_{L^{p_2}} .$$

where $L^{q_1}(\mathbb{R}^{2d}) \cdot L^{q_2}(\mathbb{R}^{2d})$ indicates the subset of $\mathcal{S}'(\mathbb{R}^d)$ of all the products fg of a function $f \in L^{q_1}(\mathbb{R}^{2d})$ with a function $g \in L^{q_2}(\mathbb{R}^{2d})$.

The generalized Hölder's formula:

$$\|V_\phi f \overline{V_\psi g}\|_q \leq \|V_\phi f\|_{L^{q_1}} \|\overline{V_\psi g}\|_{L^{q_2}}, \quad \text{for } \frac{1}{q_1} + \frac{1}{q_2} = \frac{1}{q} \tag{11}$$

proves then (10). Moreover (11) means:

$$L^{q_1}(\mathbb{R}^{2d}) \cdot L^{q_2}(\mathbb{R}^{2d}) \hookrightarrow L^q(\mathbb{R}^{2d}) ,$$

which proves (9). □

We end this section by showing that the form (7) represents a link between the spectrogram $Sp_\phi(f)$ in (3) and the Rihaczek's distribution $R(f)$ in (4).

Theorem 2.10. *Let $f \in \mathcal{S}(\mathbb{R}^d)$, and $\phi \in \mathcal{S}'(\mathbb{R}^d)$, then*

i) $Q_{\phi,\phi}(f) = Sp_\phi(f)$;
ii) $Q_{\delta,1}(f) = R(f)$.

Proof. The choice $\phi = \psi$ in (7) proves immediately the first part of the theorem. Consider now the limit cases $\phi = \delta$, the point measure $(\delta, \varphi) = \overline{\varphi(0)}$, and $\psi = 1$. For $\chi \in \mathcal{S}(\mathbb{R}^{2d})$ we have:

$$\begin{aligned}
(V_\delta f, \chi) &= (\mathcal{F}_2 \tau_a (f \otimes \overline{\delta}), \chi) = (f \otimes \overline{\delta}, {\tau_a}^{-1} {\mathcal{F}_2}^{-1} \chi) = (f, \overline{(\overline{\delta}, {\tau_a}^{-1} {\mathcal{F}_2}^{-1} \chi)}) \\
&= (f, \textstyle\int_{\mathbb{R}^d} e^{2\pi i x \omega} \chi(x-t, \omega)\, d\omega \,|_{t=0}) = (f, \textstyle\int_{\mathbb{R}^d} e^{2\pi i x \omega} \chi(x, \omega)\, d\omega) \\
&= \textstyle\int_{\mathbb{R}^{2d}} e^{-2\pi i x \omega} f(x) \overline{\chi(x, \omega)}\, dx d\omega = (e^{-2\pi i x \omega} f(x), \chi(x, \omega)) .
\end{aligned}$$

Therefore

$$V_\delta f(x,\omega) = e^{-2\pi i x\omega} f(x). \tag{12}$$

Now by relation (12) we obtain

$$V_1 g(x,\omega) = e^{-2\pi i x\omega} V_{\hat{1}}\hat{g}(\omega,-x) = e^{-2\pi i x\omega} V_\delta \hat{g}(\omega,-x) = \hat{g}(\omega).$$

We conclude that $V_\delta f \overline{V_1 g}(x,\omega) = e^{-2\pi i x\omega} f(x)\overline{\hat{g}(\omega)}$. □

We observe that it is of course possible to define explicitly a "path" between the spectrogram and the Rihaczek representation. For example, using Gaussian functions, we can consider the L^1-normalized window $\phi_\lambda(x) = \lambda^{d/2}e^{-\pi\lambda x^2}$ and set $\psi_\lambda(x) = \hat{\phi}_\lambda(x) = e^{-\pi\frac{1}{\lambda}x^2}$ with $\lambda \in [1,\infty]$, adopting the convention $\phi_\infty = \delta$, $\psi_\infty = 1$.

3. Uncertainty principles

In this section we examine various forms of the uncertainty principle. We study in particular which forms of this principle apply to the quadratic representation $Q_{\phi,\psi}$ and show that suitable L^p formulations have, as limit case, uncertainty principles for the spectrogram and for the Rihaczek representation. We begin with recalling the classical uncertainty principle.

Proposition 3.1. *For $f \in L^2(\mathbb{R})$, $a, b \in \mathbb{R}$ we have*

$$\left(\int_{\mathbb{R}} (x-a)^2 |f(x)|^2 \, dx\right)^{1/2} \left(\int_{\mathbb{R}} (\omega-b)^2 |\hat{f}(\omega)|^2 \, d\omega\right)^{1/2} \geq \frac{1}{4\pi}\|f\|_{L^2}^2.$$

As pointed out in [11] this principle is essentially an inequality for the Rihaczek representation and can actually be reformulated as:

$$\left(\int_{\mathbb{R}^2} (x-a)^2(\omega-b)^2 |Rf(x,\omega)|^2 \, dx d\omega\right)^{1/2} \geq \frac{1}{4\pi}\|f\|_{L^2}^2.$$

Another expression of the uncertainty principle for the Rihaczek representation is the well-known uncertainty principle of Donoho-Stark for which we refer to [6].

We prove next a form of the uncertainty principle for the Rihaczek representation in terms of the integral $\int_U |R(f)(z)| \, dz$ instead of $\int_U |R(f)(z)|^2 \, dz$ (as usual, $z = (x,\omega) \in \mathbb{R}^{2d}$).

Proposition 3.2. *Let $f \in L^1(\mathbb{R}^d) \cap L^\infty(\mathbb{R}^d)$ and $\epsilon \geq 0$. If $U \subset \mathbb{R}^{2d}$ is a measurable set and*

$$(1-\epsilon)\|f\|_{L^1}\|f\|_{L^\infty} \leq \int_U |Rf(z)| dz,$$

then $\mu(U) \geq 1-\epsilon$, where $\mu(U)$ is the Lebesgue measure of U.

Proof. As $|Rf(z)| = |f(x)\hat{f}(\omega)|$ we have immediately:

$$\begin{aligned}(1-\epsilon)\|f\|_{L^\infty}\|f\|_{L^1} &\le \int_U |Rf(z)|dz \\ &\le \|f\|_{L^\infty}\|\hat{f}\|_{L^\infty}\int_U dz \\ &\le \|f\|_{L^\infty}\|f\|_{L^1}\,\mu(U).\end{aligned}$$

□

For $p \in (1,\infty)$ we have the following result.

Proposition 3.3. *Let $f \in L^p(\mathbb{R}^d) \cap L^{p'}(\mathbb{R}^d)$, U be measurable with finite measure and $\epsilon \ge 0$. If*

$$(1-\epsilon)\|f\|_{L^p}\|f\|_{L^{p'}} \le \int_U |Rf(z)|dz,$$

then $\mu(U) \ge C_{p_0}^{dp_0}(1-\epsilon)^{p_0}$, where $p_0 = \min(p,p')$ and $C_{p_0} = \sqrt{\frac{(p_0')^{1/p_0'}}{(p_0)^{1/p_0}}}$.

Proof. In the proof we make use of the continuous inclusion $L^{p_2}(U_j) \subset L^{p_1}(U_j)$ for $p_1 \le p_2$, where $\|f\|_{L^{p_2}} \le \mu(U_j)^{(1/p_1-1/p_2)}\|f\|_{L^{p_1}}$ and, of the boundedness of the Fourier transform $\mathcal{F} : L^q(\mathbb{R}^d) \to L^{q'}(\mathbb{R}^d)$ with $\|\hat{f}\|_{L^{q'}} \le \left(\sqrt{q^{1/q}/q'^{1/q'}}\right)^d \|f\|_{L^q}$, which holds for $q \in [1,2]$.
We consider at first the case $p \le 2$, we have the estimates

$$\begin{aligned}(1-\epsilon)\|f\|_{L^{p'}(\mathbb{R}^d)}\|f\|_{L^p(\mathbb{R}^d)} &\le \textstyle\int_U |f(x)\hat{f}(\omega)|\,dxd\omega \\ &\le \|f\otimes\hat{f}\|_{L^{p'}(U)}\mu(U)^\alpha \\ &\le \|f\|_{L^{p'}(\mathbb{R}^d)}\|\hat{f}\|_{L^{p'}(\mathbb{R}^d)}\mu(U)^\alpha \\ &\le \left(\sqrt{p^{1/p}/p'^{1/p'}}\right)^d \|f\|_{L^{p'}(\mathbb{R}^d)}\|f\|_{L^p(\mathbb{R}^d)}\mu(U)^\alpha\end{aligned} \tag{13}$$

where $\alpha = 1 - 1/p' = 1/p$, which proves the thesis.

Analogously for $p \ge 2$ we have

$$\begin{aligned}(1-\epsilon)\|f\|_{L^{p'}(\mathbb{R}^d)}\|f\|_{L^p(\mathbb{R}^d)} &\le \textstyle\int_U |f(x)\hat{f}(\omega)|\,dxd\omega \\ &\le \|f\otimes\hat{f}\|_{L^p(U)}\mu(U)^\beta \\ &\le \|f\|_{L^p(\mathbb{R}^d)}\|\hat{f}\|_{L^p(\mathbb{R}^d)}\mu(U)^\beta \\ &\le \left(\sqrt{p'^{1/p'}/p^{1/p}}\right)^d \|f\|_{L^p(\mathbb{R}^d)}\|f\|_{L^{p'}(\mathbb{R}^d)}\mu(U)^\beta\end{aligned} \tag{14}$$

where $\beta = 1 - 1/p = 1/p'$. The thesis follows then from (13) and (14). □

Remark 3.4. *We note that we can regard Proposition 3.2 as a particular case of Proposition 3.3 if we allow $p \in [1,\infty]$, adopting, in the case $p_0 = 1$, the convention $\infty^{1/\infty} = 0^{1/0} = 1$.*

Our next goal is the investigation of possible relations between these results and corresponding uncertainty principles for our general class of representations $Q_{\phi,\psi}$. For this class we prove now the following simple uncertainty principle.

Proposition 3.5. *Consider non zero functions* $\phi \in L^p(\mathbb{R}^d)$, $\psi \in L^{p'}(\mathbb{R}^d)$ *and* $f \in L^p(\mathbb{R}^d) \cap L^{p'}(\mathbb{R}^d)$, *where* $1 \leq p \leq \infty$. *Let* U *be a measurable set in* $\mathbb{R}^{2d}$ *and* $\epsilon \geq 0$. *If*

$$(1-\epsilon)\|f\|_{L^p}\|f\|_{L^{p'}}\|\phi\|_{L^p}\|\psi\|_{L^{p'}} \leq \int_U |Q_{\psi,\phi}(f)(z)|\, dz, \tag{15}$$

then $\mu(U) \geq 1-\epsilon$.

Proof. Writing, as usual, $z=(x,\omega)$, $\phi_z(t)=e^{2\pi it\omega}\phi(t-x)$, $\psi_z(t)=e^{2\pi it\omega}\psi(t-x)$, we have

$$(1-\epsilon)\|f\|_{L^p}\|f\|_{L^{p'}}\|\phi\|_{L^p}\|\psi\|_{L^{p'}} \leq \int_U |Q_{\psi,\phi}(f)(z)|\, dz = \int_U |(f,\phi_z)(\psi_z,f)|\, dz$$

$$\leq \|f\|_{L^p}\|f\|_{L^{p'}}\|\phi\|_{L^p}\|\psi\|_{L^{p'}} \int_U dz = \|f\|_{L^p}\|f\|_{L^{p'}}\|\phi\|_{L^p}\|\psi\|_{L^{p'}}\mu(U). \tag{16}$$

□

Though its simplicity, this result is enlightening for what concerns the connections with the uncertainty principles for the spectrogram and the Rihaczek representation. Actually, for $\psi = \phi \in L^2(\mathbb{R}^d)$, it contains the well-known weak-uncertainty principle for the spectrogram (see [14]), namely:

Proposition 3.6. *For* $f,\phi \in L^2(\mathbb{R}^d)$, U *measurable set in* $\mathbb{R}^{2d}$, $\epsilon \geq 0$, *the condition*

$$(1-\epsilon)\|f\|^2_{L^2}\|\phi\|^2_{L^2} \leq \int_U |V_\phi f(z)|^2\, dz \tag{17}$$

implies $\mu(U) \geq 1-\epsilon$.

On the other hand, for $p=1$, keeping $\|\phi\|_{L^1} = \|\psi\|_{L^\infty} = 1$, we can let ϕ tend to δ and ψ tend to 1 in $\mathcal{S}'(\mathbb{R}^d)$ and we see therefore from Theorem 2.10 that the previous proposition yields, in this limit case, exactly the uncertainty principle of Proposition 3.2 for the Rihaczek representation.

Proposition 3.5 also permits to understand the substantial reason why, in the cases $1 < p < \infty$, we do not recapture the uncertainty principle of the Rihaczek representation as a limit case of that of the generalized spectrogram. Namely, if $\psi \in L^{p'}(\mathbb{R}^d)$ tends to 1 in $\mathcal{S}'(\mathbb{R}^d)$, then $\|\psi\|_{L^{p'}}$ cannot remain bounded, otherwise there would exist a subsequence weakly convergent to an element $\psi_0 \in L^{p'}(\mathbb{R}^d)$, contradicting $\psi \to 1$ in $\mathcal{S}'(\mathbb{R}^d)$.

In order that the product $\|\phi\|_{L^p}\|\psi\|_{L^{p'}}$, which appears in (15), remains bounded, it is then necessary that $\|\phi\|_{L^p} \to 0$. This implies $\phi \to 0$ in $\mathcal{S}'(\mathbb{R}^d)$, and is therefore incompatible with the condition $\phi \to \delta$. This justifies the more involved result for the Rihaczek representation which we have obtained in Proposition 3.3.

4. Marginal distributions

In this section we consider the quadratic form

$$Q_{\phi,\psi}(f,f) = V_\phi f \overline{V_\psi f},$$

and we want to study its marginal distributions, i.e.,

$$Q^{(1)}_{\phi,\psi}(x) = \int V_\phi f \overline{V_\psi f}\, d\omega, \tag{18}$$

and

$$Q^{(2)}_{\phi,\psi}(\omega) = \int V_\phi f \overline{V_\psi f}\, dx. \tag{19}$$

The marginals (18) and (19) are well defined for instance if $V_\phi f \overline{V_\psi f} \in L^1(\mathbb{R}^{2d})$, that is satisfied when q_1, q_2 in Theorem 2.9 verify $\frac{q_1 q_2}{q_1+q_2} = 1$, that means $q_1 = q_2'$. Since $q_j \geq 2$, $j = 1, 2$ we have to choose $q_1 = q_2 = 2$, and consequently $p_1 = p_2 = 2$. Then from now on we shall fix $f, \phi, \psi \in L^2(\mathbb{R}^d)$.

We start now by giving the explicit expression of (18) and (19) and then we study their average E and their standard deviation σ^2.

Proposition 4.1. *For every $f \in L^2(\mathbb{R}^d)$ and $\phi, \psi \in L^2(\mathbb{R}^d)$ we have:*

$$\int V_\phi f \overline{V_\psi f}\, d\omega = |f|^2 * \left(\tilde{\psi}\,\overline{\tilde{\phi}}\right)(x), \tag{20}$$

where as usual $\tilde{\phi}(x) = \phi(-x)$ and $\tilde{\psi}(x) = \psi(-x)$.

Proof. We start by considering $f, \phi, \psi \in \mathcal{S}(\mathbb{R}^d)$, recovering then (20) in the generical case by density arguments. By a simple change of variables we get:

$$\int V_\phi f \overline{V_\psi f}\, d\omega = \int e^{-2\pi i q\omega} f(q+s)\, \overline{f(s)}\, \psi(s-x)\, \overline{\phi(q+s-x)}\, ds\, dq\, d\omega;$$

now, by integrating at first in the ω-variable, since $\int e^{-2\pi i q\omega}\, d\omega$ is the δ distribution in the q-variable, we obtain that

$$\int V_\phi f \overline{V_\psi f}\, d\omega = \int f(s)\, \overline{f(s)}\, \psi(s-x)\, \overline{\phi(s-x)}\, ds = |f|^2 * \left(\tilde{\psi}\,\overline{\tilde{\phi}}\right)(x);$$

that is what we want to prove. □

Let us analyze now (19). From the previous proposition and the formula $V_g f(x,\omega) = e^{-2\pi i \omega x} V_{\hat{f}} \hat{g}(\omega, -x)$ we get immediately an analogous result.

Proposition 4.2. *For every $f \in L^2(\mathbb{R}^d)$ and $\phi, \psi \in L^2(\mathbb{R}^d)$ we have:*

$$\int V_\phi f \overline{V_\psi f}\, dx = |\hat{f}|^2 * \left(\tilde{\hat{\psi}}\,\overline{\tilde{\hat{\phi}}}\right)(\omega), \tag{21}$$

with the same notations as in the previous proposition.

In formula (20) we obtain the marginal with respect the x-variable, $|f(x)|^2$, if $(\tilde{\psi}\,\overline{\tilde{\phi}})(x) = \delta$, $(\delta, \varphi) = \overline{\varphi(0)}$, which is the case when

$$\phi = \delta \text{ and } \psi(0) = 1 \text{ or vice versa } \psi = \delta \text{ and } \phi(0) = 1\ .$$

In formula (21) we obtain the marginal with respect the ω-variable, $|\hat{f}(\omega)|^2$, if $(\hat{\tilde{\psi}}\,\overline{\hat{\tilde{\phi}}})(\omega) = \delta$, which happens when

$$\hat{\phi} = \delta \text{ and } \hat{\psi}(0) = 1 \text{ or vice versa } \hat{\psi} = \delta \text{ and } \hat{\phi}(0) = 1\ .$$

We have therefore:

Corollary 4.3. *For the generalized spectrogram $Q_{\phi,\psi}(f)$ both marginal distributions conditions are satisfied in the cases $\phi = \delta$, $\psi = 1$ and $\phi = 1$, $\psi = \delta$, which correspond to the Rihaczek's distribution $R(f) = e^{-2\pi i x\omega} f(x)\overline{\hat{f}(\omega)}$ and its conjugate $\overline{R(f)}$ respectively. More generally, for every multi-index α, the marginal distributions conditions are satisfied by generalized spectrograms with windows $\phi = \partial^\alpha \delta$ and $\psi = x^\alpha/\alpha!$, or vice versa $\phi = x^\alpha/\alpha!$ and $\psi = \partial^\alpha\delta$.*

We analyze now some properties of the marginal distributions (18) and (19).

Proposition 4.4. *We have:*

$$E(Q^{(1)}_{\phi,\psi}) = E(|f|^2)\,(\psi,\phi)_{L^2} - \|f\|^2_{L^2}(x\psi,\phi)_{L^2} \tag{22}$$

and

$$\begin{aligned} \sigma^2(Q^{(1)}_{\phi,\psi}) = & \int \Big[y - E(Q^{(1)}_{\phi,\psi})\Big]^2 |f(y)|^2\,dy \quad (\psi,\phi)_{L^2} \\ & - 2\int \Big[y - E(Q^{(1)}_{\phi,\psi})\Big] |f(y)|^2\,dy \quad (x\psi,\phi)_{L^2} \\ & + \|f\|^2_{L^2}\ (x^2\psi,\phi)_{L^2}. \end{aligned} \tag{23}$$

Proof. By (20) and a simple exchange of integration order we have:

$$\begin{aligned} E(Q^{(1)}_{\phi,\psi}) &= \int x\big(|f|^2 * (\tilde{\psi}\,\overline{\tilde{\phi}})(x)\big)\,dx \\ &= \int y|f(y)|^2\,\tilde{\psi}(x-y)\,\overline{\tilde{\phi}}(x-y) + (x-y)|f(y)|^2\,\tilde{\psi}(x-y)\,\overline{\tilde{\phi}}(x-y)\,dy\,dx \\ &= (\psi,\phi)_{L^2}\int y|f(y)|^2\,dy - (x\psi,\phi)_{L^2}\int |f(y)|^2\,dy, \end{aligned}$$

and so (22) is proved. Regarding (23) we observe that

$$\begin{aligned} \sigma^2(Q^{(1)}_{\phi,\psi}) &= \int \Big[x - E(Q^{(1)}_{\phi,\psi})\Big]^2 \big(|f|^2 * (\tilde{\psi}\,\overline{\tilde{\phi}})(x)\big)\,dx \\ &= \int \Big[y - E(Q^{(1)}_{\phi,\psi}) + x - y\Big]^2 |f(y)|^2\,\tilde{\psi}(x-y)\,\overline{\tilde{\phi}}(x-y)\,dy\,dx \end{aligned}$$

$$
\begin{aligned}
&= \int \Big[y - E(Q^{(1)}_{\phi,\psi})\Big]^2 |f(y)|^2 \, \tilde{\psi}(x-y)\, \overline{\tilde{\phi}}(x-y)\, dy\, dx \\
&\quad + 2\int (x-y)\Big[y - E(Q^{(1)}_{\phi,\psi})\Big] |f(y)|^2 \, \tilde{\psi}(x-y)\, \overline{\tilde{\phi}}(x-y)\, dy\, dx \\
&\quad + \int (x-y)^2 |f(y)|^2 \, \tilde{\psi}(x-y)\, \overline{\tilde{\phi}}(x-y)\, dy\, dx \\
&= \int \Big[y - E(Q^{(1)}_{\phi,\psi})\Big]^2 |f(y)|^2 \, dy \, \big(\psi,\phi\big)_{L^2} \\
&\quad - 2\int \Big[y - E(Q^{(1)}_{\phi,\psi})\Big] |f(y)|^2 \, dy \, \big(x\psi,\phi\big)_{L^2} \\
&\quad + \int |f(y)|^2 \, dy \, \big(x^2\psi,\phi\big)_{L^2},
\end{aligned}
$$

that is what we wanted to prove. □

With analogous computations results as in Proposition 4.4 can be proved also for $Q^{(2)}_{\phi,\psi}(\omega)$, namely:

Proposition 4.5. *We have:*

$$
E(Q^{(2)}_{\phi,\psi}) = E(|\hat{f}|^2)\big(\psi,\phi\big)_{L^2} - \|f\|^2_{L^2}\big(\omega\hat{\psi},\hat{\phi}\big)_{L^2} \tag{24}
$$

and

$$
\begin{aligned}
\sigma^2(Q^{(2)}_{\phi,\psi}) &= \int \Big[\eta - E(Q^{(2)}_{\phi,\psi})\Big]^2 |\hat{f}(\eta)|^2 \, d\eta \, \big(\psi,\phi\big)_{L^2} \\
&\quad - 2\int \Big[\eta - E(Q^{(2)}_{\phi,\psi})\Big] |\hat{f}(\eta)|^2 \, d\eta \, \big(\omega\hat{\psi},\hat{\phi}\big)_{L^2} \\
&\quad + \|f\|^2_{L^2}\big(\omega^2\hat{\psi},\hat{\phi}\big)_{L^2}.
\end{aligned} \tag{25}
$$

Remark 4.6. *Let us remark that in the particular case when*

$$
\big(\psi,\phi\big)_{L^2} = 1, \quad \big(x\psi,\phi\big)_{L^2} = 0, \tag{26}
$$

that is verified for example when $\tilde{\psi}(x)\overline{\tilde{\phi}}(x)$ is a Gaussian, we have from (22) that

$$
E(Q^{(1)}_{\phi,\psi}) = E(|f|^2); \tag{27}
$$

analogously, when

$$
\big(\psi,\phi\big)_{L^2} = 1, \quad \big(\omega\hat{\psi},\hat{\phi}\big)_{L^2} = 0 \tag{28}
$$

we obtain from (24)

$$
E(Q^{(2)}_{\phi,\psi}) = E(|\hat{f}|^2). \tag{29}
$$

Moreover, if (26) and

$$
\big(x^2\psi,\phi\big)_{L^2} = 0 \tag{30}
$$

are satisfied, we also have from (23) and (27) that

$$
\sigma^2(Q^{(1)}_{\phi,\psi}) = \sigma^2(|f|^2),
$$

and when the windows ϕ *and* ψ *satisfy* (28) *and*

$$\left(\omega^2\hat{\psi},\hat{\phi}\right)_{L^2}=0 \tag{31}$$

then we get from (25) *and* (29)

$$\sigma^2(Q^{(2)}_{\phi,\psi})=\sigma^2(|\hat{f}|^2)\ .$$

A simple example in the one-dimensional case of a function $\tilde{\psi}(x)\overline{\tilde{\phi}}(x)=h(x)$ *satisfying the conditions* (26) *and* (30) *is given by* $h(x)=k(|x|)$, *where*

$$k(x)=\begin{cases}\frac{3}{4x^2} & x\in(\frac{1}{2},1)\\ -\frac{3}{4x^2} & x\in(1,\frac{3}{2})\\ 0 & elsewhere\ .\end{cases}$$

Acknowledgements

The authors are thankful to the referee for his valuable remarks and suggestions.

References

[1] P. Boggiatto, G. De Donno and A. Oliaro, Weyl Quantization of Lebesgue spaces, to appear.

[2] P. Boggiatto, E. Buzano and L. Rodino, *Hypoellipticity and spectral theory*, Akademie Verlag, Berlin, 1996.

[3] L. Cohen, Time-frequency distributions – a review, *Proc. IEEE,* **77** (7) (1989), 941–981.

[4] L. Cohen, *Time-Frequency Analysis*, Prentice Hall, New Jersey, 1995.

[5] L. Cohen, The uncertainty principle for the short-time Fourier transform, *Proc. Int. Soc. Opt. Eng.* **22563** (1995), 80–90.

[6] D.L. Donoho and P.B. Stark, Uncertainty principles and signal recovery, *SIAM J. Appl. Math.*, **49** (3) (1989), 906–931.

[7] G.B. Folland, *Harmonic Analysis in Phase Space*, Princeton Univ. Press, 1989.

[8] G.B. Folland and A. Sitaram, The uncertainty principle: a mathematical survey, *J. Fourier Anal. Appl.*, **3** (3) (1989), 207–238.

[9] L. Galleani and L. Cohen, The Wigner distribution for classical systems, Physics Letters A, **302** (2002), 149–155.

[10] K. Gröchenig, *Foundations of Time-Frequency Analysis*, Birkhäuser, Boston, 2001.

[11] K. Gröchenig, Uncertainty principles for time-frequency representations. in *Advances in Gabor analysis*, Editors: H.G. Feichtinger and T. Strohmer, Birkhäuser, Boston, Boston, 2003, 11–30.

[12] A.J.A. Janssen, Proof of a conjecture on the supports of Wigner distributions, *J. Fourier Anal. Appl.*, **4** (6) (1998), 723–726.

[13] G. Kaiser, *A Friendly Guide to Wavelets*, Birkhäuser, Boston, 1994.

[14] E.H. Lieb, Integral bounds for radar ambiguity functions and Wigner distributions, *J. Math. Phys.*, **31**(3) (1990), 594–599.

[15] M.A. Shubin, *Pseudodifferential Operators and Spectral Theory*, Second Edition, Springer-Verlag, Berlin, 2001.

[16] J. Toft, Continuity properties for modulation spaces with applications to pseudo-differential calculus, I, *J. Func. Anal.* **207** (2) (2004), 399–429.

[17] J. Toft, Continuity properties for modulation spaces with applications to pseudo-differential calculus, II, *Ann. Global Anal. Geom.* **26** (2004), 73–106.

[18] E. Wigner, On the quantum correction for thermodynamic equilibrium, *Phys. Rev.* **40** (1932), 749–759.

[19] M.W. Wong, *Weyl Transforms,* Springer-Verlag, 1998.

[20] M.W. Wong, *Wavelet Transforms and Localization Operators*, Birkhäuser, Basel, 2002.

Paolo Boggiatto, Giuseppe De Donno and Alessandro Oliaro
Department of Mathematics
University of Torino
Via Carlo Alberto 10
I-10123 Torino, Italy

e-mail: `boggiatto@dm.unito.it`
e-mail: `dedonno@dm.unito.it`
e-mail: `oliaro@dm.unito.it`

Operator Theory:
Advances and Applications, Vol. 172, 251–257

A Characterization of Stockwell Spectra

M.W. Wong and Hongmei Zhu

Abstract. Signals in real applications are typically finite in duration, dynamic and non-stationary processes with frequency characteristics varying over time. This often requires techniques capable of locally analyzing and processing signals. An integral transform known as the Stockwell transform is a combination of the classic Gabor transform and the current and versatile wavelet transform. It allows more accurate detection of subtle changes and easy interpretation in the time-frequency domain. In this paper, we study the mathematical underpinnings of the Stockwell transform. We look at the Stockwell transform as a stack of simple pseudo-differential operators parameterized by frequencies and give a complete description of the Stockwell spectra.

Mathematics Subject Classification (2000). Primary 47G10, 47G30, 65R10; Secondary 92C55, 94A12.

Keywords. Stockwell transforms, pseudo-differential operators, signals, Fourier spectra, Stockwell spectra.

1. Introduction

Rapid development of information technology demands innovative mathematical tools to represent a signal in such a way that the representation can provide an accurate and easily understood description of the behavior of the signal. An optimal representation of a signal allows better designs of filters to improve the quality of signal processing and aids extraction of important information from the signal.

The classical Fourier transform forms the foundation of signal processing and modern medical imaging modalities such as computed tomography and magnetic resonance imaging. The Fourier transform $\mathcal{F}f$, also denoted by $\hat{f}$, of a function f in $L^2(\mathbb{R})$ is defined by

$$(\mathcal{F}f)(k) = \int_{-\infty}^{\infty} e^{-2\pi i t k} f(t)\, dt, \quad k \in \mathbb{R},$$

This research has been partially supported by the Natural Sciences and Engineering Research Council of Canada.

where the integral $\int_{-\infty}^{\infty}$ is understood as $\lim_{R\to\infty}\int_{-R}^{R}$ in the mean. Here, f and $\hat{f}$ describe, respectively, the temporal and frequency behavior of the signal. In general, it is difficult to observe the temporal occurrence of a particular frequency component from $\hat{f}$, and it is not easy to understand the frequency information from f. Therefore f and $\hat{f}$ are two different but equivalent representations of the same signal. The Fourier transform is ideal when the frequency characteristics of a signal do not change over time. In reality, however, most signals are non-stationary, and hence a mixed and simultaneous representation of f and $\hat{f}$ is desirable in signal processing.

Two well-developed integral transforms addressing this issue are the Gabor transform and the wavelet transform. The Gabor transform $V_g f$ of a signal f in $L^2(\mathbb{R})$, also known as the short-time Fourier transform, provides a time-frequency representation of f by applying the Fourier transform to the signal conditioned by a window g that translates over time. More precisely,

$$(V_g f)(\tau, k) = \int_{-\infty}^{\infty} e^{-2\pi itk}\overline{g(t-\tau)}f(t)\,dt, \quad \tau, k \in \mathbb{R}.$$

Analysis based on the Gabor transform is intuitive but is limited by the fixed time and frequency resolutions. Chapter 3 of the book [6] by Gröchenig contains detailed discussions on the Gabor transform.

The wavelet transform projects a signal f in $L^2(\mathbb{R})$ on a family of wavelets obtained by scaling and shifting a mother wavelet g. Mathematically, the wavelet transform $W_g f$ of f with respect to a mother wavelet g is defined by

$$(W_g f)(\tau, s) = \frac{1}{\sqrt{|s|}}\int_{-\infty}^{\infty} f(t)\,\overline{g\left(\frac{t-\tau}{s}\right)}\,dt, \quad \tau \in \mathbb{R},\ s \in \mathbb{R}\setminus\{0\}.$$

The multi-scale analysis provided by the wavelet transform gives a more accurate assessment of the local characteristics of a signal and thereby has extensive applications in many fields. Nevertheless, the time-scale representation may be difficult to interpret. Moreover, the right choice of a mother wavelet for a given problem may be a difficult task in signal processing. See, for instance, Chapter 2 of the book [3] by Daubechies, Chapter 10 of the book by [6] by Gröchenig and Chapter 18 of the book [12] by Wong for the mathematical aspects of the wavelet transform.

The search for optimal data representations leads to new development in time-frequency analysis and results in generating new integral transforms combining the merits of the Gabor transform and the wavelet transform. Among the remarkable ones is the Stockwell transform.

The Stockwell transform, first proposed by geophysicists in the paper [9] by Stockwell, Mansinha and Lowe in 1996, can be considered as a multi-scale local Fourier transform that provides a multi-resolution time-frequency data representation. It is a natural extension of the Gabor transform and gives a more accurate description of the local behavior of a signal through multi-scale analysis. The Stockwell transform has gained popularity in the signal processing community because of its applications in geophysics, oceanology, engineering and biomedicine.

See the papers [1], [2], [4], [5] and [13] for diverse applications of the Stockwell transform.

In spite of its manifold applications, the mathematics of the Stockwell transform has not been well investigated. The aim of this paper is to give, among other mathematical properties, a complete description of the range $\{Sf : f \in L^2(\mathbb{R})\}$ of the Stockwell transform. Since each element Sf in the range is simply the Stockwell spectrum of the signal f, the main result in this paper can be seen as a characterization of the Stockwell spectra.

In Section 2, we look at the Stockwell transform as a pseudo-differential operator. Pseudo-differential operators and their variants, first introduced by Kohn and Nirenberg in the seminal paper [8] in 1965 and modified by Hörmander [7] and others for problems in partial differential equations, have roots in quantization due to Weyl [10] and fertilizations with modern developments in mathematical sciences and engineering. The new way of looking at Stockwell transforms as pseudo-differential operators is promising and research in this direction is underway by the authors. A compact and self-contained account of pseudo-differential operators can be found in the book [11] by Wong.

In Section 3, we make precise some of the results in the original paper [9]. In particular, we establish the inversion formula and the Plancherel theorem for the Stockwell transform.

New results on the Stockwell transform are given in Section 4. We first introduce a new function space M, which is an interesting space in its own right. Then, in terms of the function space M, we give a characterization of the Stockwell spectra.

2. Stockwell transforms and pseudo-differential operators

Let $f \in L^2(\mathbb{R})$. Then the Stockwell transform Sf with Gaussian window of f is the function on $\mathbb{R} \times \mathbb{R}$ defined by

$$(Sf)(\tau, k) = e^{-2\pi i k\tau} \int_{-\infty}^{\infty} e^{2\pi i \tau v} e^{-2\pi^2 (v-k)^2/k^2} \hat{f}(v)\, dv, \quad \tau \in \mathbb{R},\, k \in \mathbb{R}\backslash\{0\}. \tag{2.1}$$

Thus, for each nonzero frequency k, the Stockwell transform can be thought of as the Fourier multiplier $\mathcal{F}^{-1} e^{-2\pi^2(\cdot - k)^2/k^2} \mathcal{F}$ followed by a modulation.

We have defined the Stockwell transform in the framework of Fourier multipliers or pseudo-differential operators. The following proposition gives a formula for the Stockwell transform in terms of an integral operator.

Proposition 2.1. *Let $f \in L^2(\mathbb{R})$. Then*

$$(Sf)(\tau, k) = \frac{|k|}{\sqrt{2\pi}} \int_{-\infty}^{\infty} e^{-(\tau - t)^2 k^2/2} e^{-2\pi i t k} f(t)\, dt, \quad \tau, k \in \mathbb{R}.$$

The proof of Proposition 2.1 is a simple exercise in the Fourier transform and is omitted.

By (2.1), we note that the computation of the Stockwell transform is usually implemented using the fast Fourier transform. Figure 1 illustrates the time-frequency representation computed from the Stockwell transform for a signal. The signal consists of two cosine functions with frequencies 0.1 and 0.2 Hz that are long in duration, and two cosine functions with the same frequency 0.4 Hz, which are short in duration. Due to the Heisenberg uncertainty principle, the Stockwell spectrum exhibits good frequency resolution for low frequency components and good time resolution for high frequency components. For users in the non-mathematical community, the close connections between the Fourier and Stockwell spectra make these features transparent.

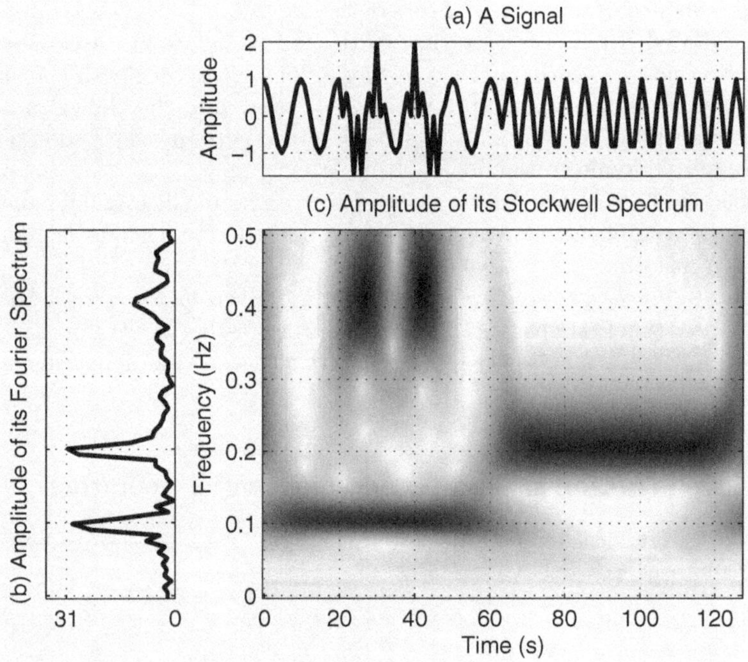

FIGURE 1. Time-frequency representation of the Stockwell transform: (a) A signal consisting of multiple frequency components (b) The amplitude of the corresponding Fourier spectrum, i.e., $|(\mathcal{F}f)(k)|$ (c) The contour plotting the amplitude of the corresponding Stockwell spectrum, i.e., $|(Sf)(\tau, k)|$

3. An inversion formula and Plancherel's theorem

For a measurable function $F : \mathbb{R} \times \mathbb{R} \to \mathbb{C}$, we define the time average AF of F to be the function on $\mathbb{R}$ by

$$(AF)(k) = \int_{-\infty}^{\infty} F(\tau, k)\, d\tau, \quad k \in \mathbb{R}. \tag{3.1}$$

Theorem 3.1. $ASf = \hat{f}, \quad f \in L^2(\mathbb{R})$.

Proof. Let f be a Schwartz function. Then, by Fubini's theorem, we get for all k in $\mathbb{R} \setminus \{0\}$,

$$\begin{aligned}(ASf)(k) &= \int_{-\infty}^{\infty} (Sf)(\tau,k)\,d\tau = \int_{-\infty}^{\infty} e^{-2\pi ik\tau}\left(\int_{-\infty}^{\infty} e^{2\pi i\tau v} e^{-2\pi^2(v-k)^2/k^2}\hat{f}(v)\,dv\right) d\tau \\ &= \int_{-\infty}^{\infty}\left(\int_{-\infty}^{\infty} e^{2\pi i\tau(v-k)}\,d\tau\right) e^{-2\pi^2(v-k)^2/k^2}\hat{f}(v)\,dv.\end{aligned}$$

Since for all k in $\mathbb{R} \setminus \{0\}$ and v in $\mathbb{R}$,

$$\int_{-\infty}^{\infty} e^{2\pi i\tau(v-k)}\,d\tau = \delta(v-k),$$

where δ is the Dirac delta function, it follows that

$$\begin{aligned}(ASf)(k) &= \int_{-\infty}^{\infty} \delta(v-k) e^{-2\pi^2(v-k)^2/k^2}\hat{f}(v)\,dv \\ &= \int_{-\infty}^{\infty} \delta(v) e^{-2\pi^2 v^2/k^2}\hat{f}(v+k)\,dv = \hat{f}(k)\end{aligned}$$

for all k in $\mathbb{R} \setminus \{0\}$. The proof is then complete by a density argument. □

Remark 3.2. The physical meaning of Theorem 3.1 can be understood visually by means of Figure 1.

We can look at Theorem 3.1 as the Stockwell inversion formula in the sense that for all f in $L^2(\mathbb{R})$,

$$f = \mathcal{F}^{-1}ASf,$$

which can be written out explicitly as

$$f(t) = \int_{-\infty}^{\infty} e^{2\pi itk}\left(\int_{-\infty}^{\infty}(Sf)(\tau,k)\,d\tau\right) dk, \quad t \in \mathbb{R}.$$

Another useful consequence of Theorem 3.1 is the Plancherel theorem for the Stockwell transform to the effect that for all f in $L^2(\mathbb{R})$,

$$\int_{-\infty}^{\infty}\left|\int_{-\infty}^{\infty}(Sf)(\tau,k)\,d\tau\right|^2 dk = \|f\|^2_{L^2(\mathbb{R})}.$$

Remark 3.3. The results in this section can be found in the paper [9] by Stockwell, Mansinha and Lowe.

4. Stockwell spectra

The main results in this paper are built on the function space M defined by

$$M = \left\{ F : \mathbb{R} \times \mathbb{R} \to \mathbb{C} : \int_{-\infty}^{\infty} \left| \int_{-\infty}^{\infty} F(\tau, k)\, d\tau \right|^2 dk < \infty \right\}.$$

Roughly speaking, M is the space of all functions F such that the time-average of F has finite energy with respect to the frequency k. In fact, we have the following theorem.

Theorem 4.1. *M is an indefinite Hilbert space in which the indefinite inner product $(\,,\,)_M$ is given by*

$$(F, G)_M = (AF, AG)_{L^2(\mathbb{R})}, \quad F, G \in M,$$

where AF is the time-average of F defined by (3.1) *and $(\,,\,)_{L^2(\mathbb{R})}$ is the inner product in $L^2(\mathbb{R})$.*

We note that $(\,,\,)_M$ is indefinite in the sense that there exist nonzero functions F in M for which $(F, F)_M = 0$.

Proof of Theorem 4.1. We only need to prove that M is complete with respect to the norm $\|\,\|_M$. Let $\{F_j\}_{j=1}^{\infty}$ be a Cauchy sequence of functions in M. Then $\{AF_j\}_{j=1}^{\infty}$ is Cauchy in $L^2(\mathbb{R})$. Since $L^2(\mathbb{R})$ is complete, $AF_j \to f$ for some f in $L^2(\mathbb{R})$ as $j \to \infty$. Therefore

$$\int_{-\infty}^{\infty} F_j(\tau, \cdot)\, d\tau \to f$$

in $L^2(\mathbb{R})$ as $j \to \infty$. By Theorem 3.1 and the Fourier inversion formula, $f = AS\check{f}$. So,

$$AF_j \to AS\check{f}$$

in $L^2(\mathbb{R})$ as $j \to \infty$. Hence $F_j \to S\check{f}$ in M as $j \to \infty$, and the proof is complete.

Theorem 4.2. *The range $R(S)$ of the Stockwell transform $S : L^2(\mathbb{R}) \to M$ is a Hilbert space with respect to the inner product $(\,,\,)_M$. In fact, S is an isometry of $L^2(\mathbb{R})$ onto $R(S)$.*

Proof. That $R(S)$ is complete follows from the proof of Theorem 4.1. To see that $(\,,\,)_M$ is a genuine inner product in $R(S)$, we only need to prove that

$$(Sf, Sf)_M = 0 \Rightarrow Sf = 0, \quad f \in L^2(\mathbb{R}).$$

But this follows from the Plancherel theorem for the Stockwell transform. That $S : L^2(\mathbb{R}) \to R(S)$ is a surjective isometry is also a consequence of the Plancherel theorem for the Stockwell transform. □

We end this paper with a characterization of $R(S)$.

Theorem 4.3. *$R(S) = M/Z$, where*

$$Z = \{F : \mathbb{R} \times \mathbb{R} \to \mathbb{C} : AF = 0\}.$$

Proof. Let $G \in M$ be such that $G \perp R(S)$. Then for all f in $L^2(\mathbb{R})$, $(Sf, G)_M = 0$. This means that

$$\int_{-\infty}^{\infty} (ASf)(k)\overline{(AG)(k)}\, dk = 0, \quad f \in L^2(\mathbb{R}),$$

or, by Theorem 3.1,

$$\int_{-\infty}^{\infty} \hat{f}(k)\overline{(AG)(k)}\, dk = 0, \quad f \in L^2(\mathbb{R}).$$

Thus, $AG = 0$ and the proof is complete. □

References

[1] R.A. Brown, H. Zhu and J.R. Mitchell, Distributed vector processing of a new local multi-scale Fourier transform for medical imaging applications, *IEEE Trans. Medical Imaging* **24** (2005), 689–691.

[2] P.K. Dash, B.K. Panigrahi and G. Panda, Power quality analysis using S-transform, *IEEE Trans. Power Delivery* **18** (2003), 406–412.

[3] I. Daubechies, *Ten Lectures on Wavelets*, SIAM, 1992.

[4] M.G. Eramian, R.A. Schincariol, L. Mansinha, and R.G. Stockwell, Generation of aquifer heterogeneity maps using two-dimensional spectral texture segmentation techniques, Math. Geology **31** (1999), 327–348.

[5] B.G. Goodyear, H. Zhu, R.A. Brown and J.R. Mitchell, Removal of phase artifacts from fMRI data using a Stockwell transform filter improves brain activity detection, Magn. Reson. Med. 51 (2004), 16–21.

[6] K. Gröchenig, *Foundations of Time-Frequency Analysis*, Birkhäuser, 2001.

[7] L. Hörmander, *The Analysis of Linear Partial Differential Operators III*, Springer-Verlag, 1985.

[8] J.J. Kohn and L. Nirenberg, An algebra of pseudo-differential operators, *Comm. Pure Appl. Math.* **18** (1965), 269–305.

[9] R.G. Stockwell, L. Mansinha and R.P. Lowe, Localization of the complex spectrum: the S transform, *IEEE Trans. Signal Processing* **44** (1996), 998–1001.

[10] H. Weyl, *The Theory of Groups and Quantum Mechanics*, Dover, 1950.

[11] M.W. Wong, *An Introduction to Pseudo-differential Operators*, Second Edition, World Scientific, 1999.

[12] M.W. Wong, *Wavelet Transforms and Localization Operators*, Birkhäuser, 2002.

[13] H. Zhu, B.G. Goodyear, M.L. Lauzon, R.A. Brown, G.S. Mayer, L. Mansinha, A.G. Law and J.R. Mitchell, A new local multiscale Fourier analysis for MRI, *Med. Phys.* **30** (2003), 1134–1141.

M.W. Wong and Hongmei Zhu
Department of Mathematics and Statistics, York University
4700 Keele Street, Toronto, Ontario M3J 1P3, Canada
e-mail: mwwong@mathstat.yorku.ca
e-mail: hmzhu@mathstat.yorku.ca

Operator Theory:
Advances and Applications, Vol. 172, 259–277

Exact and Numerical Inversion of Pseudo-differential Operators and Applications to Signal Processing

Vladimir S. Rabinovich and Steffen Roch

Abstract. A large class of time-varying filters can be described via pseudo-differential operators belonging to the Hörmander class $OPS^0_{0,0}$. The questions whether and how an input signal can be reconstructed from a known output lead to the problems of invertibility of pseudo-differential operators in that class and of (at least, numerical) solution of pseudo-differential equations. We are going to derive effective conditions for the invertibility for pseudo-differential operators with globally slowly varying symbols as well as for causal pseudo-differential operators, and we study the stability of the finite sections method with respect to time and frequency for these operators.

Mathematics Subject Classification (2000). Primary 35S05; Secondary 65N30, 94A12.

Keywords. Pseudo-differential operators, finite sections, reconstruction of signals.

1. Introduction

We start with recalling some basic definitions and facts from signal processing theory. Standard references to this field are [8, 9, 4, 22], for instance.

An *analog complex signal* (ACS for short) is a function $u : \mathbb{R} \to \mathbb{C}$. We will only consider ACS with finite energy, that is, we suppose that u belongs to the Hilbert space $L^2(\mathbb{R})$ with norm

$$\|u\|_2 := \left(\int_{\mathbb{R}} |u(t)|^2 \, dt \right)^{1/2} < \infty.$$

The authors are grateful for the support by the German Research Foundation (DFG), grant 444 MEX-112/2/05.

An *analog linear filter* is a linear mapping A which transforms every ACS u (which is then called an *input signal*) into an ACS $v := Au$, the so-called *output signal*. We will exclusively consider linear filters for which A acts as a bounded linear operator $A : L^2(\mathbb{R}) \to L^2(\mathbb{R})$.

An interesting but still easy to analyze class of linear filters is constituted by the *time-invariant filters*. They are characterized by their invariance with respect to shifts. More precisely, an analog filters A is said to be invariant with respect to shifts if

$$V_h A = A V_h \quad \text{for every } h \in \mathbb{R}$$

where the shift operators V_h on $L^2(\mathbb{R})$ acts via $(V_h u)(x) := u(x - h)$. It is well known that every bounded linear operator $A : L^2(\mathbb{R}) \to L^2(\mathbb{R})$ which is invariant with respect to shifts is in fact a convolution operator

$$(Au)(t) = \int_{\mathbb{R}} k_A(t - \tau)\, u(\tau)\, d\tau, \quad t \in \mathbb{R}, \tag{1}$$

where the kernel k_A has a bounded Fourier transform $\hat{k}_A$ in the distributional sense. Thus, $\hat{k}_A$ belongs to the space $L^\infty(\mathbb{R})$ of all essentially bounded measurable functions with norm

$$\|u\|_\infty := \operatorname{esssup}_{t \in \mathbb{R}} |u(t)| < \infty.$$

For a time-invariant filter, every input signal can be reconstructed from its output by means of the Fourier transform. Indeed, if A is the filter (1) and if

$$\operatorname{essinf}_{\omega \in \mathbb{R}} |\hat{k}_A(\omega)| > 0$$

then the input signal u is obtained from the output $v = Au$ via

$$u(t) = (A^{-1}v)(t) = \frac{1}{2\pi} \int_{\mathbb{R}} \frac{\hat{v}(\omega)}{\hat{k}_A(\omega)}\, e^{it\omega}\, dt. \tag{2}$$

A natural and important from the point of applications generalization of time-invariant filters are time-varying filters. They are described by linear operators A of the form

$$(Au)(t) = \int_{\mathbb{R}} k_A(t, t - \tau)\, u(\tau)\, d\tau, \quad t \in \mathbb{R}. \tag{3}$$

Evidently, for time-varying filters, the problem of reconstruction of the input signal from a given output is much more involved than for time-invariant filters. In particular, there is not explicit formula for the dependence of the input from the output signal. Thus, it is both necessary and natural to consider numerical methods to determine the input approximately. The present paper is devoted to this circle of problems.

The crucial observation in our approach to the problem of reconstruction of input signals is that the operator (3) can be written as the pseudo-differential operator

$$(Au)(t) = \frac{1}{2\pi} \int_{\mathbb{R}} \int_{\mathbb{R}} \sigma_A(t, \omega)\, u(\tau) e^{i(t-\tau)\omega}\, d\tau\, d\omega \tag{4}$$

with symbol

$$\sigma_A(t, \omega) := \int_{\mathbb{R}} k_A(t, \tau) \, e^{-i\tau\omega} \, d\tau.$$

The Fredholm properties of pseudo-differential operators in the Hörmander class $OPS^m_{0,0}$ have been studied in [13] (see also [15]) without imposing any additional conditions on the symbol. In particular, we derived necessary and sufficient conditions for the Fredholmness of an operator $A \in OPS^m_{0,0}$ thought of as acting from the Sobolev space $H^s(\mathbb{R}^N)$ into $H^{s-m}(\mathbb{R}^N)$, and these conditions are formulated in terms of the limit operators of A. On the other hand, for the purpose of reconstruction of the input signals, the Fredholmness is of less use. What one really needs is the (exact) invertibility of the pseudo-differential operators which model the time-varying filters rather than their Fredholmness. Thus, the main goals of the present paper are:

(a) to consider classes of pseudo-differential operators which are important for signal processing, and to derive conditions for their invertibility as well as effective formulas for their inversion (which generalize formula (2) for time-varying filters), and
(b) to consider the problem of approximation in $L^2(\mathbb{R})$ of the input signal by a sequence u_N of signals with finite support or finite spectrum.

To make the latter point more explicit, let P_N be the operator of multiplication by the characteristic function of the interval $[-N, N]$, i.e., P_N is the projection on $L^2(\mathbb{R})$ acting as

$$(P_N u)(t) := \begin{cases} u(t) & \text{if } t \in [-N, N] \\ 0 & \text{if } t \notin [-N, N]. \end{cases} \tag{5}$$

Together with the equation $Au = v$ where A is the operator (4) we consider the sequence of the equations

$$P_N A P_N u_N = v_N = P_N v, \quad N \in \mathbb{N}, \tag{6}$$

where u_N and v_N are functions with support in $[-N, N]$. Moreover, let $\hat{P}_N$ be the operator which is dual to P_N with respect to the Fourier transform, i.e.,

$$(\hat{P}_N u)(t) = \frac{1}{2\pi} \int_{-N}^{N} \hat{u}(\omega) \, e^{i\omega t} \, d\omega. \tag{7}$$

Then we also consider the sequence of the dual equations

$$\hat{P}_N A \hat{P}_N u'_N = v'_N := \hat{P}_N v. \tag{8}$$

Below we shall discuss under which conditions the solutions u_N of (6) and u'_N of (8) converge in the norm of $L^2(\mathbb{R})$ to the input signal u. From the point of view of signal processing, the more important question is that of the convergence of the u'_N to u. This is simply because the functions u'_N have finite spectra $[-N, N]$. Thus, by the well-known Shannon-Nyquist Sampling Theorem (see, for instance, [4]), the signals u'_N can be reconstructed from their samples $u'_N(nh)$ where $n \in \mathbb{Z}$ and $h \in (0, \pi/N]$.

The obtained results concerning the invertibility of pseudo-differential operators arising from problems in signal theory and concerning the applicability of the finite sections method for them with respect to the sequences (P_N) and $(\hat{P}_N)$ are new both from the point of view of signal theory and of the theory of pseudo-differential operators.

The paper is organized as follows. In Section 2 we recall some auxiliary material on pseudo-differential operators. Standard references to the theory of pseudo-differential operators are [6, 20, 21], to mention only a few. All cited facts can be also found in [17]. In Section 3 we study the invertibility of slowly varying pseudo-differential operators, which simulate slowly time-varying filters. In Section 4 we consider causal pseudo-differential operators simulating so-called causal time-varying filters. Finally, Section 5 is devoted to the stability of the finite sections methods, that is, to the problems of numerical approximation of the input signal by the sequences (u_N) and (u'_N) of signals with finite supports or with finite spectra.

2. The calculus of pseudo-differential operators

2.1. Function spaces

Let $\mathbb{N}$ and $\mathbb{N}_0$ stand for the sets of the positive and non-negative integers, respectively, and let $N \in \mathbb{N}$. For $x \in \mathbb{R}^N$, write $|x|$ for the Euclidean norm of x and set $\langle x \rangle := (1 + |x|^2)^{1/2}$. The usual operators of first order partial differentiation on $\mathbb{R}^N$ are denoted by $\partial_{x_j} := \frac{\partial}{\partial x_j}$, and for every multi-index $\alpha = (\alpha_1, \ldots, \alpha_N) \in \mathbb{N}_0$ we write $\partial^\alpha := \partial_{x_1}^{\alpha_1} \ldots \partial_{x_N}^{\alpha_N}$.

We will have to work in the following standard function spaces where Ω is an open subset of $\mathbb{R}^N$:

- $C^\infty(\Omega)$ is the linear topological space of all real-valued infinitely differentiable functions on Ω. A sequence (φ_m) in $C^\infty(\Omega)$ converges to zero if, for each compact subset K of Ω and for every multi-index $j \in \mathbb{N}_0^N$,
$$\lim_{m\to\infty} \sup_{t\in K} |(\partial^j \varphi_m)(t)| = 0.$$
- $C_0^\infty(\Omega)$ is the subspace of $C^\infty(\Omega)$ which consists of all functions with compact support. A sequence (φ_m) in $C_0^\infty(\Omega)$ converges to zero if there is a compact subset K of Ω such that $\operatorname{supp} \varphi_m \subset K$ and if, for every multi-index $j \in \mathbb{N}_0^N$,
$$\lim_{m\to\infty} \sup_{t\in K} |(\partial^j \varphi_m)(t)| = 0.$$
- $S(\mathbb{R}^N)$ is the subspace of $C^\infty(\mathbb{R}^N)$ of all functions φ with convergence defined by the family of semi-norms
$$|\varphi|_{m,j} := \sup_{t\in\mathbb{R}^N} \langle t \rangle^m |(\partial^j \varphi)(t)| < \infty$$
where $m \in \mathbb{N}_0$ and $j \in \mathbb{N}_0^N$.

- $C_b^\infty(\mathbb{R}^N)$ is the subspace of $C^\infty(\mathbb{R}^N)$ of all functions which are bounded together with all of their derivatives. Convergence in $C_b^\infty(\mathbb{R}^N)$ is defined by the family of the semi-norms

$$|\varphi|_j := \sup_{t\in\mathbb{R}^N} |(\partial^j\varphi)(t)| < \infty$$

where $j \in \mathbb{N}_0^N$.

We will use the Fourier transform F for functions on $\mathbb{R}$ in the form

$$\hat{\varphi}(\omega) = (F\varphi)(\omega) := \int_{\mathbb{R}} \varphi(t)\, e^{-it\omega}\, dt.$$

The operator $F : S(\mathbb{R}) \to S(\mathbb{R})$ acts as an isomorphism. It can be continuously extended to a bounded linear operator on $L^2(\mathbb{R})$ such that Parseval identity

$$\|\hat{\varphi}\|_2^2 = 2\pi\|\varphi\|_2^2$$

holds for all functions $\varphi \in L^2(\mathbb{R})$.

2.2. Oscillatory integrals

In what follows we will have to deal with functions on $\mathbb{R}$ and on $\mathbb{R}\times\mathbb{R}$ only. A point in $\mathbb{R}\times\mathbb{R}$ will be usually denoted by (t, ω), and we then write ∂_t and ∂_ω for the operators of partial differentiation with respect to the first and second variable, respectively. Further, $\langle D_t\rangle^2$ and $\langle D_\omega\rangle^2$ stand for the operators $I - \partial_t^2$ and $I - \partial_\omega^2$, respectively.

Definition 1. *Let $m \in \mathbb{N}_0$. A function $a \in C^\infty(\mathbb{R}\times\mathbb{R})$ belongs to the Hörmander class $S_{0,0}^m$ if, for each pair k, $l \in \mathbb{N}_0$,*

$$|a|_{k,l,m} := \sum_{\alpha\le k,\,\beta\le l}\ \sup_{(t,\,\omega)\in\mathbb{R}\times\mathbb{R}} |\partial_\omega^\alpha\,\partial_t^\beta\, a(t,\,\omega)|\,\langle\omega\rangle^{-m} < \infty.$$

A function $\chi \in S(\mathbb{R}\times\mathbb{R})$ is called a cut-off function if it is identically 1 in a neighborhood of the origin of $\mathbb{R}\times\mathbb{R}$. For each $R > 0$, set $\chi_R(x,\,\omega) := \chi(x/R,\,\omega/R)$.

Definition 2. *Let a be a measurable function on $\mathbb{R}\times\mathbb{R}$, and let χ be a cut-off function on $\mathbb{R}\times\mathbb{R}$. If the limit*

$$\lim_{R\to\infty} \frac{1}{2\pi}\int_{\mathbb{R}}\int_{\mathbb{R}} \chi_R(t,\,\omega)\, a(t,\,\omega)\, e^{-it\omega}\, dt\, d\omega$$

exists and if it is independent of the choice of χ, then this limit is called the oscillatory integral of a, *and we denote it by*

$$os\frac{1}{2\pi}\int_{\mathbb{R}}\int_{\mathbb{R}} a(t,\,\omega)\, e^{-it\omega}\, dt\, d\omega. \tag{9}$$

Proposition 3. *Let $a \in S^m_{0,0}$. Then the oscillatory integral* (9) *exists, and the following equality holds for all integers k_1, k_2 with $2k_1 > 1$ and $2k_2 > 1 + m$:*

$$os\frac{1}{2\pi}\int_{\mathbb{R}}\int_{\mathbb{R}} a(t,\,\omega)\, e^{-it\omega}\, dt\, d\omega$$
$$= \frac{1}{2\pi}\int_{\mathbb{R}}\int_{\mathbb{R}} e^{-it\omega}\langle\omega\rangle^{-2k_2}\langle D_t\rangle^{2k_2}\langle t\rangle^{-2k_1}\langle D_\omega\rangle^{2k_1} a(t,\,\omega)\, dt\, d\omega.$$

Moreover, for all $2k_1 > 1$ and $2k_2 > 1 + m$, the estimate

$$\left| os\frac{1}{2\pi}\int_{\mathbb{R}}\int_{\mathbb{R}} a(t,\,\omega)e^{-it\omega}\, dt\, d\omega \right| \le C|a|_{2k_1,\,2k_2,\,m} \tag{10}$$

holds with a constant C independent of a (but depending on k_1 and k_2).

Proposition 4. *Let $a \in C^\infty(\mathbb{R})$, and suppose there are an $m \in \mathbb{N}_0$ and constants C_j such that*

$$|a^{(j)}(t)| \le C_j\langle t\rangle^m$$

for all $j \in \mathbb{N}_0$. Then, for all $t \in \mathbb{R}$,

$$os\frac{1}{2\pi}\int_{\mathbb{R}}\int_{\mathbb{R}} a(t+\tau)\, e^{-i\tau\omega}\, d\tau\, d\omega = a(t). \tag{11}$$

2.3. Pseudo-differential operators

Let $a \in S^m_{0,0}$. The operator $Op(a)$ defined on $S(\mathbb{R})$ by

$$(Op(a)u)(t) := \frac{1}{2\pi}\int_{\mathbb{R}} a(t,\,\omega)\, \hat{u}(\omega)\, e^{it\omega}\, d\omega, \quad t \in \mathbb{R}, \tag{12}$$

is called the *pseudo-differential operator* with symbol a. The class of all pseudo-differential operators with symbol in $S^m_{0,0}$ is denoted by $OPS^m_{0,0}$.

Proposition 5. *Let $a \in S^m_{0,0}$. Then $Op(a)$ is bounded on $S(\mathbb{R})$.*

Let $a \in S^m_{0,0}$ and $u \in S(\mathbb{R})$. It follows from (12) that

$$(Op(a)u)(t) = \frac{1}{2\pi}\int_{\mathbb{R}}\int_{\mathbb{R}} a(t,\,\omega)\, u(t+\tau)\, e^{-i\tau\omega}\, d\tau\, d\omega.$$

One easily checks that this integral can be considered as the oscillatory integral

$$(Op(a)u)(t) = os\frac{1}{2\pi}\int_{\mathbb{R}}\int_{\mathbb{R}} a(t,\,\omega)\, u(t+\tau)\, e^{-i\tau\omega}\, d\tau\, d\omega. \tag{13}$$

This observation offers another way to define the pseudo-differential operator $Op(a)$ as the oscillatory integral (13) depending on the parameter t. Note that (13) makes sense also for functions u in $C^\infty_b(\mathbb{R})$. Thus, this alternative definition allows one to consider the pseudo-differential operator $Op(a) \in OPS^m_{0,0}$ as being defined on $C^\infty_b(\mathbb{R})$.

Proposition 6. *Let $a \in S^m_{0,0}$. Then the pseudo-differential operator $Op(a)$ defined by* (13) *acts boundedly on $C^\infty_b(\mathbb{R})$.*

Let A be a linear operator acting from $C_b^\infty(\mathbb{R})$ into $C_b^\infty(\mathbb{R})$. The *formal symbol* of A is defined by

$$\sigma_A(t,\,\omega) = e^{-it\omega}A(e^{it\omega}), \quad (t,\,\omega) \in \mathbb{R}\times\mathbb{R}. \tag{14}$$

Proposition 7. *Let $A = Op(a) \in OPS^m_{0,0}$. Then the formal symbol σ_A of A coincides with a.*

Basically, this proposition states that each time-varying filter $A \in OPS^m_{0,0}$ can be reconstructed from its action on the system of the harmonic signals $\{e^{it\omega}\}_{\omega\in\mathbb{R}}$.

Proposition 8 (Calderón-Vaillancourt). *Let $a \in S^0_{0,0}$. Then $Op(a)$ is a bounded operator on $L^2(\mathbb{R})$, and*

$$\|Op(a)\|_{\mathcal{B}(L^2(\mathbb{R}))} \le C|a|_{2,2,0}.$$

Here and in what follows, $\mathcal{B}(X)$ stands for the Banach algebra of all bounded linear operators on X.

Proposition 9. *Let $a \in S^0_{0,0}$. Then $Op(a)$ is a compact operator on $L^2(\mathbb{R})$ if and only if*

$$\lim_{(t,\,\omega)\to\infty} a(t,\,\omega) = 0.$$

The next result describes the symbol of the composition of pseudo-differential operators.

Proposition 10. *Let $A = Op(a) \in OPS^{m_1}_{0,0}$ and $B = Op(b) \in OPS^{m_2}_{0,0}$. Then $AB \in OPS^{m_1+m_2}_{0,0}$ and*

$$\sigma_{AB}(t,\,\omega) = os\frac{1}{2\pi}\int_{\mathbb{R}}\int_{\mathbb{R}} a(t,\,\omega+\eta)\,b(t+\tau,\,\omega)\,e^{-i\tau\eta}\,d\tau\,d\eta. \tag{15}$$

Moreover,

$$|\sigma_{AB}|_{l_1,\,l_2,\,m_1+m_2} \le C(l_1,\,l_2)|a|_{2k_1+l_1,\,l_2,\,m_1}|b|_{l_1,\,2k_2+l_2,\,m_2} \tag{16}$$

for all integers k_1, k_2 with $2k_1 > m_1+1$ and $2k_2 > m_2+1$ and for all non-negative integers l_1, l_2.

2.4. Fredholmness of pseudo-differential operators

For α, $\beta \in \mathbb{Z}$, consider the unitary operators V_α and E_β on $L^2(\mathbb{R})$ defined by

$$(V_\alpha u)(t) := u(t-\alpha) \quad \text{and} \quad (E_\beta u)(t) := e^{i\beta t}u(t), \quad t\in\mathbb{R},$$

and set $U_{(\alpha,\,\beta)} := V_\alpha E_\beta$. Clearly, the $U_{(\alpha,\,\beta)}$ are also unitary on $L^2(\mathbb{R})$. Let A be a bounded operator on $L^2(\mathbb{R})$, and let

$$h : \mathbb{N}\to\mathbb{Z}\times\mathbb{Z},\ m\mapsto h(m) = (\alpha(m),\,\beta(m))$$

be a sequence which tends to infinity in the sense that $|h(m)|\to\infty$ as $m\to\infty$. The operator $A_h \in \mathcal{B}(L^2(\mathbb{R}))$ is called the *limit operator* of A with respect to the sequence h if

$$A_h = \text{s-lim}_{m\to\infty}U^*_{h(m)}AU_{h(m)}, \quad A^*_h = \text{s-lim}_{m\to\infty}U^*_{h(m)}A^*U_{h(m)},$$

where $\text{s-lim}_{m\to\infty} A_m$ denotes the strong limit of the sequence (A_m) of operators in $\mathcal{B}(L^2(\mathbb{R}))$. We denote the set of all limit operators of A by $\sigma_{op}(A)$.

Let $A = Op(a) \in OPS^0_{0,0}$ and let h be as above. Then

$$U^*_{h(m)} A U_{h(m)} = Op(a_{h(m)}) \quad \text{with} \quad a_{h(m)}(t,\, \omega) := a(t + \alpha(m),\, \omega + \beta(m)).$$

The sequence h has a subsequence g such that the sequence $(a_{g(m)})_{m\in\mathbb{N}}$ converges in the topology of $C^\infty(\mathbb{R}\times\mathbb{R})$ to a certain function a_g. One can show that a_g belongs to $S^0_{0,0}$ and that $A_g := Op(a_g)$ is just the limit operator of A defined by the sequence g.

Recall that a bounded linear operator acting on a Banach space X is called a *Fredholm operator* if $\ker A := \{x \in X : Ax = 0\}$ and $\operatorname{coker} A := X/(AX)$ are finite-dimensional spaces. This means that the equations $Ax = 0$ and $A^*y = 0$ possess only a finite number of linear independent solutions in the spaces X and X^*, respectively, and the equation $Ax = f$ is solvable if and only if $y_j(f) = 0$ for $j = 1, \ldots, m$ where the functionals $y_1, \ldots, y_m$ form a basis of $\ker A^*$. The integer

$$\operatorname{ind}(A) := \dim\ker A - \dim\ker A^*$$

is called the Fredholm index of the Fredholm operator A.

The following theorem is the main result of [13] (see also Theorems 4.3.9 and 4.3.15 in [17]). It states necessary and sufficient conditions for operator $A = Op(a) \in OPS^0_{0,0}$ to be Fredholm on the space $L^2(\mathbb{R})$.

Theorem 11. *The operator $A = Op(a) \in OPS^0_{0,0}$ is Fredholm on $L^2(\mathbb{R})$ if and only if all limit operators of A are invertible on $L^2(\mathbb{R})$.*

Our next goal is to single out a special class of pseudo-differential operators for which the Fredholm criterion from Theorem 11 can be made more explicit and effective. We say that a symbol $a \in S^0_{0,0}$ is *slowly oscillating at infinity with respect to t* if

$$\lim_{t\to\infty} \sup_{\omega\in\mathbb{R}} |\partial_t a(t,\, \omega)| = 0, \tag{17}$$

and we call $a \in S^0_{0,0}$ *slowly oscillating at infinity with respect to ω* if

$$\lim_{\omega\to\infty} \sup_{t\in\mathbb{R}} |\partial_\omega a(t,\, \omega)| = 0. \tag{18}$$

Finally, $a \in S^0_{0,0}$ is said to be *slowly oscillating at infinity with respect to both variables $(t,\, \omega)$* if conditions (17) and (18) hold simultaneously.

Proposition 12. *Let $a \in S^0_{0,0}$ be slowly oscillating at infinity with respect to t (resp. ω) and let $Op(a_h)$ be the limit operator of $Op(a)$ corresponding to the sequence $h = (\alpha,\, \beta) : \mathbb{N} \to \mathbb{Z}\times\mathbb{Z}$ with $\alpha(m) \to \infty$ (resp. $\beta(m) \to \infty$). Then the symbol a_h of $Op(a_h)$ is independent of t (resp. ω):*

$$a_h(\omega) = \lim_{(\alpha(m),\, \beta(m))\to\infty} a(\alpha(m),\, \beta(m) + \omega)$$

resp.

$$a_h(t) = \lim_{(\alpha(m),\, \beta(m))\to\infty} a(\alpha(m) + t,\, \beta(m)).$$

Theorem 11 and Proposition 12 imply the following result.

Theorem 13. *Let the symbol a be slowly oscillating at infinity with respect to both variables* (t, ω). *Then* $Op(a) : L^2(\mathbb{R}) \to L^2(\mathbb{R})$ *is a Fredholm operator if and only if*

$$\lim_{R\to\infty} \inf_{|t|+|\omega|>R} |a(t, \omega)| > 0.$$

3. Invertibility of pseudo-differential operators with globally slowly varying symbols

For $a \in S^0_{0,0}$ and $\varepsilon > 0$, define a_ε by $a_\varepsilon(t, \omega) := a(\varepsilon t, \omega)$. Clearly, if ε tends to zero, then the variation of a_ε with respect to t becomes slower and slower.

Theorem 14. *Let* $a \in S^0_{0,0}$ *satisfy*

$$\inf_{(t,\,\omega)\in\mathbb{R}^2} |a(t, \omega)| > 0,$$

and set $a_\varepsilon(t, \omega) := a(\varepsilon t, \omega)$ *for* $\varepsilon > 0$. *Then there exists* $\varepsilon_0 > 0$ *such that the pseudo-differential operator* $Op(a_\varepsilon) : L^2(\mathbb{R}) \to L^2(\mathbb{R})$ *is invertible for all* $\varepsilon \in (0, \varepsilon_0)$.

Proof. Set $b_\varepsilon(t, \omega) := a^{-1}(\varepsilon t, \omega)$. Via (15) we obtain that $Op(b_\varepsilon)Op(a_\varepsilon) = Op(c_\varepsilon)$ where

$$\begin{aligned}
c_\varepsilon(t, \omega) &= os\frac{1}{2\pi}\int\int_{\mathbb{R}^2} a(\varepsilon t, \omega+\eta)\, a^{-1}(\varepsilon(t+\tau), \omega)\, e^{-i\tau\eta}\, d\tau\, d\eta \\
&= \left(os\frac{1}{2\pi}\int\int_{\mathbb{R}^2} a(\varepsilon t, \omega+\eta)\, e^{-i\tau\eta}\, d\tau\, d\eta\right) a^{-1}(\varepsilon t, \omega) \\
&\quad + os\frac{1}{2\pi}\int\int_{\mathbb{R}^2} a(\varepsilon t, \omega+\eta)(a^{-1}(\varepsilon(t+\tau), \omega) - a^{-1}(\varepsilon t, \omega))\, e^{-i\tau\eta}\, d\tau\, d\eta \\
&=: 1 + l_\varepsilon(t, \omega)
\end{aligned}$$

with $l_\varepsilon(t, \omega)$ being equal to

$$os\frac{\varepsilon i}{2\pi}\int\int_{\mathbb{R}^2}\int_0^1 \partial_\eta a(\varepsilon t, \omega+\eta)\partial_\tau a^{-1}(\varepsilon(t+\theta\tau), \omega)\, e^{-i\tau\eta}\, d\theta\, d\tau\, d\eta.$$

Applying estimate (16) and Proposition 8 one obtains

$$\|Op(l_\varepsilon)\|_{\mathcal{B}(L^2(\mathbb{R}))} \le C(a)\varepsilon$$

where the constant $C(a)$ depends on a only. Hence, there exists an $\varepsilon_0 > 0$ such that $\|Op(l_\varepsilon)\| < 1$ for all $\varepsilon \in (0, \varepsilon_0)$. For $\varepsilon \in (0, \varepsilon_0)$, the operator

$$(I + Op(l_\varepsilon))^{-1} Op(b_\varepsilon) = \sum_{j=0}^{\infty} Op(l_\varepsilon)^j\, Op(b_\varepsilon)$$

is a left inverse for $Op(a_\varepsilon)$. In the same way one gets the existence of a right inverse operator for $Op(a_\varepsilon)$ whenever $\varepsilon > 0$ is sufficiently small. □

Theorem 15. *Let $a \in S^0_{0,0}$ satisfy*

$$\inf_{(t,\,\omega)\in\mathbb{R}^2} |a(t,\,\omega)| > 0,$$

and set $a_\varepsilon(t,\,\omega) := a(t,\,\varepsilon\omega)$ for $\varepsilon > 0$. Then there exists $\varepsilon_0 > 0$ such that the pseudo-differential operator $Op(a_\varepsilon) : L^2(\mathbb{R}) \to L^2(\mathbb{R})$ is invertible for all $\varepsilon \in (0, \varepsilon_0)$.

The proof runs similarly to the proof of Theorem 14.

4. Invertibility of causal pseudo-differential operators

4.1. Invertibility on weighted spaces

Causal pseudo-differential operators are realizations of casual filter, that is, of filters A owning the property that if $u \in L^2(\mathbb{R})$ and $u(t) = 0$ for $t < 0$, then $(Au)(t) = 0$ for $t < 0$. Casual time-varying filters admit the following representation

$$(Au)(t) = \int_0^t k_A(t,\,t-\tau)\,u(\tau)\,d\tau, \quad t > 0.$$

Hence, the support of the function $\tau \mapsto k_A(t,\,\tau)$ is located on the positive semi-axis. Consequently, its Fourier transform

$$\hat{k}_A(t,\,z) = \int_0^\infty k_A(t,\,\tau)e^{-i\tau z}\,d\tau$$

(understood, if necessary, in the sense of the distributions) has an analytic extension into the lower half-plane $\Pi_- := \{z \in \mathbb{C} : \operatorname{Im} z < 0\}$. Thus, the consideration of causal time-varying filters leads in a natural way to the study of so-called causal pseudo-differential operators, i.e., of pseudo-differential operators with symbols $(t, \omega) \mapsto a(t,\,\omega)$ admitting an analytic extension with respect to ω into the lower half-plane Π_-. Let $\mathbb{R}_+$ stand for the set of the positive real numbers.

Definition 16. *Let $S^0(\Pi_-)$ denote the set of all symbols $a \in C^\infty(\mathbb{R}_+ \times \mathbb{R})$ for which the function $(t,\,\omega) \mapsto a(t,\,\omega)$ possesses an analytic extension with respect to ω into the half-plane Π_- and for which*

$$|a|_{jk} := \sup_{t\in\mathbb{R}_+,\,z\in\Pi_-} \left|\partial_t^j \partial_z^k a(t,\,z)\right| < \infty \tag{19}$$

for all $j,\,k \geq 0$ and

$$\lim_{\Pi_-\ni z\to\infty} \sup_{t\in\mathbb{R}_+} |\partial_z a(t,\,z)| = 0. \tag{20}$$

Let further $OPS^0(\Pi_-)$ denote the set of all pseudo-differential operators with symbols in $S^0(\Pi_-)$.

Let $L^2(\mathbb{R}_+,\,e^{-ht})$ stand for the Banach space of all measurable functions on $\mathbb{R}_+$ with norm

$$\|u\|^2_{L^2(\mathbb{R}_+,\,e^{-ht})} := \int_{\mathbb{R}_+} |u(t)|^2 e^{-2ht}\,dt < \infty.$$

Proposition 17. *Let $a \in S^0(\Pi_-)$. Then $Op(a)$ is bounded on $L^2(\mathbb{R}_+, e^{-ht})$ for every $h \geq 0$, and*

$$\|Op(a)\|_{\mathcal{B}(L^2(\mathbb{R}_+, e^{-ht}))} \leq C \sum_{j \leq 2,\, k \leq 2} |a|_{jk}.$$

Proof. Let k_A denote the kernel of the operator $A := Op(a)$ understood in the sense of distributions. The analyticity of the function $z \mapsto a(t, z)$ in the lower complex half-plane and the estimate (19) imply that the support of the kernel $k_A(t, \tau)$ with respect to τ is located on the real interval $[0, \infty)$. Hence, if u is a distribution in $\mathcal{D}'(\mathbb{R})$ with support in $[0, \infty)$ and if the convolution of $\tau \mapsto k_A(t, \tau)$ with u (with respect to τ) is well defined, then this convolution is supported in $[0, \infty)$ again. Further,

$$\begin{aligned}
\|Op(a)\|_{\mathcal{B}(L^2(\mathbb{R}_+, e^{-ht}))} &\leq \|e^{-ht} Op(a) e^{ht}\|_{\mathcal{B}(L^2(\mathbb{R}))} \\
&= \|Op(a(t, \omega - ih))\|_{\mathcal{B}(L^2(\mathbb{R}))} \\
&\leq |\partial_t^j \partial_\omega^k a(t, w - ih)| \\
&\leq C \sum_{j \leq 2,\, k \leq 2} |a|_{jk}
\end{aligned}$$

by the Calderón-Vaillancourt theorem (Proposition 8) and by (19). This estimate yields the boundedness of A as well as the norm estimate. □

Theorem 18. *Let $a \in S^0(\Pi_-)$ and assume that*

$$\inf_{t \in \mathbb{R}_+,\, z \in \Pi_-} |a(t, z)| > 0. \tag{21}$$

Then there exists an $h_0 > 0$ such that the operator $Op(a) : L^2(\mathbb{R}_+, e^{-ht}) \to L^2(\mathbb{R}_+, e^{-ht})$ is invertible for all $h \geq h_0$.

Proof. Let $b := 1/a$. Condition (21) implies that $b \in S^0(\Pi_-)$. Set $L := Op(b)$. Via the composition formula (15) we get $LOp(a) = Op(c)$ where

$$c(t, \omega) = os \frac{1}{2\pi} \int_{\mathbb{R}} \int_{\mathbb{R}} b(t, \omega + \eta)\, a(t + \tau, \omega) e^{-i\tau\eta}\, d\tau\, d\eta. \tag{22}$$

Applying the Lagrange formula

$$b(t, \omega + \eta) = b(t, \omega) + \eta \int_0^1 b'_\omega(t, \omega + \theta\eta)\, d\theta$$

we further obtain from (22) that $c = 1 + r$ where

$$r(t, \omega) = os \frac{i}{2\pi} \int_{\mathbb{R}} \int_{\mathbb{R}} \left(\int_0^1 b'_\omega(t, \omega + \theta\eta)\, a'_t(t + \tau, \omega)\, e^{-i\tau\eta} \right) d\theta\, d\tau\, d\eta.$$

It follows from the definition of oscillatory integrals that $r \in S^0(\Pi_-)$. Moreover, condition (20) yields that

$$\lim_{h \to +\infty} \sup_{t \in \mathbb{R}_+,\, \omega \in \mathbb{R}} |\partial_t^k \partial_\omega^j r(t, \omega - ih)| = 0 \tag{23}$$

for all non-negative integers k, j. Hence, Proposition 17 together with (23) imply that

$$\begin{aligned}\|Op(r)\|_{\mathcal{B}(L^2(\mathbb{R}_+,\, e^{-ht}))} &= \|e^{-ht}Op(r)e^{ht}\|_{\mathcal{B}(L^2(\mathbb{R}_+))}\\ &= \|Op(r(t,\, \omega - ih))\|_{\mathcal{B}(L^2(\mathbb{R}_+))} \to 0.\end{aligned}$$

if $h \to +\infty$. Thus, there exists an $h_0 > 0$ such that, for all $h \geq h_0$,

$$\|Op(r)\|_{\mathcal{B}(L^2(\mathbb{R}_+,\, e^{-ht}))} < 1.$$

Hence, $I + Op(r)$ is invertible, and the operator

$$L := (I + Op(r))^{-1} Op(b)$$

is a left inverse of $Op(a)$ in the space $L^2(\mathbb{R}_+,\, e^{-ht})$ for every $h \geq h_0$ if only $h_0 > 0$ is large enough. In the same way, one gets the existence a right inverse of $Op(a)$. Thus, $L = Op(a)^{-1}$. □

4.2. Invertibility on $L^2(\mathbb{R}_+)$

Our next goal is effective conditions for the invertibility of causal pseudo-differential operators on the (unweighted) Hilbert space $L^2(\mathbb{R}_+)$.

Theorem 19. *Let $a \in S^0(\Pi_-)$ satisfy the conditions*

$$\lim_{t \to +\infty} \sup_{z \in \Pi_-} |\partial_t a(t,\, z)| = 0 \tag{24}$$

and

$$\inf_{t \in \mathbb{R}_+,\, z \in \Pi_-} |a(t,\, z)| > 0. \tag{25}$$

Then $Op(a) : L^2(\mathbb{R}_+) \to L^2(\mathbb{R}_+)$ is invertible.

Proof. Let L be as in the proof of Theorem 18. Then $LOp(a) = I + Op(r)$ where $r \in S^0(\Pi_-)$. Condition (24) implies

$$\lim_{t + |\omega| \to \infty} r(t,\, \omega - ih) = 0$$

for every $h > 0$. Hence, and by Proposition 9, $Op(r)$ acts as a compact operator on $L^2(\mathbb{R}_+,\, e^{-th})$ for every $h \geq 0$. Consequently, L is a left regularizator of $Op(a)$ on $L^2(\mathbb{R}_+,\, e^{-th})$ for every $h \geq 0$. In the same way one checks that L is also a right regularizator of $Op(a)$ on $L^2(\mathbb{R}_+,\, e^{-th})$ for every $h \geq 0$. Thus, the operator $Op(a)$ considered as acting from $L^2(\mathbb{R}_+,\, e^{-th})$ into $L^2(\mathbb{R}_+,\, e^{-th})$ is Fredholm. But then the operator

$$Op(a)_{\langle h \rangle} := e^{-th} Op(a) e^{th} : L^2(\mathbb{R}_+) \to L^2(\mathbb{R}_+)$$

is a Fredholm operator, too. It follows from the definition of the class $S^0(\Pi_-)$ and from Proposition 8 that the operators $Op(a)_{\langle h \rangle}$ depend continuously on the parameter $h \geq 0$ in the norm topology of $\mathcal{B}(L^2(\mathbb{R}_+))$. Hence, being a continuous function on the set of all Fredholm operators, the index of the Fredholm operators $Op(a)_{\langle h \rangle}$ is independent of h. Since the operator $Op(a)_{\langle h \rangle}$ is invertible for h large enough, this implies that the index of $Op(a)_{\langle h \rangle}$ is zero for all $h \geq 0$.

We prove that the kernel of $Op(a)$ is trivial. Indeed, suppose that $u \in L^2(\mathbb{R}_+) \cap \ker Op(a)$. Then $e^{-th}u \in L^2(\mathbb{R}_+)$ for every $h \geq 0$. By Theorem 18, there exists an $h > 0$ for which $Op(a)_{\langle h \rangle}$ is an invertible operator. The equality

$$Op(a)_{\langle h \rangle} e^{-th} u = (e^{-th} Op(a) e^{th}) e^{-th} u = e^{-th} Op(a) u = 0$$

and the invertibility of $Op(a)_{\langle h \rangle}$ imply that $e^{-th}u = 0$. Hence, $u = 0$, and since the index of $Op(a) : L^2(\mathbb{R}_+) \to L^2(\mathbb{R}_+)$ is zero, this operator is invertible. □

5. Finite sections of pseudo-differential operators

Let $A \in \mathcal{B}(L^2(\mathbb{R}))$ be an invertible operator. Hence, the equation

$$Au = v \tag{26}$$

has a unique solution u for every right-hand side $v \in L^2(\mathbb{R})$. Let $P_N \in \mathcal{B}(L^2(\mathbb{R}))$ be the operator of multiplication by the characteristic function of the interval $[-N, N]$. Together with equation (26), we consider the sequence of its finite sections

$$P_N A P_N u_N = P_N v, \quad N \in \mathbb{N}. \tag{27}$$

The crucial question is whether the equations (27) possess unique solutions for sufficiently large N and whether the sequence (u_N) of these solutions converges in the norm of $L^2(\mathbb{R})$ to the solution u of equation (26). It is well known (see for instance [5, 3, 17]) that the convergence $\|u - u_N\|_2 \to 0$ is guaranteed if the operator $A : L^2(\mathbb{R}) \to L^2(\mathbb{R})$ is invertible and if the sequence $(P_N A P_N)$ is *stable*, i.e., if the operators $P_N A P_N : \operatorname{im} P_N \to \operatorname{im} P_N$ are invertible for sufficiently large N, say for $N \geq N_0$, and if the norms of their inverses are uniform bounded,

$$\sup_{N \geq N_0} \|(P_N A P_N)^{-1}\| < \infty.$$

In this case, we also say that the finite sections method with respect to the projections P_N is stable. In the same way, one defines the stability of finite sections method with respect to the sequence of the projections $\hat{P}_N$ defined by (7).

A general approach to the finite sections method via limit operators is proposed in [15, 16], see also Chapter 6 in [17]. This approach usually yields results of the following form: The finite sections method $(P_N A P_N)$ is stable if and only if the operator A is invertible and if a certain associated family of operators (related with the limit operators of A) is uniformly invertible. In any concrete situation, the discussion of the invertibility of the operators in this auxiliary family requires an additional (in general, large) amount of work. Only in some instances, one can state the additional conditions in a more explicit way (see, e.g., Theorem 6.2.8 in [17] for the finite sections of convolution type operators).

Below we shall discuss this problem for pseudo-differential operators. The obtained general Theorems 22 and 23, holding for (almost) arbitrary pseudo-differential operators, are certainly of their own interest. But for some special classes of symbols (viz. the multiplicatively slowly oscillating symbols), we will be able to formulate the extra conditions in a completely explicit manner, which yields a complete and final solution of the problem of the stability of the finite section method for these operators.

One should also mention that the obtained results are well known for some special classes of pseudo-differential operators with symbols which behave sufficiently well at infinity (see, for example, [18, 19] where the finite sections method is considered for operators which are constituted by operators of multiplication and of convolution by piecewise continuous functions).

The formulation of our results for the stability of the finite sections method with respect to the sequences (P_N) and $(\hat{P}_N)$ requires a modification of the notion of a limit operator. To each operator $A \in \mathcal{B}(L^2(\mathbb{R}))$, we now associate two families of limit operators. These families are defined with respect to the families $\mathcal{V} := \{V_\alpha\}_{\alpha\in\mathbb{R}}$ and $\mathcal{E} := \{E_\beta\}_{\beta\in\mathbb{R}}$ of unitary operators where V_α and E_β are as in Section 2.5.

Definition 20. *Let $A \in \mathcal{B}(L^2(\mathbb{R}))$, and let the sequence $h : \mathbb{N} \to \mathbb{Z}$ tend to $+\infty$ or $-\infty$.*

(a) *An operator A_h is called the limit operator of A with respect to the family $\mathcal{V}$ and to the sequence h if*

$$\lim_{k\to\infty} \|(V^*_{h(k)} A V_{h(k)} - A_h)P_N\| = \lim_{k\to\infty} \|(V^*_{h(k)} A^* V_{h(k)} - A^*_h)P_N\| = 0 \tag{28}$$

for every $N \in \mathbb{N}$. We denote the set of all limit operators of A with respect to the family $\mathcal{V}$ which correspond to sequences h tending to $+\infty$ (resp. $-\infty$) by $\sigma^{\mathcal{V}}_{op,+}(A)$ (resp. $\sigma^{\mathcal{V}}_{op,-}(A)$).

(b) *An operator A_h is called the limit operator of A with respect to the family $\mathcal{E}$ and to the sequence h if*

$$\lim_{k\to\infty} \|(E^*_{h(k)} A E_{h(k)} - A_h)\hat{P}_N\| = \lim_{k\to\infty} \|(E^*_{h(k)} A^* E_{h(k)} - A^*_h)\hat{P}_N\| = 0 \tag{29}$$

for every $N \in \mathbb{N}$. We denote the set of all limit operators of A with respect to the family $\mathcal{E}$ which correspond to sequences h tending to $+\infty$ (resp. $-\infty$) by $\sigma^{\mathcal{E}}_{op,+}(A)$ (resp. $\sigma^{\mathcal{E}}_{op,-}(A)$).

We denote by $\tilde{S}$ the closure in the topology of $S^0_{0,0}$ of the set of all functions of the form

$$a(t,\omega) = \sum_{n=1}^{N} a_n(t) b_n(\omega) \quad \text{with } a_n,\, b_n \in C_b^\infty(\mathbb{R}).$$

The class of all pseudo-differential operators with symbols in $\tilde{S}$ will be denoted by $OP\tilde{S}$.

Proposition 21. *Let $a \in \tilde{S}$. Then every sequence h tending to ∞ possesses a subsequence g for which the limit operator of $Op(a)$ with respect to the family $\mathcal{V}$ exists. The same statement holds with respect to the family $\mathcal{E}$.*

We denote by $P_+ \in \mathcal{B}(L^2(\mathbb{R}))$ the operator of multiplication by the characteristic function of the real interval $(0, \infty)$ and by $\hat{P}_+$ its dual with respect to the Fourier transform, that is, $\hat{P}_+ := FP_+F^{-1}$. Further, let $P_- := I - P_+$ and $\hat{P}_- := I - \hat{P}_+$, and write $H^2_+ := \hat{P}_+(L^2(\mathbb{R}))$ and $H^2_- := \hat{P}_-(L^2(\mathbb{R}))$ for the corresponding Hardy spaces.

The following theorems state necessary and sufficient conditions for the stability of the finite sections method with respect to the sequences (P_N) and $(\hat{P}_N)$, respectively.

Theorem 22. *Let $A := Op(a) \in OP\tilde{S}$ be an invertible operator on $L^2(\mathbb{R})$. Then the finite sections method for A with respect to the family $(P_N)_{N\in\mathbb{N}}$ is stable if and only if*

(a) *the Toeplitz operators $P_+A_hP_+ : L^2(\mathbb{R}_+) \to L^2(\mathbb{R}_+)$ are invertible for every limit operator $A_h \in \sigma^{\mathcal{V}}_{op,-}(A)$ and if*

(b) *the Toeplitz operators $P_-A_hP_- : L^2(\mathbb{R}_-) \to L^2(\mathbb{R}_-)$ are invertible for every limit operator $A_h \in \sigma^{\mathcal{V}}_{op,+}(A)$.*

Theorem 23. *Let $A := Op(a) \in OP\tilde{S}$ be an invertible operator on $L^2(\mathbb{R})$. Then the finite sections method for A with respect to the family $(\hat{P}_N)_{N\in\mathbb{N}}$ is stable if and only if*

(a) *the Toeplitz operators $\hat{P}_+A_h\hat{P}_+ : H^2_+ \to H^2_+$ are invertible for every limit operator $A_h \in \sigma^{\mathcal{E}}_{op,-}(A)$ and if*

(b) *the Toeplitz operators $\hat{P}_-A_h\hat{P}_- : H^2_- \to H^2_-$ are invertible for every limit operator $A_h \in \sigma^{\mathcal{E}}_{op,+}(A)$.*

The proofs of these results follow (with evident modifications) as in [14].

Theorems 22 and 23 reduce the problem of stability of the finite sections method to the problem of invertibility of pseudo-differential operators in $OP\tilde{S}$ and of invertibility of the Toeplitz operators $P_+A_hP_+$, $P_-A_hP_-$ and $\hat{P}_+A_h\hat{P}_+$, $\hat{P}_-A_h\hat{P}_-$. We are not aware of any effective invertibility criteria for general operators of this form. However, below we will consider certain subclasses of operators in $OP\tilde{S}$ which are both important in signal processing and for which we will be able to derive effective conditions of their invertibility and, thus, for the stability of the related finite sections methods.

Consider Theorem 22 for operators with symbols $(t, \omega) \mapsto a(t, \omega)$ which are slowly oscillating with respect to $t \to \infty$. In this case, the limit operators of $A := Op(a)$ have the form $A_h = Op(a_h)$ where the symbol a_h is independent of t and belongs to $C_b^\infty(\mathbb{R})$. Hence, $P_+Op(a_h)P_+$ and $P_-Op(a_h)P_-$ are Toeplitz operators on the positive and negative half-line, respectively, with symbols a_h in $C_b^\infty(\mathbb{R})$. Necessary and sufficient conditions for the invertibility of the Toeplitz operator $P_+Op(a_h)P_+$ on $L^2(\mathbb{R}_+)$ are well known in case the symbol a_h possesses one-sided limits

$$a_h^\pm := \lim_{\omega\to\pm\infty} a_h(\omega)$$

at infinity (consult [2, 5], for example). We are going to consider the more general class of symbols which are multiplicatively slowly oscillating at infinity in the following sense.

Definition 24. *A function $a \in C_b(\mathbb{R})$ is* multiplicatively slowly oscillating at infinity *if*

$$\lim_{\lambda \to +\infty} \sup_{\omega \in [a,\, b]} |a(\lambda\omega) - a(\omega)| = 0$$

for every interval $[a,\, b]$ in $(0,\, \infty) \cup (-\infty,\, 0)$.

The class of all multiplicatively slowly oscillating functions is (with respect to pointwise defined operations and the supremum norm) a C^*-subalgebra of $C_b(\mathbb{R})$. We denote it by $C_b^{sl}(\mathbb{R})$. Note that this algebra indeed contains all functions in $C_b(\mathbb{R})$ which possess one-sided limits at $\pm\infty$.

Let $a \in C_b^{sl}(\mathbb{R})$. Clearly, the Toeplitz operators $\hat{P}_+ a \hat{P}_+ : H_+^2 \to H_+^2$ and $\hat{P}_- a \hat{P}_- : H_-^2 \to H_-^2$ are Fredholm operators if and only if the operators $A_\pm := a\hat{P}_\pm + \hat{P}_\mp : L^2(\mathbb{R}) \to L^2(\mathbb{R})$ are Fredholm operators (this holds, of course, for every $a \in L^\infty$). Let

$$(S_{\mathbb{R}} u)(\omega) = \frac{1}{\pi i} \int_{-\infty}^{+\infty} \frac{u(\eta)}{\eta - \omega}\, d\eta$$

stand for the Cauchy singular integral operator on $\mathbb{R}$. Then $\hat{P}_\pm = (I \pm S_{\mathbb{R}})/2$ and

$$A_\pm = \frac{a+1}{2} I \pm \frac{a-1}{2} S_{\mathbb{R}}.$$

Hence, the operators $A_\pm$ can be considered as singular integral operators on $L^2(\mathbb{R})$ with multiplicatively slowly oscillating coefficients. Singular integral operators with multiplicatively slowly oscillating coefficients acting on $L^p(\Gamma,\, w)$, with $p \in (1,\, \infty)$, Γ is a Carleson curve and w being a Muckenhoupt weight, have been intensively studied in [1, 10, 12, 11]. Following these papers, we associate to the singular integral operator

$$A := aI + bS_{\mathbb{R}} : L^2(\mathbb{R}) \to L^2(\mathbb{R})$$

with $a,\, b \in C_b^{sl}(\mathbb{R})$ the pair

$$\sigma_A(\omega,\, \lambda) := (a(\omega) + b(\omega) sgn(\lambda),\, q_A(\omega,\, \lambda)), \quad (\omega,\, \lambda) \in \mathbb{R}_+ \times \mathbb{R},$$

where $q_A(\omega,\, \lambda)$ is the 2×2 matrix

$$q_A(\omega,\, \lambda) = \begin{pmatrix} a_+(\omega) + b_+(\omega)s(\lambda) & -b_+(\omega)n(\lambda) \\ b_-(\omega)n(\lambda) & a_-(\omega) - b_-(\omega)s(\lambda) \end{pmatrix}$$

with

$$s(\lambda) := \coth \pi(\lambda + i/2), \quad n(\lambda) := (\sinh \pi(\lambda + i/2))^{-1}$$

and

$$a_\pm(\omega) := a(\pm\omega), \quad b_\pm(\omega) := b(\pm\omega).$$

Theorem 25. *The singular integral operator $A = aI + bS_{\mathbb{R}} : L^2(\mathbb{R}) \to L^2(\mathbb{R})$ is Fredholm if and only if*

$$\inf_{\omega\in\mathbb{R}} |a(\omega) \pm b(\omega)| > 0 \tag{30}$$

and

$$\liminf_{\omega\to\infty} \inf_{\lambda\in\mathbb{R}} |q_A(\omega, \lambda)| > 0. \tag{31}$$

If conditions (30) *and* (31) *are satisfied, then the index of A is equal to*

$$-\lim_{R\to+\infty} \frac{1}{2\pi} \left(\left[\arg \frac{a(\omega)+b(\omega)}{a(\omega)-b(\omega)} \right]_{\omega=-R}^{R} + [\arg\det q_A(R, \lambda)]_{\lambda=-\infty}^{\infty} \right).$$

In particular, the matrices $q_{A_\pm}(\omega, \lambda)$ for the operators $A_\pm$ are equal to

$$\frac{1}{2}\begin{pmatrix} (a_+(\omega)+1) \pm (a_+(\omega)-1)s(\lambda) & \mp(a_+(\omega)-1)n(\lambda) \\ \pm(a_-(\omega)-1)n(\lambda) & (a_-(\omega)+1) \mp (a_-(\omega)-1)s(\lambda) \end{pmatrix},$$

and their determinants are

$$\frac{1}{2}\left[(a_+(\omega)+a_-(\omega)) \pm (a_+(\omega)-a_-(\omega)) \coth \pi(\lambda + \frac{i}{2}) \right] =: M_\pm(a, \omega, \lambda).$$

Hence, in our context, Theorem 25 specializes as follows.

Theorem 26. *Let $a \in C_b^{sl}(\mathbb{R})$. Then the Toeplitz operator $\hat{P}_\pm a \hat{P}_\pm : H_\pm^2 \to H_\pm^2$ is Fredholm if and only if*

$$\inf_{\omega\in\mathbb{R}} |a(\omega)| > 0 \tag{32}$$

and

$$\liminf_{\omega\to+\infty} \inf_{\lambda\in\mathbb{R}} |M_\pm(a, \omega, \lambda)| > 0. \tag{33}$$

If conditions (32) *and* (33) *are satisfied, then the index of $\hat{P}_\pm a \hat{P}_\pm$ is equal to*

$$-\lim_{R\to\infty} \frac{1}{2\pi} \left([\arg a(\omega)]_{\omega=-R}^{R} - [\arg M_\pm(a, R, \lambda)]_{\lambda=-\infty}^{+\infty} \right). \tag{34}$$

In case the limits $a_\pm := \lim_{\omega\to\pm\infty} a(\omega)$ exist, Theorem 26 reduces to a well-known result in the Toeplitz operator theory. We formulate it for the operator $\hat{P}_+ a \hat{P}_+$.

Let $\tilde{a}(\mathbb{R})$ denote the closed and naturally oriented curve which results from the essential range of $a : \mathbb{R} \to \mathbb{C}$ by filling in the line segment $[a_-, a_+]$ between the one-sided limits a_- and a_+. If $0 \notin \tilde{a}(\mathbb{R})$, let $\operatorname{wind}(\tilde{a}, 0)$ refer to the winding number of the closed curve $\tilde{a}(\mathbb{R})$ with respect to 0. The following result was discovered by many people, including Calderón, Spitzer, Widom, Devinatz, Gohberg, Krupnik, and Simonenko. For detailed references, see [2].

Theorem 27. *Let $a \in C_b(\mathbb{R})$ be a function for which the one-sided limits $a_\pm := \lim_{\omega\to\pm\infty} a(\omega)$ at infinity exist. Then $\hat{P}_+ a \hat{P}_+ : H_+^2 \to H_+^2$ is a Fredholm operator if and only if $0 \notin \tilde{a}(\mathbb{R})$. In this case,*

$$\operatorname{ind}(\hat{P}_+ a \hat{P}_+) = \operatorname{wind}(\tilde{a}, 0).$$

Corollary 28. *Let $a \in C_b^{sl}(\mathbb{R})$. Then $\hat{P}_+ a \hat{P}_+ : H_+^2 \to H_+^2$ is an invertible operator if and only if conditions* (32) *and* (33) *hold and if the index of $\hat{P}_+ a \hat{P}_+$ which is given by* (34) *is equal to zero.*

Proof. A well-known lemma by Coburn and Simonenko (see, for instance, Theorem 2.38 in [2]) states that every Fredholm Toeplitz operator has a trivial kernel or a trivial cokernel. Hence, the conditions (32) and (33) together with the vanishing of the index already imply the invertibility of $\hat{P}_+ a \hat{P}_+$. □

The operator $P_\pm Op(a) P_\pm$ on $L^2(\mathbb{R}_\pm)$ is unitarily equivalent to the Toeplitz operator $\hat{P}_\pm a \hat{P}_\pm$ acting on $H_\pm^2$. Hence, Theorem 26 has the following corollary.

Corollary 29. *Let $a \in C_b^{sl}(\mathbb{R})$. Then $P_\pm Op(a) P_\pm : L^2(\mathbb{R}_\pm) \to L^2(\mathbb{R}_\pm)$ is a Fredholm operator if and only if conditions* (32) *and* (33) *are satisfied. In this case, the index of $P_\pm Op(a) P_\pm$ is given by* (34). *Moreover, this operator is invertible if and only if it is a Fredholm operator with index zero.*

References

[1] A. Böttcher, Yu.I. Karlovich and V.S. Rabinovich, Mellin pseudodifferential operators with slowly varying symbols and singular integrals on Carleson curves with Muckenhoupt weights, *Manuscripta Math.* **95** (1998), 363–376.

[2] A. Böttcher and B. Silbermann, *Analysis of Toeplitz Operators*, Akademie-Verlag, Berlin, 1989 and Springer-Verlag, Berlin, 1990.

[3] A. Böttcher and B. Silbermann, *Introduction to Large Truncated Toeplitz Matrices*, Springer-Verlag, New York, 1999.

[4] P. Brémaud, *Mathematical Principles of Signal Processing: Fourier and Wavelet Analysis*, Springer-Verlag, New York, Berlin, Heidelberg, 2002.

[5] I. Gohberg and I. Feldman, *Convolution Equations and Projection Methods for Their Solution*, Nauka, Moskva, 1971 (Russian, Engl. transl.: Amer. Math. Soc. Transl. of Math. Monographs 41, Providence, R.I., 1974.

[6] L. Hörmander, *The Analysis of Linear Partial Differential Operators, Vol. I-IV*, Springer-Verlag, Berlin, Heidelberg, New York, Tokyo, 1983/1985.

[7] M. Lindner, V.S. Rabinovich and S. Roch, Finite sections of band operators with slowly oscillating coefficients, *Linear Algebra Appl.* **390** (2004), 19–26.

[8] A.V. Oppenheim and R.W. Schafer, *Digital Signal Processing*, Prentice-Hall, Inc., 1975.

[9] A. Papoulis, *Signal Analysis*, McGraw-Hill, New York, 1984.

[10] V.S. Rabinovich, Singular integral operators on a composed contour with oscillating tangent and pseudodifferential Mellin operators, *Soviet Math. Dokl.* **44** (1992), 791–796.

[11] V.S. Rabinovich, Algebras of singular integral operators on complicated contours with nodes that are of logarithmic whirl points, *Izv. Ross. Akad. Nauk, Ser. Mat.* **60** (1996), 6, 169–200 (Russian, Engl. transl.: Izv. Math. **60** (1996), 1261–1292).

[12] V.S. Rabinovich, Mellin pseudodifferential operators techniques in the theory of singular integral operators on some Carleson curves, in Operator Theory: Advances and Applications, Vol. 102, Birkhäuser, Basel, 1998, 201–218.

[13] V.S. Rabinovich and S. Roch, Wiener algebras of operators, and applications to pseudodifferential operators, *J. Anal. Appl.* **23** (2004), 437–482.

[14] V.S. Rabinovich, S. Roch and B. Silbermann, Fredholm theory and finite section method for band-dominated operators, *Integral Equations Operator Theory* **30** (1998), 452–495.

[15] V.S. Rabinovich, S. Roch and B. Silbermann, Band-dominated operators with operator-valued coefficients, their Fredholm properties and finite sections, *Integral Equations Operator Theory* **40** (2001), 342–381.

[16] V.S. Rabinovich, S. Roch and B. Silbermann, Algebras of approximation sequences: Finite sections of band-dominated operators, *Acta Applicande Mathematicae* **65** (2001), 315–332.

[17] V.S. Rabinovich, S. Roch and B. Silbermann, *Limit Operators and Their Applications in Operator Theory*, Birkhäuser, Basel, 2004.

[18] S. Roch, Finite sections of operators generated by convolutions, in *Seminar Analysis, Operator Equations and Numerical Analysis* 1987/88, *Berlin 1988*, 118–138.

[19] S. Roch, P.A. Santos and B. Silbermann, Finite section method in some algebras of multiplication and convolution operators and a flip, *J. Anal. Appl.* **16** (1997), 575–606.

[20] M.A. Shubin, *Pseudodifferential Operators and Spectral Theory*, Second Edition, Springer-Verlag, Berlin, Heidelberg, New York, Tokyo, 2001.

[21] M.E. Taylor, *Pseudodifferential Operators*, Princeton University Press, Princeton, New Jersey, 1981.

[22] M. Vetterli and J. Kovačevič, *Wavelets and Sub-Band Coding*, Prentice-Hall, Englewood Cliffs, N.J., 1995.

Vladimir S. Rabinovich
Instituto Politecnico Nacional
ESIME Zacatenco
Avenida IPN
Mexico, D. F. 07738, Mexico
e-mail: vladimir.rabinovich@gmail.com

Steffen Roch
Department of Mathematics
Technical University of Darmstadt
Schlossgartenstrasse 7
D-64289 Darmstadt, Germany
e-mail: roch@mathematik.tu-darmstadt.de

Operator Theory:
Advances and Applications, Vol. 172, 279–295

On the Product of Localization Operators

Elena Cordero and Karlheinz Gröchenig

Abstract. We provide examples of the product of two localization operators. As a special case, we study the composition of Gabor multipliers. The results highlight the instability of this product and underline the necessity of expressing it in terms of asymptotic expansions.

Mathematics Subject Classification (2000). Primary 35S05; Secondary 47G30.

Keywords. Localization operators, modulation spaces, time-frequency analysis, Weyl calculus, Wigner distribution, short-time Fourier transform.

We study the problem of calculating or estimating the product of two localization operators. The motivation comes either from signal analysis or pseudo-differential operator theory. On the one hand, in signal analysis the problem of finding a filter that has the same effect as two filters arranged in series amounts to the computation of the product of two localization operators, see [8, 9] and references therein.

On the other side, composition of pseudodifferential operators by means of symbolic calculus gives rise to asymptotic expansions, mainly employed in PDEs (see, e.g., [14, 16]). Outcomes are regularity properties of partial differential operators and the construction of an approximate inverse (so-called parametrix).

Since localization operators are a sub-class of pseudodifferential operators, looking for asymptotic expansions of the localization operator product appears to be natural as well. Applications can be found in the framework of PDEs and Quantum Mechanics [1, 6, 7, 15].

In this paper, we survey the known approaches to this problem and provide concrete examples of the composition of localization operators. Indeed, very few cases allow the product to be written as a localization operator as well, consequently the class of localization operators is not closed under composition. Thus the product is unstable with respect to composition. This instability highlights the importance of a symbolic calculus for localization operators [6].

K. Gröchenig was supported by the Marie-Curie Excellence Grant MEXT-CT 2004-517154.

We present localization operators using language and tools from time-frequency analysis. First, the definition of the short-time Fourier transform is required.

Given a function f on $\mathbb{R}^d$ and a point (x,ω) of the phase space $\mathbb{R}^{2d}$, the operators of translation and modulation are defined to be

$$T_x f(t) = f(t-x) \quad \text{and} \quad M_\omega f(t) = e^{2\pi i \omega t} f(t)\,. \tag{1}$$

We often combine translations and modulations into *time-frequency shift* (phase-space shifts in physical terminology). Set $z = (x,\omega) \in \mathbb{R}^{2d}$, then the general time-frequency shift is defined by

$$\pi(z) = M_\omega T_x. \tag{2}$$

Associated to time-frequency shifts is an important time-frequency representation, the *short-time Fourier transform* (STFT), also well known as coherent state transform, Gabor transform and windowed Fourier transform. The STFT of a function or distribution f with respect to a fixed non-zero *window* function g is given by

$$V_g f(x,\omega) = \int_{\mathbb{R}^d} f(t)\, \overline{g(t-x)}\, e^{-2\pi i \omega t}\, dt = \langle f, M_\omega T_x g\rangle = \langle f, \pi(z) g\rangle\,, \tag{3}$$

whenever the integral or the brackets $\langle\cdot,\cdot\rangle$ (expressing a sesquilinear form) are well defined. The short-time Fourier transform can be defined on many pairs of distribution spaces and test functions. For instance, $V_g f$ maps $L^2(\mathbb{R}^d) \times L^2(\mathbb{R}^d)$ into $L^2(\mathbb{R}^{2d})$ and $\mathcal{S}(\mathbb{R}^d) \times \mathcal{S}(\mathbb{R}^d)$ into $\mathcal{S}(\mathbb{R}^{2d})$. Furthermore, $V_g f$ can be extended to a map from $\mathcal{S}'(\mathbb{R}^d) \times \mathcal{S}'(\mathbb{R}^d)$ into $\mathcal{S}'(\mathbb{R}^{2d})$ [12, p. 41]. The short-time Fourier transform is the appropriate tool for defining localization operators, as we shall see presently.

Let a be a *symbol* on the time-frequency plane $\mathbb{R}^{2d}$ and choose two windows φ_1, φ_2 on $\mathbb{R}^d$, then the localization operator $A_a^{\varphi_1,\varphi_2}$ is defined as

$$A_a^{\varphi_1,\varphi_2} f(t) = \int_{\mathbb{R}^{2d}} a(x,\omega) V_{\varphi_1} f(x,\omega) M_\omega T_x \varphi_2(t)\, dx d\omega\,. \tag{4}$$

Taking the inner product with a test function g, the definition of $A_a^{\varphi_1,\varphi_2}$ can be written in a weak sense, namely,

$$\langle A_a^{\varphi_1,\varphi_2} f, g\rangle = \langle a V_{\varphi_1} f, V_{\varphi_2} g\rangle = \langle a, \overline{V_{\varphi_1} f}\, V_{\varphi_2} g\rangle. \tag{5}$$

If $a \in \mathcal{S}'(\mathbb{R}^{2d})$ and $\varphi_1, \varphi_2 \in \mathcal{S}(\mathbb{R}^d)$, then (5) is a well-defined continuous operator from $\mathcal{S}(\mathbb{R}^d)$ to $\mathcal{S}'(\mathbb{R}^d)$. If $\varphi_1(t) = \varphi_2(t) = 2^{d/4} e^{-\pi t^2}$, then $A_a = A_a^{\varphi_1,\varphi_2}$ is well known as (anti-)Wick operator and the mapping $a \to A_a^{\varphi_1,\varphi_2}$ is interpreted as a quantization rule [2, 8, 15, 16, 19].

Both the exact and the asymptotic product of localization operators rely upon the connection with the Weyl calculus. Namely, a localization operator $A_a^{\varphi_1,\varphi_2}$ can be represented as a Weyl transform. Here we need to refer to another time-frequency representation, the cross-Wigner distribution $W(g,f)$ of the functions g, f defined below (18).

The Weyl transform L_σ of $\sigma \in \mathcal{S}'(\mathbb{R}^{2d})$ is then defined by

$$\langle L_\sigma f, g\rangle = \langle \sigma, W(g,f)\rangle, \qquad f, g \in \mathcal{S}(\mathbb{R}^d). \tag{6}$$

Every continuous operator from $\mathcal{S}(\mathbb{R}^d)$ to $\mathcal{S}'(\mathbb{R}^d)$ can be represented as a Weyl transform, and a calculation in [3, 11, 16] reveals that

$$A_a^{\varphi_1,\varphi_2} = L_{a*W(\varphi_2,\varphi_1)}, \tag{7}$$

so, the (Weyl) symbol of $A_a^{\varphi_1,\varphi_2}$ is given by

$$\sigma = a * W(\varphi_2, \varphi_1)\,. \tag{8}$$

The composition of the Weyl operators L_σ and L_τ, with symbols σ and τ belonging to suitable function spaces, can be expressed in the Weyl form [11, Chap. 2.3]

$$L_\sigma L_\tau = L_{\sigma\natural\tau}, \tag{9}$$

where $\sigma\natural\tau$, the twisted multiplication of the symbols σ and τ, is given by

$$\sigma\natural\tau(\zeta) = 2^{2d} \iint_{\mathbb{R}^{2d}} \sigma(z)\tau(w)e^{4\pi i[\zeta-w,\zeta-z]}\, dzdw \tag{10}$$

and the brackets $[\cdot,\cdot]$ express the *symplectic form* on $\mathbb{R}^{2d}$

$$[(z_1,z_2),(\zeta_1,\zeta_2)] = z_1\zeta_2 - z_2\zeta_1, \qquad z=(z_1,z_2),\ \zeta=(\zeta_1,\zeta_2)\in\mathbb{R}^{2d}.$$

Thus, the composition of Weyl operators, whenever possible, defines a bilinear form (the twisted multiplication) on the corresponding symbols.

The product of localization operators will be studied along the following steps in Sections 3–5:

(i) Rewrite the two localization operators in terms of Weyl transforms (7);
(ii) Use the product formula for Weyl symbols (9), (10) to compute the Weyl symbol of their product;
(iii) Express, whenever possible, the resulting operator as a localization operator. In view of (8) this amounts to a deconvolution problem.

Notation We define $t^2 = t\cdot t$, for $t\in\mathbb{R}^d$, and $xy = x\cdot y$ is the scalar product on $\mathbb{R}^d$. The Schwartz class is denoted by $\mathcal{S}(\mathbb{R}^d)$, the space of tempered distributions by $\mathcal{S}'(\mathbb{R}^d)$. We use the brackets $\langle f,g\rangle$ to denote the extension to $\mathcal{S}'(\mathbb{R}^d)\times\mathcal{S}(\mathbb{R}^d)$ of the inner product $\langle f,g\rangle = \int f(t)\overline{g(t)}dt$ on $L^2(\mathbb{R}^d)$. The Fourier transform is normalized to be $\hat f(\omega) = \mathcal{F}f(\omega) = \int f(t)e^{-2\pi i t\omega}dt$. Given a continuous positive function (so-called weight function) m and $1\le p\le\infty$, we define as $L^p_m(\mathbb{R}^d)$ the space of all (Lebesgue) measurable functions on $\mathbb{R}^d$ such that the norm $\|f\|_{L^p_m} := \left(\int_{\mathbb{R}^d}|f(x)|^p m(x)^p\,dx\right)^{1/p}$ is finite (obvious changes for $p=\infty$). For $1\le p,q\le\infty$, the mixed-norm space $L^{p,q}(\mathbb{R}^{2d})$ is the Banach space of all (Lebesgue) measurable functions on $\mathbb{R}^{2d}$ satisfying

$$\|F\|_{L^{p,q}} := \left(\int_{\mathbb{R}^d}\left(\int_{\mathbb{R}^d}|F(x,\omega)|^p\,dx\right)^{1/p} d\omega\right)^{1/q},$$

again with obvious modifications whether $p=\infty$ or $q=\infty$.

Throughout the paper, we shall use the notation $A \lesssim B$ to indicate $A \leq cB$ for a suitable constant $c > 0$, whereas $A \asymp B$ if $A \leq cB$ and $B \leq kA$, for suitable $c, k > 0$.

1. Product formulae

In this section we briefly review two approaches to handle the product of localization operators.

1.1. Exact product

We reformulate the result of [8, 9] according to the notation of [11, 12].

We consider the window functions $\varphi_1(t) = \varphi_2(t) = \varphi(t) = 2^{d/4}e^{-\pi t^2}$, $t \in \mathbb{R}^d$. In this case, the Wigner distribution of the Gaussian φ is a Gaussian as well. Precisely, we have

$$W(\varphi, \varphi)(z) = 2^d e^{-2\pi z^2}, \quad z \in \mathbb{R}^{2d}. \tag{11}$$

If we compute the Weyl symbol σ of the operator $A_a^{\varphi,\varphi}$ we obtain $\sigma(\zeta) = 2^d(a * e^{-2\pi z^2})(\zeta)$, $z, \zeta \in \mathbb{R}^{2d}$. In order to express the product in the form of a localization operator, we rewrite the factors in a Weyl form, as detailed in the end of the previous section. Secondly, we come back to localization operators by means of relation (8). Given the Weyl symbols $\sigma, \tau \in \mathcal{S}(\mathbb{R}^{2d})$, we are interested in the Fourier transform of the twisted multiplication $\sigma \sharp \tau$, that is

$$\mathcal{F}(\sigma \sharp \tau)(\zeta) = \hat{\sigma} \natural \hat{\tau}(\zeta),$$

where the twisted convolution $\natural$ is given by

$$\hat{\sigma} \natural \hat{\tau}(\zeta) = \iint_{\mathbb{R}^{2d}} \hat{\sigma}(z)\hat{\tau}(\zeta - z)e^{\pi i[z,\zeta]}\, dz. \tag{12}$$

For any $f, g \in \mathcal{S}(\mathbb{R}^{2d})$, we define the $\natural^\flat$ product by

$$f \natural^\flat g(\zeta) = \iint_{\mathbb{R}^{2d}} f(z)g(z - \zeta)e^{\pi(z\zeta + i[z,\zeta])}\, e^{-\pi z^2}\, dz, \tag{13}$$

then the product of localization operators is given by the following formula.

Theorem 1.1. *Let $a, b \in \mathcal{S}(\mathbb{R}^{2d})$. If there exists a symbol $c \in \mathcal{S}'(\mathbb{R}^{2d})$ so that*

$$\hat{c} = 2^{-2d}\hat{a} \natural^\flat \hat{b}, \tag{14}$$

then we have

$$A_a^{\varphi,\varphi} A_b^{\varphi,\varphi} = A_c^{\varphi,\varphi}.$$

The proof is a straightforward consequence of relations (8) and (12). Indeed, one rewrites $A_a^{\varphi,\varphi} A_b^{\varphi,\varphi}$ in Weyl form and uses relation (12) for the product. The result is the Weyl operator L_μ, where the Fourier transform of μ is given by

$$\begin{aligned}\hat{\mu}(\zeta) &= [\mathcal{F}(a * W(\varphi,\varphi))\natural\mathcal{F}(b * W(\varphi,\varphi))](\zeta)\\ &= 2^{-2d}\iint_{\mathbb{R}^{2d}} \hat{a}(z)\hat{b}(z-\zeta)e^{-(\pi/2)z^2}e^{-(\pi/2)(\zeta-z)^2}e^{\pi i[z,\zeta]}\,dz\\ &= 2^{-2d}e^{-(\pi/2)\zeta^2}\iint_{\mathbb{R}^{2d}} \hat{a}(z)\hat{b}(z-\zeta)e^{\pi(z\zeta+i[z,\zeta])}\,e^{-\pi z^2}\,dz.\end{aligned}$$

Hence, we have

$$\hat{\mu}(\zeta) = \hat{c}(\zeta)(e^{-(\pi/2)\zeta^2}) = \mathcal{F}(c * W(\varphi,\varphi))(\zeta),$$

where $\hat{c}$ is given by relation (14).

The stability of the product when localization symbols live in the linear space spanned by the Gaussian functions is proven in [9, Thm. 2.1]. Here we reformulate the result using our terminology and we shall prove it using the Weyl connection and the twisted multiplication, instead of the $\natural^b$ one (Section 5).

Theorem 1.2. *Let $\varphi_i(t) = \varphi(t) = 2^{d/4}e^{-\pi t^2}$, $i = 1,\dots,4$, $t \in \mathbb{R}^d$. Consider the symbols*

$$a(z) = \sum_{k=1}^{m} C_k e^{-2\pi d_k z^2}, \quad b(z) = \sum_{j=1}^{l} C'_j e^{-2\pi d'_j z^2}, \quad z \in \mathbb{R}^{2d}, \tag{15}$$

where $C_1,\dots,C_m; C'_1,\dots,C'_l$ are complex numbers while $d_1,\dots,d_m; d'_1,\dots,d'_m$ are positive real numbers. Then $A_a^{\varphi,\varphi}A_b^{\varphi,\varphi} = A_c^{\varphi,\varphi}$, with

$$c(z) = \sum_{k=1}^{m}\sum_{j=1}^{l} C_k C'_j e^{-2\pi r_{k,j} z^2}, \quad z \in \mathbb{R}^{2d},$$

with $r_{k,j} = d_k + d'_j + 2d_k d'_j$.

1.2. Asymptotic product

Asymptotic expansions realize the composition of two localization operators as a sum of localization operators plus a controllable remainder. Versions of such a symbolic calculus are developed in [1, 7, 15, 6]. Most of them, as we observed in the introduction, were mainly motivated by PDEs and energy estimates, and therefore used smooth symbols that are defined by differentiability properties, such as Hörmander or Shubin classes. For applications in quantum mechanics and signal analysis, alternative notions of smoothness – "smoothness in phase-space" or quantitative measures of "time-frequency concentration" – have turned out to be useful. This point of view is pursued in [6], and we shall present the corresponding results.

The starting point is the following composition formula for two localization operators derived in [7]:

$$A_a^{\varphi_1,\varphi_2}A_b^{\varphi_3,\varphi_4} = \sum_{|\alpha|=0}^{N-1}\frac{(-1)^{|\alpha|}}{\alpha!}A_{a\partial^\alpha b}^{\Phi_\alpha,\varphi_2} + E_N. \tag{16}$$

The essence of this formula is that the product of two localization operators can be written as a sum of localization operators, with new windows Φ_α suitably defined, and a remainder term E_N, which is "small".

In the spirit of the classical symbolic calculus, this formula was derived in [7, Thm. 1.1] for *smooth* symbols belonging to some Shubin class $S^m(\mathbb{R}^{2d})$ and for windows in the Schwartz class $\mathcal{S}(\mathbb{R}^d)$.

In [6] the validity of (16) is established on the modulation spaces. The innovative features of this extension are highlighted below (we do not give here detailed statements and proofs).

(i) *Rough symbols.* While in (16) the symbol b must be N-times differentiable, the symbol a only needs to be locally bounded. The classical results in symbolic calculus require both symbols to be smooth.

(ii) *Growth conditions on symbols.* The symbolic calculus in (16) can handle symbols with ultra-rapid growth (as long as it is compensated by a decay of b or vice versa). For instance, a may grow subexponentially as $a(z) \sim e^{\alpha|z|^\beta}$ for $\alpha > 0$ and $0 < \beta < 1$. This goes far beyond the usual polynomial growth and decay conditions.

(iii) *General window classes.* A precise description of the admissible windows φ_j in (16) is provided. Usually only the Gaussian $e^{-\pi t^2}$ or Schwartz functions are considered as windows.

(iv) *Size of the remainder term.* Norm estimates for the size of the remainder term E_N are derived. They depend explicitly on the symbols a, b and the windows φ_j.

(v) *The Fredholm property of localization operators.* By choosing $N = 1$, $\varphi_1 = \varphi_2 = \varphi$ with $\|\varphi\| = 1$, $a(z) \neq 0$, and $b = 1/a$, the composition formula (16) yields the following important special case:

$$A_a^{\varphi,\varphi} A_{1/a}^{\varphi,\varphi} = A_1^{\varphi,\varphi} + R = \mathrm{I} + R\,. \tag{17}$$

If the symbol a belongs to $L^\infty_m(\mathbb{R}^{2d})$ and $|a| \asymp 1/m$, and the first partial derivative satisfies $(\partial_j a)m \in L^\infty$ and vanishes at infinity for $j = 1, \dots, 2d$, then R is shown to be a compact operator. Besides, $A_a^{\varphi,\varphi}$ is proven to be a Fredholm operator between suitable modulation spaces. This result works even for ultra-rapidly growing symbols such as $a(z) = e^{\alpha|z|^\beta}$ for $\alpha > 0$ and $0 < \beta < 1$.

2. Concepts of time-frequency analysis

We first present the tools and properties from time-frequency analysis that we shall use in the following sections.

2.1. The (cross-)Wigner distribution and the Weyl calculus

The *cross-Wigner distribution* $W(f,g)$ of $f, g \in L^2(\mathbb{R}^d)$ is defined to be

$$W(f,g)(x,\omega) = \int f(x+\frac{t}{2})\overline{g(x-\frac{t}{2})}e^{-2\pi i\omega t}\,dt. \tag{18}$$

The quadratic expression $Wf = W(f,f)$ is usually called the Wigner distribution of f.

The Wigner distribution $W(f,g)$ is defined on many pairs of Banach or topological vector spaces. For instance, they both map $L^2(\mathbb{R}^d) \times L^2(\mathbb{R}^d)$ into $L^2(\mathbb{R}^{2d})$ and $\mathcal{S}(\mathbb{R}^d) \times \mathcal{S}(\mathbb{R}^d)$ into $\mathcal{S}(\mathbb{R}^{2d})$. Furthermore, they can be extended to a map from $\mathcal{S}'(\mathbb{R}^d) \times \mathcal{S}'(\mathbb{R}^d)$ into $\mathcal{S}'(\mathbb{R}^{2d})$.

We first report a crucial property of the (cross-)Wigner distribution (for proofs, see [12, Ch. 4] and [11]).

Lemma 2.1. *For $\lambda = (u,\eta), \mu = (v,\gamma) \in \mathbb{R}^{2d}$ and $z = (x,\omega) \in \mathbb{R}^{2d}$ we have*

$$\begin{aligned} W(\pi(\lambda)f, \pi(\mu)g)(x,\omega) &= e^{\pi i(u+v)(\eta-\gamma)} e^{2\pi i x(\eta-\gamma)} e^{-2\pi i\omega(u-v)} \\ &\qquad \times W(f,g)(z - \frac{\lambda+\mu}{2}). \end{aligned} \tag{19}$$

In particular, if $\lambda = \mu$, relation (19) *becomes*

$$W(\pi(\lambda)f, \pi(\lambda)g)(z) = W(f,g)(z-\lambda). \tag{20}$$

Since the (cross-)Wigner distribution of Schwartz functions $f, g \in \mathcal{S}(\mathbb{R}^d)$ is a Schwartz function on $\mathbb{R}^{2d}$, its partial derivatives are well defined and may be expressed explicitly as linear combinations of (cross-)Wigner distributions of the functions $t^{\gamma_1}\partial^{\delta_1} f$ and $t^{\gamma_2}\partial^{\delta_2} g$.

Lemma 2.2. *Let $(x,\omega) \in \mathbb{R}^{2d}$, $f, g \in \mathcal{S}(\mathbb{R}^d)$, $\alpha = (\alpha_1, \alpha_2), \beta = (\beta_1, \beta_2) \in \mathbb{Z}_+^d \times \mathbb{Z}_+^d$, then*

$$\partial_x^{\alpha_1}\partial_\omega^{\alpha_2} W(f,g)(x,\omega) = (-2\pi i)^{|\alpha_2|} \sum_{\beta \le \alpha} \binom{\alpha}{\beta} (-1)^{|\beta_2|} W(t^{\alpha_2-\beta_2}\partial^{\alpha_1-\beta_1} f, t^{\beta_2}\partial^{\beta_1} g)(x,\omega). \tag{21}$$

Proof. Using the product formula for derivatives, the first partial derivative with respect to the *time* variable x_j, with $j = 1,\dots,d$, is given by

$$\partial_{x_j} W(f,g)(x,\omega) = W(\partial_{t_j} f, g)(x,\omega) + W(f, \partial_{t_j} g)(x,\omega),$$

and, by induction or by Leibniz' formula, we obtain

$$\partial_x^\alpha W(f,g)(x,\omega) = \sum_{\beta_1 \le \alpha_1} \binom{\alpha_1}{\beta_1} W(\partial^{\alpha_1-\beta_1} f, \partial^{\beta_1} g)(x,\omega). \tag{22}$$

The first partial derivative with respect to the *frequency* variable ω_j, with $j = 1,\dots,d$, is

$$\partial_{\omega_j} W(f,g)(x,\omega) = \int_{\mathbb{R}^d} (-2\pi i t_j) e^{-2\pi i t\omega} f\left(x + \frac{t}{2}\right) \overline{g\left(x - \frac{t}{2}\right)}\, dt. \tag{23}$$

The cross-Wigner distributions of the functions $t_j f, g$ and $f, t_j g$, respectively, are given by

$$W(t_j f, g)(x,\omega) = x_j W(f,g)(x,\omega) + \frac{1}{2}\int_{\mathbb{R}^d} t_j e^{-2\pi i t\omega} f\left(x+\frac{t}{2}\right)\overline{g\left(x-\frac{t}{2}\right)}\,dt\,,$$

$$W(f, t_j g)(x,\omega) = x_j W(f,g)(x,\omega) - \frac{1}{2}\int_{\mathbb{R}^d} t_j e^{-2\pi i t\omega} f\left(x+\frac{t}{2}\right)\overline{g\left(x-\frac{t}{2}\right)}\,dt.$$

Subtracting the latter from the former and using (23) we obtain

$$\partial_{\omega_j} W(f,g)(x,\omega) = (-2\pi i)[W(t_j f, g)(x,\omega) - W(f, t_j g)(x,\omega)].$$

As for the time case, the induction process yields

$$\partial_\omega^{\alpha_2} W(f,g)(x,\omega) = (-2\pi i)^{|\alpha_2|} \sum_{\beta_2 \le \alpha_2} \binom{\alpha_2}{\beta_2} (-1)^{\beta_2} W(t^{\alpha_2-\beta_2} f, t^{\beta_2} g)(x,\omega). \tag{24}$$

By combining (22) and (24) we obtain (21). □

2.2. Modulation spaces

The modulation space norm measures the joint time-frequency distribution of a $f \in \mathcal{S}'$. For their basic properties we refer, for instance, to [12, Ch. 11–13] and the original literature quoted there. The introduction of the general theory of modulation spaces is beyond our scope, therefore we shall limit ourselves to draw on their unweighted version.

Given a non-zero window $g \in \mathcal{S}(\mathbb{R}^d)$, and $1 \le p,q \le \infty$, the *modulation space* $M^{p,q}(\mathbb{R}^d)$ consists of all tempered distributions $f \in \mathcal{S}'(\mathbb{R}^d)$ such that $V_g f \in L^{p,q}(\mathbb{R}^{2d})$ (mixed-norm spaces). The norm on $M^{p,q}$ is

$$\|f\|_{M^{p,q}} = \|V_g f\|_{L^{p,q}} = \left(\int_{\mathbb{R}^d}\left(\int_{\mathbb{R}^d} |V_g f(x,\omega)|^p\,dx\right)^{q/p} d\omega\right)^{1/q}.$$

If $p = q$, we write M^p instead of $M^{p,p}$.

Modulation spaces $M^{p,q}$ are Banach spaces whose definition is independent of the choice of the window g. Moreover, if $g \in M^1 \setminus \{0\}$, then $\|V_g f\|_{L^{p,q}}$ is an equivalent norm for $M^{p,q}(\mathbb{R}^d)$ (see [12, Thm. 11.3.7]).

We recall that $M^2(\mathbb{R}^d) = L^2(\mathbb{R}^d)$ and, among the weighted modulation spaces one can encounter Sobolev spaces and Shubin-Sobolev spaces. Furthermore, the space of tempered distribution $\mathcal{S}'$ is recovered as unions of suitable weighted modulation spaces.

2.3. Convolution relations and Wigner estimate

In view of the relation between the multiplier a and the Weyl symbol (8), we shall use convolution relations between modulation spaces and some properties of the Wigner distribution.

Convolution relations for modulation spaces, studied in [5, 18], yield the following unweighted version.

Proposition 2.3. *Let* $1 \le p, q, r, s, t \le \infty$. *If*

$$\frac{1}{p} + \frac{1}{q} - 1 = \frac{1}{r}, \qquad \text{and} \qquad \frac{1}{t} + \frac{1}{t'} = 1,$$

then

$$M^{p,st}(\mathbb{R}^d) * M^{q,st'}(\mathbb{R}^d) \hookrightarrow M^{r,s}(\mathbb{R}^d) \tag{25}$$

with norm inequality $\|f * h\|_{M^{r,s}} \lesssim \|f\|_{M^{p,st}} \|h\|_{M^{q,st'}}$.

The modulation space norm of a cross-Wigner distribution may be controlled by the window norms, as expressed below (see [5]).

Proposition 2.4. *If* $\varphi_1, \varphi_2 \in M^1(\mathbb{R}^d)$ *we have* $W(\varphi_2, \varphi_1) \in M^1(\mathbb{R}^{2d})$, *with*

$$\|W(\varphi_2, \varphi_1)\|_{M^1} \lesssim \|\varphi_1\|_{M^1} \|\varphi_2\|_{M^1}. \tag{26}$$

The modulation space $M^{\infty,1}$ is the so-called Sjöstrand class and deserves quite an attention when studying Weyl operators. In particular, Sjöstrand in [17] proved that, if the Weyl symbol σ belongs to $M^{\infty,1}$, the corresponding Weyl operator L_σ is bounded on $L^2(\mathbb{R}^d)$. Besides, if $\sigma, \tau \in M^{\infty,1}$, and $L_\mu = L_\sigma L_\tau$, then $\mu \in M^{\infty,1}$; thus $M^{\infty,1}$ is a Banach algebra of pseudo-differential operators. In [13] the previous result is recaptured by using time-frequency analysis techniques. In particular, the Banach algebra property follows from the continuity of the twisted multiplication (see [13, Thm. 4.2]):

Theorem 2.5. *The modulation space* $M^{\infty,1}$ *is a Banach* $*$*-algebra with respect to twisted multiplication* $\sharp$ *and the involution* $\sigma \to \bar{\sigma}$. *In particular, the* $M^{\infty,1}$*-norm with respect to a window Wigner distribution* $W(\varphi, \varphi)$, *with* $\varphi \in \mathcal{S}$, *is given by*

$$\|\sigma \sharp \tau\|_{M^{\infty,1}} \le C_\varphi \|\sigma\|_{M^{\infty,1}} \|\tau\|_{M^{\infty,1}}, \quad \forall \sigma, \tau \in M^{\infty,1}. \tag{27}$$

The continuity of the twisted multiplication on $M^{\infty,1} \times M^{\infty,1}$ will be employed when composing Gabor multipliers (Section 4).

3. Examples of well-localized products

We shall provide few examples of products of localization operators. To this aim, we first need some results on Weyl operators.

Every rank one linear operator acting on $L^2(\mathbb{R}^d)$ can be interpreted as Weyl operator. The characterization of its Weyl symbol is given in [12, Lemma 14.6.3]:

Lemma 3.1. *Given* $h, k \in L^2(\mathbb{R}^d)$ *and set* $\sigma = W(h, k)$. *Then* L_σ *is the rank one operator*

$$L_\sigma f = \langle f, k \rangle h, \quad f \in L^2(\mathbb{R}^d). \tag{28}$$

If we express the product of two rank one operators as a Weyl transform, we obtain immediately a formula for the twisted multiplication of cross-Wigner distributions.

Lemma 3.2. *Given $\varphi_1, \varphi_2, \psi_1, \psi_2 \in L^2(\mathbb{R}^d)$ we have*

$$W(\varphi_1, \varphi_2) \sharp W(\psi_1, \psi_2) = \langle \psi_1, \varphi_2 \rangle W(\varphi_1, \psi_2) \tag{29}$$

Proof. Instead of the explicit formula (10), we use Lemma 3.1 and compute the product of the rank one operators $L_{W(\varphi_1,\varphi_2)} L_{W(\psi_1,\psi_2)}$. Let f be in $L^2(\mathbb{R}^d)$, then

$$\begin{aligned} L_{W(\varphi_1,\varphi_2)} L_{W(\psi_1,\psi_2)} f &= \langle L_{W(\psi_1,\psi_2)} f, \varphi_2 \rangle \varphi_1 \\ &= \langle \langle f, \psi_2 \rangle \psi_1, \varphi_2 \rangle \varphi_1 \\ &= \langle \psi_1, \varphi_2 \rangle \langle f, \psi_2 \rangle \varphi_1 \\ &= L_{\langle \psi_1, \varphi_2 \rangle W(\varphi_1,\psi_2)} f. \end{aligned}$$

Hence, the Weyl operator $L_{W(\varphi_1,\varphi_2)\sharp W(\psi_1,\psi_2)} = L_{W(\varphi_1,\varphi_2)} L_{W(\psi_1,\psi_2)}$ possesses the Weyl symbol claimed in (29). □

With these tools, we can now compute the product of two localization operators whose symbols have minimal support.

Proposition 3.3. *Let $1 \le j \le d$, and consider the distributions with support at the origin $a = \partial_{x_j} \delta$, $b = \delta$. For every $\varphi_k \in \mathcal{S}(\mathbb{R}^d)$, $k = 1, \dots, 4$, we have*

$$A^{\varphi_1,\varphi_2}_{\partial_{x_j}\delta} A^{\varphi_3,\varphi_4}_{\delta} = A^{\varphi_3,\partial_j\varphi_2}_{\langle \varphi_4,\varphi_1\rangle \delta} + A^{\varphi_3,\varphi_2}_{\langle \varphi_4, \partial_j \varphi_1\rangle \delta}. \tag{30}$$

Proof. Rewriting the composition of two localization operators as a Weyl transform (7), we reduce ourselves to compute the twisted multiplication of the corresponding Weyl symbols. Using Lemma 3.1 and (29), the desired result follows. Namely,

$$\begin{aligned} &[(\partial_{x_j}\delta) * W(\varphi_2,\varphi_1)] \sharp [\delta * W(\varphi_4,\varphi_3)] \\ &= [\delta * \partial_{x_j} W(\varphi_2,\varphi_1)] \sharp W(\varphi_4,\varphi_3) \\ &= [W(\partial_j\varphi_2,\varphi_1) + W(\varphi_2,\partial_j\varphi_1)] \sharp W(\varphi_4,\varphi_3) \\ &= W(\partial_j\varphi_2,\varphi_1) \sharp W(\varphi_4,\varphi_3) + W(\varphi_2,\partial_j\varphi_1) \sharp W(\varphi_4,\varphi_3) \\ &= \langle \varphi_4,\varphi_1\rangle W(\partial_j\varphi_2,\varphi_3) + \langle \varphi_4,\partial_j\varphi_1\rangle W(\varphi_2,\varphi_3), \\ &= \delta * [\langle \varphi_4,\varphi_1\rangle W(\partial_j\varphi_2,\varphi_3)] + \delta * [\langle \varphi_4,\partial_j\varphi_1\rangle W(\varphi_2,\varphi_3)]. \end{aligned}$$ □

The product above is no longer a single localization operator, in this sense the composition is unstable. However, the product in (30) is still a sum of two localization operators, and both have symbols localized at the origin If we choose the window with the optimal time-frequency localization, i.e., the normalized Gaussian $\varphi(t) = 2^{d/4} e^{-\pi t^2}$ and set $\varphi_1 = \varphi_4 = \varphi$, then formula (30) reduces to

$$A^{\varphi,\varphi_2}_{\partial_{x_j}\delta} A^{\varphi_3,\varphi}_{\delta} = A^{\varphi_3,\partial_j\varphi_2}_{\delta}, \quad \forall\, \varphi_2,\varphi_3 \in \mathcal{S} \tag{31}$$

because $\langle \varphi_4, \partial_j \varphi_1 \rangle = 0$ and $\langle \varphi, \varphi \rangle = 1$.

Next, choosing the partial derivatives $\partial_{\omega_j}\delta$, $j = 1, \dots, d$, as symbols, we obtain a similar formula.

Proposition 3.4. *Let $1 \leq j \leq d$, and consider the distributions with support at the origin $a = \partial_{\omega_j}\delta$, $b = \delta$. For every $\varphi_k \in \mathcal{S}(\mathbb{R}^d)$, $k = 1,\dots,4$, we have*

$$A^{\varphi_1,\varphi_2}_{\partial_{\omega_j}\delta} A^{\varphi_3,\varphi_4}_{\delta} = A^{\varphi_3,t_j\varphi_2}_{\langle\varphi_4,\varphi_1\rangle\delta} + A^{\varphi_3,\varphi_2}_{\langle\varphi_4,t_j\varphi_1\rangle\delta}. \tag{32}$$

The proof is similar to the one of Proposition 3.3. Again, if $\varphi_1 = \varphi_4 = \varphi = 2^{d/4}e^{-\pi t^2}$, then product is is stable, and

$$A^{\varphi,\varphi_2}_{\partial_{\omega_j}\delta} A^{\varphi_3,\varphi}_{\delta} = A^{\varphi_3,t_j\varphi_2}_{\delta}, \quad \forall\, \varphi_2,\varphi_3 \in \mathcal{S}.$$

If we increase the order of the derivative of the symbol, the product of two localization operators is never a single localization operator, and the stability of the product is definitely lost, as is shown the the following observation. Nevertheless, the supports of the symbols are all localized at the origin.

Proposition 3.5. *Let $\alpha = (\alpha_1,\alpha_2), \beta = (\beta_1,\beta_2) \in \mathbb{Z}^d_+ \times \mathbb{Z}^d_+$, $\varphi_i \in \mathcal{S}(\mathbb{R}^d)$, $i = 1,\dots,4$. Then the product of localization operators whose symbols are derivatives of the delta distribution is given by*

$$A^{\varphi_1,\varphi_2}_{(\partial_x^{\alpha_1}\partial_\omega^{\alpha_2}\delta)} A^{\varphi_3,\varphi_4}_{(\partial_x^{\beta_1}\partial_\omega^{\beta_2}\delta)} = (-2\pi i)^{|\alpha_2+\beta_2|} \sum_{\substack{\gamma\leq\alpha \\ \nu\leq\beta}} A^{(t^{\nu_2}\partial^{\nu_1}\varphi_3),(t^{\alpha_2-\nu_2}\partial^{\alpha_1-\nu_1}\varphi_2)}_{c_{\alpha,\beta,\gamma,\nu}\delta}, \tag{33}$$

where $c_{\alpha,\beta,\gamma,\nu} = \binom{\alpha}{\gamma}\binom{b}{\nu}(-1)^{|\gamma_2+\nu_2|}\langle t^{\beta_2-\nu_2}\partial^{\beta_1-\nu_1}\varphi_4, t^{\gamma_2}\partial^{\gamma_1}\varphi_1\rangle.$

Again, the proof relies on the same tools as for Proposition 3.3 and therefore we shall omit it. If the derivative order is greater than one, we highlight that neither the Gaussian choice for the windows could help us to have a single localization operator in the right-hand side. In fact, the brackets $\langle t^{\beta_2-\nu_2}\partial^{\beta_1-\nu_1}\varphi, t^{\gamma_2}\partial^{\gamma_1}\varphi\rangle$ do not vanish if $|\beta - \nu + \gamma| \in 2\mathbb{N}$.

4. Product of Gabor multipliers

In this section we shall study the product of Gabor multipliers, for a survey on the topic we refer to [10]. Consider a time-frequency lattice Λ in $\mathbb{R}^{2d}$, for instance, $\Lambda = \alpha\mathbb{Z}^d \times \beta\mathbb{Z}^d$, $\alpha,\beta \in \mathbb{R}$ and set $\mathbf{a} = (a_\lambda)_{\lambda\in\Lambda}$; moreover, choose two non-zero window functions $\varphi_1,\varphi_2 \in L^2(\mathbb{R}^d)$. Then, a *Gabor multiplier* $G_{\mathbf{a}}$ associated to the triple $(\varphi_1,\varphi_2,\Lambda)$ and with symbol $\mathbf{a}$ is given by

$$G_{\mathbf{a}} f = \sum_{\lambda\in\Lambda} a_\lambda \langle f, \pi(\lambda)\varphi_1\rangle \pi(\lambda)\varphi_2, \quad f \in L^2(\mathbb{R}^d). \tag{34}$$

The domain of a Gabor multiplier can even be a subspace of tempered distribution rather than simply a space of functions. The distribution/function space of f in (34) depends on the decay and smoothness properties of the pair (φ_1,φ_2) of dual windows and on the decay of the symbol $\mathbf{a}$.

Gabor multipliers are special cases of localization operators, as we shall see presently.

Lemma 4.1. *Let Λ be a lattice in $\mathbb{R}^{2d}$, $(a_\lambda)_{\lambda\in\Lambda} \in \ell^\infty(\Lambda)$, $\varphi_1, \varphi_2 \in M^1(\mathbb{R}^d)$. Consider the mapping $i : \ell^\infty(\Lambda) \to M^\infty(\mathbb{R}^{2d})$ defined by $i(\mathbf{a}) = \sum_{\lambda\in\Lambda} a_\lambda \delta_\lambda$, then we obtain*

$$G_{\mathbf{a}} = A^{\varphi_1,\varphi_2}_{i(\mathbf{a})} . \tag{35}$$

Proof. An easy computation shows that $i(\mathbf{a}) \in M^\infty(\mathbb{R}^{2d})$. More precisely, choose a window $g \in \mathcal{S}(\mathbb{R}^d)$ with compact support such that, for some constants $A, B > 0$ we have $A \le \sum_{\mu\in\Lambda^\perp} |\hat{g}(\xi - \mu)|^2 \le B < \infty$ for all $\xi \in \mathbb{R}^d$ (as usual $\Lambda^\perp$ denotes the dual lattice of Λ). A result in approximation theory then implies that $\sup_{x\in\mathbb{R}^d} |\sum_{\lambda\in\Lambda} a_\lambda g(x-\lambda)| \asymp \|a\|_\infty$. Using this observation, we obtain

$$\begin{aligned} \|i(\mathbf{a})\|_{M^\infty} &= \sup_{x,\xi\in\mathbb{R}^d} |\langle \sum_{\lambda\in\Lambda} a_\lambda \delta_\lambda, M_\xi T_x g\rangle| \\ &= \sup_{x,\xi\in\mathbb{R}^d} |\sum_{\lambda\in\Lambda} a_\lambda e^{-2\pi i\lambda\xi} g(x-\lambda)| \\ &\asymp \sup_{\lambda\in\Lambda} |a_\lambda| = \|\mathbf{a}\|_\infty . \end{aligned}$$

Consequently,

$$\| \sum_{\lambda\in\Lambda} a_\lambda \delta_\delta \|_{M^\infty} \asymp \|\mathbf{a}\|_\infty \qquad \forall \mathbf{a} \in \ell^\infty(\Lambda) . \tag{36}$$

By assumption $\varphi_1, \varphi_2 \in M^1(\mathbb{R}^d)$, we then appeal to [5, Theorem 3.2] to deduce the boundedness of the localization operator $A^{\varphi_1,\varphi_2}_{i(\mathbf{a})}$ on $L^2(\mathbb{R}^d)$. Using the weak definition (5), we observe that, for every $f, g \in \mathcal{S}(\mathbb{R}^d)$,

$$\begin{aligned} \langle A^{\varphi_1,\varphi_2}_{i(\mathbf{a})} f, g\rangle &= \langle i(\mathbf{a}), \overline{V_{\varphi_1} f}\, V_{\varphi_2} g\rangle \\ &= \sum_{\lambda\in\Lambda} a_\lambda \langle \delta_\lambda, \overline{V_{\varphi_1} f}\, V_{\varphi_2} g\rangle \\ &= \sum_{\lambda\in\Lambda} a_\lambda V_{\varphi_1} f(\lambda) \overline{V_{\varphi_2} g(\lambda)} \\ &= \sum_{\lambda\in\Lambda} a_\lambda \langle f, \pi(\lambda)\varphi_1\rangle \langle \pi(\lambda)\varphi_2, g\rangle . \end{aligned}$$

That is, the localization operator $A^{\varphi_1,\varphi_2}_{i(\mathbf{a})}$ coincides with the Gabor multiplier $\sum_{\lambda\in\Lambda} a_\lambda \langle \cdot, \pi(\lambda)\varphi_1\rangle \pi(\lambda)\varphi_2$.

Finally, we note that for $f, g \in L^2(\mathbb{R}^d)$ we have $(\langle f, \pi(\lambda)\varphi_1\rangle) \in \ell^2(\Lambda)$ with a norm estimate $\|\langle f, \pi(\lambda)\varphi_1\rangle_{\lambda\in\Lambda}\|_2 \lesssim \|f\|_2$, and likewise $(\langle \pi(\lambda)\varphi_2, g\rangle) \in \ell^2(\Lambda)$, thus $(\langle f, \pi(\lambda)\varphi_1\rangle\langle \pi(\lambda)\varphi_2, g\rangle) \in \ell^1(\Lambda)$ and the ℓ^1-norm is bounded by $\|f\|_2 \|g\|_2$. Consequently, the series defining $\langle A^{\varphi_1,\varphi_2}_{i(\mathbf{a})} f, g\rangle$ converges absolutely. This implies that the partial sums $S_N = \sum_{\lambda\in\Lambda:|\lambda|\le N} a_\lambda \langle \cdot, \pi(\lambda)\varphi_1\rangle \pi(\lambda)\varphi_2$ converge in the strong operator topology to $A^{\varphi_1,\varphi_2}_{i(\mathbf{a})}$. □

Since Gabor multipliers are a special case of localization operators, we may compute their product via Weyl calculus. We shall see that the result is a Gabor multiplier plus a remainder term. The remainder operator is no more expressible in terms of Gabor multipliers but it can be recaptured as series of suitable localization operators.

Proposition 4.2. *Let Λ be a time-frequency lattice in $\mathbb{R}^{2d}$. Consider $\varphi_i \in M^1$, $i = 1,\dots,4$, two sequences $\mathbf{a},\mathbf{b} \in \ell^\infty(\Lambda)$ and the* Gabor multipliers *$G_{\mathbf{a}}$ and $G_{\mathbf{b}}$ associated to the triple $(\varphi_1,\varphi_2,\Lambda)$ and $(\varphi_3,\varphi_4,\Lambda)$. Set $\mathbf{c} = (a_\lambda b_\lambda)_\lambda$, and let $G_{\mathbf{c}}$ be the* Gabor multiplier *associated to the triple $(\varphi_3,\varphi_2,\Lambda)$. Then,*

$$G_{\mathbf{a}}G_{\mathbf{b}} = \langle\varphi_4,\varphi_1\rangle G_{\mathbf{c}} + \sum_{\substack{\lambda,\mu\in\Lambda\\ \lambda\neq\mu}} A^{\pi(\mu)\varphi_3,\pi(\lambda)\varphi_2}_{\langle\pi(\mu)\varphi_4,\pi(\lambda)\varphi_1\rangle a(\lambda)b(\mu)\delta}. \tag{37}$$

Proof. In virtue of (35), we rewrite the Gabor multipliers $G_{\mathbf{a}}, G_{\mathbf{b}}$ as localization operators and then use the Weyl connection as in the preceding proofs. Precisely, if we name σ and τ the Weyl symbols of $G_{\mathbf{a}}$ and $G_{\mathbf{b}}$, respectively, their expression is given by

$$\sigma = \left(\sum_{\lambda\in\Lambda} a_\lambda\delta_\lambda\right) * W(\varphi_2,\varphi_1), \qquad \tau = \left(\sum_{\lambda\in\Lambda} b_\lambda\delta_\lambda\right) * W(\varphi_4,\varphi_3).$$

Next, we need to compute $\sigma \sharp \tau$. In this framework, Proposition 2.4 guarantees $W(\varphi_2,\varphi_1), W(\varphi_4,\varphi_3) \in M^1(\mathbb{R}^{2d})$, whereas Proposition 2.3 provides the continuity of the convolution acting from $M^\infty \times M^1$ into $M^{\infty,1}$. Getting the preceding results all together, we observe that the Weyl symbols σ and τ belong to $M^{\infty,1}(\mathbb{R}^{2d})$. Thus, the corresponding operators L_σ and L_τ are bounded operators on $L^2(\mathbb{R}^d)$ and the same for their product $L_{\sigma\sharp\tau}$, with $\sigma\sharp\tau \in M^{\infty,1}$. In the following, we shall calculate $\sigma\sharp\tau$ explicitly, using the boundedness properties of the convolution and twisted multiplication listed above. Relation (20) will be repeatedly used in the sequel,

$$\begin{aligned}
\sigma\sharp\tau &= \left[\left(\sum_{\lambda\in\Lambda} a_\lambda\delta_\lambda\right) * W(\varphi_2,\varphi_1)\right] \sharp \left[\left(\sum_{\mu\in\Lambda} b_\mu\delta_\mu\right) * W(\varphi_4,\varphi_3)\right]\\
&= \left[\sum_{\lambda\in\Lambda} (a_\lambda\delta_\lambda * W(\varphi_2,\varphi_1))\right] \sharp \left[\sum_{\mu\in\Lambda} (b_\mu\delta_\mu * W(\varphi_4,\varphi_3))\right]\\
&= \left[\sum_{\lambda\in\Lambda} a_\lambda W(\pi(\lambda)\varphi_2,\pi(\lambda)\varphi_1)\right] \sharp \left[\sum_{\lambda\in\Lambda} b_\mu W(\pi(\mu)\varphi_4,\pi(\mu)\varphi_3)\right]\\
&= \sum_{\lambda,\mu\in\Lambda} a_\lambda b_\mu [W(\pi(\lambda)\varphi_2,\pi(\lambda)\varphi_1) \sharp W(\pi(\mu)\varphi_4,\pi(\mu)\varphi_3)]\\
&= \langle\varphi_4,\varphi_1\rangle \sum_{\lambda\in\Lambda} a_\lambda b_\lambda W(\pi(\lambda)\varphi_2,\pi(\lambda)\varphi_3) +\\
&\qquad + \sum_{\lambda,\mu\in\Lambda,\ \lambda\neq\mu} a_\lambda b_\mu \langle\pi(\mu)\varphi_4,\pi(\lambda)\varphi_1\rangle W(\pi(\lambda)\varphi_2,\pi(\mu)\varphi_3).
\end{aligned}$$

$$= \langle \varphi_4, \varphi_1 \rangle \sum_{\lambda\in\Lambda} (a_\lambda b_\lambda)\, \delta_\lambda * W(\varphi_2, \varphi_3) +$$

$$+ \sum_{\lambda,\mu\in\Lambda,\ \lambda\neq\mu} a_\lambda b_\mu \,\langle \pi(\mu)\varphi_4, \pi(\lambda)\varphi_1 \rangle W(\pi(\lambda)\varphi_2, \pi(\mu)\varphi_3).$$

We notice that the convergence of the series in the line above is assumed to be in the $M^{\infty,1}$-norm. Passing to the corresponding Weyl operator we obtain the desired result. □

In the above argument we have assumed that all series converge in norm, in particular that $\sum_\lambda a_\lambda \delta_\lambda$ converges in the M^∞-norm. According to (36) this is the case if and only if $\mathbf{a} \in c_0(\Lambda)$, i.e., if $\lim_{|\lambda|\to\infty} |a_\lambda| = 0$. For arbitrary bounded sequences we now give an alternative argument.

Second proof of Proposition 4.2. Assume first that $\mathbf{a}$ and $\mathbf{b}$ have finite support, so that there are no issues about the convergence of the series involved. Then we calculate directly that

$$\begin{aligned}
G_a G_b f &= \sum_{\lambda\in\Lambda}\sum_{\mu\in\Lambda} a_\lambda b_\mu \langle f, \pi(\mu)\varphi_3\rangle \langle \pi(\mu)\varphi_4, \pi(\lambda)\varphi_1\rangle \pi(\lambda)\varphi_2 \\
&= \sum_{\mu=\lambda} + \sum_{\mu\neq\lambda} \cdots \\
&= \sum_{\lambda\in\Lambda} a_\lambda b_\lambda \langle \varphi_4, \varphi_1\rangle \langle f, \pi(\lambda)\varphi_3\rangle \pi(\lambda)\varphi_2 + \\
&+ \sum_{\lambda,\mu\in\Lambda,\mu\neq\lambda} a_\lambda b_\mu \langle f, \pi(\mu)\varphi_3\rangle \langle \pi(\mu)\varphi_4, \pi(\lambda)\varphi_1\rangle \pi(\lambda)\varphi_2 \\
&= \langle \varphi_4, \varphi_1\rangle G_c + \sum_{\lambda,\mu\in\Lambda,\lambda\neq\mu} A^{\pi(\mu)\varphi_3,\pi(\lambda)\varphi_2}_{\langle \pi(\mu)\varphi_4,\pi(\lambda)\varphi_1\rangle a(\lambda) b(\mu)\delta}.
\end{aligned}$$

If $\mathbf{a}, \mathbf{b} \in \ell^\infty(\lambda)$, we consider the partial sums

$$G_{\mathbf{a}_M} f = \sum_{\lambda\in\Lambda,|\lambda|\leq M} a_\lambda \langle f, \pi(\lambda)\varphi_1\rangle \pi(\lambda)\varphi_2$$

and

$$G_{\mathbf{b_N}} f = \sum_{\mu\in\Lambda,|\leq N} b_\mu \langle f, \pi(\mu)\varphi_1\rangle \pi(\mu)\varphi_2.$$

Since $G_{\mathbf{a}_M}$ and $G_{\mathbf{b}_N} f$ converge strongly to G_a and G_b, the above argument carries over to general bounded sequences $\mathbf{a}$ and $\mathbf{b}$. □

5. Gaussian functions as symbols

In this section we prove Theorem 1.2. Instead of using the product formula (12), as done in [9], we use the techniques of the preceding two sections. The result is a mere consequence of the Gaussian nice behavior under convolution and twisted products.

Lemma 5.1. *Let $a, b > 0$, then*

(i) *Gaussian convolution:*

$$[e^{-\pi a t^2} * e^{-\pi b t^2}](x) = (a+b)^{-d/2} e^{-\pi \frac{ab}{a+b} x^2}, \qquad t, x \in \mathbb{R}^d. \tag{38}$$

(ii) *Gaussian twisted multiplication:*

$$[e^{-2\pi a z^2} \sharp e^{-2\pi b z^2}](\zeta) = (1+ab)^{-d} e^{-2\pi \frac{a+b}{1+ab} \zeta^2}, \qquad z, \zeta \in \mathbb{R}^{2d}. \tag{39}$$

Proof. (i) The semigroup property of Gaussians is well known, see for instance [11], and follows by an easy calculation

(ii) We use the definition of $\sharp$ in (10) and make a direct computation using Gaussian integrals. All integrals converge absolutely and exchanging the order of integration is justified by Fubini's Theorem.

$$\begin{aligned}
[e^{-2\pi a z^2} \sharp e^{-2\pi b z^2}](\zeta) &= 2^{2d} \iint_{\mathbb{R}^{2d}} e^{-2\pi a z^2} e^{-2\pi b w^2} e^{4\pi i [\zeta - w, \zeta - z]}\, dz dw \\
&= 2^{2d} \int_{\mathbb{R}^{2d}} \left(\int_{\mathbb{R}^{2d}} e^{-2\pi a z^2} e^{-4\pi i [\zeta - w, z]}\, dz \right) e^{-2\pi b w^2} e^{-4\pi i [w, \zeta]} dw.
\end{aligned}$$

To begin with, we compute the interior integral with respect to the variable z. Namely, if $z = (z_1, z_2)$, $w = (w_1, w_2)$, $\zeta = (\zeta_1, \zeta_2) \in \mathbb{R}^{2d}$,

$$\begin{aligned}
\int_{\mathbb{R}^{2d}} e^{-2\pi a z^2} e^{-4\pi i [\zeta - w, z]}\, dz &= \left(\int_{\mathbb{R}^d} e^{-2\pi a z_1^2} e^{-2\pi i (-2(\zeta_2 - w_2) z_1)}\, dz_1 \right) \\
&\quad \cdot \left(\int_{\mathbb{R}^d} e^{-2\pi a z_2^2} e^{-2\pi i (2(\zeta_1 - w_1) z_2)}\, dz_2 \right) \\
&= (2a)^{-d} e^{-\frac{2\pi}{a} (\zeta - w)^2}.
\end{aligned}$$

Then, we have

$$\begin{aligned}
[e^{-2\pi a z^2} \sharp e^{-2\pi b z^2}](\zeta) &= \left(\frac{2}{a}\right)^d \int_{\mathbb{R}^{2d}} e^{-\frac{2\pi}{a}(\zeta - w)^2} e^{-2\pi b w^2} e^{-4\pi i [w, \zeta]}\, dw \\
&= \left(\frac{2}{a}\right)^d e^{-\frac{2\pi}{a}\zeta^2} \int_{\mathbb{R}^{2d}} e^{-2\pi \frac{1+ab}{a} w^2} e^{\frac{4\pi i}{a} \zeta w} e^{-4\pi i [w, \zeta]} dw \\
&= \left(\frac{2}{a}\right)^d e^{-\frac{2\pi}{a}\left(1 - \frac{1}{1+ab}\right)\zeta^2} \int_{\mathbb{R}^{2d}} e^{-2\pi \left[\left(\frac{1+ab}{a}\right)^{1/2} w - \frac{\zeta}{(a(1+ab))^{1/2}} \right]^2} \\
&\quad \cdot\ e^{-4\pi i [w, \zeta]}\, dw.
\end{aligned}$$

Splitting up the variables again, the twisted multiplication reduces to

$$\begin{aligned}
[e^{-2\pi a z^2} \sharp e^{-2\pi b z^2}](\zeta) &= \left(\frac{2}{a}\right)^d e^{-\frac{2\pi}{a}\left(1 - \frac{1}{1+ab}\right)\zeta^2} \int_{\mathbb{R}^d} T_{\frac{\zeta_1}{(a(1+ab))^{1/2}}}\, e^{-2\pi \frac{(1+ab)}{a} w_1^2} \\
&\quad \cdot\ e^{-2\pi i (2\zeta_2) w_1} dw_1 \int_{\mathbb{R}^d} T_{\frac{\zeta_2}{(a(1+ab))^{1/2}}}\, e^{-2\pi \frac{(1+ab)}{a} w_2^2} e^{-2\pi i (-2\zeta_1) w_2} dw_2
\end{aligned}$$

$$= \left(\frac{2}{a}\right)^d e^{-\frac{2\pi}{a}\left(1-\frac{1}{1+ab}\right)\zeta^2} \int_{\mathbb{R}^d} e^{-2\pi\frac{(1+ab)}{a}w_1^2} e^{-2\pi i(2\zeta_2)w_1}\, dw_1$$
$$\cdot \int_{\mathbb{R}^d} e^{-2\pi\frac{(1+ab)}{a}w_2^2} e^{-2\pi i(-2\zeta_1)w_2}\, dw_2$$
$$= (1+ab)^{-d} e^{-2\pi\frac{a+b}{1+ab}\zeta^2},$$

as desired. □

Corollary 5.2. *Let* $\varphi(t) = 2^{d/4}e^{-\pi t^2}$ *anc* $c > 0$. *Then,*

$$(e^{-2\pi c z^2} * W(\varphi,\varphi))(\zeta) = (c+1)^{-d}\, e^{-2\pi\frac{c}{c+1}\zeta^2}, \quad z,\zeta \in \mathbb{R}^{2d}. \tag{40}$$

Proof. It is a straightforward consequence of the relations (11) and (38). □

Now we have all we need to prove the main result of this section.

Proof of Theorem 1.2. We use relation (7) and the bilinearity of both the convolution and twisted multiplication. This yields

$$[a * W(\varphi,\varphi)] \,\sharp\, [b * W(\varphi,\varphi)](\zeta) = \sum_{k=1}^{m}\sum_{j=1}^{l} C_k C'_j p_{k,j}(\zeta),$$

with $p_{k,j}(\zeta) := [e^{-2\pi d_k z^2} * W(\varphi,\varphi)] \,\sharp\, [e^{-2\pi d'_j z^2} * W(\varphi,\varphi)](\zeta)$. The convolution products are achieved by (40) and the outcomes are

$$p_{k,j}(\zeta) = (d_k+1)^{-d}(d'_j+1)^{-d}(e^{-2\pi\frac{d_k}{d_k+1}z^2} \,\sharp\, e^{-2\pi\frac{d'_j}{d'_j+1}z^2})(\zeta).$$

Finally, we compute the twisted multiplication using (39) and we get

$$\begin{aligned} p_{k,j}(\zeta) &= (d_k+1)^{-d}(d'_j+1)^{-d}(e^{-2\pi\frac{d_k}{d_k+1}z^2} \,\sharp\, e^{-2\pi\frac{d'_j}{d'_j+1}z^2})(\zeta) \\ &= (d_k+d'_j+2d_kd'_j+1)^{-d}\, e^{-2\pi\frac{d_k+d'_j+2d_kd'_j}{d_k+d'_j+2d_kd'_j+1}\zeta^2} \\ &= [e^{-2\pi(d_k+d'_j+2d_kd'_j)z^2} * W(\varphi,\varphi)](\zeta), \end{aligned}$$

where in the last equality we used relation (40) backwards. □

Acknowledgment

This work was started while both authors were visiting the Erwin Schrödinger Institute in Vienna. Its hospitality and great working conditions are gratefully acknowledged.

References

[1] H. Ando and Y. Morimoto, Wick calculus and the Cauchy problem for some dispersive equations. *Osaka J. Math.* **39** (1) (2002), 123–147.

[2] F.A. Berezin, Wick and anti-Wick symbols of operators, *Mat. Sb. (N.S.)* **86** (128) (1971), 578–610.

[3] P. Boggiatto, E. Cordero, and K. Gröchenig, Generalized Anti-Wick operators with symbols in distributional Sobolev spaces. *Integral Equations Operator Theory* **48** (2004), 427–442.

[4] E. Cordero, Gelfand-Shilov windows for weighted modulation spaces, *Integral Transforms Spec. Funct.*, to appear.

[5] E. Cordero and K. Gröchenig, Time-frequency analysis of localization operators. *J. Funct. Anal.* **205** (1) (2003), 107–131.

[6] E. Cordero and K. Gröchenig. Symbolic calculus and Fredholm property for localization operators, *J. Fourier Anal. Appl.* **12** (4) (2006), 371–392.

[7] E. Cordero and L. Rodino, Wick calculus: a time-frequency approach, *Osaka J. Math.* **42** (1) (2005), 43–63.

[8] J. Du and M.W. Wong, A product formula for localization operators, *Bull. Korean Math. Soc.* **37** (1) (2000), 77–84.

[9] J. Du and M.W. Wong, Gaussian functions and Daubechies operators, *Integral Equations Operator Theory* **38** (1) (2000), 1–8.

[10] H.G. Feichtinger and K. Nowak, A first survey of Gabor multipliers, in *Advances in Gabor Analysis*, Editors: H.G. Feichtinger and T. Strohmer, Birkhäuser, Boston, 2002, 99–128.

[11] G.B. Folland, *Harmonic Analysis in Phase Space*, Princeton Univ. Press, Princeton, NJ, 1989.

[12] K. Gröchenig, *Foundations of Time-Frequency Analysis*, Birkhäuser, Boston, 2001.

[13] K. Gröchenig, Composition and spectral invariance of pseudodifferential operators on modulation spaces, *J. Anal. Math.*, *J. Anal. Math.* **98** (2006), 65–82.

[14] L. Hörmander, *The Analysis of Linear Partial Differential Operators III*, Springer-Verlag, Berlin, 1994.

[15] N. Lerner, The Wick calculus of pseudo-differential operators and energy estimates, in *New Trends in Microlocal Analysis*, Springer, 1997, 23–37.

[16] M.A. Shubin, *Pseudodifferential Operators and Spectral Theory*, Second Edition, Springer-Verlag, Berlin, 2001.

[17] J. Sjöstrand, An algebra of pseudodifferential operators, *Math. Res. Lett.* **1** (2) (1994), 185–192.

[18] J. Toft, Continuity properties for modulation spaces with applications to pseudo-differential calculus, I, *J. Funct. Anal.* **207** (2) (2004), 399–429.

[19] M.W. Wong, *Wavelet Transforms and Localization Operators*, Birkhäuser, Basel, 2002.

Elena Cordero
Department of Mathematics
University of Torino, Italy
e-mail: `elena.cordero@unito.it`

Karlheinz Gröchenig
Faculty of Mathematics
University of Vienna, Austria
e-mail: `karlheinz.groechenig@univie.ac.at`

Operator Theory:
Advances and Applications, Vol. 172, 297–312

Gelfand–Shilov Spaces, Pseudo-differential Operators and Localization Operators

Marco Cappiello, Todor Gramchev and Luigi Rodino

Abstract. We present new results concerning pseudo-differential operators in the function spaces $S^\mu_\nu(\mathbb{R}^n)$ of Gelfand and Shilov. In particular we discuss $S^\mu_\nu(\mathbb{R}^n)$-regularity of solutions to SG-elliptic pseudo-differential equations, allowing lower order semilinear perturbations. The results apply to SG-elliptic partial differential equations with polynomial coefficients. We also study the action of Weyl operators and localization operators on $S^\mu_\mu(\mathbb{R}^n)$.

Mathematics Subject Classification (2000). Primary 35S05; Secondary 47G30.

Keywords. Ultradifferentiable functions, Gelfand–Shilov spaces, pseudo-differential operators, time-frequency analysis.

1. Introduction

In [14] I.M. Gelfand and G.E. Shilov defined general classes of ultradifferentiable test functions, including as relevant case $S^\mu_\nu(\mathbb{R}^n), \mu \geq 0, \nu \geq 0$, space of all functions $f \in C^\infty(\mathbb{R}^n)$ satisfying for suitable positive constants C, ε the estimates in $\mathbb{R}^n$

$$|\partial^\alpha_x f(x)| \leq C^{|\alpha|+1}(\alpha!)^\mu e^{-\varepsilon|x|^{1/\nu}}. \tag{1.1}$$

Hence $S^\mu_\nu(\mathbb{R}^n)$ is a subspace of the Schwartz space $\mathcal{S}(\mathbb{R}^n)$. In fact, motivation in [14] was to introduce general ultradistributions spaces $(S^\mu_\nu)'(\mathbb{R}^n)$, containing the Schwartz dual $\mathcal{S}'(\mathbb{R}^n)$; in the corresponding functional frame, existence and uniqueness was proved for the (non-Kowalewskian) Cauchy problem for heat equation and other parabolic equations with constant coefficients. It is easy to verify that $f \in S^\mu_\nu(\mathbb{R}^n)$ if and only if there exists a positive constant C such that

$$\sup_{\alpha,\beta\in\mathbb{Z}^n_+}\sup_{x\in\mathbb{R}^n} C^{-|\alpha|-|\beta|}(\alpha!)^{-\mu}(\beta!)^{-\nu}|x^\beta\partial^\alpha_x f(x)| < +\infty. \tag{1.2}$$

We have $S^\mu_\nu(\mathbb{R}^n) \neq \{0\}$ iff $\mu+\nu > 1$ or $\mu+\nu = 1$ and $\mu\nu > 0$, $S^\mu_\nu(\mathbb{R}^n) \hookrightarrow S^{\mu'}_{\nu'}(\mathbb{R}^n)$ if $\mu \leq \mu'$, $\nu \leq \nu'$ and the Fourier transform acts as an isomorphism interchanging the indices μ and ν

$$\mathcal{F} : S^{\mu}_{\nu}(\mathbb{R}^n) \longrightarrow S^{\nu}_{\mu}(\mathbb{R}^n). \tag{1.3}$$

In particular, we obtain that $\mathcal{F}$ is an automorphism of $S^{\mu}_{\mu}(\mathbb{R}^n)$ for $\mu \geq 1/2$, and $S^{\mu}_{\mu}(\mathbb{R}^n)$, $1/2 \leq \mu < 1$, is a subset of the restrictions of the entire functions on $\mathbb{C}^n$ of exponential type $1/\mu$ (cf. [14], [20] and the references therein). There are equivalent ways of defining $S^{\mu}_{\nu}(\mathbb{R}^n)$ as inductive limit of scales of Banach spaces, cf. [9] and Section 3 below.

Gelfand–Shilov spaces, namely the above classes $S^{\mu}_{\nu}(\mathbb{R}^n)$, played an important role in the development of the general theory of partial differential equations, after [14]. Recently, they have been also considered as functional framework for pseudo-differential operators and localization operators. We want to present some new results in this line, improving in part preceding contributions of the authors and their collaborators ([3], [5], [6], [7], [8], [10], [15]). We first recall that in [5] Fourier integral operators in $\mathbb{R}^n$ were treated, of the form

$$(2\pi)^{-n} \int e^{i\varphi(x,\xi)} p(x,\xi)\hat{u}(\xi)d\xi,$$

where $\hat{u}$ denotes the Fourier transform of u. Suitable assumptions on the phase $\varphi(x,\xi)$ and the symbol $p(x,\xi)$ grant a calculus on S^{μ}_{ν} classes. In the pseudo-differential case $\varphi(x,\xi) = \langle x,\xi\rangle$, these results were applied in [6] to the study of the class of the SG-elliptic equations, having as a model

$$P = (1 + |x|^{2k})(-\Delta + 1) + L_1(x, D), \tag{1.4}$$

where L_1 is a first order operator with polynomial coefficients of degree $2k-1$, $k \geq 1$. The main concern in [6], deriving from problems in Quantum Mechanics (see [1]) was the exponential decay of the solutions, expressed by the index ν in the estimates (1.1). In the next Section 2 we present a (light) generalization of the calculus in [6]. To be precise, we fix attention on the linear problem

$$Pu = f, \tag{1.5}$$

where f belongs to the Gelfand–Shilov classes and P is an SG-elliptic differential operator with polynomial coefficients or, more generally, an SG elliptic pseudo-differential operator. Namely, we assume that the symbol $p(x,\xi)$ of P satisfies the following estimates

$$\left|D^{\alpha}_{\xi} D^{\beta}_{x} p(x,\xi)\right| \leq C^{|\alpha|+|\beta|+1}(\alpha!)^{\nu}(\beta!)^{\mu}\langle\xi\rangle^{m_1-|\alpha|}\langle x\rangle^{m_2-|\beta|} \tag{1.6}$$

for every $(x,\xi) \in \mathbb{R}^{2n}$ and $\alpha, \beta \in \mathbb{Z}^n_+$, and a global ellipticity condition of the form

$$|p(x,\xi)| \geq C_o\langle\xi\rangle^{m_1}\langle x\rangle^{m_2} \quad \textit{for} \quad |x| + |\xi| \geq R > 0. \tag{1.7}$$

For example (1.6) and (1.7) are satisfied by P in (1.4) with $m_1 = 2, m_2 = 2k$. The result of [6] was that for $f \in S^{\theta}_{\theta}(\mathbb{R}^n)$ in (1.5), with $\theta \geq \mu+\nu-1, \mu \geq 1, \nu \geq 1$, under the further assumption $\theta > 1$, all the solutions $u \in (S^{\theta}_{\theta})'(\mathbb{R}^n)$ are in fact in $S^{\theta}_{\theta}(\mathbb{R}^n)$. In Section 2 we shall allow $f \in S^{\mu'}_{\nu'}(\mathbb{R}^n)$, with $\mu' > 1, \nu' > 1, \min\{\mu',\nu'\} \geq \mu+\nu-1$, and for $u \in (S^{\mu'}_{\nu'})'(\mathbb{R}^n)$ conclude that $u \in S^{\mu'}_{\nu'}(\mathbb{R}^n)$, see Corollary 2.8 below. This result is more general than [6], but it does not eliminate the drawback of the assumption $\theta > 1$. So for example, for SG-elliptic operators satisfying (1.6) with $\mu = \nu = 1$, in particular for P in (1.4), the solutions of the homogeneous equation $Pu = 0$ are proved by Corollary 2.8 to belong to $S_\nu^\mu(\mathbb{R}^n)$ for any $\mu > 1, \nu > 1$, hence in particular

$$|u(x)| \leq Ce^{-\varepsilon|x|^{1/\nu}} \tag{1.8}$$

with $1/\nu < 1$. However the expected optimal result is $u \in S_1^1(\mathbb{R}^n)$, that implies $\nu = 1$ in (1.8), according to the known results for the one-dimensional case. The fact is that in [6] and Section 2 below the technique does not give satisfactory results for analytic symbols since it involves Gevrey cut-off functions of order μ and ν, which must be assumed both strictly greater than 1.

On the other hand, in a recent paper [7], the authors proved some kinds of S_1^1-estimates for SG-elliptic differential operators with polynomial coefficients. Here we want to improve and generalize these results to pseudo-differential operators, see the next Theorem 3.1. To do this, we use an iterative scheme developed in [3], [15] for studying the S_1^1 regularity of travelling waves, cf. [4]. This method requires some technical assumptions we shall give in Section 3. In Section 4, we show that for a self-adjoint differential operator with polynomial coefficients of the form

$$P(x, D) = \sum_{\substack{|\alpha| \leq m_1 \\ |\beta| \leq m_2}} c_{\alpha\beta} x^\beta D_x^\alpha \tag{1.9}$$

satisfying (1.7) these conditions are fulfilled. On the other hand, the iterative scheme mentioned above allows to prove analogous results also for semilinear perturbations of (1.5) of the form

$$Pu = f + F[u] \tag{1.10}$$

where the nonlinear term F is of the form

$$F[u] = \sum F_{\gamma jk} x^j u^k (D^\gamma u)^{h_\gamma}, \tag{1.11}$$

with $j, k \in \mathbb{Z}_+, 0 \leq k \leq M$ for some $M, |j| \leq m_2 - 1, 0 < |\gamma| \leq m_o$ for some $m_o < m_1, h_\gamma < m_1/|\gamma|$. Namely, if P in (1.10) is of the form (1.9), under the only assumption of SG-ellipticity we get solutions in $S_\nu^\mu(\mathbb{R}^n)$ if $f \in S_\nu^\mu(\mathbb{R}^n)$, for any $\mu \geq 1, \nu \geq 1$, hence $u \in S_1^1(\mathbb{R}^n)$ if $f = 0$, see Theorem 4.5 below. This applies in particular to P in (1.4). The method has been tested already in [7] on some examples of SG-elliptic operators and in [8] on globally elliptic operators of Shubin type (cf. [24]) for semilinear autonomous equations.

In Section 5 finally, in a different direction, we consider localization operators and Weyl pseudo-differential operators on Gelfand–Shilov spaces. We do not state new results here, but just observe that localization operators with windows in $S_\mu^\mu(\mathbb{R}^n)$, as well as generic Weyl operators, define continuous maps from $S_\mu^\mu(\mathbb{R}^n)$ to $(S_\mu^\mu)'(\mathbb{R}^n)$, provided the symbol is in $(S_\mu^\mu)'(\mathbb{R}^{2n})$, see Propositions 5.1 and 5.2 below. This is an easy consequence of the results of [17], [25] concerning the $S_\nu^\mu(\mathbb{R}^n)$-regularity of the short-time Fourier transform and Wigner transform.

We obtain in this way a very general setting for localization operators, which was already adopted in [10] in the case $\mu = 1$, to characterize ultradistribution symbols giving L^2-bounded operators.

2. Pseudo-differential operators and Gelfand–Shilov spaces

In this section, we briefly describe the action of SG pseudo-differential operators with symbols defined by (1.6) on the spaces $S^\mu_\nu(\mathbb{R}^n), \mu \geq 1, \nu \geq 1$ and give some results about composition of operators and parametrices. The statements in the sequel provide a generalization of the calculus developed in [6] for $\mu = \nu > 1$ following essentially the same approach. We shall omit the proofs which are just a repetition of [6], and only detail the new part.
Let $m = (m_1, m_2) \in \mathbb{R}^2$ and let μ, ν be real numbers such that $\mu \geq 1, \nu \geq 1$.

Definition 2.1. We shall denote by $\Gamma^m_{\nu,\mu} = \Gamma^m_{\nu,\mu}(\mathbb{R}^{2n})$ the space of all functions $p(x,\xi) \in C^\infty(\mathbb{R}^{2n})$ satisfying the estimates (1.6). We shall use a separate notation for the class of analytic symbols $\Gamma^m_{1,1}$ denoting it by Γ^m_a.

Given $p \in \Gamma^m_{\nu,\mu}$, we can consider the pseudo-differential operator defined as standard by

$$Pu(x) = p(x,D)u(x) = (2\pi)^{-n} \int_{\mathbb{R}^n} e^{i\langle x,\xi\rangle} p(x,\xi)\hat{u}(\xi)d\xi, \quad u \in \mathcal{S}(\mathbb{R}^n), \tag{2.1}$$

where $\hat{u}$ denotes the Fourier transform of u. We shall denote by $OP\Gamma^m_{\nu,\mu}$ the space of all operators of the form (2.1), with symbol in $\Gamma^m_{\nu,\mu}$.

Theorem 2.2. *Given $p \in \Gamma^m_{\nu,\mu}$, the operator P defined by (2.1) is linear and continuous from $S^{\mu'}_{\nu'}$ into itself for any μ', ν' with $\mu' \geq \mu, \nu' \geq \nu$. Furthermore, P can be extended to a linear and continuous map from $(S^{\mu'}_{\nu'})'$ into itself.*

Proof. For any $\alpha, \beta \in \mathbb{Z}^n_+$ and for any positive integer N, we can write:

$$x^\alpha D^\beta_x Pu(x) = (2\pi)^{-n} x^\alpha \sum_{\beta_1+\beta_2=\beta} \binom{\beta}{\beta_1} \int_{\mathbb{R}^n} e^{i\langle x,\xi\rangle} \xi^{\beta_1} D^{\beta_2}_x p(x,\xi)\hat{u}(\xi)d\xi$$

$$= (2\pi)^{-n} x^\alpha \langle x\rangle^{-2N} \sum_{\beta_1+\beta_2=\beta} \binom{\beta}{\beta_1} \int_{\mathbb{R}^n} e^{i\langle x,\xi\rangle} (1-\Delta_\xi)^N [\xi^{\beta_1} D^{\beta_2}_x p(x,\xi)\hat{u}(\xi)]d\xi.$$

Choosing $N = \left[\frac{|\alpha|+m_2}{2}\right] + 1$, by (1.3) and by standard factorial inequalities, we obtain

$$\langle x\rangle^{|\alpha|-2N} \left|(1-\Delta_\xi)^N [\xi^{\beta_1} D^{\beta_2}_x p(x,\xi)\hat{u}(\xi)]\right| \leq C^{|\alpha|+|\beta|+1} (\alpha!)^\nu (\beta_1!)^{\mu'} (\beta_2!)^\mu e^{-a\langle\xi\rangle^{1/\mu'}}$$

for some positive constants C, a. Then, by the conditions $\mu' \geq \mu, \nu' \geq \nu$, it follows that P is continuous from $S^{\mu'}_{\nu'}$ into itself. In order to extend P on the dual space

$(S^{\mu'}_{\nu'})'$, we observe that, for $u, v \in S^{\mu'}_{\nu'}$,

$$\int_{\mathbb{R}^n} Pu(x)v(x)dx = \int_{\mathbb{R}^n} \hat{u}(\xi)p_v(\xi)d\xi$$

where

$$p_v(\xi) = (2\pi)^{-n} \int_{\mathbb{R}^n} e^{i\langle x,\xi\rangle} p(x,\xi)v(x)dx$$

Furthermore, by the same argument of the first part of the proof, it follows that the map $v \to p_v$ is linear and continuous from $S^{\mu'}_{\nu'}$ to $S^{\nu'}_{\mu'}$. Then, we can define, for $u \in (S^{\mu'}_{\nu'})'$,

$$Pu(v) = \hat{u}(p_v), \quad v \in S^{\mu'}_{\nu'}.$$

This is a linear continuous map from $(S^{\mu'}_{\nu'})'$ into itself and it extends P.
For $t \geq 0$, denote by Q_t the set

$$Q_t = \{(x,\xi) \in \mathbb{R}^{2n} : \langle\xi\rangle < t \quad and \quad \langle x\rangle < t\}$$

and

$$Q_t^e = \mathbb{R}^{2n} \setminus Q_t.$$

Definition 2.3. We denote by $FS^m_{\nu,\mu}$ the space of all formal sums $\sum_{j\geq 0} p_j$ such that $p_j \in C^\infty(\mathbb{R}^{2n})$ for all $j \geq 0$ and there exist positive constants B, C such that $\forall j \geq 0$:

$$\left|D^\alpha_\xi D^\beta_x p_j(x,\xi)\right| \leq C^{|\alpha|+|\beta|+2j+1}(\alpha!)^\nu(\beta!)^\mu(j!)^{\mu+\nu-1}\cdot \cdot\langle\xi\rangle^{m_1-|\alpha|-j}\langle x\rangle^{m_2-|\beta|-j} \tag{2.2}$$

for all $\alpha, \beta \in \mathbb{Z}^n_+$ and for all $(x,\xi) \in Q^e_{Bj^{\mu+\nu-1}}$.

Definition 2.4. We say that two sums $\sum_{j\geq 0} p_j, \sum_{j\geq 0} q_j \in FS^m_{\nu,\mu}$ are equivalent if there exist positive constants B, C such that for every $N = 1, 2, \ldots$

$$\left|D^\alpha_\xi D^\beta_x \sum_{j<N}(p_j - q_j)\right| \leq C^{|\alpha|+|\beta|+2N+1}(\alpha!)^\nu(\beta!)^\mu(N!)^{\mu+\nu-1}\cdot \cdot\langle\xi\rangle^{m_1-|\alpha|-N}\langle x\rangle^{m_2-|\beta|-N}$$

for all $\alpha, \beta \in \mathbb{Z}^n_+$ and for all $(x,\xi) \in Q^e_{BN^{\mu+\nu-1}}$. In this case, we write $\sum_{j\geq 0} p_j \sim \sum_{j\geq 0} q_j$.

Proposition 2.5. *Let $p \in \Gamma^0_{\nu,\mu}$ such that $p \sim 0$ and let μ', ν' be real numbers with $\min\{\mu',\nu'\} \geq \mu+\nu-1$. Then the operator P is $S^{\mu'}_{\nu'}$-regularizing, i.e., it extends to a continuous linear map from $(S^{\mu'}_{\nu'})'$ into $S^{\mu'}_{\nu'}$. In particular, if $p \in \Gamma^0_a$ and $p \sim 0$ in FS^0_a, then P is S^1_1-regularizing.*

Proof. Denote $\theta = \min\{\mu', \nu'\}$. Since $S_\theta^\theta(\mathbb{R}^n) \subset S_{\nu'}^{\mu'}(\mathbb{R}^n)$, by duality it is sufficient to show that P is S_θ^θ-regularizing to prove the assertion. Arguing as in [6], it is easy to prove that if $p \sim 0$, then $p \in S_\theta^\theta(\mathbb{R}^{2n})$. Then, the assertion follows invoking Proposition 2.11 in [5].

As standard we can identify $\Gamma^m_{\nu,\mu}$ with a subspace of $FS^m_{\nu,\mu}$ by associating to each symbol $p \in \Gamma^m_{\nu,\mu}$ a sum $\sum_{j\geq 0} p_j \in FS^m_{\nu,\mu}$, with $p_0 = p$ and $p_j = 0 \quad \forall j \geq 1$. In the opposite direction, by Theorem 2.14 in [5], starting from a sum $\sum_{j\geq 0} p_j \in FS^m_{\nu,\mu}$, we can find a symbol $p \in \Gamma^m_{\nu,\mu}$ such that $p \sim \sum_{j\geq 0} p_j$ in $FS^m_{\nu,\mu}$. However, the arguments used to prove this statement involve Gevrey cut-off functions of order μ and ν. This leads to assume the additional restriction $\mu > 1, \nu > 1$. From now on in this section, we shall assume $\mu > 1, \nu > 1$. The next results hold also for analytic symbols of Γ^m_a only considering it as a subclass of $\Gamma^m_{\nu,\mu}$ for any choice of $\mu > 1, \nu > 1$.

We conclude this section recalling and stating more precisely the composition theorem for SG operators in Gelfand–Shilov spaces. We address to [6] for the proof. Construction of parametrices and regularity theorem follow then as standard.

Theorem 2.6. *Let $p \in \Gamma^m_{\nu,\mu}, q \in \Gamma^{m'}_{\nu,\mu}$ for some $m, m' \in \mathbb{R}^2$. Then, there exists a symbol $s \in \Gamma^{m+m'}_{\nu,\mu}$ such that $p(x,D)q(x,D) = s(x,D) + R$ for some $S_{\nu'}^{\mu'}$-regularizing operator R, where* $\min\{\mu', \nu'\} \geq \mu + \nu - 1$.

Theorem 2.7. *Let $p \in \Gamma^m_{\nu,\mu}$ be SG-elliptic, i.e., let (1.7) be satisfied. Then there exists $q \in \Gamma^{-m}_{\nu,\mu}$ such that $p(x,D)q(x,D) = I + R_1, q(x,D)p(x,D) = I + R_2$, for some $S_{\nu'}^{\mu'}$-regularizing operators R_1, R_2, with* $\min\{\mu', \nu'\} \geq \mu + \nu - 1$.

Corollary 2.8. *Let $p \in \Gamma^m_{\nu,\mu}, \mu \geq 1, \nu \geq 1$ be SG-elliptic. Let $\mu' > 1, \nu' > 1$ satisfy* $\min\{\mu', \nu'\} \geq \mu + \nu - 1$. *Then, every solution $u \in (S_{\nu'}^{\mu'})'(\mathbb{R}^n)$ of the equation $Pu = f \in S_{\nu'}^{\mu'}(\mathbb{R}^n)$ belongs to $S_{\nu'}^{\mu'}(\mathbb{R}^n)$.*

3. Semilinear equations

In this section, we prove the S_ν^μ estimates for the solutions of the equation (1.10). To do this, we first need to introduce some scales of Sobolev norms defining $S_\nu^\mu(\mathbb{R}^n)$. Consider $s \in \mathbb{R}$. Let us fix $\mu \geq 1, \nu \geq 1$ and define, for $\delta, \varepsilon > 0$:

$$|||u|||_{\varepsilon,\delta} = \sum_{\alpha,\beta\in\mathbb{Z}^n_+} \| [u^{(\alpha,\beta)}]^{\mu,\nu}_{\varepsilon,\delta} \|_s, \tag{3.1}$$

where

$$[u^{(\alpha,\beta)}]^{\mu,\nu}_{\varepsilon,\delta}(x) = \frac{\varepsilon^{|\alpha|}\delta^{|\beta|}}{(\alpha!)^\mu(\beta!)^\nu} x^\beta \partial_x^\alpha u(x).$$

Equivalence with the definition from (1.2) is easily proved by standard factorial inequalities and Sobolev embedding estimates. Namely, for any fixed $s \in \mathbb{R}$, $u \in S^\mu_\nu(\mathbb{R}^n)$ iff $|||u|||_{\varepsilon,\delta} < \infty$ for some positive δ, ε. We also define the partial sums

$$S_N^{\varepsilon,\delta}[u] = \sum_{\substack{\alpha,\beta\in\mathbb{Z}^n_+ \\ |\alpha|+|\beta|\le N}} \|[u^{(\alpha,\beta)}]^{\mu,\nu}_{\varepsilon,\delta}\|_s, \quad N \in \mathbb{Z}_+ \tag{3.2}$$

and

$$E_N^\varepsilon[u] = \sum_{\substack{\alpha\in\mathbb{Z}^n_+ \\ |\alpha|\le N}} \frac{\varepsilon^{|\alpha|}}{(\alpha!)^\mu} \|\partial^\alpha_x u\|_s \, .$$

Let $P \in OP\Gamma^m_{\nu,\mu}$ be SG-elliptic, i.e., let (1.7) be satisfied. In the subsequent proof we need the existence of $\lambda \in \mathbb{C} \setminus \operatorname{spec}(P)$. This hypothesis is always satisfied if, for example, P is selfadjoint, what we shall assume in the following for simplicity. Furthermore, in view of the results in [11], [12], [19], [23] on SG pseudo-differential calculus, if $m_1 > 0, m_2 > 0$ the operator $P - \lambda$ is also SG-elliptic and

$$(P-\lambda)^{-1} \circ x^q \partial^p_x : H^s(\mathbb{R}^n) \mapsto H^s(\mathbb{R}^n) \tag{3.3}$$

is continuous for all $p, q \in \mathbb{Z}^n_+, |p| \le m_1, |q| \le m_2$ and for every $s \in \mathbb{R}$. We will assume the following condition on the operator $(P-\lambda)^{-1}$:

$$\left\|(P-\lambda)^{-1}[P, x^\beta\partial^\alpha_x]v\right\|_s \le (\alpha!)^\mu(\beta!)^\nu \sum_{\substack{\gamma\le\alpha,\sigma\le\beta \\ |\gamma+\sigma|<|\alpha+\beta|}} \frac{C_1^{|\alpha-\gamma|+|\beta-\sigma|}}{(\gamma!)^\mu(\sigma!)^\nu} \|x^\sigma\partial^\gamma_x v\|_s \tag{3.4}$$

for every $s \in \mathbb{R}, \alpha, \beta \in \mathbb{Z}^n_+$ and for a positive constant C_1 independent of α, β. We shall prove in Section 4 that SG-ellipticity actually implies the technical condition (3.4) for partial differential operators P with polynomial coefficients (1.9). We are not able to prove the same for general operators P, but address to [3], [15] for examples of pseudo-differential operators satisfying (3.4).

Theorem 3.1. *Let $P \in OP\Gamma^m_{\nu,\mu}, m = (m_1, m_2), m_1 > 0, m_2 > 0, \mu \ge 1, \nu \ge 1$ be SG-elliptic, selfadjoint and satisfy the condition (3.4). Let $f \in S^\mu_\nu(\mathbb{R}^n)$. Assume F to be of the form (1.11) and let $u \in H^{s+m_o}(\mathbb{R}^n)$ for some $s > n/2$ be a solution of (1.10). Then $u \in S^\mu_\nu(\mathbb{R}^n)$.*

To prove Theorem 3.1 we need a preliminary result.

Proposition 3.2. *Let $P \in OP\Gamma^m_{\nu,\mu}$ be as in Theorem 3.1. Fix $\lambda \notin \operatorname{spec}(P)$. Let $s > n/2$. Assume moreover that $F = x^j u^k(\partial^\gamma_x u)^r$ for some $k, r \in \mathbb{Z}_+$ and $j, \gamma \in \mathbb{Z}^n_+$ with $r|\gamma| \le m_1 - 1, 0 \le j \le m_2 - 1$. Then, there exists a positive constant $C > 0$ such that for every $N \in \mathbb{Z}_+, \varepsilon, \delta \in (0,1)$*

$$\sum_{\substack{\alpha,\beta\in\mathbb{Z}^n_+ \\ |\alpha|+|\beta|\le N}} \frac{\varepsilon^{|\alpha|}\delta^{|\beta|}}{(\alpha!)^\mu(\beta!)^\nu} \left\|(P-\lambda)^{-1} \circ x^\beta\partial^\alpha_x F\right\|_s \le C\|u\|^{k+r}_{s+|\gamma|}$$

$$+C\left((\varepsilon+\delta)S^{\varepsilon,\delta}_{N-1}[u](E^{\varepsilon}_{N-1}[u])^{k+r}+\delta S^{\varepsilon,\delta}_{N-1}[u]\cdot\|u\|_s^{k-1}\cdot\|u\|^r_{s+|\gamma|}\right). \tag{3.5}$$

Proof. First fix $\alpha\in\mathbb{Z}^n_+$ with $|\alpha|\le r|\gamma|$. By (3.3), the operator $(P-\lambda)^{-1}\circ\partial_x^\alpha\circ x^j$ is bounded on $H^s(\mathbb{R}^n)$ for every $s\in\mathbb{R}$, then applying Schauder's lemma we have

$$\sum_{\substack{|\alpha|\le N\\|\alpha|\le r|\gamma|}}\frac{\varepsilon^{|\alpha|}}{(\alpha!)^\mu}\left\|(P-\lambda)^{-1}\circ\partial_x^\alpha\left(x^ju^k(\partial_x^\gamma u)^r\right)\right\|_s\le C_1^{k+r+1}\|u\|_s^k\cdot\|u\|^r_{s+|\gamma|} \tag{3.6}$$

for some positive constant C_1 depending only on m_1,m_2. On the other hand, if $|\alpha|>r|\gamma|$, then we can write $\alpha=(\alpha-\theta)+\theta$ with $|\theta|=r|\gamma|+1\le m_1$ and then

$$\partial_x^\alpha\left(x^ju^k(\partial_x^\gamma u)^r\right)=\partial_x^\theta\partial_x^{\alpha-\theta}\left(x^ju^k(\partial_x^\gamma u)^r\right)$$

$$=\partial_x^\theta\Bigg(\sum_{\substack{\sigma_0+\sigma_1+\dots+\sigma_k+\rho_1+\dots+\rho_r=\alpha-\theta\\ \sigma_0\le j}}\frac{(\alpha-\theta)!}{\sigma_1!\dots\sigma_k!\rho_1!\dots\rho_r!}\binom{j}{\sigma_0}$$

$$\times\, x^{j-\sigma_0}\partial_x^{\sigma_1}u\cdot\dots\cdot\partial_x^{\sigma_k}u\cdot\partial_x^{\rho_1+\gamma}u\cdot\dots\cdot\partial_x^{\rho_r+\gamma}u\Bigg).$$

By (3.3), the operators $(P-\lambda)^{-1}\circ\partial_x^\theta\circ x^{j-\sigma_0}$ are bounded in $H^s(\mathbb{R}^n)$. Then, applying Schauder's lemma and observing that

$$\frac{1}{(\alpha)!^\mu}\cdot\frac{(\alpha-\theta)!}{\sigma_1!\dots\sigma_k!\rho_1!\dots\rho_r!}\le\frac{C}{(\sigma_1!)^\mu\dots(\sigma_k!)^\mu(\rho_1+\gamma)!^\mu\dots(\rho_r+\gamma)!^\mu},$$

for some constant C depending only on r,γ, we obtain

$$\frac{\varepsilon^{|\alpha|}}{(\alpha!)^\mu}\left\|(P-\lambda)^{-1}\circ\partial_x^\alpha\left(x^ju^k(\partial_x^\gamma u)^r\right)\right\|_s$$

$$\le C\varepsilon\sum_{\substack{\sigma_0+\sigma_1+\dots+\sigma_k+\rho_1+\dots+\rho_r=\alpha-\theta\\ \sigma_0\le j}}\frac{\varepsilon^{|\sigma_1|}}{(\sigma_1!)^\mu}\|\partial_x^{\sigma_1}u\|_s\cdot\dots\cdot\frac{\varepsilon^{|\sigma_k|}}{(\sigma_k!)^\mu}\|\partial_x^{\sigma_k}u\|_s$$

$$\times\frac{\varepsilon^{|\rho_1+\gamma|}}{(\rho_1+\gamma)!^\mu}\|\partial_x^{\rho_1+\gamma}u\|_s\cdot\dots\cdot\frac{\varepsilon^{|\rho_r+\gamma|}}{(\rho_r+\gamma)!^\mu}\|\partial_x^{\rho_r+\gamma}u\|_s.$$

then

$$\sum_{r|\gamma|<|\alpha|\le N}\frac{\varepsilon^{|\alpha|}}{(\alpha!)^\mu}\left\|(P-\lambda)^{-1}\circ\partial_x^\alpha\left(x^ju^k(\partial_x^\gamma u)^r\right)\right\|_s\le C_2^{k+r+1}\varepsilon\left(E^\varepsilon_{N-1}[u]\right)^{k+r}. \tag{3.7}$$

for some constant C_2 depending only on m_1,m_2. From (3.6) and (3.7), we deduce that

$$\sum_{|\alpha|\le N}\frac{\varepsilon^{|\alpha|}}{(\alpha!)^\mu}\left\|(P-\lambda)^{-1}\circ\partial_x^\alpha\left(x^ju^k(\partial_x^\gamma u)^r\right)\right\|_s$$

$$\le C_3^{k+r+1}\left(\varepsilon\left(E^\varepsilon_{N-1}[u]\right)^{k+r}+\|u\|_s^k\cdot\|u\|^r_{s+|\gamma|}\right). \tag{3.8}$$

Now, let $\alpha, \beta \in \mathbb{Z}^n_+$, with $\beta_i \neq 0$ for some $i \in \{1, \dots, n\}$ and $|\alpha| \geq r|\gamma| + 1$. We can decompose

$$\begin{aligned}
x^\beta \partial_x^\alpha \left(x^j u^k (\partial_x^\gamma u)^r \right) &= x_i x^{\beta - e_i} \partial_x^\theta \partial_x^{\alpha-\theta} \left(x^j u^k (\partial_x^\gamma u)^r \right) \\
&= x_i \partial_x^\theta \left(x^j \left(x^{\beta - e_i} \partial_x^{\alpha-\theta} \left(u^k (\partial_x^\gamma u)^r \right) \right) \right) \\
&\quad + x_i \left[x^{\beta-e_i} \partial_x^{\alpha-\theta}, \partial_x^\theta x^j \right] \left(u^k (\partial_x^\gamma u)^r \right) \\
&= x_i \partial_x^\theta \left(x^j \left(x^{\beta - e_i} \partial_x^{\alpha-\theta} \left(u^k (\partial_x^\gamma u)^r \right) \right) \right) \\
&\quad + \sum_{\substack{0 \neq \rho \leq \alpha \\ \rho \leq j}} \frac{\alpha!}{(\alpha-\rho)!} \binom{j}{\rho} x^{j-\rho+\beta} \partial_x^{\alpha-\rho} \left(u^k (\partial_x^\gamma u)^r \right) \\
&\quad + \sum_{\substack{0 \neq \sigma \leq \theta \\ \sigma \leq j+\beta-e_i}} \frac{(j+\beta-e_i)!}{(j+\beta-e_i-\sigma)!} \binom{\theta}{\sigma} x^{j-\sigma+\beta} \partial_x^{\alpha-\sigma} \left(u^k (\partial_x^\gamma u)^r \right).
\end{aligned}$$

Now, the operators $(P-\lambda)^{-1} \circ x_i \partial_x^\theta \circ x^j$, $(P-\lambda)^{-1} \circ x^{j-\rho}$ and $(P-\lambda)^{-1} \circ x^{j-\sigma}$ are bounded in $H^s(\mathbb{R}^n)$, then, applying Leibniz rule and arguing as for (3.7), we obtain

$$\sum_{\substack{|\alpha|+|\beta| \leq N \\ r|\gamma| < |\alpha|, \beta \neq 0}} \frac{\varepsilon^{|\alpha|} \delta^{|\beta|}}{(\alpha!)^\mu (\beta!)^\nu} \left\| (P-\lambda)^{-1} \circ x^\beta \partial_x^\alpha \left(x^j u^k (\partial_x^\gamma u)^r \right) \right\|_s$$
$$\leq C_4^{k+r+1} (\varepsilon + \delta) S_{N-1}^{\varepsilon,\delta}[u] (E_{N-1}^\varepsilon[u])^{k+r-1}. \tag{3.9}$$

Finally, arguing as before, we derive the estimate

$$\sum_{\substack{|\alpha|+|\beta| \leq N \\ |\alpha| \leq r|\gamma|, \beta \neq 0}} \frac{\varepsilon^{|\alpha|} \delta^{|\beta|}}{(\alpha!)^\mu (\beta!)^\nu} \left\| (P-\lambda)^{-1} \circ x^\beta \partial_x^\alpha \left(x^j u^k (\partial_x^\gamma u)^r \right) \right\|_s$$
$$\leq C_4^{k+r+1} (\varepsilon + \delta) S_{N-1}^{\varepsilon,\delta}[u] \|u\|_s^{k-1} \|u\|_{s+|\gamma|}^r. \tag{3.10}$$

The estimates (3.8), (3.9), (3.10) directly yield (3.5).

Proof of Theorem 3.1. Now, if u is a solution of (1.10), then

$$Pu - \lambda u = -\lambda u + f + F.$$

Hence, for every $\alpha, \beta \in \mathbb{Z}^n_+$, we have

$$(P-\lambda) x^\beta \partial_x^\alpha u = -\lambda x^\beta \partial_x^\alpha u + x^\beta \partial_x^\alpha f + \left[P, x^\beta \partial_x^\alpha \right] u + x^\beta \partial_x^\alpha F. \tag{3.11}$$

from which we get

$$S_N^{\varepsilon,\delta}[u] = \sum_{\substack{\alpha,\beta \in \mathbb{Z}^n_+ \\ |\alpha|+|\beta| \leq N}} \left[(P-\lambda)^{-1} ([f^{(\alpha,\beta)}]^{\mu,\nu}_{\varepsilon,\delta}) - \lambda (P-\lambda)^{-1} ([u^{(\alpha,\beta)}]^{\mu,\nu}_{\varepsilon,\delta}) \right.$$
$$\left. + \frac{\varepsilon^{|\alpha|} \delta^{|\beta|}}{(\alpha!)^\mu (\beta!)^\nu} (P-\lambda)^{-1} \left[P, x^\beta \partial_x^\alpha \right] u + (P-\lambda)^{-1} ([F^{(\alpha,\beta)}]^{\mu,\nu}_{\varepsilon,\delta}) \right]. \tag{3.12}$$

To conclude the proof, it is sufficient to show that $S_N^{\varepsilon,\delta}[u]$ converges for some $\varepsilon, \delta > 0$. The assumption (3.4) implies that

$$\sum_{\substack{\alpha,\beta\in\mathbb{Z}_+^n \\ |\alpha|+|\beta|\le N}} \frac{\varepsilon^{|\alpha|}\delta^{|\beta|}}{(\alpha!)^{\mu}(\beta!)^{\nu}} \left\|(P-\lambda)^{-1}\left[P, x^{\beta}\partial_x^{\alpha}\right]u\right\|_s \le C(\varepsilon+\delta)S_{N-1}^{\varepsilon,\delta}[u] \tag{3.13}$$

for every $\varepsilon, \delta \in (0,1)$ and for some positive constant C. Moreover, (3.3) yields that for some $C_1 > 0$

$$\|\lambda(P-\lambda)^{-1}([u^{(\alpha,\beta)}]_{\varepsilon,\delta}^{\mu,\nu})\|_s \le C_1 \frac{\delta}{(\beta_\ell)^{\nu}} \|[u^{(\alpha,\beta-e_\ell)}]_{\varepsilon,\delta}^{\mu,\nu}\|_s \tag{3.14}$$

if $\beta_\ell \ge 1$ for some $\ell \in \{1,\dots,n\}$ and

$$\|\lambda(P-\lambda)^{-1}([u^{(\alpha,0)}]_{\varepsilon,\delta}^{\mu,\nu})\|_s \le C_1 \frac{\varepsilon}{(\alpha_j)^{\mu}} \|[u^{(\alpha-e_j,0)}]_{\varepsilon,\delta}^{\mu,\nu}\|_s \tag{3.15}$$

if $\alpha_j \ge 1$ for some $j \in \{1,\dots,n\}$. Hence, from (3.13), (3.14), (3.15) and Proposition 3.2, we deduce that

$$S_N^{\varepsilon,\delta}[u] \le C\Big(\|u\|_{s+m_o}^{k+h} + |||f|||_{\varepsilon,\delta} + (\delta+\varepsilon)S_{N-1}^{\varepsilon,\delta}[u]$$
$$+(\delta+\varepsilon)\left(S_{N-1}^{\varepsilon,\delta}[u]\right)^{k+h} + \delta S_{N-1}^{\varepsilon,\delta}[u]\|u\|_s^{k-1}\|u\|_{s+m_o}^{h}\Big). \tag{3.16}$$

for every $\varepsilon, \delta \in (0,1)$ and for some positive constant C and positive integers h, k, m_o. Then, iterating (3.16) and choosing δ and ε sufficiently small, we obtain that $S_N^{\varepsilon,\delta}[u]$ is bounded and hence it converges.

Remark 3.3. We proved Theorem 3.1 under the assumption $m_2 > 0$. The S_ν^μ-estimates for u solution of (1.10) in the case $m_2 = 0$ have been already proved in [15]. We recall that this case presents some additional difficulties and requires stronger assumptions on the a priori H^s regularity of u. Namely, one must assume $\langle x\rangle^\tau u \in H^{s+m_o}$ for some $s > n/2, \tau > 0$. Moreover, in this case, P has not compact resolvent in general. In [15], the operator P is assumed to be invertible.

4. Commutator identities for differential operators

In this section, we introduce some commutator identities. These allow to show easily that the assumption (3.4) is fulfilled in the case of SG-elliptic differential operators with polynomial coefficients, then Theorem 3.1 applies to this class of operators.

Lemma 4.1. *Let $\alpha, \beta, \rho, \sigma \in \mathbb{Z}_+^n$ and $u \in \mathcal{S}(\mathbb{R}^n)$. Then the following identity holds:*

$$x^{\beta}\partial_x^{\alpha}(x^{\sigma}\partial_x^{\rho}u) = \sum_{\substack{\gamma\le\alpha \\ \gamma\le\sigma}} \frac{\alpha!}{(\alpha-\gamma)!}\binom{\sigma}{\gamma} x^{\sigma-\gamma}x^{\beta}\partial_x^{\rho}(\partial_x^{\alpha-\gamma}u). \tag{4.1}$$

Proof. We have, by the Leibniz rule,

$$x^\beta \partial_x^\alpha (x^\sigma \partial_x^\rho u) = x^\beta \sum_{\substack{\gamma\leq\alpha\\ \gamma\leq\sigma}} \binom{\alpha}{\gamma} \partial_x^\gamma (x^\sigma) \partial_x^{\alpha-\gamma+\rho} u$$
$$= \sum_{\substack{\gamma\leq\alpha\\ \gamma\leq\sigma}} \frac{\alpha!}{(\alpha-\gamma)!} \binom{\sigma}{\gamma} x^{\sigma-\gamma} x^\beta \partial_x^{\alpha-\gamma+\rho} u. \tag{4.2}$$

□

Next, we need another commutator identity.

Lemma 4.2. *Let $\beta, \rho \subset \mathbb{Z}_+^n$ and $w \in \mathcal{S}(\mathbb{R}^n)$. Then the following identity holds*

$$x^\beta \partial_x^\rho w = \sum_{\substack{j\leq\beta\\ j\leq\rho}} \frac{\beta!}{(\beta-j)!} \binom{\rho}{j} (-1)^{|j|} \partial_x^{\rho-j} (x^{\beta-j} w). \tag{4.3}$$

Proof. Denote by $\mathcal{F}^{-1}$ the inverse Fourier transform

$$\mathcal{F}^{-1}_{\xi\to x} \hat{w} = (2\pi)^{-n} \int_{\mathbb{R}^n} \exp(ix\xi) \hat{w}(\xi)\, d\xi \tag{4.4}$$

and recall that $\mathcal{F}_{x\to\xi} \left(x^\beta \partial_x^\rho w\right)(\xi) = i^{|\beta|+|\rho|} \partial_\xi^\beta (\xi^\rho \hat{w}(\xi))$. We have, by (4.4) and the Leibniz rule

$$x^\beta \partial_x^\rho w(x) = i^{|\beta|+|\rho|} \mathcal{F}^{-1}_{\xi\to x} \left(\partial_\xi^\beta (\xi^\rho \hat{w}(\xi))\right)$$
$$= i^{|\beta|+|\rho|} \sum_{\substack{j\leq\beta\\ j\leq\rho}} \frac{\beta!}{(\beta-j)!} \binom{\rho}{j} \mathcal{F}^{-1}_{\xi\to x} (\xi^{\rho-j} \partial_\xi^{\beta-j} \hat{w}(\xi))$$
$$- \sum_{\substack{j\leq\beta\\ j\leq\rho}} \frac{\beta!}{(\beta-j)!} \binom{\rho}{j} (-1)^{|j|} \partial_x^{\rho-j} (x^{\beta-j} w(x)). \tag{4.5}$$

□

As an immediate consequence of Lemma 4.1 and Lemma 4.2, we are able to show the following commutator identity.

Lemma 4.3. *Let $\alpha, \beta, \rho, \sigma \in \mathbb{Z}_+^n$. Then the following commutator decomposition is true*

$$[x^\beta \partial_x^\alpha, x^\sigma \partial_x^\rho] u = x^\beta \partial_x^\alpha (x^\sigma \partial_x^\rho u) - x^\sigma \partial_x^\rho (x^\beta \partial_x^\alpha u) \tag{4.6}$$
$$= {\sum_{\substack{\gamma\leq\alpha\\ \gamma\leq\sigma}}}' \frac{\alpha!}{(\alpha-\gamma)!} \binom{\sigma}{\gamma} {\sum_{\substack{j\leq\beta\\ j\leq\rho}}}' \frac{\beta!}{(\beta-j)!} \binom{\rho}{j} (-1)^{|j|} x^{\sigma-\gamma} \partial_x^{\rho-j} (x^{\beta-j} \partial_x^{\alpha-\gamma} u)$$

where $\sum'\sum'$ means that $|\gamma| + |j| > 0$.

Proposition 4.4. *Let P be a selfadjoint SG elliptic operator of the form (1.9) with $m_1 > 0, m_2 > 0$ and let $\lambda \in \mathbb{C} \setminus \operatorname{spec}(P)$. Then, for every $\alpha, \beta \in \mathbb{Z}_+^n$ and for every $s \in \mathbb{R}$ the condition (3.4) holds with $\mu = \nu = 1$.*

Proof. By Lemma 4.3, we can write:

$$(P-\lambda)^{-1}[P, x^\beta \partial_x^\alpha]u = \sum_{\substack{|\rho|\le m \\ |\sigma|\le m_2}} \tilde{c}_{\rho\sigma} \sum_{\substack{\gamma\le\alpha \\ \gamma\le\sigma}}{}' \frac{\alpha!}{(\alpha-\gamma)!}\binom{\sigma}{\gamma} \sum_{\substack{j\le\beta \\ j\le\rho}}{}' \frac{\beta!}{(\beta-j)!}\binom{\rho}{j}(-1)^{|j|}$$
$$\times (P-\lambda)^{-1} x^{\sigma-\gamma}\partial_x^{\rho-j}(x^{\beta-j}\partial_x^{\alpha-\gamma}u) \tag{4.7}$$

with $\tilde{c}_{\rho\sigma} = (-i)^{|\rho|}c_{\rho\sigma}$. Using the smoothing property (3.3), we deduce that the operator $(P-\lambda)^{-1}\circ x^{\sigma-\gamma}\partial_x^{\rho-j}$ is bounded on $H^s(\mathbb{R}^n)$ for every $s\in\mathbb{R}$. Then

$$\left\|(P-\lambda)^{-1}[P, x^\beta\partial_x^\alpha]u\right\|_s$$
$$\le C \sum_{\substack{|\rho|\le m \\ |\sigma|\le m_2}} \sum_{\substack{\gamma\le\alpha \\ \gamma\le\sigma}}{}' \frac{\alpha!}{(\alpha-\gamma)!}\binom{\sigma}{\gamma} \sum_{\substack{j\le\beta \\ j\le\rho}}{}' \frac{\beta!}{(\beta-j)!}\binom{\rho}{j} \|x^{\beta-j}\partial_x^{\alpha-\gamma}u\|_s$$
$$\le C_1 \sum_{\gamma\le\alpha}{}' \frac{\alpha!}{(\alpha-\gamma)!} \sum_{j\le\beta}{}' \frac{\beta!}{(\beta-j)!}\|x^{\beta-j}\partial_x^{\alpha-\gamma}u\|_s \tag{4.8}$$

which gives (3.4), since $|\gamma|+|j|>0$ in (4.8).

To be explicit, we state again the result we have obtained.

Theorem 4.5. *Consider the semilinear partial differential equation*

$$\sum_{\substack{|\alpha|\le m_1 \\ |\beta|\le m_2}} c_{\alpha\beta}x^\beta D_x^\alpha u = f + F[u]. \tag{4.9}$$

We assume that the linear part is self-adjoint and SG-elliptic, that is the constants $c_{\alpha\beta}\in\mathbb{R}$ are fixed in such a way that

$$\left|\sum_{\substack{|\alpha|\le m_1 \\ |\beta|\le m_2}} c_{\alpha\beta}x^\beta\xi^\alpha\right| \ge C_o\langle x\rangle^{m_2}\langle\xi\rangle^{m_1} \quad \textit{for} \quad |x|+|\xi|\ge R>0.$$

The nonlinearity F is of the form (1.11) and $f\in S_\nu^\mu(\mathbb{R}^n), \mu\ge 1, \nu\ge 1$. Let $u\in H^{s+m_1-1}(\mathbb{R}^n)$, for some $s>n/2$, be a solution of (4.9), then $u\in S_\nu^\mu(\mathbb{R}^n)$. In particular, if $f=0$, then $u\in S_1^1(\mathbb{R}^n)$.

5. Localization operators, Weyl operators and Gelfand–Shilov spaces: Some remarks

Arguing first at a formal level, we recall the definition of localization operator in terms of time-frequency representation. Building blocks are the translation and modulation operators, with $t, x, \omega\in\mathbb{R}^n$:

$$T_x f(t) = f(t-x), \qquad M_\omega f(t) = e^{2\pi i\omega t}f(t). \tag{5.1}$$

For a fixed non-zero function φ, called "window", the short time Fourier transform (STFT) of f with respect to φ is given by

$$V_\varphi f(x,\omega) = \langle f, M_\omega T_x \varphi \rangle = \int f(t)\overline{\varphi(t-x)}e^{-2\pi i \omega t}dt. \tag{5.2}$$

(here and in the following we use the normalization of Fourier integrals with a factor 2π in the phase, which is standard in time-frequency analysis). For φ_1, φ_2 with $\langle \varphi_1, \varphi_2 \rangle = 1$, the following inversion formula is valid:

$$f(t) = \int V_{\varphi_1} f(x,\omega) M_\omega T_x \varphi_2(t) dx d\omega. \tag{5.3}$$

Given a function $a(x,\omega)$, we then define the localization operator $A_a^{\varphi_1,\varphi_2}$, with symbol a and windows φ_1, φ_2 :

$$A_a^{\varphi_1,\varphi_2} f(t) = \int a(x,\omega) V_{\varphi_1} f(x,\omega) M_\omega T_x \varphi_2(t) dx d\omega \tag{5.4}$$

or, in the weak sense,

$$\langle A_a^{\varphi_1,\varphi_2} f, g \rangle = \langle a, \overline{V_{\varphi_1} f} V_{\varphi_2} g \rangle. \tag{5.5}$$

A precise functional setting for the above formal computations is obtained by taking f, g, φ and φ_1, φ_2 in the Schwartz space $\mathcal{S}(\mathbb{R}^n)$. Since V_φ is a continuous map from $\mathcal{S}(\mathbb{R}^n)$ to $\mathcal{S}(\mathbb{R}^{2n})$, we may give sense to (5.5) for any $a \in \mathcal{S}'(\mathbb{R}^n)$. Note that in (5.5) we read $\langle \cdot, \cdot \rangle$ as the extension to $\mathcal{S}(\mathbb{R}^n), \mathcal{S}'(\mathbb{R}^n)$ of the L^2-inner product, i.e., $\langle f, g \rangle = (f, \overline{g})$. So we get $A_a^{\varphi_1,\varphi_2} : \mathcal{S}(\mathbb{R}^n) \to \mathcal{S}'(\mathbb{R}^n)$. Similar settings are obtained by fixing less regular windows φ_1, φ_2 and more general symbols in such a way that an extension of L^2-inner product in (5.5) makes sense, see [16], [26]. In the opposite direction, we may take more regular windows, belonging to Gelfand–Shilov spaces, $\varphi_1, \varphi_2 \in S_\mu^\mu(\mathbb{R}^n), \mu \geq 1/2$. This allows to consider even more general symbols than $a \in \mathcal{S}'(\mathbb{R}^{2n})$.

Proposition 5.1. *Let $\varphi_1, \varphi_2 \in S_\mu^\mu(\mathbb{R}^n), \mu \geq 1/2$, and let a be an element of the dual space $(S_\mu^\mu)'(\mathbb{R}^n)$. Then (5.5) defines a linear map*

$$A_a^{\varphi_1,\varphi_2} : S_\mu^\mu(\mathbb{R}^n) \to (S_\mu^\mu)'(\mathbb{R}^n). \tag{5.6}$$

Proof. In fact, it follows from the results in [17] that for windows $\varphi \in S_\mu^\mu(\mathbb{R}^n), \mu \geq 1/2$, the STFT V_φ maps continuously $S_\mu^\mu(\mathbb{R}^n)$ into $S_\mu^\mu(\mathbb{R}^{2n})$. Since $S_\mu^\mu(\mathbb{R}^n)$ is an algebra, then

$$\overline{V_{\varphi_1} f} V_{\varphi_2} g \in S_\mu^\mu(\mathbb{R}^{2n}).$$

By L^2-duality, the right-hand side of (5.5) is well defined for $a \in (S_\mu^\mu)'(\mathbb{R}^{2n})$ and we get (5.6).

For more details concerning the setting given by Proposition 5.1, we refer to [25] in the case $\mu > 1$ and to [10] in the case $\mu = 1$. In particular, in [10] it was proved that for windows $\varphi_1, \varphi_2 \in S_1^1(\mathbb{R}^n)$, every ultradistribution with compact support $a \in \mathcal{E}'_\mu(\mathbb{R}^{2n}), \mu > 1$, cf. [22], considered as a symbol, defines a trace class, hence L^2-bounded, localization operator. The properties of the operators from

Proposition 5.1 in the case $1/2 \leq \mu < 1$ are rather unexplored, but for the case

$$\varphi_1(t) = \varphi_2(t) = 2^{n/4} e^{-\pi|t|^2} \tag{5.7}$$

coming back to [2]. With the choice of φ_1, φ_2 in (5.7), the operator $A_a^{Wick} = A_a^{\varphi_1,\varphi_2}$ comes in [2] as a result of a quantization procedure and it is used in the PDE context, see [24]. Since $e^{-\pi|t|^2} \in S_{1/2}^{1/2}(\mathbb{R}^n)$, we have for $a \in (S_{1/2}^{1/2})'(\mathbb{R}^{2n})$:

$$A_a^{Wick} : S_{1/2}^{1/2}(\mathbb{R}^n) \to (S_{1/2}^{1/2})'(\mathbb{R}^n). \tag{5.8}$$

Let us pass now to consider Weyl pseudo-differential operators L_σ with symbol σ, defined by

$$L_\sigma f(x) = \int \sigma\left(\frac{x+y}{2}, \omega\right) f(y) e^{2\pi i(x-y)\omega} dy d\omega.$$

Connected with Weyl operators is the Wigner transform of a pair of functions f, g, defined by

$$W(f,g)(x,\omega) = \int f\left(x + \frac{t}{2}\right) \overline{g\left(x - \frac{t}{2}\right)} e^{-2\pi i t\omega} dt. \tag{5.9}$$

In fact, an easy formal computation shows that

$$\langle L_\sigma f, g\rangle = \langle \sigma, W(f,g)\rangle. \tag{5.10}$$

Taking (5.10) as definition of L_σ, we have the following analogous to Proposition 5.1.

Proposition 5.2. *If $\sigma \in (S_\mu^\mu)'(\mathbb{R}^n), \mu \geq 1/2$, then*

$$L_\sigma : S_\mu^\mu(\mathbb{R}^n) \to (S_\mu^\mu)'(\mathbb{R}^n). \tag{5.11}$$

Proof. Basic fact is that the Wigner transform maps $S_\mu^\mu(\mathbb{R}^n) \otimes S_\mu^\mu(\mathbb{R}^n)$ continuously into $S_\mu^\mu(\mathbb{R}^{2n})$. See for the proof Theorem 3.8 in [25], or else observe that $W(f,g)(x,\omega)$ can be seen as STFT, namely:

$$W(f,g)(x,\omega) = 2^n e^{4\pi i x\omega} V_{\tilde{g}} f(2x, 2\omega), \tag{5.12}$$

where $\tilde{g}(t) = g(-t)$ (see [16] Lemma 4.3.1). In (5.12) we may then appeal to the results of [17] used to prove Proposition 5.1. The right-hand side of (5.10) is then well defined for $\sigma \in (S_\mu^\mu)'(\mathbb{R}^{2n})$ and we get (5.11).

There is a simple connection between Weyl symbols and localization symbols. Namely, $A_a^{\varphi_1,\varphi_2} = L_\sigma$ for

$$\sigma = a * W(\varphi_2, \varphi_1), \tag{5.13}$$

see for example [13]. In the Gelfand–Shilov spaces, taking $a \in (S_\mu^\mu)'(\mathbb{R}^{2n}), \mu \geq 1/2$, we assume $\varphi_1, \varphi_2 \in S_\mu^\mu(\mathbb{R}^n)$ to apply Proposition 5.1. Hence, $W(\varphi_1, \varphi_2) \in S_\mu^\mu(\mathbb{R}^{2n})$ and from (5.13) and the convolution properties in [14], we get $\sigma \in (S_\mu^\mu)'(\mathbb{R}^{2n})$. So we may apply Proposition 5.2 as well, to obtain

$$A_a^{\varphi_1,\varphi_2} = L_\sigma : S_\mu^\mu(\mathbb{R}^n) \to (S_\mu^\mu)'(\mathbb{R}^n). \tag{5.14}$$

In particular, if we assume $\varphi_1 = \varphi_2 = 2^{n/4}e^{-\pi|t|^2}$, by observing that in this case $W(x,\omega) = 2^n e^{-2\pi(|x|^2+|\omega|^2)}$, we have

$$\sigma = a * 2^n e^{-2\pi(|x|^2+|\omega|^2)}, \tag{5.15}$$

cf. [24]. Summing up, starting from $a \in (S^{1/2}_{1/2})'(\mathbb{R}^{2n})$, we may consider $A^{Wick}_a : S^{1/2}_{1/2}(\mathbb{R}^n) \to (S^{1/2}_{1/2})'(\mathbb{R}^n)$ and write

$$A^{Wick}_a = L_\sigma$$

where the Weyl symbol $\sigma \in (S^{1/2}_{1/2})'(\mathbb{R}^{2n})$ is given by (5.15). Note however that the convolution (5.15) has a strong smoothing effect, in particular $a \in (S^{1/2}_{1/2})'(\mathbb{R}^{2n})$ implies the analyticity of σ, cf. [14]. So, given a Weyl symbol $\sigma \in (S^{1/2}_{1/2})'(\mathbb{R}^{2n})$ or even $\sigma \in \mathcal{S}'(\mathbb{R}^{2n})$, we cannot find in general $a \in (S^{1/2}_{1/2})'(\mathbb{R}^{2n})$ such that $L_\sigma = A^{Wick}_a$. The study of the convolution equation (5.15) in appropriate spaces of distributions is in fact a challenging problem, see for example [18] and the references there.

References

[1] S. Agmon, *Lectures on Exponential Decay of Second-Order Elliptic Equations: Bounds on Eigenfunctions of N-body Schrödinger Operators*, Princeton University Press, Princeton, 1982.

[2] F.A. Berezin, Wick and anti-Wick symbols of operators, *Mat. Sb. (N.S.)* **86** (1971), 578–610.

[3] H.A. Biagioni and T. Gramchev, Fractional derivative estimates in Gevrey spaces, global regularity and decay for solutions to semilinear equations in $\mathbb{R}^n$, *J. Diff. Equations,* **194** (2003), 140–165.

[4] J. Bona and Y. Li, Decay and analyticity of solitary waves, *J. Math. Pures Appl.* **76** (1997), 377–430.

[5] M. Cappiello, Fourier integral operators of infinite order and applications to SG-hyperbolic equations, *Tsukuba J. Math.* **28** (2004), 311–361.

[6] M. Cappiello and L. Rodino, SG-pseudo-differential operators and Gelfand–Shilov spaces, *Rocky Mountain J. Math.* **36** (2006), no. 4, 1117–1148.

[7] M. Cappiello, T. Gramchev and L. Rodino, Exponential decay and regularity for SG-elliptic operators with polynomial coefficients, in *Hyperbolic Problems and Regularity Questions*, Editors: M. Padula and L. Zanghirati, Birkhäuser, Basel, to appear.

[8] M. Cappiello, T. Gramchev and L. Rodino, *Superexponential decay and holomorphic extensions for semilinear equations with polynomial coefficients*, J. Funct. Anal. **237** (2006), 634–654.

[9] J. Chung, S.Y. Chung and D. Kim, Characterization of the Gelfand–Shilov spaces via Fourier transforms, *Proc. Amer. Math. Soc.* **124** (1996), 2101–2108.

[10] E. Cordero, S. Pilipovic, L. Rodino and N. Teofanov, Localization operators and exponential weights for modulation spaces, *Mediterranean. J. Math.* **2** (2005), 381–394.

[11] H.O. Cordes, *The Technique of Pseudodifferential Operators*, Cambridge Univ. Press, 1995.

[12] Y.V. Egorov and B.-W. Schulze, *Pseudo-differential Operators, Singularities, Applications*, Birkhäuser, Basel, 1997.

[13] G.B. Folland, *Harmonic Analysis in Phase Space*, Princeton Univ. Press, Princeton, 1989.

[14] I.M. Gel'fand and G.E. Shilov, *Generalized Functions II*, Academic Press, New York, 1968.

[15] T. Gramchev, Perturbative Methods in Scales of Banach Spaces: Applications for Gevrey Regularity of Solutions to Semilinear Partial Differential Equations, *Rend. Sem. Mat. Univ. Pol. Torino* **61** (2003), 101–134.

[16] K. Gröchenig, *Foundations of Time-Frequency Analysis*, Birkhäuser, Boston 2001.

[17] K. Gröchenig and G. Zimmermann, Spaces of test functions via the STFT, *J. Function Spaces Appl.* **2** (2004), 25–53.

[18] B.C. Hall, Harmonic analysis with respect to heat kernel measure, *Bull. Amer. Math. Soc.* **38** (2000), 43–78.

[19] C. Parenti, Operatori pseudodifferenziali in $\mathbb{R}^n$ e applicazioni, *Ann. Mat. Pura Appl.* **93** (1972), 359–389.

[20] S. Pilipovic, Tempered ultradistributions, *Boll. Unione Mat. Ital., VII. Ser., B,* **2** (1988), 235–251.

[21] V.S. Rabinovich, Exponential estimates for eigenfunctions of Schrödinger operators with rapidly increasing and discontinuous potentials, *Contemporary Math.* **364** (2004), 225–236.

[22] L. Rodino, *Linear Partial Differential Operators in Gevrey Spaces*, World Scientific Publishing Co., Singapore, 1993.

[23] E. Schrohe, Spaces of weighted symbols and weighted Sobolev spaces on manifolds, in *Pseudodifferential Operators*, Editors: H.O. Cordes, B. Gramsch and H. Widom, Springer, 1987, 360–377.

[24] M. Shubin, *Pseudodifferential Operators and Spectral theory*, Second Edition, Springer-Verlag, Berlin, 2001.

[25] N. Teofanov, Ultradistributions and time-frequency analysis, in *Pseudo-differential Operators and Related Topics*, Editors: P. Boggiatto, L. Rodino, J. Toft, M.W. Wong, Birkhäuser, Basel, 2006, 173–191.

[26] M.W. Wong, *Wavelet Transforms and Localization Operators*, Birkäuser, Basel, 2002.

Marco Cappiello
Dipartimento di Matematica, Università di Ferrara
Via Machiavelli 35, I-44100 Ferrara, Italy
e-mail: `marco.cappiello@unife.it`

Todor Gramchev
Dipartimento di Matematica e Informatica, Università di Cagliari
Via Ospedale 72, I-09124 Cagliari, Italy
e-mail: `todor@unica.it`

Luigi Rodino
Dipartimento di Matematica, Università di Torino
Via Carlo Alberto 10, I-10123 Torino, Italy
e-mail: `luigi.rodino@unito.it`

Operator Theory:
Advances and Applications, Vol. 172, 313–328

Continuity and Schatten Properties for Toeplitz Operators on Modulation Spaces

Joachim Toft

Abstract. Let $M^{p,q}_{(\omega)}$ be the modulation space with parameters p,q and weight function ω. We prove that if $p,q,p_1,p_2,q_1,q_2 \in [1,\infty]$, ω_1, ω_2, ω and h_1,h_2 are appropriate, and $a \in M^{p,q}_{(\omega)}$, then the Toeplitz operator $\mathrm{Tp}_{h_1,h_2}(a) : M^{p_1,q_1}_{(\omega_1)} \to M^{p_2,q_2}_{(\omega_2)}$ is continuous. If in addition $p_1 = p_2 = q_1 = q_2 = 2$, then we present sufficient conditions on p, q, h_1 and h_2 in order for $\mathrm{Tp}_{h_1,h_2}(a)$ should be a Schatten-von Neumann operator of certain degree.

Mathematics Subject Classification (2000). Primary 47B10, 47B35, 47B37: Secondary 35S05, 42B35.

Keywords. Toeplitz operators, localization operators, anti-Wick operators, modulation spaces, Schatten classes.

0. Introduction

In this paper we discuss continuity properties for Toeplitz operators (in the literature, the terms localization operators or anti-Wick operators also occur) in background of modulation space theory. More precisely, we consider Toeplitz operators where the corresponding window functions (or coherent state) and symbols belong to appropriate modulation spaces, and discuss continuity for such operators when acting on modulation spaces. At the same time we clarify some details and correct some mistakes in [26], and show that stronger results can be obtained by a slight improvement of the techniques there.

Furthermore, if such operators act on modulation spaces of Hilbert type, then a detailed study of continuity and compactness is performed in terms of Schatten-von Neumann classes. In particular we investigate trace-class and Hilbert-Schmidt properties. In Section 1 we discuss general properties for such classes, and recall from [28, 29] embedding relations between such operators in Weyl calculus and modulation spaces. Thereafter we apply, in Section 2, a standard technique to analyze Toeplitz operators in background of Weyl calculus, using the fact that the

Weyl symbol of a Toeplitz operator is equal to a convolution between the Toeplitz symbol and a Wigner distribution (see [2–4,17,21,24–29]). In that end, we are able to extend certain continuity and Schatten-von Neumann results in [2, 4, 25, 26].

Toeplitz operators have been applied in quantum physics (see, e.g., [18] and the references therein). They were introduced in time-frequency analysis in [6] by Daubechies. In this field, Toeplitz operators can be used as a tool for filtering signals. A convenient way to measure the time-frequency content of these signals then can be done by means of modulation spaces. It might therefore not be surprising that Toeplitz operators have been considered in background of modulation space theory before (see, e.g., [2–5, 8, 25, 26]). As a consequence of the investigations in the present paper we are able to present a general result which essentially contains the most of the results in the latter papers.

The basic theory of modulation spaces were established by Feichtinger and Gröchenig. (See [9, 10] and and the references therein.) Roughly speaking, for an appropriate weight function ω, the modulation space $M^{p,q}_{(\omega)}$ is obtained by imposing a mixed $L^{p,q}_{(\omega)}$-norm on the short-time Fourier transform of a tempered distribution. The non-weighted (or classical) modulation space $M^{p,q}$ is then obtained by choosing $\omega = 1$. By using modulation spaces it might be easy to obtain information concerning growth and decay properties, as well as certain localization and regularity properties for distributions.

In order to describe our results more in detail we recall the definition of modulation spaces, and start to consider appropriate conditions on the involved weight functions. Assume that $0 < \omega, v \in L^\infty_{\mathrm{loc}}(\mathbf{R}^m)$. Then recall that ω is called *v-moderate* if there is a constant C such that $\omega(x_1 + x_2) \le C\omega(x_1)v(x_2)$ holds for every $x_1, x_2 \in \mathbf{R}^m$, and if ω is v-moderate for some polynomial v, then ω is called polynomial moderate. (See [13].) As in [25] we let $\mathscr{P}(\mathbf{R}^m)$ be the set of all polynomial moderate functions on $\mathbf{R}^m$.

Next assume that $\chi \in \mathscr{S}(\mathbf{R}^m) \setminus 0$, $p, q \in [1, \infty]$ and that $\omega \in \mathscr{P}(\mathbf{R}^{2m})$, and let $\tau_x\chi(y) = \chi(y - x)$ when $x, y \in \mathbf{R}^m$. (We use the same notation for the usual functions and distribution spaces as in, e.g., [16].) Then the modulation space $M^{p,q}_{(\omega)}(\mathbf{R}^m)$ consists of all $f \in \mathscr{S}'(\mathbf{R}^m)$ such that

$$\begin{aligned} \|f\|_{M^{p,q}_{(\omega)}} &= \|f\|_{M^{p,q,\chi}_{(\omega)}} \\ &\equiv \Big(\int \Big(\int |\mathscr{F}(f\,\tau_x\chi)(\xi)\omega(x,\xi)|^p\,dx \Big)^{q/p} d\xi \Big)^{1/q} < \infty, \end{aligned} \tag{0.1}$$

with the obvious modifications when $p = \infty$ and/or $q = \infty$. Here $\mathscr{F}$ denotes the Fourier transform on $\mathscr{S}'(\mathbf{R}^m)$, which takes the form

$$\mathscr{F}f(\xi) = \widehat{f}(\xi) = (2\pi)^{-m/2} \int f(x)e^{-i\langle x,\xi\rangle}\,dx$$

when $f \in \mathscr{S}(\mathbf{R}^m)$, and the function $(x, \xi) \mapsto \mathscr{F}(f\,\tau_x\chi)(\xi)$ in (0.1) is called the *short-time Fourier transform* of f with respect to the *window function* χ.

For functions and distributions on the phase space $\mathbf{R}^{2m}$, we sometimes use the symplectic Fourier transform (see (1.1) below) instead of the Fourier transform $\mathscr{F}$ above, in the definition of modulation space norm. Then we use the notation $\widetilde{M}^{p,q}_{(\omega)}(\mathbf{R}^{2m})$ and $\|\cdot\|_{\widetilde{M}^{p,q}_{(\omega)}}$ instead of $M^{p,q}_{(\omega)}(\mathbf{R}^{2m})$ and $\|\cdot\|_{M^{p,q}_{(\omega)}}$ respectively.

Another "advanced Fourier transform", similar to the short-time Fourier transform, is the Wigner distribution which we shall discuss now. Assume that $f_1, f_2 \in \mathscr{S}(\mathbf{R}^m)$. Then

$$W_{f_1,f_2}(x,\xi) = (2\pi)^{-m/2} \int f_1(x-y/2)\overline{f_2(x+y/2)}e^{i\langle y,\xi\rangle}\,dy \tag{0.2}$$

is the Wigner distribution of f_1 and f_2. The map $(f_1, f_2) \mapsto W_{f_1,f_2}$ is continuous from $\mathscr{S}(\mathbf{R}^m) \times \mathscr{S}(\mathbf{R}^m)$ to $\mathscr{S}(\mathbf{R}^{2m})$, which extends to a continuous map from $\mathscr{S}'(\mathbf{R}^m) \times \mathscr{S}'(\mathbf{R}^m)$ to $\mathscr{S}'(\mathbf{R}^{2m})$. (See [11].) Furthermore, some extensions to continuity in terms of modulation spaces can be found in [25, 26].

Next assume that $a \in \mathscr{S}(\mathbf{R}^{2n})$, and that $h_1, h_2 \in \mathscr{S}(\mathbf{R}^n)$. Then the Toeplitz operator, with symbol a and window functions h_1 and h_2, is defined by the formula

$$(\mathrm{Tp}_{h_1,h_2}(a)f_1, f_2) = (a(2\,\cdot\,)W_{f_1,h_1}, W_{f_2,h_2}) \tag{0.3}$$

when $f_1, f_2 \in \mathscr{S}(\mathbf{R}^m)$. The definition extends in several ways (see, e.g., [2–4, 24–29]).

In Section 2 we discuss continuity for Toeplitz operators acting on modulation spaces when the window functions and operator symbols belong to modulation spaces. In particular we find appropriate conditions on the weight functions ω, ω_j, ν_j and the exponents p, q, p_j, q_j, r_j, s_j for $j = 1, 2$, in order for the map

$$\mathrm{Tp}_{h_1,h_2}(a) \,:\, M^{p_1,q_1}_{(\omega_1)}(\mathbf{R}^m) \to M^{p_2,q_2}_{(\omega_2)}(\mathbf{R}^m) \tag{0.4}$$

should be continuous when $a \in \widetilde{M}^{p,q}_{(\omega)}(\mathbf{R}^{2m})$ and $h_j \in M^{r_j,s_j}_{(\nu_j)}(\mathbf{R}^m)$ for $j = 1, 2$.

As it is seen above, the general situation involves at least ten different exponents and five weight functions attached to the modulation spaces. The interplay between them is rather complicated and is regulated with certain equalities and inequalities. Here we present some examples of possible conditions, and consequences of our results. At the same time we get a link about the generality of our investigations.

First assume first that $r_j = s_j = 1$ and $q = \infty$ here above. Then it turns out that appropriate conditions for the remaining exponents are

$$\frac{1}{p_2} - \frac{1}{p_1} = \frac{1}{q_2} - \frac{1}{q_1} = \frac{1}{p}. \tag{0.5}$$

If instead $r_j = s_j = 2$, then appropriate conditions are

$$\frac{1}{p_2} - \frac{1}{p_1} = \frac{1}{q_2} - \frac{1}{q_1} = \frac{1}{p} + \frac{1}{q} - 1, \quad q \le p_2, q_2 \le p. \tag{0.5$'$}$$

In order to establish appropriate conditions for the involved weight functions, it is convenient to include weight functions κ, κ_j and v_j for $j = 1, 2$, such that ω_1 is v_1-moderate and ω_2 is $\check{v}_2$-moderate. Here and in what follows we let $\check{f}(x) = f(-x)$

for every function or distribution f. Then we sometimes assume that the involved weight functions satisfy

$$\begin{aligned} \nu_j = \kappa_j v_j, \quad \omega(X,Y) &= \frac{\omega_2(X)}{\omega_1(X)\kappa(2Y)}, \quad \text{for} \quad j = 1,2, \\ \text{and} \quad \kappa(X+Y) &\le C\kappa_1(-X)\kappa_2(Y), \end{aligned} \tag{0.6}$$

for some constant C.

As a consequence of our general result we have the following result.

Proposition 0.1. *Assume that $p, p_j, q_j \in [1,\infty]$ for $j = 1,2$ satisfy (0.5). Also assume that (0.6) is fulfilled, for some $\kappa, \kappa_j, v_j, \nu_j, \omega_j \in \mathscr{P}(\mathbf{R}^{2m})$ such that ω_1 is v_1-moderate and ω_2 is $\check{v}_2$-moderate. If $h_j \in M^{1,1}_{(\nu_j)}(\mathbf{R}^m)$ for $j = 1,2$ and $a \in \widetilde{M}^{p,\infty}_{(\omega)}(\mathbf{R}^{2m})$, then the map (0.4) is continuous.*

We note that if we choose $p = \infty$, $\omega_1 = \omega_2 = 1$ and $\kappa(X) = \kappa_j(X) = \langle X\rangle^s$ for some $s \ge 0$, then Proposition 0.1 agrees with Theorem 3.2 in [4]. Here and in what follows we let $\langle x\rangle = (1+|x|^2)^{1/2}$. If instead $\omega_1 = \omega_2$ and $\kappa(X) = \kappa_j(X) = 1$, then Proposition 0.1 agrees with Corollary 4.2 in [4]. Furthermore, from embedding properties between Lebesgue spaces and modulation spaces (cf., e.g., [27]), it follows that Proposition 3.3 in [2] in the case $q_0 \ge p_0$ is an immediate consequence of Proposition 0.1 in the case $\kappa = \kappa_j = 1$. (See also Corollary 2.7 and Remark 2.11 below.)

If, more generally, $h_j \in M^{2,2}_{(\nu_j)}$, then our results give the following.

Proposition 0.1′. *Assume that $p, q, p_j, q_j \in [1,\infty]$ for $j = 1,2$ satisfy (0.5)′. Also assume that (0.6) is fulfilled, for some $\kappa, \kappa_j, v_j, \nu_j, \omega_j \in \mathscr{P}(\mathbf{R}^{2m})$ such that ω_1 is v_1-moderate and ω_2 is $\check{v}_2$-moderate. If $h_j \in M^{2,2}_{(\nu_j)}(\mathbf{R}^m)$ for $j = 1,2$ and $a \in \widetilde{M}^{p,q}_{(\omega)}(\mathbf{R}^{2m})$, then the map (0.4) is continuous.*

If we let $p_1 = p_2 = q_1 = q_2 = 2$, then the operator $\mathrm{Tp}_{h_1,h_2}(a)$ in Proposition 0.1 and Proposition 0.1′ acts between the Hilbert spaces $\mathscr{H}_1 = M^{2,2}_{(\omega_1)}$ and $\mathscr{H}_2 = M^{2,2}_{(\omega_2)}$. In this case, a more detailed compactness investigation in terms of Schatten-von Neumann classes is performed in Section 3 (see Section 1 for strict definitions). These investigations are based on certain results in [29] in the present volume, where similar questions are considered for pseudo-differential operators. (See also [28].) As a consequence of these investigations, we are able to prove a general result for Toeplitz operators, which in particular implies that the following two propositions hold. Here and in what follows we let p' denote the conjugate exponent of $p \in [1,\infty]$, i.e., $1/p + 1/p' = 1$.

Proposition 0.2. *Assume that $p \in [1,\infty]$. Also assume that (0.6) is fulfilled, for some $\kappa, \kappa_j, v_j, \nu_j, \omega_j \in \mathscr{P}(\mathbf{R}^{2m})$ such that ω_1 is v_1-moderate and ω_2 is $\check{v}_2$-moderate. If $h_j \in M^{1,1}_{(\nu_j)}(\mathbf{R}^m)$ for $j = 1,2$ and $a \in \widetilde{M}^{p,\infty}_{(\omega)}(\mathbf{R}^{2m})$, then*

$$\mathrm{Tp}_{h_1,h_2}(a) \in \mathscr{I}_p(M^{2,2}_{(\omega_1)}, M^{2,2}_{(\omega_2)}).$$

Proposition 0.2$'$. *Assume that $p, q \in [1,\infty]$ satisfy $q \le \min(p,p')$. Also assume that (0.6) is fulfilled for some $\kappa, \kappa_j, v_j, \nu_j, \omega_j \in \mathscr{P}(\mathbf{R}^{2m})$ such that ω_1 is v_1-moderate and ω_2 is $\check{v}_2$-moderate. If $h_j \in M^{2,2}_{(\nu_j)}(\mathbf{R}^m)$ for $j = 1,2$ and $a \in \widetilde{M}^{p,q}_{(\omega)}(\mathbf{R}^{2m})$, then $\mathrm{Tp}_{h_1,h_2}(a) \in \mathscr{I}_p(M^{2,2}_{(\omega_1)}, M^{2,2}_{(\omega_2)})$.*

1. Preliminaries

In the first part of the section we discuss basic properties for modulation spaces, Wigner distributions and Weyl operators in pseudo-differential calculus. The proofs are in the most cases omitted since they can be found in [7,9,10,13,24–26]. In the second part we discuss Schatten-von Neumann properties and recall some facts which can be found in [1,22,25–29].

We start to present some notations which are used. If $\mathscr{H}$ is a Hilbert space, then its scalar product is denoted by $(\,\cdot\,,\,\cdot\,)_{\mathscr{H}}$, or $(\,\cdot\,,\,\cdot\,)$ when there are no confusions about the Hilbert space structure.

The duality between a topological vector space and its dual is denoted by $\langle\,\cdot\,,\,\cdot\,\rangle$. For admissible a and b in $\mathscr{S}'(\mathbf{R}^m)$, we set $(a,b) = \langle a, \overline{b}\rangle$, and it is obvious that $(\,\cdot\,,\,\cdot\,)$ on L^2 is the usual scalar product. Furthermore, the restriction of $(\,\cdot\,,\,\cdot\,)$ to $\mathscr{S} \times \mathscr{S}$ is the (standard) sesqui-linear form on $\mathscr{S}$.

Next assume that $\mathscr{B}_1$ and $\mathscr{B}_2$ are topological spaces. Then $\mathscr{B}_1 \hookrightarrow \mathscr{B}_2$ means that $\mathscr{B}_1$ is continuously embedded in $\mathscr{B}_2$. In the case that $\mathscr{B}_1$ and $\mathscr{B}_2$ are Banach spaces, $\mathscr{B}_1 \hookrightarrow \mathscr{B}_2$ is equivalent to $\mathscr{B}_1 \subseteq \mathscr{B}_2$ and $\|x\|_{\mathscr{B}_2} \le C\|x\|_{\mathscr{B}_1}$, for some constant $C > 0$ which is independent of $x \in \mathscr{B}_1$.

Assume that $\omega \in \mathscr{P}(\mathbf{R}^{2m})$, $p, q \in [1,\infty]$, and that $\chi \in \mathscr{S}(\mathbf{R}^m) \setminus 0$. Then recall that the *modulation space* $M^{p,q}_{(\omega)}(\mathbf{R}^m)$ is the set of all $f \in \mathscr{S}'(\mathbf{R}^m)$ such that (0.1) holds. Obviously, this is equivalent to that the short-time Fourier transform $\mathscr{F}(f\,\tau_x\chi)(\xi)$ belongs to $L^{p,q}_{(\omega)}(\mathbf{R}^{2m})$, the set of all $F \in L^1_{loc}(\mathbf{R}^{2m})$ such that

$$\|F\|_{L^{p,q}_{(\omega)}} \equiv \Big(\int \Big(\int |F(x,\xi)\omega(x,\xi)|^p \, dx \Big)^{q/p} d\xi \Big)^{1/q}$$

is finite.

If $\omega = 1$, then the notation $M^{p,q}$ is used instead of $M^{p,q}_{(\omega)}$. Moreover we set $M^p_{(\omega)} = M^{p,p}_{(\omega)}$ and $M^p = M^{p,p}$.

The convention of indicating weight functions with parenthesis is also used in other situations. For example, if $\omega \in \mathscr{P}(\mathbf{R}^m)$, then $L^p_{(\omega)}(\mathbf{R}^m)$ is the set of all measurable functions f on $\mathbf{R}^m$ such that $f\omega \in L^p(\mathbf{R}^m)$, i.e., such that $\|f\|_{L^p_{(\omega)}} \equiv \|f\omega\|_{L^p}$ is finite.

Next we consider the Fourier transform of functions and distributions defined on $\mathbf{R}^{2m}$. By interpreting $\mathbf{R}^{2m}$ as the phase space consisting of all pairs (x,ξ), where $x \in \mathbf{R}^m$ and $\xi \in \mathbf{R}^m$ are dual variables, it is sometimes convenient to use the symplectic Fourier transform $\mathscr{F}_\sigma$ which is given by

$$(\mathscr{F}_\sigma f)(y,\eta) = \widehat{f}(y,\eta) \equiv \pi^{-m} \iint f(x,\xi) e^{2i(\langle x,\eta\rangle - \langle y,\xi\rangle)}\, dx d\xi, \tag{1.1}$$

when $f \in L^1(\mathbf{R}^{2m})$. Then it follows that $\mathscr{F}_\sigma$ is a homeomorphism on $f \in \mathscr{S}(\mathbf{R}^{2m})$ which extends to a homeomorphism on $f \in \mathscr{S}'(\mathbf{R}^{2m})$ and to a unitary map on $L^2(\mathbf{R}^{2m})$, since similar facts hold for $\mathscr{F}$.

We use the notation $\widetilde{M}^{p,q}_{(\omega)}$, $\widetilde{M}^{p}_{(\omega)}$, $\widetilde{M}^{p,q}$ and $\widetilde{M}^{p}$ instead of $M^{p,q}_{(\omega)}$, $M^{p}_{(\omega)}$, $M^{p,q}$ and M^p respectively, when $\mathscr{F}_\sigma$ is used instead of $\mathscr{F}$, in the definition of modulation spaces of distributions on $\mathbf{R}^{2m}$. Note that $\widetilde{M}^{p,q}_{(\omega)}$ here does not agree with $\widetilde{M}^{p,q}_{(\omega)}$ in [29], since $\mathscr{F}_\sigma$ does not agree with the phase space Fourier transform which is used in the definition of $\widetilde{M}^{p,q}_{(\omega)}$ in [29].

The following proposition is a consequence of well-known facts in [7, 13].

Proposition 1.1. *Assume that $p,q,p_j,q_j \in [1,\infty]$ for $j=1,2$, and that $\omega,\omega_1,\omega_2,v \in \mathscr{P}(\mathbf{R}^{2m})$ are such that ω is v-moderate. Then the following are true:*

1. *if $\chi \in M^1_{(v)}(\mathbf{R}^m)\setminus 0$, then $f \in M^{p,q}_{(\omega)}(\mathbf{R}^m)$ if and only if (0.1) holds, i.e., $M^{p,q}_{(\omega)}(\mathbf{R}^m)$ is independent of the choice of χ. Moreover, $M^{p,q}_{(\omega)}$ is a Banach space under the norm in (0.1), and different choices of χ give rise to equivalent norms;*
2. *if $p_1 \le p_2$, $q_1 \le q_2$ and $\omega_2 \prec \omega_1$, then*
$$\mathscr{S}(\mathbf{R}^m) \hookrightarrow M^{p_1,q_1}_{(\omega_1)}(\mathbf{R}^m) \hookrightarrow M^{p_2,q_2}_{(\omega_2)}(\mathbf{R}^m) \hookrightarrow \mathscr{S}'(\mathbf{R}^m);$$
3. *the L^2 product $(\,\cdot\,,\,\cdot\,)$ from $\mathscr{S}$ extends to a continuous map from $M^{p,q}_{(\omega)}(\mathbf{R}^m)\times M^{p',q'}_{(1/\omega)}(\mathbf{R}^m)$ to $\mathbf{C}$. On the other hand, if $\|a\| = \sup|(a,b)|$, where the supremum is taken over all $b \in M^{p',q'}_{(1/\omega)}(\mathbf{R}^m)$ such that $\|b\|_{M^{p',q'}_{(1/\omega)}} \le 1$, then $\|\cdot\|$ and $\|\cdot\|_{M^{p,q}_{(\omega)}}$ are equivalent norms;*
4. *if $p,q<\infty$, then $\mathscr{S}(\mathbf{R}^m)$ is dense in $M^{p,q}_{(\omega)}(\mathbf{R}^m)$. The dual space of $M^{p,q}_{(\omega)}(\mathbf{R}^m)$ can be identified with $M^{p',q'}_{(1/\omega)}(\mathbf{R}^m)$ through the form $(\,\cdot\,,\,\cdot\,)_{L^2}$. Moreover, $\mathscr{S}(\mathbf{R}^m)$ is weakly dense in $M^{\infty}_{(\omega)}(\mathbf{R}^m)$.*

Proposition 1.1 (1) permits us be rather vague about to the choice of $\chi \in M^1_{(v)}\setminus 0$ in (0.1). For example, if $C>0$ is a constant and Ω is a subset of $\mathscr{S}'$, then $\|a\|_{M^{p,q}_{(\omega)}} \le C$ for every $a \in \Omega$, means that the inequality holds for some choice of $\chi \in M^1_{(v)}\setminus 0$ and every $a\in\Omega$. Evidently, for any other choice of $\chi \in M^1_{(v)}\setminus 0$, a similar inequality is true although C may have to be replaced by a larger constant, if necessary.

It is convenient to let $\mathcal{M}^{p,q}_{(\omega)}(\mathbf{R}^m)$ ($\widetilde{\mathcal{M}}^{p,q}_{(\omega)}(\mathbf{R}^{2m})$) be the completion of $\mathscr{S}(\mathbf{R}^m)$ ($\mathscr{S}(\mathbf{R}^{2m})$) under the norm $\|\cdot\|_{M^{p,q}_{(\omega)}}$ ($\|\cdot\|_{\widetilde{M}^{p,q}_{(\omega)}}$). Then $\mathcal{M}^{p,q}_{(\omega)} \subseteq M^{p,q}_{(\omega)}$ with equality if and only if $p < \infty$ and $q < \infty$. It follows that most of the properties which are valid for the usual modulation spaces also hold for $\mathcal{M}^{p,q}_{(\omega)}(\mathbf{R}^m)$.

Remark 1.2. Assume that p, q, q_1, $q_2 \in [1,\infty]$. If $q_1 \le \min(p,p')$ and $q_2 \ge \max(p,p')$, then $M^{p,q_1} \subseteq L^p \subseteq M^{p,q_2}$. In particular, $M^2 = L^2$. (See, e.g., [25,26].)

Remark 1.3. Assume that $p_j, q_j \in [1,\infty]$ and $\omega_j \in \mathscr{P}(\mathbf{R}^{2m})$ for $0 \le j \le 2$, satisfy

$$\frac{1}{p_1} + \frac{1}{p_2} = 1 + \frac{1}{p_0}, \qquad \frac{1}{q_1} + \frac{1}{q_2} = \frac{1}{q_0},$$

and

$$\omega_0(x_1 + x_2, \xi) \le C\omega_1(x_1,\xi)\omega_2(x_2,\xi),$$

for some constant C independent on $x_1, x_2, \xi \in \mathbf{R}^m$. Then Theorem 5.5 in [27] shows that there is a canonical way to extend the the usual convolution $*$ on $\mathscr{S}$ to a continuous multiplication from $M^{p_1,q_1}_{(\omega_1)} \times M^{p_2,q_2}_{(\omega_2)}$ to $M^{p_0,q_0}_{(\omega_0)}$, and if $f_j \in M^{p_j,q_j}_{(\omega_j)}$ for $j = 1,\dots,2$, then

$$\|f_1 * f_2\|_{M^{p_0,q_0}_{(\omega_0)}} \le C\|f_1\|_{M^{p_1,q_1}_{(\omega_1)}}\|f_2\|_{M^{p_2,q_2}_{(\omega_2)}}, \tag{1.2}$$

for some constant C, independent of f_1 and f_2. (See also [4].)

Next we recall the definition of Weyl operators. Assume that $a \in \mathscr{S}(\mathbf{R}^{2m})$. Then the Weyl operator $a^w(x,D)$ with respect to the symbol a is the continuous operator on $\mathscr{S}(\mathbf{R}^m)$, defined by the formula

$$(a^w(x,D)f)(x) = (2\pi)^{-m} \iint a((x+y)/2,\xi)f(y)e^{i\langle x-y,\xi\rangle}\,dyd\xi.$$

The definition of $a^w(x,D)$ extends to any $a \in \mathscr{S}'(\mathbf{R}^{2m})$, and then $a^w(x,D)$ is continuous from $\mathscr{S}(\mathbf{R}^m)$ to $\mathscr{S}'(\mathbf{R}^m)$. This is a consequence of the relation

$$(a^w(x,D)f,g) = (2\pi)^{-m/2}(a, W_{g,f}), \tag{1.3}$$

when $a \in \mathscr{S}(\mathbf{R}^{2m})$ and $f,g \in \mathscr{S}(\mathbf{R}^m)$, and the fact that the map $(f,g) \mapsto W_{f,g}$ is continuous from $\mathscr{S}(\mathbf{R}^m) \times \mathscr{S}(\mathbf{R}^m)$ to $\mathscr{S}(\mathbf{R}^{2m})$. (See [11,16].)

Next we discuss Schatten-von Neumann classes of linear operators from a Hilbert space $\mathscr{H}_1$ to another Hilbert space $\mathscr{H}_2$. Such operator classes were introduced by R. Schatten in [19] in the case $\mathscr{H}_1 = \mathscr{H}_2$. (See also [12,22].) The general situation when $\mathscr{H}_1$ is not necessarily equal to $\mathscr{H}_2$ have thereafter been discussed in [1,20]. Here we give a short introduction, based on [28,29].

For any Hilbert space $\mathscr{H}$, we let $\mathrm{ON}(\mathscr{H})$ be the set of orthonormal sequences in $\mathscr{H}$. Assume that $T : \mathscr{H}_1 \to \mathscr{H}_2$ is linear, and that $p \in [1,\infty]$. Then set

$$\|T\|_{\mathscr{I}_p} = \|T\|_{\mathscr{I}_p(\mathscr{H}_1,\mathscr{H}_2)} \equiv \sup\Big(\sum |(Tf_j, g_j)_{\mathscr{H}_2}|^p\Big)^{1/p} \tag{1.4}$$

(with obvious modifications when $p = \infty$). Here the supremum is taken over all $(f_j) \in \mathrm{ON}(\mathscr{H}_1)$ and $(g_j) \in \mathrm{ON}(\mathscr{H}_2)$. Now recall that $\mathscr{I}_p = \mathscr{I}_p(\mathscr{H}_1,\mathscr{H}_2)$, the Schatten-von Neumann class of order p, is the Banach space which consists of all linear operators T from $\mathscr{H}_1$ to $\mathscr{H}_2$ such that $\|T\|_{\mathscr{I}_p(\mathscr{H}_1,\mathscr{H}_2)}$ is finite. Obviously,

$\mathscr{I}_\infty(\mathscr{H}_1,\mathscr{H}_2)$ consists of all linear and continuous operators from $\mathscr{H}_1$ to $\mathscr{H}_2$. If $\mathscr{H}_1=\mathscr{H}_2$, then the shorter notation $\mathscr{I}_p(\mathscr{H}_1)$ is used instead of $\mathscr{I}_p(\mathscr{H}_1,\mathscr{H}_2)$. We also let $\mathscr{I}_\sharp(\mathscr{H}_1,\mathscr{H}_2)$ be the set of all linear and compact operators from $\mathscr{H}_1$ to $\mathscr{H}_2$, and equip this space with the norm $\|\cdot\|_{\mathscr{I}_\infty}$ as usual. The spaces $\mathscr{I}_1(\mathscr{H}_1,\mathscr{H}_2)$ and $\mathscr{I}_2(\mathscr{H}_1,\mathscr{H}_2)$ are called the sets of trace-class operators and Hilbert-Schmidt operators respectively. These definitions agree with the old ones when $\mathscr{H}_1=\mathscr{H}_2$, and in this case the norms $\|\cdot\|_{\mathscr{I}_1}$ and $\|\cdot\|_{\mathscr{I}_2}$ agree with the trace-class norm and Hilbert-Schmidt norm respectively.

We refer to [1,12,20,22,28,29] for more properties concerning Schatten-von Neumann classes on general Hilbert spaces.

Next we consider Schatten-von Neumann classes in background of Weyl calculus. Let $p\in[1,\infty]$ and $\omega_1,\omega_2\in\mathscr{P}(\mathbf{R}^{2m})$ be fixed. Then let $s_p^w(\omega_1,\omega_2)$ be the set of all $a\in\mathscr{S}'(\mathbf{R}^{2m})$ such that $a^w(x,D)$ belongs to $\mathscr{I}_p(M^2_{(\omega_1)},M^2_{(\omega_2)})$. Also let $s_\sharp^w(\omega_1,\omega_2)$ be the set of all $a\in\mathscr{S}'(\mathbf{R}^{2m})$ such that $a^w(x,D)\in\mathscr{I}_\sharp(M^2_{(\omega_1)},M^2_{(\omega_2)})$. We let $s_p^w(\omega_1,\omega_2)$ and $s_\sharp^w(\omega_1,\omega_2)$ be equipped by the norms

$$\|a\|_{s_p^w}=\|a\|_{s_p^w(\omega_1,\omega_2)}\equiv\|a^w(x,D)\|_{\mathscr{I}_p(M^2_{(\omega_1)},M^2_{(\omega_2)})}$$

and $\|\cdot\|_{s_\infty^w}$ respectively.

Since any linear and continuous operator from $\mathscr{S}(\mathbf{R}^m)$ to $\mathscr{S}'(\mathbf{R}^m)$ is equal to $a^w(x,D)$, for a unique $a\in\mathscr{S}'(\mathbf{R}^{2m})$ (cf. the introduction or [16]), it follows that the map $a\mapsto a^w(x,D)$ is an isometric homeomorphism from $s_p^w(\omega_1,\omega_2)$ to $\mathscr{I}_p(M^2_{(\omega_1)},M^2_{(\omega_2)})$ when $p\in[1,\infty]$, and from $s_\sharp^w(\omega_1,\omega_2)$ to $\mathscr{I}_\sharp(M^2_{(\omega_1)},M^2_{(\omega_2)})$. Consequently, most of the properties for the $\mathscr{I}_p$ spaces which are listed earlier in this section carry over to the s_p^w-spaces. For example, $s_p^w(\omega_1,\omega_2)$ is a Banach space which increases with the parameter $p\in[1,\infty]$.

Next we discuss continuity for Wigner distributions in terms of modulation spaces. The following result is essentially a restatement of Theorem 5.1 in [26] and Proposition 4.1 in [29].

Proposition 1.4. *Assume that $p_j,q_j,p,q\in[1,\infty]$ such that $p\le p_j,q_j\le q$, for $j=1,2$, and that*

$$1/p_1+1/p_2=1/q_1+1/q_2=1/p+1/q. \tag{1.5}$$

Also assume that $\omega_1,\omega_2\in\mathscr{P}(\mathbf{R}^{2m})$, and that $\omega\in\mathscr{P}(\mathbf{R}^{2m}\oplus\mathbf{R}^{2m})$ satisfy

$$\omega(X,Y)\le C\omega_1(X-Y)\omega_2(X+Y).$$

Then the map $(f_1,f_2)\mapsto W_{f_1,f_2}$ from $\mathscr{S}'(\mathbf{R}^m)\times\mathscr{S}'(\mathbf{R}^m)$ to $\mathscr{S}'(\mathbf{R}^{2m})$ restricts to a continuous mapping from $M^{p_1,q_1}_{(\omega_1)}(\mathbf{R}^m)\times M^{p_2,q_2}_{(\omega_2)}(\mathbf{R}^m)$ to $\widetilde{M}^{p,q}_{(\omega)}(\mathbf{R}^{2m})$, and for some constant C it holds

$$\|W_{f_1,f_2}\|_{\widetilde{M}^{p,q}_{(\omega)}}\le C\|f_1\|_{M^{p_1,q_1}_{(\omega_1)}}\|f_2\|_{M^{p_2,q_2}_{(\omega_2)}} \tag{1.6}$$

when $f_1,f_2\in\mathscr{S}'(\mathbf{R}^m)$.

We finish this section by recalling continuity and Schatten-von Neumann properties for Weyl operators in [25–29]. The following result is a restatement of Theorem 4.2 in [29] in the case of Weyl operators. A similar result, containing a mistake is also available in [26].

Theorem 1.5. *Assume that* $p, q, p_j, q_j \in [1, \infty]$ *for* $j = 1, 2$, *satisfy* $(0.5)'$. *Also assume that* $\omega \in \mathscr{P}(\mathbf{R}^{2m} \oplus \mathbf{R}^{2m})$ *and* $\omega_1, \omega_2 \in \mathscr{P}(\mathbf{R}^{2m})$ *satisfy*

$$\frac{\omega_2(X - Y)}{\omega_1(X + Y)} = \omega(X, Y). \tag{1.7}$$

If $a \in M^{p,q}_{(\omega)}(\mathbf{R}^{2m})$, *then* $a^w(x, D)$ *from* $\mathscr{S}(\mathbf{R}^m)$ *to* $\mathscr{S}'(\mathbf{R}^m)$ *extends uniquely to a continuous map from* $M^{p_1,q_1}_{(\omega_1)}(\mathbf{R}^m)$ *to* $M^{p_2,q_2}_{(\omega_2)}(\mathbf{R}^m)$.

Moreover, if in addition $a \in \widetilde{\mathcal{M}}^{p,q}_{(\omega)}$, *then*

$$a^w(x, D) : M^{p_1,q_1}_{(\omega_1)} \to M^{p_2,q_2}_{(\omega_2)}$$

is compact.

The next result concerns Schatten-von Neumann properties for Weyl operators with symbols in appropriate modulation spaces. For the proof we refer to [28, 29] in the general case, and to [25] in the unweighted case (see also [14]).

Theorem 1.6. *Assume that* $p, q, p_j, q_j \in [1, \infty]$ *for* $j = 1, 2$, *satisfy*

$$p_1 \le p \le p_2, \quad q_1 \le \min(p, p') \quad \text{and} \quad q_2 \ge \max(p, p'). \tag{1.8}$$

Also assume that $\omega \in \mathscr{P}(\mathbf{R}^{2m} \oplus \mathbf{R}^{2m})$ *and* $\omega_1, \omega_2 \in \mathscr{P}(\mathbf{R}^{2m})$ *satisfy* (1.7). *Then*

$$\widetilde{M}^{p_1,q_1}_{(\omega)}(\mathbf{R}^{2m}) \hookrightarrow s^w_p(\omega_1, \omega_2) \hookrightarrow M^{p_2,q_2}_{(\omega)}(\mathbf{R}^{2m}). \tag{1.9}$$

2. Continuity for Toeplitz operators

In this section, we extend certain continuity properties in Section 5 in [26] concerning Toeplitz operators, when the operator symbols belong to appropriate classes of modulation spaces.

We start by recalling the definition of Toeplitz operators. Assume that $h_1, h_2 \in \mathscr{S}(\mathbf{R}^m) \setminus 0$ are fixed. For any $a \in \mathscr{S}(\mathbf{R}^{2m})$, the Toeplitz operator $\mathrm{Tp}_{h_1,h_2}(a)$ on $\mathscr{S}(\mathbf{R}^m)$, with respect to h_1, h_2 and symbol a is defined by (0.3). Obviously, $\mathrm{Tp}_{h_1,h_2}(a)$ extends to a continuous map from $\mathscr{S}'(\mathbf{R}^m)$ to $\mathscr{S}(\mathbf{R}^m)$. As in [2–4, 21, 24–30] we study Toeplitz operators in background of Weyl calculus, using the fact that if $a \in \mathscr{S}(\mathbf{R}^{2m})$ and $h_1, h_2 \in \mathscr{S}'(\mathbf{R}^m)$, then

$$\begin{gathered} \mathrm{Tp}_{h_1,h_2}(a) = (a * u_{h_1,h_2})^w(x, D), \\ \text{where} \quad u_{h_1,h_2}(X) = (2\pi)^{-m/2} W_{h_2,h_1}(-X). \end{gathered} \tag{2.1}$$

If, more generally, $a \in \mathscr{S}'(\mathbf{R}^{2m})$ and $h_1, h_2 \in \mathscr{S}'(\mathbf{R}^m)$ are distributions such that $a * u_{h_1,h_2}$ makes sense as an element in $\mathscr{S}'(\mathbf{R}^{2m})$, then (2.1) is taken as the definition of the Toeplitz operator $\mathrm{Tp}_{h_1,h_2}(a)$. We refer to [2–4, 21, 24, 26, 30–32] for more facts about Toeplitz operators.

Next we discuss $\mathrm{Tp}_{h_1,h_2}(a)$ when h_1, h_2 and a belong to certain types of modulation spaces. In the introduction it was stated that Proposition 0.1 and Proposition 0.1′ are special cases of more general results. The first result is Proposition 2.1 below, which in some sense is the most general one. Here the exponentials for the involved modulation spaces are given by $p, q, p_0, q_0, r_j, s_j \in [1,\infty]$ for $j = 1, 2$, and they should satisfy

$$\begin{gathered}\frac{1}{r_1}+\frac{1}{r_2}=\frac{1}{s_1}+\frac{1}{s_2}=1-\frac{1}{p}-\frac{1}{q}+\frac{1}{p_0}+\frac{1}{q_0}\\ 0\le\frac{1}{p}-\frac{1}{p_0}\le\min\Big(\frac{1}{r_j'},\frac{1}{s_j'}\Big)\\ 0\le\frac{1}{q_0}-\frac{1}{q}\le\min\Big(\frac{1}{r_j},\frac{1}{s_j}\Big),\quad j=1,2.\end{gathered}\tag{2.2}$$

It is also convenient to let $\mathscr{L}^w_{p,q}(\omega)$ be the set of all operators of the type $a^w(x,D)$ such that $a \in \widetilde{M}^{p,q}_{(\omega)}(\mathbf{R}^{2m})$, where the topology for $\mathscr{L}^w_{p,q}(\omega)$ is inherited from $\widetilde{M}^{p,q}_{(\omega)}(\mathbf{R}^{2m})$.

Proposition 2.1. *Assume that $r_j, s_j, p, p_0, q, q_0 \in [1,\infty]$ for $j = 1, 2$ satisfy (2.2). Also assume that $\nu_j \in \mathscr{P}(\mathbf{R}^{2m})$, $h_j \in M^{r_j,s_j}_{(\nu_j)}(\mathbf{R}^m)$, $j = 1, 2$, and that $\omega, \omega_0 \in \mathscr{P}(\mathbf{R}^{2m} \oplus \mathbf{R}^{2m})$ satisfy*

$$\omega_0(X_1+X_2,Y)\le C\omega(X_1,Y)\,\check{\nu}_1(Y+X_2)\nu_2(Y-X_2)\quad X_1,X_2,Y\in\mathbf{R}^{2m},$$

for some constant $C > 0$. Then the map $a \mapsto \mathrm{Tp}_{h_1,h_2}(a)$ from $\mathscr{S}(\mathbf{R}^{2m})$ to the set of linear and continuous operators from $\mathscr{S}(\mathbf{R}^m)$ to $\mathscr{S}'(\mathbf{R}^m)$ extends to a continuous map from $\widetilde{M}^{p,q}_{(\omega)}(\mathbf{R}^{2m})$ to $\mathscr{L}^w_{p_0,q_0}(\omega_0)$.

Proof. By (2.2) it is possible to find $p_1, q_1 \in [1,\infty]$ such that $p_1 \le r_j, s_j \le q_1$ for $j = 1, 2$ and

$$1/p-1/p_0=1-1/p_1,\quad 1/q_0-1/q=1/q_1,\quad 1/r_1+1/r_2=1/p_1+1/q_1.$$

Hence Proposition 1.4 implies that $u_{h_1,h_2} \in \widetilde{M}^{p_1,q_1}_{(\omega_1)}$, where $\omega_1(X,Y) = \check{\nu}_1(Y+X)\nu_2(Y-X)$. If $a \in \widetilde{M}^{p,q}_{(\omega)}(\mathbf{R}^{2m})$, then Remark 1.3 gives

$$a*u_{h_1,h_2}\in\widetilde{M}^{p,q}_{(\omega)}*\widetilde{M}^{p_1,q_1}_{(\omega_1)}\subseteq\widetilde{M}^{p_0,q_0}_{(\omega_0)},$$

and the result follows from (2.1). The proof is complete. □

Next we combine Proposition 2.1 with Theorem 1.5 in order to obtain more explicit continuity results for $\mathrm{Tp}_{h_1,h_2}(a)$ when acting on modulation spaces. More precisely we establish continuity properties of the form

$$\|\mathrm{Tp}_{h_1,h_2}(a)\|_{M^{p_1,q_1}_{(\omega_1)}\to M^{p_2,q_2}_{(\omega_2)}}\le C\|a\|_{\widetilde{M}^{p,q}_{(\omega)}}\|h_1\|_{M^{r_1,s_1}_{(\nu_1)}}\|h_2\|_{M^{r_2,s_2}_{(\nu_2)}}.\tag{2.3}$$

It turns out that an appropriate condition for the involving weight functions is

$$\sup_Z \Big(\frac{\omega_2(X-2Y-Z)}{\omega_1(X-Z)\nu_1(Z)\nu_2(2Y+Z)} \Big) \le C\omega(X,Y). \tag{2.4}$$

and appropriate conditions for the exponents p, q, p_j, q_j, r_j, s_j is that they should satisfy (2.2) for some $p_0, q_0 \in [1,\infty]$, and

$$\frac{1}{p_2} - \frac{1}{p_1} = \frac{1}{q_2} - \frac{1}{q_1} = \frac{1}{p_0} + \frac{1}{q_0} - 1, \quad q_0 \le p_2, q_2 \le p_0. \tag{0.5)$''$ (}$$

Theorem 2.2. *Assume that $p, q, p_j, q_j, r_j, s_j \in [1,\infty]$ for $j = 0,1,2$ satisfy (0.5)$''$ and (2.2). Also assume that $\nu_j, \omega_j \in \mathscr{P}(\mathbf{R}^{2m})$ for $j = 1,2$ and $\omega \in \mathscr{P}(\mathbf{R}^{2m} \oplus \mathbf{R}^{2m})$ satisfy (2.4) for some constant $C > 0$ independent of $X, Y \subset \mathbf{R}^{2m}$. If $h_j \in M^{r_j,s_j}_{(\nu_j)}(\mathbf{R}^m)$ for $j = 1,2$ and $a \in \widetilde{M}^{p,q}_{(\omega)}(\mathbf{R}^{2m})$, then the map (0.4) is continuous, and (2.3) holds for some constant C which is independent of $a \in \widetilde{M}^{p,q}_{(\omega)}$ and $h_j \in M^{r_j,s_j}_{(\nu_j)}$ for $j = 1,2$.*

Proof. Let

$$\omega_0(X,Y) = \omega_2(X-Y)/\omega_1(X+Y).$$

Then it follows from (2.4) that

$$\omega_0(X_1+X_2,Y) \le C\omega(X_1,Y)\,\check{\nu}_1(X_2+Y)\nu_2(Y-X_2),$$

for some constant C which is independent of $X_1, X_2, Y \in \mathbf{R}^{2m}$. Hence $\mathrm{Tp}_{h_1,h_2}(a) \in \mathscr{L}^w_{p_0,q_0}(\omega_0)$ by Proposition 2.1. It follows now from the definition of ω_0, Theorem 1.5, (2.2) and (0.5) that $\mathrm{Tp}_{h_1,h_2}(a)$ is continuous from $M^{p_1,q_1}_{(\omega_1)}$ to $M^{p_2,q_2}_{(\omega_2)}$. The proof is complete. □

As a consequence of Theorem 2.2 we have the following result.

Proposition 0.1$''$. *Assume that $p, q, p_j, q_j, r_j, s_j \in [1,\infty]$ for $j = 0,1,2$ satisfy (0.5)$''$ and (2.2). Also assume that (0.6) is fulfilled, for some $\kappa, \kappa_j, v_j, \nu_j, \omega_j \in \mathscr{P}(\mathbf{R}^{2m})$ such that ω_1 is v_1-moderate and ω_2 is $\check{v}_2$-moderate. If $h_j \in M^{r_j,s_j}_{(\nu_j)}(\mathbf{R}^m)$ for $j = 1,2$ and $a \in \widetilde{M}^{p,q}_{(\omega)}(\mathbf{R}^{2m})$, then the map (0.4) is continuous.*

Proof. The assertion is an immediate consequence of Theorem 2.2, since the condition that ω_1 is ν_1-moderate and ω_2 is $\check{\nu}_2$-moderate immediately gives (2.4). The proof is complete. □

Theorem 2.2 may be used to give other results, parallel to Proposition 0.1$''$. For example, if

$$\begin{aligned} &\omega(X,Y) = (v_1(X)v_2(X)\kappa(2Y))^{-1}, \\ \text{and}\quad &\kappa(X+Y) \le C\kappa_1(X)\kappa_2(Y), \end{aligned} \qquad (0.6)'$$

are imposed on the involved weight functions, then we get.

Proposition 2.3. *Assume that $p, q, p_j, q_j, r_j, s_j \in [1, \infty]$ for $j = 0, 1, 2$ satisfy (0.5)$''$ and (2.2). Also assume that (0.6)$'$ is fulfilled, for some $\kappa, \kappa_j, v_j, \omega_j \in \mathscr{P}(\mathbf{R}^{2m})$ such that ω_j is v_j-moderate for $j = 1, 2$. If $h_j \in M^{r_j,s_j}_{(\omega_j\kappa_j)}(\mathbf{R}^m)$ for $j = 1, 2$ and $a \in \widetilde{M}^{p,q}_{(\omega)}(\mathbf{R}^{2m})$, then the map*

$$\mathrm{Tp}_{h_1,h_2}(a) : M^{p_1,q_1}_{(v_1)} \to M^{p_2,q_2}_{(1/v_2)}$$

is continuous.

Proof. The result follows if we let $\omega_1 = v_1$, $\omega_2 = 1/v_2$ and $\nu_j = \omega_j\kappa_j$ in (2.4). The proof is complete. □

Remark 2.4. If $\kappa = \kappa_1 = \kappa_2 = 1$, then Proposition 2.3 is equivalent to Proposition 0.1$''$. In fact, it follows from (0.2) that $\overline{W_{f,g}} = W_{g,f}$ for every $f, g \in \mathscr{S}'$. This in turn implies that if $a \in \mathscr{S}'(\mathbf{R}^{2m})$ and $f_j, h_j \in \mathscr{S}'(\mathbf{R}^m)$, then $(\mathrm{Tp}_{h_1,h_2}(a)f_1, f_2)$ is well defined if and only if $(\mathrm{Tp}_{f_2,f_1}(a)h_2, h_1)$ is well defined, and then

$$(\mathrm{Tp}_{h_1,h_2}(a)f_1, f_2) = (\mathrm{Tp}_{f_2,f_1}(a)h_2, h_1).$$

It is now straightforward to control that the asserted equivalence follows from the latter identification formula and Proposition 1.1.

Remark 2.5. Theorem 2.2 can be used to obtain other results, similar to Proposition 0.1$''$ and Proposition 2.3. For example, the proof of Proposition 0.1$''$ gives the following more general result. Assume that $\theta, \theta_1, \theta_2 \in \mathscr{P}(\mathbf{R}^{2m})$ satisfies

$$\theta(X + Y) \le C\frac{\theta_2(-Y)}{\theta_1(X)}, \quad X, Y \in \mathbf{R}^{2m},$$

for some constant C, and that the hypothesis in Proposition 0.1$''$ holds, except that the conditions on ω in (0.6) is replaced by

$$\omega(X, Y) = \frac{\omega_2(X)\theta(2Y)}{\omega_1(X)\kappa(2Y)}.$$

Then

$$\mathrm{Tp}_{h_1,h_2}(a) : M^{p_1,q_1}_{(\omega_1\theta_1)}(\mathbf{R}^m) \to M^{p_2,q_2}_{(\omega_2\theta_2)}(\mathbf{R}^m)$$

is continuous.

Next we present a consequence of Proposition 0.1$''$. Here and in what follows we let $\mathscr{P}_0(\mathbf{R}^m)$ be the set of all $\omega \in \mathscr{P}(\mathbf{R}^m) \cap C^\infty(\mathbf{R}^m)$ such that $\partial^\alpha\omega/\omega \in L^\infty$ for every multi-index α.

Corollary 2.6. *Assume that $p, q, p_j, q_j, r_j, s_j \in [1, \infty]$ for $j = 0, 1, 2$ satisfy (0.5)$''$ and (2.2). Also assume that $v_j, \omega_j \in \mathscr{P}_0(\mathbf{R}^{2m})$ such that ω_1 is v_1-moderate and ω_2 is $\check{v}_2$-moderate such that*

$$\omega(X, Y) = \frac{\omega_2(X)}{\omega_1(X)} \tag{2.5}$$

is fulfilled. If $h_j \in M^{r_j,s_j}_{(v_j)}(\mathbf{R}^m)$ for $j = 1, 2$ and $\omega(X, D)a \in \widetilde{M}^{p,q}(\mathbf{R}^{2m})$, then the map (0.4) is continuous.

Proof. The result follows immediately from the fact that $a \in \widetilde{M}^{p,q}_{(\omega)}$, if and only if $\omega(X,D)a \in \widetilde{M}^{p,q}$ (see, e.g., Theorem 3.2 in [27]), and Proposition 0.1″. □

Corollary 2.7. *Assume that $p, p_j, q_j \in [1,\infty]$ for $j = 0,1,2$ satisfy (0.5) and (2.2). Also assume that $v_j, \omega_j \in \mathscr{P}_0(\mathbf{R}^{2m})$ such that ω_1 is v_1-moderate and ω_2 is $\check{v}_2$-moderate such that (2.5) is fulfilled. If $h_j \in M^{1,1}_{(v_j)}(\mathbf{R}^m)$ for $j = 1,2$ and $\omega(X,D)a \in L^p(\mathbf{R}^{2m})$, then the map (0.4) is continuous.*

Proof. The result follows immediately by letting $r_j = s_j = 1$ in Corollary 2.6 and using the embedding $L^p \subseteq M^{p,q}$ when $q = \max(p,p')$ (see Remark 1.2). □

Remark 2.8. By straightforward computations it follows that if it in addition is assumed that $a \in \widetilde{\mathcal{M}}^{p,q}_{(\omega)}(\mathbf{R}^{2m})$ and $h_j \in \mathcal{M}^{r_j,s_j}_{(\nu_j)}(\mathbf{R}^m)$ for $j = 1,2$ in Theorem 2.2, Proposition 0.1″ or Proposition 2.6, then (0.4) is compact.

Remark 2.9. In [4], Cordero and Gröchenig present a converse of Corollary 0.1 in the case $p = q = \infty$ and $r_j = s_j = 1$ for $j = 1,2$, and the weight functions satisfy some extra conditions. (Cf. Theorem 4.3 in [4].)

Remark 2.10. For each $\omega \in \mathscr{P}(\mathbf{R}^m)$, it exists an element $\omega_0 \in \mathscr{P}_0(\mathbf{R}^m)$ which is equivalent to ω, i.e., there exists a constant C such that $C^{-1}\omega \le \omega_0 \le C\omega$ holds. In fact, it suffices to let $\omega_0 = \omega * \varphi$, for some $0 \le \varphi \in \mathscr{S}(\mathbf{R}^m) \setminus 0$. (Cf. Lemma 1.2 in [26].) It follows from this equivalence that the conditions $\omega_j, v_j \in \mathscr{P}_0(\mathbf{R}^{2m})$ in Corollary 2.7 can be replaced by $\omega_j, v_j \in \mathscr{P}(\mathbf{R}^{2m})$.

Remark 2.11. From the equivalence property in Remark 2.10 it also follows that Proposition 3.3 in [2] can be formulated as:

Assume that $p, q, p_j, q_j \in [1,\infty]$ for $j = 1,2$ are such that

$$\frac{1}{p_2} - \frac{1}{p_1} = \frac{1}{p}, \qquad \frac{1}{q_2} - \frac{1}{q_1} = \frac{1}{q}.$$

Also assume that $\omega_j, v_j \in \mathscr{P}_0(\mathbf{R}^{2m})$ for $j = 1,2$ are such that ω_1 is v_1-moderate and ω_2 is $\check{v}_2$-moderate, $a\omega_2/\omega_1 \in L^{p,q}(\mathbf{R}^{2m})$ and that $h_j \in M^{1,1}_{(v_j)}$ for $j = 1,2$. Then the map (0.4) is continuous.

We note that Proposition 1.1 (2) implies that this result is a consequence of Corollary 2.7 when $p \le q$. On the other hand, if instead $q < p$, then Proposition 1.1 (2) shows that Corollary 2.7 is a consequence of Proposition 3.3 in [2].

3. Schatten-von Neumann properties for Toeplitz operators

In this section, we give examples on how the results in the previous section can be combined with Theorem 1.6 in order to establish Schatten-von Neumann properties for Toeplitz operators acting on modulation spaces of Hilbert types. This means that $p_1 = p_2 = q_1 = q_2 = 2$ in Proposition 2.1, Theorem 2.2 and Proposition 0.1″, and that (0.4) takes the form

$$\operatorname{Tp}_{h_1,h_2}(a) : M^{2,2}_{(\omega_1)} \to M^{2,2}_{(\omega_2)}. \tag{0.4$'$}$$

By combining Theorem 1.6 with Proposition 2.1 we obtain.

Proposition 3.1. *Assume that $r_j, s_j, p, p_0, q, q_0 \in [1,\infty]$ for $j = 1,2$ satisfy (2.2) and $q_0 \le \min(p_0, p_0')$. Also assume that $\nu_j \in \mathscr{P}(\mathbf{R}^{2m})$, $h_j \in M^{r_j,s_j}_{(\nu_j)}(\mathbf{R}^m)$, $j = 1,2$, and that $\omega, \omega_0 \in \mathscr{P}(\mathbf{R}^{2m} \oplus \mathbf{R}^{2m})$ satisfy*

$$\omega_0(X_1 + X_2, Y) \le C\omega(X_1, Y)\,\check{\nu}_1(Y + X_2)\nu_2(Y - X_2) \quad X_1, X_2, Y \in \mathbf{R}^{2m},$$

for some constant $C > 0$. Then $a \mapsto \operatorname{Tp}_{h_1,h_2}(a)$ from $\mathscr{S}(\mathbf{R}^{2m})$ to the set of linear and continuous operators from $\mathscr{S}(\mathbf{R}^m)$ to $\mathscr{S}'(\mathbf{R}^m)$ extends to a continuous map from $\widetilde{M}^{p,q}_{(\omega)}(\mathbf{R}^{2m})$ to $\mathscr{I}_{p_0}(M^{2,2}_{(\omega_1)}, M^{2,2}_{(\omega_2)})$.

If we instead combine Theorem 1.6 with Theorem 2.2, then we get the following.

Theorem 3.2. *Assume that $p, q, q_0, r_j, s_j \in [1,\infty]$ for $j = 1,2$ satisfy (2.2) and $q_0 \le \min(p, p')$. Also assume that $\nu_j, \omega_j \in \mathscr{P}(\mathbf{R}^{2m})$ for $j = 1,2$ and $\omega \in \mathscr{P}(\mathbf{R}^{2m} \oplus \mathbf{R}^{2m})$ satisfy (2.4) for some constant $C > 0$ independent of $X, Y \in \mathbf{R}^{2m}$. If $h_j \in M^{r_j,s_j}_{(\nu_j)}$ for $j = 1,2$ and $a \in \widetilde{M}^{p,q}_{(\omega)}(\mathbf{R}^{2m})$, then $\operatorname{Tp}_{h_1,h_2}(a) \in \mathscr{I}_p(M^{2,2}_{(\omega_1)}, M^{2,2}_{(\omega_2)})$.*

The next result is a consequence of Proposition 0.1″ and Proposition 3.1.

Proposition 0.2″. *Assume that $p, q, , q_0, r_j, s_j \in [1,\infty]$ for $j = 1,2$ satisfy (2.2) and $q_0 \le \min(p, p')$. Also assume (0.6) is fulfilled for some $\kappa, \kappa_j, v_j, \nu_j, \omega_j \in \mathscr{P}(\mathbf{R}^{2m})$ such that ω_1 is v_1-moderate and ω_2 is $\check{v}_2$-moderate. If $h_j \in M^{r_j,s_j}_{(\nu_j)}(\mathbf{R}^m)$ for $j = 1,2$ and $a \in \widetilde{M}^{p,q}_{(\omega)}(\mathbf{R}^{2m})$, then $\operatorname{Tp}_{h_1,h_2}(a) \in \mathscr{I}_p(M^{2,2}_{(\omega_1)}, M^{2,2}_{(\omega_2)})$.*

The next result is a consequence of Proposition 2.3 and Proposition 3.1.

Proposition 3.3. *Assume that $p, q, q_0, r_j, s_j \in [1,\infty]$ for $j = 0,1,2$ satisfy (2.2) and $q_0 \le \min(p, p')$. Also assume (0.6)′ is fulfilled, for some $\kappa, \kappa_j, v_j, \omega_j \in \mathscr{P}(\mathbf{R}^{2m})$ such that ω_j is v_j-moderate for $j = 1,2$. If $h_j \in M^{r_j,s_j}_{(\omega_j\kappa_j)}$ for $j = 1,2$ and $a \in \widetilde{M}^{p,q}_{(\omega)}(\mathbf{R}^{2m})$, then $\operatorname{Tp}_{h_1,h_2}(a) \in \mathscr{I}_p(M^{2,2}_{(v_1)}, M^{2,2}_{(1/v_2)})$.*

Finally we remark that more Schatten-von Neumann results can be obtained by combining Proposition 3.1 with other results in Section 2. We leave the details for the reader.

References

[1] M.S. Birman and M.Z. Solomyak, Estimates for the singular numbers of integral operators (Russian), Uspehi Mat. Nauk. **32**, (1977), 17–84.

[2] P. Boggiatto, Localization operators with L^p symbols on modulation spaces, in *Advances in Pseudo-differential Operators,* Editors: R. Ashino, P. Boggiatto and M.W. Wong, Birkhäuser, Basel 2004, 149–163.

[3] P. Boggiatto, E. Cordero and K. Gröchenig, Generalized anti-Wick operators with symbols in distributional Sobolev spaces, *Integral Equations Operator Theory*, **48** (2004), 427–442.

[4] E. Cordero and K. Gröchenig, Time-frequency analysis of localization operators, *J. Funct. Anal.* **205** (2003), 107–131.

[5] E. Cordero, S. Pilipović, L. Rodino and N. Teofanov, Localization operators and exponential weights for modulation spaces, *Mediterranean J. Math.* **2** (2005), 381–394.

[6] I. Daubechies, Time-frequency localization operators: a geometric phase space approach, *IEEE Trans. Inform. Theory* **34** (1988), 605–612.

[7] H. G. Feichtinger, Modulation spaces on locally compact abelian groups, in *Wavelets and their Applications*, Editors: M. Krishna, R. Radha and S. Thangavelu, Allied Publishers, 2003, 99–140.

[8] C. Fernández and A. Galbis, Compactness of time-frequency localization operators on $L^2(\mathbf{R}^d)$, *J. Funct. Anal.* **233** (2006), 307–340.

[9] H.G. Feichtinger and K.H. Gröchenig, *Banach spaces related to integrable group representations and their atomic decompositions, I, J. Funct. Anal.* **86** (1989), 335–350.

[10] H.G. Feichtinger and K.H. Gröchenig, Banach spaces related to integrable group representations and their atomic decompositions, II, *Monatsh. Math.* **108** (1989), 129–148.

[11] G.B. Folland, *Harmonic Analysis in Phase Space*, Princeton University Press, Princeton, 1989.

[12] I.C. Gohberg and M.G. Krein, *Introduction to the Theory of Linear Non-Selfadjoint Operators in Hilbert Space (Russian),* Izdat. Nauka, Moscow, 1965.

[13] K. Gröchenig, *Foundations of Time-Frequency Analysis*, Birkhäuser, Boston, 2001.

[14] K. Gröchenig and C. Heil *Modulation spaces and pseudo-differential operators, Integral Equations Operator Theory* **34** (1999), 439–457.

[15] K. Gröchenig, Modulation spaces as symbol classes for pseudodifferential operators, in *Wavelets and their Applications*, Editors: M. Krishna, R. Radha and S. Thangavelu, Allied Publishers, 2003, 151–170.

[16] L. Hörmander, *The Analysis of Linear Partial Differential Operators I, III*, Springer-Verlag, Berlin, Heidelberg, New York, Tokyo, 1983, 1985.

[17] Z. He and M.W. Wong, Localization operators associated to square integrable group representations, *Panamer. Math. J.* **6** (1996), 93–104.

[18] E.H. Lieb, Integral bounds for radar ambiguity functions and Wigner distributions, *J. Math. Phys.* **31** (1990), 594–599.

[19] R. Schatten, *Norm Ideals of Completely Continuous Operators*, Springer, Berlin, 1960.

[20] B.W. Schulze and N.N. Tarkhanov, Pseudodifferential operators with operator-valued symbols, in *Israel Math. Conf. Proc.* **16**, 2003.

[21] M.A. Shubin, *Pseudodifferential Operators and Spectral Theory*, Springer-Verlag, Berlin, 1987.

[22] B. Simon, *Trace Ideals and their Applications*, Cambridge University Press, Cambridge, London, New York, Melbourne, 1979.

[23] N. Teofanov, *Ultramodulation spaces and pseudodifferential operators*, Endowment Andrejević, Beograd, 2003.

[24] J. Toft, *Continuity and Positivity Problems in Pseudo-differential Calculus,* Thesis, Department of Mathematics, University of Lund, Lund, 1996.

[25] J. Toft, Continuity properties for modulation spaces with applications to pseudo-differential calculus, I, *J. Funct. Anal.*, **207** (2004), 399–429.

[26] J. Toft, Continuity properties for modulation spaces with applications to pseudo-differential calculus, II, *Ann. Global Anal. Geom.* **26** (2004), 73–106.

[27] J. Toft, Convolution and embeddings for weighted modulation spaces, in *Advances in Pseudo-differential Operators,* Editors: R. Ashino, P. Boggiatto and M.W. Wong, Birkhäuser, Basel, 2004, 165–186.

[28] J. Toft, Continuity and Schatten-von Neumann properties for pseudo-differential operators on modulation spaces, *Research Report*, Växjö University, Växjö, 2005.

[29] J. Toft, Continuity and Schatten properties for pseudo-differential operators on modulation spaces, in *Modern Trends in Pseudo-differential Operators,* Editors: J. Toft, M.W. Wong and H. Zhu, Birkhäuser, Basel, this Volume, 173–206.

[30] M.W. Wong, *Weyl Transforms*, Springer-Verlag, 1998.

[31] M.W. Wong, Localization operators on the Weyl-Heisenberg group, in *Geometry, Analysis and applications,* Editor: R.S. Pathak,, World Scientific, 2001, 303–314.

[32] M.W. Wong *Wavelet transform and localization operators*, Birkhäuser, Basel, 2002.

Joachim Toft
Department of Mathematics and Systems Engineering
Växjö University, Sweden
e-mail: `joachim.toft@vxu.se`

Operator Theory:
Advances and Applications, Vol. 172, 329–342

Microlocalization within Some Classes of Fourier Hyperfunctions

Richard D. Carmichael, Atsuhiko Eida and Stevan Pilipović

Abstract. New presheaves of hyperfunction spaces with the growth estimates with respect to $|x| \to \infty$ and $y \to 0$ in a cone Γ are introduced. Then it is shown that the Laplace transform is a bijective mapping of the space of tempered ultradistributions on $\mathbf{R}^n$ of non-quasianalytic class onto the corresponding hyperfunction space of sections over $\mathbf{D}^n$, the compactification of $\mathbf{R}^n$. Microlocalization of tempered ultradistributions at $(x_0\infty, \xi_0)$ is introduced as well as a new microlocalization within some classes of hyperfunctions.

Mathematics Subject Classification (2000). Primary 32A40, 32A45, 46F15.

Keywords. Fourier hyperfunctions, tempered ultradistributions, analytic wavefront.

1. Introduction

Two of the authors recently studied the microlocal decomposition of several classes of Sato's hyperfunctions. They showed in [4] that the sheaves of certain classes of microfunctions, e.g., $\mathcal{C}^*$, $\mathcal{C}^{d,*}$, are supple; whereas the sheaf of microfunctions itself is flabby (cf. [19], [1]). One of the authors, S. Pilipović, has introduced in [16] and [13] the spaces of tempered ultradifferentiable functions and tempered ultradistributions which are globally expressed for the usage with Fourier transforms on $\mathbf{R}^n$. These spaces $\mathcal{S}^*(\mathbf{R}^n)$, $\mathcal{S}'^*(\mathbf{R}^n)$ are topologically isomorphic to themselves through the Fourier transform. On the other hand, T. Kawai [10] introduced the spaces of Fourier hyperfunctions which are homologically constructed on $\mathbf{D}^n$, the compactification of $\mathbf{R}^n$ with infinity. These spaces are also topologically isomorphic to themselves through the Fourier transform.

The difference between the tempered ultradistributions and the Fourier hyperfunctions is that the spaces of Fourier hyperfunctions can be locally expressed from the viewpoint of sheaf. Here the authors will apply the ideas of Fourier hyperfunctions to the new presheaf of spaces of tempered ultradistributions of Roumieu type $\mathcal{S}^{\{t\}}(\Omega)$, Ω open in $\mathbf{D}^n$, and will proceed to the microlocalization as Kaneko did ([8], [7]). The authors will do this by introducing a presheaf $\mathcal{S}b^s$, $s \in (0,1)$, which is used for the definition of the wave front at $(x_0\infty, \xi) \in \mathbf{D}^n \times \mathbf{R}^n \backslash \{0\}$. The authors will also introduce a sheaf $\mathcal{DSC}^s$, a subsheaf of $\mathcal{R}^s$, and prove the softness of $\mathcal{D}b^s/\mathcal{P}^s$, $\mathcal{DRC}^s$ and $\mathcal{DSC}^s$.

In order to point out the main idea of the paper recall that Sato had constructed hyperfunctions using analytic functions without a growth estimate and Kawai [10], Zharinov [21], Kaneko [8], [9], Saburi [18] and others had introduced various Fourier hyperfunctions by the means of appropriate growth rate at infinity with respect to real variable. In our analysis of introduced hyperfunction spaces bounds appear depending also on the imaginary variable as it approaches to the boundary.

2. Tempered ultradistributions

We refer to [11] and [12] for the definitions of spaces of ultradifferentiable functions and ultradistributions of Beurling and Roumieu type. Here we recall the definition of $\mathcal{E}^{\{t\}}(\Omega)$, where Ω is open in $\mathbf{R}^n$ and $t > 1$:

$$\mathcal{E}^{\{t\}}(\Omega) = \operatorname*{indlim}_{K\subset\subset\Omega}\ \operatorname*{indlim}_{h\to\infty} \mathcal{E}^{t,h}(K),$$

where

$$\mathcal{E}^{t,h}(K) = \{\varphi \in C^\infty(K); ||\varphi||_{K,h} < \infty,$$

$$||\varphi||_{K,h} = \sup\{\frac{h^{|\alpha|}|\varphi^{(\alpha)}(x)|}{\alpha!^t}; x \in K, \alpha \in \mathbf{N}_0^n\}.$$

($K \subset\subset \Omega$ means that compact sets K run over a set of compact sets exhausting Ω.)

Let $\mathbf{D}^n = \mathbf{R}^n \sqcup S_\infty^{n-1}$ be the directional compactification of $\mathbf{R}^n$ with the natural topology. A fundamental system of neighborhoods of $a\infty \in S_\infty^{n-1}$ is given by sets of the form

$$(\Gamma \cap \{|x| > R\}) \sqcup (\Gamma \cap S^{n-1})_\infty,$$

where Γ runs over open cones containing a and $R \gg 0$.

Let $t > 1$ and $h > 0$. Let Ω be an open set of $\mathbf{D}^n$. A space of smooth functions φ on $\mathbf{R}^n \cap \Omega$ with compact supports contained in Ω such that

$$\sigma_{h,2}(\varphi) := \left(\sum_{\alpha,\beta\in\mathbf{N}_0^n} \int_{\mathbf{R}^n\cap\Omega} \left| \frac{h^{|\alpha+\beta|}}{\alpha!^t\beta!^t}(1+|x|^2)^{|\beta|/2}\varphi^{(\alpha)}(x)\right|^2 dx \right)^{1/2} < \infty,$$

equipped with the topology induced by the norm $\sigma_{h,2}$, is denoted by $\tilde{\mathcal{S}}_2^h(\Omega)$. Denote by $\mathcal{S}_2^h(\Omega)$ its completion. The strong dual of

$$\mathcal{S}^{\{t\}}(\Omega) = \operatorname*{indlim}_{h\to 0} \mathcal{S}_2^h(\Omega),$$

$\mathcal{S}'^{\{t\}}(\Omega)$, is called the space of tempered ultradistributions of Roumieu type. In case $\Omega = \mathbf{D}^n$, we have $\mathcal{S}'^{\{t\}}(\mathbf{D}^n) = \mathcal{S}'^{\{t\}}(\mathbf{R}^n)$, the space of tempered ultradistributions. For the properties of $\mathcal{S}'^{\{t\}}(\mathbf{R}^n)$ as well as of Beurling type spaces, we refer to [13] and [17].

It is clear that a function $\varphi \in \mathcal{S}^{\{t\}}(\Omega)$ can be extended out of Ω. More precisely, there is a $\Phi \in \mathcal{S}^{\{t\}}(\Omega_\varepsilon)$, where $\Omega_\varepsilon = \{x \in \mathbf{D}^n; d(x,\Omega) < \varepsilon\}$, $\varepsilon > 0$, such that $\operatorname{supp}\varphi \subset \Omega_\varepsilon$ and $\Phi_{|\Omega} = \varphi$.

We refer to [2] for the extension of ultradifferentiable functions of appropriate classes forming Whitney's jets on a closed set.

Using the part of the proof of Theorem 2 in [17] we have that an element $f \in \mathcal{D}'^{\{t\}}(\mathbf{R}^n)$ is in $\mathcal{S}'^{\{t\}}(\mathbf{R}^n)$ if and only if it is of the form

$$\langle f, \varphi\rangle = \int_{\mathbf{R}^n} F(t)P(\partial)\varphi(t)dt, \ \varphi \in \mathcal{S}^{\{t\}}(\mathbf{R}^n), \tag{1}$$

where P is an ultradifferential operator of $\{t\}$ – class ([11]) and a function F satisfies $F(x)e^{-k|x|^s} \in L^2(\mathbf{R}^n)$, $s = 1/t$.

Let Ω be an arbitrary open set of $\mathbf{D}^n$. Then by the same proof as for $\Omega = \mathbf{D}^n$, we have the representation

$$\langle f, \phi\rangle = \int_{\mathbf{R}^n\cap\Omega} F(t)P(\partial)\phi(t)dt, \phi \in \mathcal{S}^{\{t\}}(\Omega), \tag{2}$$

where $P(\partial)$ is an ultradifferential operator of $\{t\}$ – class and F is a function such that $F(x)e^{-k|x|^s} \in L^2(\mathbf{R}^n \cap \Omega)$ for every $k > 0$. The representations in (1) and (2) are not unique.

We will show how the representation (2) can be used for the definition of the extension of an $f \in \mathcal{S}'^{\{t\}}(\Omega)$ to an element $f_{ext} \in \mathcal{S}'^{\{t\}}(\mathbf{R}^n)$. With the notation as in (2), put

$$F_{ext} = F \ \text{ on } \ \mathbf{R}^n \cap \Omega, \ \ F_{ext} = 0 \ \text{ on } \ \mathbf{R}^n \setminus (\mathbf{R}^n \cap \Omega)$$

and define $f_{ext} \in \mathcal{S}'^{\{t\}}(\mathbf{R}^n)$ by

$$\langle f_{ext}, \varphi\rangle = \int_{\mathbf{R}^n} F_{ext}(t)P(\partial)\varphi(t)dt, \ \varphi \in \mathcal{S}^{\{t\}}(\mathbf{R}^n). \tag{3}$$

We have

$$f_{ext|\Omega} = f,$$

where the left side is defined by

$$\langle f_{ext}, \varphi\rangle = \int_{\mathbf{R}^n\cap\Omega} F_{ext}(t)P(\partial)\varphi(t)dt, \ \varphi \in \mathcal{S}^{\{t\}}(\Omega).$$

Let Ω_1, Ω_2 be arbitrary open sets of $\mathbf{D}^n$ and $\Omega_1 \subset \Omega_2$. We define the restriction mapping $r_{\Omega_2,\Omega_1} : \mathcal{S}'^{\{t\}}(\Omega_2) \to \mathcal{S}'^{\{t\}}(\Omega_1)$ as follows. Let $f \in \mathcal{S}'^{\{t\}}(\Omega_2)$ be of the form (2) (with Ω_1 instead of Ω).

Then $r_{\Omega_2,\Omega_1}(f)$ is defined by

$$\langle r_{\Omega_2,\Omega_1}(f), \varphi\rangle = \int_{\mathbf{R}^n\cap\Omega_1} F(t)P(\partial)\varphi(t)dt, \ \varphi \in \mathcal{S}^{\{t\}}(\Omega_1).$$

This definition does not depend on the representatives F and $P(\partial)$ in (2) since compactly supported elements of $\mathcal{S}^{\{t\}}(\Omega_1)$ make a dense set in $\mathcal{S}^{\{t\}}(\Omega_1)$ and they are contained also in $\mathcal{S}^{\{t\}}(\Omega_2)$.

The given explanation proves the next proposition:

Proposition 1. *Let $\Omega_1 \subset \Omega_2 \subset \Omega_3$ be open sets of $\mathbf{D}^n$ and let $f \in \mathcal{S}^{\{t\}}(\Omega_3)$. Then*

$$r_{\Omega_3,\Omega_1}(f) = r_{\Omega_2,\Omega_1}(r_{\Omega_3,\Omega_2} f).$$

Thus we have constructed a presheaf $\Omega \to \mathcal{S}'^{\{t\}}(\Omega)$, Ω runs through all open sets in $\mathbf{D}^n$. It is not a sheaf. In fact, the first sheaf property related to the support is satisfied but the second sheaf property (gluing of sections) is not satisfied. The corresponding associated sheaf is denoted by $\Omega \to \mathcal{D}'^{\{t\}}(\Omega)$, $\Omega \in \mathbf{D}^n$. If $\Omega \subset \mathbf{R}^n$, then it is the Roumieu space of ultradistributions. This is a consequence of the fact that any f in the associated sheaf has the representation of an ultradistribution of Roumieu type over any open bounded set G such that $G \subset\subset \Omega \subset \mathbf{R}^n$. We denote the associated sheaf by $\mathcal{D}'^{\{t\}}$.
Recall that the Fourier transform is an isomorphism of $\mathcal{S}^{\{t\}}(\mathbf{R}^n)$ onto itself.

3. Sheaves of Fourier hyperfunctions of type $[s, 0]$ ([8])

Let $\mathcal{O}$ be the sheaf of holomorphic functions on $\mathbf{C}^n$. Kaneko has introduced in [8] the sheaves $\tilde{\mathcal{O}}^{s,\delta}$, $s > 0, \delta \in \mathbf{R}$. We need such sheaves with $s \in (0,1)$ and $\delta = 0$. Recall, for open $U \subset \mathbf{D}^n + \sqrt{-1}\mathbf{R}^n$,

$$\tilde{\mathcal{O}}^{s,0}(U) = \tilde{\mathcal{O}}^{s}(U)$$

$$= \{F(z) \in \mathcal{O}(U \cap \mathbf{C}^n);\ \forall K \subset\subset U, \forall \varepsilon > 0,\ \sup_{z \in K \cap \mathbf{C}^n} |F(z)| e^{-\varepsilon |z|^s} < \infty\},$$

where $K \subset\subset U$ means that the closure of K in $\mathbf{D}^n + \sqrt{-1}\mathbf{R}^n$ is a compact set contained in U.

Let Ω be an open subset of $\mathbf{D}^n$ and Γ be an open cone in $\mathbf{R}^n$. Then

$$\tilde{\mathcal{O}}^{s}(\Omega + \sqrt{-1}\Gamma 0) := \operatorname*{indlim}_{U} \tilde{\mathcal{O}}^{s}(U),$$

U running over infinitesimal wedges with edge Ω and opening Γ (for the definition of an infinitesimal wedge we refer to [7], Definition 2.2.9).

The sheaf $\mathcal{P}^s$ on $\mathbf{D}^n$, is defined by $\tilde{\mathcal{O}}^s|_{\mathbf{D}^n}$ and the sheaf of Fourier hyperfunctions on $\mathbf{D}^n$ is defined by $\mathcal{Q}^s = \mathcal{H}^n_{\mathbf{D}^n}(\tilde{\mathcal{O}}^s)$ ([8]).

For every open set $\Omega \subset \mathbf{D}^n$, set $\mathcal{Q}^s(\Omega) = H^n_{\Omega}(\mathbf{C}^n, \tilde{\mathcal{O}}^s)$. This cohomological expression can be translated in the following way: $\mathcal{Q}^s(\Omega) = X/Y$, where

$$X = \sum_{\Gamma} \tilde{\mathcal{O}}^s(\Omega + \sqrt{-1}\Gamma 0),\ \Gamma \text{ varying over all open convex cones,}$$

and Y is a $\mathbf{C}-$ linear subspace of X generated by all the elements of the form

$$F_1(z) + F_2(z) - F_3(z), F_j(z) \in \tilde{\mathcal{O}}^s(\Omega + \sqrt{-1}\Gamma_j 0), j = 1, 2, 3,$$

such that $\Gamma_3 \subset \Gamma_1 \cap \Gamma_2$ and $F_1(z) + F_2(z) = F_3(z)$ on a common domain of definition.

We use the same notation $\sum_{j=1}^{N} F_j(x + \sqrt{-1}\Gamma_j 0)$ for an element in X and the corresponding class in X/Y. Note that

$$\sum_{j=1}^{N} F_j(x + \sqrt{-1}\Gamma_j 0) \sim 0 \quad \text{if and only if there exists}$$

$H_{jk}(z) \in \tilde{\mathcal{O}}^s(\Omega + \sqrt{-1}(\Gamma_j + \Gamma_k)0)$ with $H_{jk}(z) = -H_{kj}(z), j, k, = 1, \ldots, N$, and

$$F_j(z) = \sum_{k=1}^{N} H_{jk}(z), \; j = 1, \ldots, N, \text{ on a common domain.}$$

The singular spectrum of an $f \in \mathcal{Q}^s(\Omega)$ is defined as follows: f is micro-analytic at $(a, \xi) \in \Omega \times \; S^{n-1}$ if there exists a local boundary value expression $f = \sum_{j=1}^{r} F_j(x + \sqrt{-1}\Gamma_j 0)$ on a neighborhood of a such that $\Gamma_j \cap \{\langle y, \xi\rangle < 0\} \neq \emptyset$ for all j. The complement of the set of points where f is microlocally analytic is called the singular spectrum of f and denoted by $SS^s f$. The corresponding (analytic) singular support is defined by $\operatorname{singsupp}_a^s f = \pi(SS^s f)$ (Refer to [7] and [8] for details.)

Next, the sheaf $\mathcal{R}^s$ of Fourier microfunctions on $\mathbf{D}^n \times S^{n-1}$ is defined as the one associated to the presheaf

$$\mathbf{D}^n \times S^{n-1} \supset \Omega \times \Lambda \to \mathcal{Q}^s(\Omega)/\{f \in \mathcal{Q}^s(\Omega); SS^s(f) \cap \Omega \times \Delta = \emptyset\}.$$

Let $\pi : \mathbf{D}^n \times S^{n-1} \to \mathbf{D}^n$ be a natural (first) projection.

Kaneko has proved a generalized version of Sato's fundamental exact sequence:

$$0 \to \mathcal{P}^s \to \mathcal{Q}^s \to \pi_* \mathcal{R}^s \to 0.$$

The sheaves $\mathcal{Q}^s, \mathcal{Q}^s/\mathcal{P}^s$, and $\mathcal{R}^s$ are flabby and $H^k(\Omega, \mathcal{P}^s) = 0, k \geq 1$, for real open $\Omega \subset \mathbf{D}^n$.

So we have a global exact sequence

$$0 \to \mathcal{P}^s(\Omega) \to \mathcal{Q}^s(\Omega) \to \pi_* \mathcal{R}^s(\Omega \times S^{n-1}) \to 0.$$

If $s = 1$, we omit the superindex. Is this case $\mathcal{Q}$ and $\mathcal{R}$ are well-known spaces for Fourier hyperfunctions and microfunctions. We have the topological isomorphisms through the Fourier transform:

$$\mathcal{F} : \mathcal{Q}(\mathbf{D}^n) \xrightarrow{\sim} \mathcal{Q}(\mathbf{D}^n), \; \mathcal{F} : \mathcal{R}(\mathbf{D}^n \times S^{n-1}) \xrightarrow{\sim} \mathcal{R}(\mathbf{D}^n \times S^{n-1}).$$

4. Tempered hyperfunctions

In the sequel we will use notation $s \in (0,1)$ and $t = 1/s$.

Definition 1. Let Ω be an open set in $\mathbf{D}^n$ and Γ be an open convex cone in $\mathbf{R}^n$. $\tilde{\mathcal{O}}^s_\tau(\Omega + \sqrt{-1}\Gamma 0)$ is the inductive limit of spaces of functions F analytic in an appropriate infinitesimal wedge with the basis Ω and the opening Γ such that

$$\begin{aligned}&(\forall K \subset\subset \Omega)\ (\forall \Delta \subset\subset \Gamma)(\forall a > 0)(\forall b > 0)(\exists c > 0)\\ &|F(x + \sqrt{-1}y)| < c\ \exp(a|x|^s + b|y|^{s/(1-s)}),\\ &x + \sqrt{-1}y \in (\mathbf{R}^n \cap K) + \sqrt{-1}\Delta, |y| < 1/c.\end{aligned} \tag{4}$$

$\mathcal{S}b^s(\Omega)$ is a $\mathbf{C}$-vector space defined as X'/Y', where X' and Y' are defined (as were X and Y in Section 3) by

$$X' = \sum_\Gamma \tilde{\mathcal{O}}^s_\tau(\Omega + i\Gamma 0),\ \Gamma \text{ varying over all open convex cones;}$$

and Y' is a $\mathbf{C}$-linear subspace of X' generated by all elements of the form

$$F_1(z) + F_2(z) - F_3(z),\ \ F_j(z) \in \tilde{\mathcal{O}}^s_\tau(\Omega + \sqrt{-1}\Gamma_j 0),\ j = 1,2,3,$$

such that $\Gamma_3 \subset \Gamma_1 \cap \Gamma_2$ and that $F_1(z) + F_2(z) = F_3(z)$ on a common domain of definition.

X' consists of sums of the form

$$\sum_{j=1}^N F_j(z),\ \ F_j \in \tilde{\mathcal{O}}^s_\tau(\Omega + \sqrt{-1}\Gamma_j 0). \tag{5}$$

The corresponding equivalence class to (5) is denoted by $\sum_{j=1}^N F_j(x + \sqrt{-1}\Gamma_j 0)$ and, as for Fourier hyperfunctions, we will use the same notation for the elements of X' and the corresponding classes.

5. Laplace transform

For a cone $\gamma \subset \mathbf{R}^n, \gamma^0$, $\operatorname{int}\gamma$ and $\bar{\gamma}$ denote the polar cone, the interior of γ and the closure of γ, respectively. Recall that a cone is said to be acute if it does not contain straight lines. In this case $\operatorname{int}(\gamma^0) \neq \emptyset$.

We use the definition of the Laplace transform with the weight function

$$\exp\sqrt{-1}(x + \sqrt{-1}y)t$$

which involves $ty = \sum t_i y_i \geq 0$ for a t belonging to a cone and y belonging to a polar cone. Thus, for an $f \in \mathcal{S}'^{\{t\}}(\mathbf{R}^n)$, such that $\operatorname{supp} f \subset \gamma$, its Laplace transformation $\mathcal{L}f$ is defined by

$$\mathcal{L}f(z) = \int_\gamma f(t)(\exp\sqrt{-1}(x + \sqrt{-1}y)t)dt,\ x + \sqrt{-1}y \in \mathbf{R}^n + \sqrt{-1}\operatorname{int}(\gamma^0),$$

and in this domain it is a holomorphic function.

Isomorphism $j : \mathcal{S}'^{\{t\}}(\mathbf{R}^n) \mapsto \mathcal{S}b^s(\mathbf{D}^n)$.

With the notation $\Gamma = \operatorname{int}(\gamma^0)$ and $f \in \mathcal{S}'^{\{t\}}(\mathbf{R}^n)$, $\operatorname{supp} f \subset \gamma$, we have proved in [3] that $F = \mathcal{L}f \in \mathcal{O}(\mathbf{R}^n + \sqrt{-1}\Gamma)$ satisfies

$$(\forall a > 0)(\forall b > 0)(\forall \varepsilon > 0)(\exists M > 0)$$

$$|F(x + \sqrt{-1}y)| \leq M \ \exp(a|x|^s + b|y|^{s/(s-1)} + \varepsilon|y|), \ \ x + \sqrt{-1}y \in \mathbf{R}^n + \sqrt{-1}\Gamma \tag{6}$$

and

$$F(x + \sqrt{-1}\Gamma_i 0) = bvF = (2\pi)^n \mathcal{F}^{-1}(f)(x) \in \mathcal{S}'^{\{t\}}(\mathbf{R}^n) \ i = 1, \ldots, p. \tag{7}$$

Let $K \subset \mathbf{R}^n$ be a compact set such that $0 \in \operatorname{int} K$; and let $\gamma_i, i = 1, \ldots, p$, be open convex cones such that $\cup_{i=1}^p \gamma_i = \mathbf{R}^n \setminus \{0\}$. Assume that cones $\overline{\gamma}_i, i = 1, \ldots, p$, are acute. Let $\eta_0, \eta_1, \ldots, \eta_p \in C^\infty$ be a partition of unity such that $\operatorname{supp} \eta_0 \subset K$, and $\operatorname{supp} \eta_i \subset \gamma_i, i = 1, \ldots, p$. Let $f \in \mathcal{S}'^{\{t\}}(\mathbf{R}^n)$. Clearly, $\tilde{f} = (2\pi)^{-n}\mathcal{F}f \in \mathcal{S}'^{\{t\}}(\mathbf{R}^n)$. We make the decomposition

$$\tilde{f} = \tilde{f}_0 + \sum_{i=1}^p \tilde{f}_i,$$

$$\tilde{f}_0 = \tilde{f}\eta_0 \in \mathcal{E}'^{\{t\}}(\mathbf{R}^n), \ \tilde{f}_i = \tilde{f}\eta_i \in \mathcal{S}'^{\{t\}}(\mathbf{R}^n), \operatorname{supp} \tilde{f}_i \subset \gamma_i, i = 1, \ldots, p.$$

We define the Laplace transform by

$$\mathcal{L}f := F_0 + \sum_{i=1}^p F_i, \tag{8}$$

where

$$F_0(z) = \mathcal{L}\tilde{f}_0(z), z \in \mathbf{C}^n; \ F_i(z) \ = \mathcal{L}\tilde{f}_i(z), \operatorname{Im} z \in \Gamma_i \ = \operatorname{int}(\gamma_i^0), i = 1, \ldots, p.$$

By the Paley-Wiener theorem we know that F_0 is an entire function. Moreover, it is an element of $\tilde{\mathcal{O}}^s_\tau(\Omega + \sqrt{-1}\Gamma 0)$, for any open cone Γ. By (6) and (7) we have that $F_i \in \mathcal{O}(\mathbf{R}^n + \sqrt{-1}\Gamma_i)$ satisfies

$$(\forall a > 0)(\forall b > 0)(\forall \varepsilon > 0)(\exists M > 0)$$

$$|F_i(x + \sqrt{-1}y)| \leq M \ \exp(a|x|^s + b|y|^{s/(s-1)} + \varepsilon|y|), \ \ x + \sqrt{-1}y \in \mathbf{R}^n + \sqrt{-1}\Gamma_i$$

and

$$F_i(x + \sqrt{-1}\Gamma_i 0) = bvF_i = (2\pi)^n \mathcal{F}^{-1}(\tilde{f}\eta_i)(x), \ i = 1, \ldots, p.$$

So by (8), we let f correspond to the formal sum $\sum_{i=0}^p F_i(x + \sqrt{-1}\Gamma_i 0)$, where (4) holds in corresponding cones.

We refer to [3] for the proof of (6); for tempered distributions supported by a cone the corresponding proof is given in [20].

In the first part of the next theorem we prove that $\mathcal{L} : \mathcal{S}'^{\{t\}}(\mathbf{R}^n) \to \mathcal{S}b^s(\mathbf{D}^n)$ does not depend on the given decomposition. So, for this purpose, let $T, \delta_j, j = 1, \ldots, r$ be another covering of $\mathbf{R}^n$ and $\nu_j, j = 0, \ldots, r$, be a corresponding partition of unity with the same properties as above. We have $\tilde{f} = \tilde{g}_0 + \sum_1^r \tilde{g}_j$, where $\tilde{g}_0 = \tilde{f}\nu_0 \in \mathcal{E}'^{\{t\}}(\mathbf{R}^n), \tilde{g}_j = \tilde{f}\nu_j \in \mathcal{S}'^{\{t\}}(\mathbf{R}^n)$, $\operatorname{supp} \tilde{g}_j \subset \delta_j$. Let $G_j = \mathcal{L}(\tilde{g}_j)$ $j = 0, \ldots, r$.

Theorem 1.

(i) *With the above notation,*

$$\sum_{i=0}^{p} F_i - (\sum_{j=0}^{r} G_j) \in Y'.$$

(ii) *Let*

$$\jmath(f) := (2\pi)^{-n}\mathcal{L}\mathcal{F}(f),\ f \in \mathcal{S}'^{\{t\}}(\mathbf{R}^n).$$

Then $\jmath$ is an algebraic isomorphism of $\mathcal{S}'^{\{t\}}(\mathbf{R}^n)$ onto $\mathcal{S}b^s(\mathbf{D}^n)$ and

$$bv(\jmath(f)) := \sum_{i=0}^{p} \lim_{y_i \to 0, y_i \in \Gamma_i} F_i(x + \sqrt{-1}y_i)$$

equals f in the sense of convergence in $\mathcal{S}'^{\{t\}}(\mathbf{R}^n)$.

Proof. (i) For every $i = 0, 1, \ldots, p$,

$$F_i(z) = \mathcal{L}(\sum_{j=0}^{r} \tilde{f}_i\nu_j)(z),\ \operatorname{Im} z \in \Gamma_i = \operatorname{int}(\gamma_i^0).$$

This implies $F_i - \sum_{j=0}^{r} \mathcal{L}(\tilde{f}_i\nu_j) \in Y'$ and

$$\begin{aligned}\sum_{i=0}^{p} F_i - \sum_{i=0}^{p}\sum_{j=0}^{r} \mathcal{L}(\nu_j\eta_i\tilde{f}) &= \sum_{i=0}^{p} F_i - \sum_{j=0}^{r}\sum_{i=0}^{p} \mathcal{L}(\eta_i\tilde{g}_j)\\ &= \sum_{i=0}^{p} F_i - \sum_{j=0}^{r} \mathcal{L}(\tilde{g}_j) = \sum_{i=0}^{p} F_i - \sum_{j=0}^{r} G_j \in Y'.\end{aligned}$$

This implies assertion (i).

(ii) Let $f \in \mathcal{S}'^{\{t\}}(\mathbf{R}^n)$. Again, we put $\tilde{f} = (2\pi)^{-n}\mathcal{F}(f)$ and use the decomposition $\tilde{f} = \tilde{f}\eta_0 + \sum_{i=1}^{p} \tilde{f}\eta_i$, using the partition of unity $\eta_i, i = 1, \ldots, p$. With $\Gamma_i = \operatorname{int}(\gamma_i^0)$, $i = 1, \ldots, p$, $F_i = \mathcal{L}(\tilde{f}\eta_i) \in \tilde{\mathcal{O}}_\tau^s(\mathbf{R}^n + \sqrt{-1}\Gamma_i 0)$ and by (7), we have

$$F_i(x + \sqrt{-1}\Gamma_i 0) = (bvF_i)(x) = (2\pi)^n \mathcal{F}^{-1}(\tilde{f}\eta_i)(x) \in \mathcal{S}'^{\{t\}}(\mathbf{R}^n),\ i = 1, \ldots, p,$$

$$\sum_{i=0}^{p} F_i(x + \sqrt{-1}\Gamma_i 0) = \sum_{i=0}^{p} (bvF_i)(x) = f(x).$$

So $\jmath(f)$ is the formal sum $\sum_{i=0}^{p} F_i(x + \sqrt{-1}\Gamma_i 0)$ such that (6) holds in corresponding cones.

Let $f, g \in \mathcal{S}'^{\{t\}}(\mathbf{R}^n)$ so that

$$\jmath(f) - \jmath(g) \in Y'.$$

By definition, any element in Y' has the boundary value equaling 0, in $\mathcal{S}'^{\{t\}}(\mathbf{R}^n)$ (the sum of limits when the imaginary part approaches to 0 in corresponding cones). This implies that

$$bv(j(f)) - bv(j(g)) = f - g = 0$$

and this implies the injectivity of j.

Let us show that the mapping $j : \mathcal{S}'^{\{t\}}(\mathbf{R}^n) \to \mathcal{S}b^s(\mathbf{D}^n)$ is surjective. Let $f \in \mathcal{S}b^s(\mathbf{D}^n)$ be determined by a single element $F(x+\sqrt{-1}\Gamma 0)$ which satisfies (4). We will show that the boundary value of F as $y \to 0,\ y \in \Gamma$, in the sense of convergence in $\mathcal{S}'^{\{t\}}(\mathbf{R}^n)$, is an element $\tilde{f} \in \mathcal{S}'^{\{t\}}(\mathbf{R}^n)$. Then, by the proof of injectivity it follows that this f is the image of $\tilde{f}$, $j(\tilde{f}) = f$.

So let us find the boundary value of F in $\mathcal{S}'^{\{t\}}(\mathbf{R}^n)$.

The idea of proof is similar to the proof of the corresponding assertions in $\mathcal{D}'(\mathbf{R}^n)$ ([7], Theorem 3.1.15). It is based on the almost analytic extension in which appropriate "cut off" functions appear.

As in [13], let κ be a non-negative function which belongs to $\mathcal{S}^{\{t\}}(\mathbf{R})$ such that $\operatorname{supp}\kappa \subset [-2,2]$, $\kappa\mid_{[-1,1]}= 1$ and let

$$\kappa_p(y) = \kappa(4yp^{t-1}/h),\quad y > 0,\ p \in \mathbf{N}_0$$

(remember, $t = 1/s > 1$).

Let φ be a smooth function such that $s_h(\varphi) < \infty$. The almost analytic extension of φ is defined by

$$\Phi(z) = \sum_{p\in\mathbf{N}_0^n} \frac{\varphi^{(p)}(x)}{p!}(\sqrt{-1}y)^p \kappa_p(y),\quad z = x+\sqrt{-1}y \in \mathbf{C}^n,$$

where

$$p! = p_1!\dots p_n!,\ (\sqrt{-1}y)^p = (\sqrt{-1}y_1)^{p_1}\dots(\sqrt{-1}y_n)^{p_n},$$

$$\kappa_p(y) = \kappa_{p_1}(y_1)\dots\kappa_{p_n}(y_n).$$

This is a smooth function in $\mathbf{R}^{2n}$.

Let $0 \neq Y \in \Gamma$, $\mid Y \mid < \gamma$, $Y = (Y_1,\dots,Y_n), Y_i \neq 0,\ i = 1,\dots,n$. (If some $Y_i = 0$, the proof is the same.) Using

$$\frac{1}{4}h/p^{t-1} \le Y_i y \le \frac{1}{2}h/p^{t-1},\ \text{ on the support of } \kappa_p(yY),\ p \in \mathbf{N}_0,$$

$$\max\{|\kappa(y)|,\ |\kappa'(y)|\} \le C,\ \ y \in \mathbf{R},$$

it follows that there are $C_h > 0$ and $H_h > 0$ such that

$$\sup_{i=1,\dots,n}\{\exp(h\mid x\mid^s + \frac{1}{H_h|y|^{s/(1-s)}})\mid \frac{\partial}{\partial\bar{z}_i}\Phi(x+\sqrt{-1}yY)\mid\} \le C_h\sigma_{h,2}(\varphi), \tag{9}$$

$$e^{h|x|^s}\mid \Phi(x+\sqrt{-1}yY)\mid \le C_h\sigma_{h,2}(\varphi),\ \ x \in \mathbf{R}^n, y \in [0,1).$$

Put $Z_Y = \{z = x + \sqrt{-1}yY;\quad x \in \mathbf{R}^n,\quad 0 \le y \le 1\}$. Let ψ be an analytic function and θ, $\frac{\partial}{\partial \bar{z}_i}\theta$, $i = 1, \ldots, n$, be continuous functions in a neighborhood of Z_Y. Assume that $\psi\theta$ is integrable in this neighborhood and that

$$| (\psi\theta)(\ldots, a_{i_1} + \sqrt{-1}tY_{i_1}, \ldots, a_{i_m} + \sqrt{-1}tY_{i_m}, \ldots) | \to 0 \quad \text{as} \quad a \to \infty,$$

uniformly in $y \in [0,1]$, and $x_j \in \mathbf{R}$, $j \neq i_1, \ldots, i_m$,

$$\text{where } | a_{i_j} | = a,\ \ a > 0,\ \ i_1, \ldots, i_m \in \{1, \ldots, n\},\ \ 1 \le m \le n.$$

Then,

$$\int_{\mathbf{R}^n} \theta(x_1, \ldots, x_n)\psi(x_1, \ldots, x_n)dx_1, \ldots, dx_n = \int_{\mathbf{R}^n} \theta(x + \sqrt{-1}Y)\psi(x + \sqrt{-1}Y)dx$$

$$+2\sqrt{-1}\sum_{i=1}^{n} Y_i \int_0^1 \int_{\mathbf{R}^n} \frac{\partial \theta}{\partial \bar{z}_i}(x + \sqrt{-1}yY)\psi(x + \sqrt{-1}yY)dydx.$$

The proof of this assertion follows from Stokes' formula on

$$Z_{Y_a} = \{z = x + \sqrt{-1}yY;\ -a \le x_i \le a,\ i = 1, \ldots, n,\ 0 \le y \le 1\}$$

letting $a \to \infty$.

There exists a constant $C > 0$ such that

$$y \le C \mid u + yY \mid,\quad u \in \Gamma,\ y > 0 \quad ([7], \text{ p. } 66).$$

Let $u \in \Gamma$, $\varphi \in \mathcal{S}^{\{t\}}(\mathbf{R}^n)$. Then functions

$$\psi(z) = F(z + \sqrt{-1}u),\ \theta(z) = \Phi(z),\ z \in Z_Y,$$

satisfy conditions given above. This follows from the assumptions and (9). Letting $u \to 0$, $u \in \Gamma$, we have

$$\langle F(x + \sqrt{-1}0), \varphi(x)\rangle = \int_{\mathbf{R}^n} F(x + \sqrt{-1}Y)\Phi(x + \sqrt{-1}Y)dx$$

$$+2\sqrt{-1}\sum_{i=1}^{n} Y_i \int_0^1 \int_{\mathbf{R}^n} \frac{\partial}{\partial \bar{z}_i}\Phi(x + \sqrt{-1}yY)F(x + \sqrt{-1}Yy)dydx.$$

Moreover, for some $C > 0$ there holds

$$| \langle F(x + \sqrt{-1}0), \varphi(x)\rangle | \le C\sigma_{h,2}(\varphi),\ \varphi \in \mathcal{S}^{\{t\}}.$$

This completes the proof of Theorem 1.

Isomorphism $j : \mathcal{S}'^{\{t\}}(\Omega) \mapsto \mathcal{S}b^s(\Omega), \Omega \subset \mathbf{D}^n$. Let $f \in \mathcal{S}'^{\{t\}}(\Omega)$. Then we define f_{ext} as in Section 2 through (2) and (3). By Theorem 1 we have a unique element $j(f_{\text{ext}}) \in \mathcal{S}b^s(\mathbf{D}^n)$. The arguments of Theorem 1 imply that

$$bvj(f_{\text{ext}})_{|\Omega} = f_{\text{ext}|\Omega} = f.$$

Thus we have the next theorem:

Theorem 2. *The mapping j is an isomorphism of $\mathcal{S}'^{\{t\}}(\Omega)$ onto $\mathcal{S}b^s(\Omega)$.*

6. Tempered hyperfunctions (Continuation)

Since the mapping j is an isomorphism of $\mathcal{S}'^{\{t\}}(\Omega)$ onto $\mathcal{S}b^s(\Omega)$, $\Omega \subset \mathbf{D}^n$ and $\Omega \to \mathcal{S}'^{\{t\}}(\Omega)$, is a presheaf, we obtain the presheaf $\Omega \mapsto \mathcal{S}b^s(\Omega)$. We denote the associated sheaf as $\mathcal{D}b^s$.

Note that the canonical mapping $\mathcal{S}b^s(\Omega) \to \mathcal{D}b^s(\Omega) = \Gamma(\Omega, \mathcal{D}b^s)$ is injective. If $f \in \mathcal{S}'^{\{t\}}(\Omega)$, we denote by jf the corresponding element of $\mathcal{D}b^s(\Omega)$.

The singular spectrum of an $f \in \mathcal{S}b^s(\Omega)$, denoted by $SS^s_\tau f$, is defined as the complement of $\mathcal{S}b^s_\tau$-microanalytic points, where we define: f is $\mathcal{S}b^s_\tau$-microanalytic at $(a, \xi) \in \Omega \times S^{n-1}$ if there exists a local boundary value expression of the form $f = \sum_{j=1}^{r} F_j(x + \sqrt{-1}\Gamma_j 0)$ $(F_j \in \tilde{\mathcal{O}}^s_\tau(\Omega + \sqrt{-1}\Gamma_j 0))$ on a neighborhood of a such that $\Gamma_j \cap \{\langle y, \xi\rangle < 0\} \neq \emptyset$ for all j.

We define

$$\operatorname{singsupp}^s_\tau f = \pi(SS^s_\tau f), \ f \in \mathcal{S}b^s(\Omega).$$

The canonical mapping $k : \mathcal{D}b^s \to \mathcal{Q}^s$ defined by:

$$\sum_{j=1}^{N} F_j(x + \sqrt{-1}\Gamma_j 0) \mapsto \sum_{j=1}^{N} F_j(x + \sqrt{-1}\Gamma_j 0)$$

is injective. So we have the injective mapping $k : \mathcal{S}'^{\{t\}}(\Omega) \to \Gamma(\Omega, \mathcal{Q}))$ The next assertions are consequences of definitions.

Theorem 3. 1) *$\tilde{\mathcal{D}}b^s := k(\mathcal{D}b^s)$ is a soft subsheaf of $\mathcal{Q}^s$.*

2) *Let $f \in \mathcal{S}b^s(\Omega)$. Then $SS^s_\tau f \supset SS^s k(f)$. In particular, if we define $\operatorname{singsupp}^s_a f := \pi(SS^s k(f))$, then $\operatorname{singsupp}^s_\tau f \supset \operatorname{singsupp}^s_a f$.*

7. Microlocalization

With the notation $\tilde{\mathcal{D}}b^s = k(\mathcal{D}b^s)$, we define quotient sheaves $\mathcal{D}b^s/\mathcal{P}^s$ and $\mathcal{Q}^s/\tilde{\mathcal{D}}b^s$ by

$$0 \to \mathcal{P}^s \to \mathcal{D}b^s \to \mathcal{D}b^s/\mathcal{P}^s \to 0,$$

$$0 \to \tilde{\mathcal{D}}b^s \to \mathcal{Q}^s \to \mathcal{Q}^s/\tilde{\mathcal{D}}b^s \to 0.$$

Theorem 4. *For every open $\Omega \subset \mathbf{D}^n$,*

$$\mathcal{D}b^s/\mathcal{P}^s(\Omega) = \mathcal{D}b^s(\Omega)/\mathcal{P}^s(\Omega),$$

$$\mathcal{Q}^s/\tilde{\mathcal{D}}b^s(\Omega) = \mathcal{Q}^s(\Omega)/\tilde{\mathcal{D}}b^s(\Omega).$$

Proof. The first assertion follows from $H^k(\Omega, \mathcal{P}^s) = 0, k \geq 1, \Omega \subset \mathbf{D}^n$. The softness of $\tilde{\mathcal{D}}b^s$ implies $H^k(\Omega, \tilde{\mathcal{D}}b^s) = 0, k \geq 1, \Omega \subset \mathbf{D}^n$ and this implies the second assertion.

Denote by ρ and $\tilde{\rho}$ the natural mappings

$$\rho : \mathcal{D}b^s(\Omega) \to \mathcal{D}b^s(\Omega)/\mathcal{P}^s(\Omega), \ \tilde{\rho} : \tilde{\mathcal{D}}b^s(\Omega) \to \tilde{\mathcal{D}}b^s(\Omega)/\mathcal{P}^s(\Omega)$$

and define the s-analytic singular support of an $f \in \mathcal{S}^{\{t\}}(\mathbf{R}^n)$ by

$$\operatorname{singsupp}_a^s f := \operatorname{supp} \rho(j(f)).$$

The proof of the next theorem follows from the definitions.

Proposition 2. *Let $f \in \mathcal{S}'^{\{t\}}(\mathbf{R}^n)$. Then*

$$\operatorname{singsupp}_a^s f = \operatorname{supp} \tilde{\rho}(k(f)), \quad \mathbf{R}^n \cap \operatorname{singsupp}_a^s f = \operatorname{singsupp}_a f,$$

where $\operatorname{singsupp}_a f$ *denotes the analytic singular support of f.*

The isomorphism j enables us to define the analytic singular support and the analytic wave front set of an $f \in \mathcal{S}'^{\{t\}}(\mathbf{R}^n)$ at $x_0\infty$ and $(x_0\infty, \xi_0)$, respectively, where $(x_0\infty, \xi_0)$ denotes a point on the cosphere bundle at $x_0 \in S_\infty^{n-1}$ with the direction at component $\xi_0 \in \mathbf{R}^n \setminus \{0\}$.

Let $f \in \mathcal{S}'^{\{t\}}(\mathbf{R}^n)$. Then we define:

$$(x_0\infty, \xi) \notin SS_\tau^s f \text{ iff } (x_0\infty, \xi) \notin SS_\tau^s j(f).$$

This leads to:

$$x_0\infty \notin \operatorname{singsupp}_\tau^s f \quad \text{if} \quad x_0\infty \notin \operatorname{singsupp}_\tau^s j(f).$$

Proposition 3. *Let $f \in \mathcal{S}'^{\{t\}}(\mathbf{R}^n)$. Then*

$$x_0\infty \notin \operatorname{singsupp}_\tau^s f \Longleftrightarrow \forall \xi \in \mathbf{R}^n \setminus \{0\}, (x_0\infty, \xi) \notin SS_\tau^s j(f).$$

We denote by Sp^s the canonical surjective spectrum mapping

$$Sp^s \, : \, \pi^{-1}\mathcal{Q}^s \to \mathcal{R}^s$$

which gives the sheaf homomorphisms

$$\pi^{-1}\tilde{\mathcal{D}}b^s \to \mathcal{R}^s.$$

This leads us to the definitions of new microfunction spaces and new microlocalizations of hyperfunctions in $\mathcal{Q}^s$.

Definition 2.
(a) $\mathcal{DRC}^s := Sp^s(\pi^{-1}\tilde{\mathcal{D}}b^s)$.
(b) $\mathcal{DSC}^s$ is defined by the sequence

$$0 \to \mathcal{DRC}^s \to \mathcal{R}^s \to \mathcal{DSC}^s \to 0.$$

(c) $Sp^{1,s} \, : \, \pi^{-1}\mathcal{Q}^s \to \mathcal{DSC}^s$ is defined as the composition of the mapping Sp^s and the canonical embedding $\mathcal{R}^s \to \mathcal{DSC}^s$.
(d) Let $u \in \mathcal{Q}^s$ Then

$$SS^{1,s}(u) := \operatorname{supp} Sp^{1,s} u.$$

One can easily prove the exactness of the next sequence.

Theorem 5.
$$0 \to \mathcal{P}^s \to \tilde{\mathcal{D}}b^s \to \pi_* \mathcal{DRC}^s \to 0.$$

Theorem 6. *Sheaves $\mathcal{D}b^s/\mathcal{P}^s$, $\mathcal{DRC}^s$ and $\mathcal{DSC}^s$ are soft sheaves.*

Proof. We will only show the last part of the theorem.

By $\mathcal{D}b^s/\mathcal{P}^s(\Omega) = \mathcal{D}b^s(\Omega)/\mathcal{P}^s(\Omega)$ and $\mathcal{Q}^s/\mathcal{P}^s(\Omega) = \mathcal{Q}^s(\Omega)/\mathcal{P}^s(\Omega)$, $\Omega \subset \mathbf{D}^n$, it follows that for every open $V \subset \mathbf{D}^n \times S_\infty^{n-1}$

$$\Gamma(V, \mathcal{R}^s) \to \Gamma(V, \mathcal{DSC}^s) \to 0 \text{ is exact;}$$

and the softness of $\mathcal{DSC}^s$ follows from the diagram

$$\begin{array}{ccccc} \Gamma(V,\mathcal{R}^s) & \to & \Gamma(V,\mathcal{DSC}^s) & \to & 0 \quad \text{exact} \\ \downarrow & & \downarrow & & \\ \Gamma(W,\mathcal{R}^s) & \to & \Gamma(W,\mathcal{DSC}^s) & \to & 0 \quad \text{exact} \\ \downarrow & & & & \\ 0 & & & & \\ \text{exact,} & & & & \end{array}$$

where V, W are open in $\mathbf{D}^n \times S_\infty^{n-1}$ and $W \subset V$.

References

[1] G. Bengel and P. Schapira, Décomposition mircrolocal analytique des distributions, *Ann. Inst. Fourier, Grenoble,* **29** (1979), 101–124.

[2] A. Bonet, R.W. Braun, R. Meise and B.A. Taylor, Whitney's extension theorem for nonquasianalytic classes of ultradifferentiable functions, *Studia Mathematica,* **99** (1991), 155–184.

[3] R. Carmichael and S. Pilipović, On the convolution and Laplace transformation in the space of Beurling–Gevrey tempered ultradistributions, *Math. Nachr.,* **158** (1992), 119–132.

[4] A. Eida and S. Pilipović, On the microlocal decomposition of some classes of ultradistributions, *Math. Proc. Cambridge Phil. Soc.* **125** (1999), 455–461.

[5] A. Eida and S. Pilipović, Suppleness of some quotient spaces of microfunctions, *Rocky Mountain J. Math.* **30** (2000), 165–177.

[6] L. Hörmander, Uniqueness theorems and wave front sets for solutions of linear differential equations with analytic coefficients, *Comm. Pure Appl. Math.* **24** (1971), 617–704.

[7] A. Kaneko, *Introduction to Hyperfunctions,* Kluwer Acad. Press, Dordrecht, Boston, London, 1988.

[8] A. Kaneko, On the global existence of real analytic solutions of linear partial differential equations on unbounded domains, *J. Fac. Sci. Univ. Tokyo Sect. IA Math.* **32** (1985), 319–372.

[9] A. Kaneko, On the flabbiness of the sheaf of Fourier microfunctions, *Sci. Papers of the College of Arts and Sci.* **36** (1986), 1–14.

[10] T. Kawai, On the theory of Fourier hyperfunctions and its applications to partial differential equations with constant coefficients, *J. Fac. Sci. Univ. Tokyo Sect. IA Math.* **17** (1970), 467–517.

[11] H. Komatsu, Microlocal analysis in Gevrey classes and in convex domains, in *Lecture Notes in Mathematics* 1495 Springer-Verlag, 1991, 161–236.

[12] H. Komatsu, Ultradistributions, I: Structure theorems and a characterization, *J. Fac. Sci. Univ. Tokyo. Sect. IA Math.* **20** (1973), 25–105.

[13] D. Kovačević and S. Pilipović, Structural properties of the space of tempered ultradistributions, in *Complex Analysis and Generalized Functions*, Varna, 1991, 169–184.

[14] P. Laubin, Front d'onde analytique et décomposition microlocale des distributions, *Ann. Inst. Fourier, Grenoble,* **33** (1983), 179–199.

[15] A. Melin and J. Sjöstrand, Fourier integral operators with complex phase functions and parametrix for an interior boundary value problem, *Comm. Partial Differential Equations* **1** (1976), 283–311.

[16] S. Pilipović, Microlocal analysis of ultradistributions, *Proc. Amer. Math. Soc.* **126** (1998), 105–113.

[17] S. Pilipović, Characterization of bounded sets in spaces of ultradistributions, *Proc. Amer. Math. Soc.* **120** (1994), 1191–1206.

[18] Y. Saburi, Fundamental properties of modified Fourier hyperfunctions, *Tokyo J. Math.* **8** (1985), 231– 273.

[19] M. Sato, T. Kawai and M. Kashiwara, Microfunctions and pseudo-differential equations in *Lecture Notes in Mathematics* 287 Springer-Verlag, 1973, 265-529.

[20] V. S. Vladimirov, *Generalized Functions in Mathematical Physics*, Mir, Moscow, 1979.

[21] V.V. Zharinov, The Laplace transform of Fourier hyperfunctions and other similar cases of analytic functionals, Part I in *Teoret. Mat. Fiz.* **33** (1977), 291–309, Part II in *Teoret. Mat. Fiz.* **37** (1978), 12–29.

Richard D. Carmichael
Department of Mathematics
Wake Forest University
Winston-Salem
NC 27109-7388, USA
e-mail: carmicha@wfu.edu

Atsuhiko Eida
Department of Engineering Science
Tokyo Engineering University
1401-1, Katakura
Hachioji, Tokyo 192, Japan
e-mail: eida@cs.teu.ac.jp

Stevan Pilipović
Institute of Mathematics
University of Novi Sad
Trg D. Obradovica 4
21000 Novi Sad, Yugoslavia
e-mail: pilipovic@im.ns.ac.yu

Operator Theory: Advances and Applications

Your Specialized Publisher in Mathematics
Birkhäuser

OT 158: Eiderman, V.Ya. / Samokhin, M.V. (Eds.), Selected Topics in Complex Analysis (2005). ISBN 3-7643-7251-6

OT 157: Alpay, D. / Vinnikov, V. (Eds.), Operator Theory, Systems Theory and Scattering Theory: Multidimensional Generalizations (2005). ISBN 3-7643-7212-5

OT 156: Ebenfelt, P. / Gustafsson, B. / Khavinson, D. / Putinar, M. (Eds.), Quadrature Domains and Their Applications. The Harold S. Shapiro Anniversary Volume (2005). ISBN 3-7643-7145-5

OT 155: Ashino, R. / Boggiatto, P. / Wong, M.W. (Eds.), Advances in Pseudo-Differential Operators (2004). ISBN 3-7643-7140-4

OT 154: Janas, J. / Kurasov, P. / Naboko, S. (Eds.), Spectral Methods for Operators of Mathematical Physics (2004). ISBN 3-7643-7133-1

OT 153: Gaspar, D. / Gohberg, I. / Timotin, D. / Vasilescu, F.H. / Zsido, L. (Eds.), Recent Advances in Operator Theory, Operator Algebras, and their Applications (2004). ISBN 3-7643-7127-7

OT 152: Eidelman, S.D. / Ivasyshen, S.D. / Kochubei, A.N., Analytic Methods in the Theory of Differential and Pseudo-differential Equations of Parabolic Type (2004). ISBN 3-7643-7115-3

OT 151: Gil, J.B. / Krainer, T. / Witt, I. (Eds.), Aspects of Boundary Problems in Analysis and Geometry (2004). Subseries **A**dvances in **P**artial **D**ifferential **E**quations ISBN 3-7643-7069-6

OT 150: Rabinovich, V. / Roch, S. / Silbermann, B., Limit Operators and their Applications in Operator Theory (2004). ISBN 3-7643-7081-5

OT 149: Ball, J.A. / Helton, J.W. / Klaus, M. / Rodman, L. (Eds.), Current Trends in Operator Theory and its Applications (2004). ISBN 3-7643-7067-X

OT 148: Ashyralyev, A. / Sobolevskii, P.E., New Difference Schemes for Partial Differential Equations (2004). ISBN 3-7643-7054-8

OT 147: Gohberg, I. / dos Santos, A.F. / Speck, F.-O. / Teixeira, F.S. / Wendland, W. (Eds.), Operator Theoretical Methods and Applications to Mathematical Physics. The Erhard Meister Memorial Volume (2004). ISBN 3-7643-6634-6

OT 146: Kopachevsky, N.D. / Krein, S.G., Operator Approach to Linear Problems of Hydrodynamics. Volume 2: Nonself-adjoint Problems for Viscous Fluids (2003). ISBN 3-7643-2190-3

OT 145: Albeverio, S. / Demuth, M. / Schrohe, E. / Schulze, B.-W. (Eds.), Nonlinear Hyperbolic Equations, Spectral Theory, and Wavelet Transformations (2003). Subseries **A**dvances in **P**artial **D**ifferential **E**quations ISBN 3-7643-2168-7

OT 144: Belitskii, G. / Tkachenko, V., One-dimensior Functional Equations (2003). ISBN 3-7643-0084-1

OT 143: Alpay, D. (Ed.), Reproducing Kernel Spaces an Applications (2003). ISBN 3-7643-0068-X

OT 142: Böttcher, A. / dos Santos, A.F. / Kaashoek Brites Lebre, A. / Speck, F.-O. (Eds.), Singular Integral Factorization and Applications (2003). ISBN 3-7643-694

OT 141: dos Santos, A.F. / Gohberg, I. / Manojlovi Factorization and Integrable Systems. Proceedings of the School, Faro, Portugal, 2000 (2003). ISBN 3-7643-6938-

OT 140: Ellis, R. / Gohberg, I. Orthogonal Systems an Convolution Operators (2002). ISBN 3-7643-6929-9

OT 139: Müller, V. Spectral Theory of Linear Operators a Spectral Systems in Banach Algebras (2003). ISBN 3-764

OT 138: Albeverio, S. / Demuth, M. / Schrohe, E. / S B.-W. (Eds.). Parabolicity, Volterra Calculus, and Conical Singularities (2002). Subseries **A**dvances in **P**artial **D**iffere **E**quations. ISBN 3-7643-6906-X

OT 137: Dybin, V. / Grudsky, S.M. Introduction to the Toeplitz Operators with Infinite Index (2002) ISBN 3-7643-6906-X

OT 136: Wong, M.W. Wavelet Transforms and Localizati Operators (2002). ISBN 3-7643-6789-X

OT 135: Böttcher, A. / Gohberg, I. / Junghanns, P. (E Toeplitz Matrices, Convolution Operators, and Integral Equ The Bernd Silbermann Anniversary Volume (2002) ISBN 3-7643-6877-2

OT 134: Alpay, D. / Gohberg, I. / Vinnikov, V. (Eds.). Interpolation Theory, Systems Theory and Related Topics. T Dym Anniversary Volume (2002). ISBN 3-7643-6762-8

OT 133: Krall, A.M. Hilbert Space, Boundary Value Probl Orthogonal Polynomials (2002). ISBN 3-7643-6701-6

OT 132: Albeverio, S. / Elander, N. / Everitt, W.N. / K P. (Eds.). Operator Methods in Ordinary and Partial Differer Equations. S. Kovalevsky Symposium, University of Stockho 2000 (2002). ISBN 3-7643-6790-3

OT 131: Böttcher, A. / Karlovich, Y.I. / Spitkovsky, I.M Convolution Operators and Factorization of Almost Periodic Functions (2002). ISBN 3-7643-6672-9

OT 130: Gohberg, I. / Langer, H. (Eds.). Linear Operator Matrices (2001). ISBN 3-7643-6655-9

OT 129: Borichev, A.A. / Nikolski, N.K. (Eds.). Systems, Approximation, Singular Integral Operators, and Related To (2001). ISBN 3-7643-6645-1

GPSR Compliance

The European Union's (EU) General Product Safety Regulation (GPSR) is a set of rules that requires consumer products to be safe and our obligations to ensure this.

If you have any concerns about our products, you can contact us on ProductSafety@springernature.com

In case Publisher is established outside the EU, the EU authorized representative is:

Springer Nature Customer Service Center GmbH
Europaplatz 3
69115 Heidelberg, Germany

Batch number: 10578039

Printed by Printforce, the Netherlands